Chemometrics in Chromatography

CHROMATOGRAPHIC SCIENCE SERIES

A Series of Textbooks and Reference Books

Editor:
Nelu Grinberg

Founding Editor:
Jack Cazes

1. Dynamics of Chromatography: Principles and Theory, J. Calvin Giddings
2. Gas Chromatographic Analysis of Drugs and Pesticides, Benjamin J. Gudzinowicz
3. Principles of Adsorption Chromatography: The Separation of Nonionic Organic Compounds, Lloyd R. Snyder
4. Multicomponent Chromatography: Theory of Interference, Friedrich Helfferich and Gerhard Klein
5. Quantitative Analysis by Gas Chromatography, Josef Novák
6. High-Speed Liquid Chromatography, Peter M. Rajcsanyi and Elisabeth Rajcsanyi
7. Fundamentals of Integrated GC-MS (in three parts), Benjamin J. Gudzinowicz, Michael J. Gudzinowicz, and Horace F. Martin
8. Liquid Chromatography of Polymers and Related Materials, Jack Cazes
9. GLC and HPLC Determination of Therapeutic Agents (in three parts), Part 1 edited by Kiyoshi Tsuji and Walter Morozowich, Parts 2 and 3 edited by Kiyoshi Tsuji
10. Biological/Biomedical Applications of Liquid Chromatography, edited by Gerald L. Hawk
11. Chromatography in Petroleum Analysis, edited by Klaus H. Altgelt and T. H. Gouw
12. Biological/Biomedical Applications of Liquid Chromatography II, edited by Gerald L. Hawk
13. Liquid Chromatography of Polymers and Related Materials II, edited by Jack Cazes and Xavier Delamare
14. Introduction to Analytical Gas Chromatography: History, Principles, and Practice, John A. Perry
15. Applications of Glass Capillary Gas Chromatography, edited by Walter G. Jennings
16. Steroid Analysis by HPLC: Recent Applications, edited by Marie P. Kautsky
17. Thin-Layer Chromatography: Techniques and Applications, Bernard Fried and Joseph Sherma
18. Biological/Biomedical Applications of Liquid Chromatography III, edited by Gerald L. Hawk
19. Liquid Chromatography of Polymers and Related Materials III, edited by Jack Cazes
20. Biological/Biomedical Applications of Liquid Chromatography, edited by Gerald L. Hawk
21. Chromatographic Separation and Extraction with Foamed Plastics and Rubbers, G. J. Moody and J. D. R. Thomas
22. Analytical Pyrolysis: A Comprehensive Guide, William J. Irwin
23. Liquid Chromatography Detectors, edited by Thomas M. Vickrey
24. High-Performance Liquid Chromatography in Forensic Chemistry, edited by Ira S. Lurie and John D. Wittwer, Jr.
25. Steric Exclusion Liquid Chromatography of Polymers, edited by Josef Janca
26. HPLC Analysis of Biological Compounds: A Laboratory Guide, William S. Hancock and James T. Sparrow
27. Affinity Chromatography: Template Chromatography of Nucleic Acids and Proteins, Herbert Schott

Chemometrics in Chromatography

Edited by
Łukasz Komsta
Yvan Vander Heyden
Joseph Sherma

CRC Press
Taylor & Francis Group
Boca Raton London New York

CRC Press is an imprint of the
Taylor & Francis Group, an **informa** business

CRC Press
Taylor & Francis Group
6000 Broken Sound Parkway NW, Suite 300
Boca Raton, FL 33487-2742

First issued in paperback 2020

ISBN-13: 978-0-367-57223-5 (pbk)
ISBN-13: 978-1-4987-7253-2 (hbk)

Library of Congress Cataloging-in-Publication Data

Names: Komsta, Łukasz, editor. | Vander Heyden, Yvan, editor. | Sherma, Joseph, editor.
Title: Chemometrics in chromatography / editors, Łukasz Komsta, Yvan Vander Heyden, Joseph Sherma.
Description: Boca Raton : CRC Press, 2018. | Series: Chromatographic science series | Includes bibliographical references.
Identifiers: LCCN 2017037096 | ISBN 9781498772532 (hardback : alk. paper)
Subjects: LCSH: Chemometrics. | Chromatographic analysis.
Classification: LCC QD75.4.C45 C48448 2018 | DDC 543.01/5195--dc23
LC record available at https://lccn.loc.gov/2017037096

Visit the Taylor & Francis Web site at
http://www.taylorandfrancis.com

and the CRC Press Web site at
http://www.crcpress.com

We dedicate this book to Prof. Yi-Zeng Liang, an excellent chemometrician and chromatographer who also contributed to this book. He passed away on October 14, 2016, after suffering from an incurable disease. The chapter he coauthored for this book is one of his last scientific contributions.

Yi-Zeng Liang was professor at the Central South University in Changsha, China, where he was working at the Research Center for Modernization of Chinese Medicines. His chromatographic work thus was often related to the analysis of traditional Chinese medicines, for which fingerprint chromatograms were developed from herbal samples. The entire fingerprint profiles then were multivariately analyzed using chemometric methods.

Prof. Liang was the best known Chinese chemometrician for many years. In 1997, he organized the first Chinese Conference on Chemometrics, together with his Norwegian friend Olav Kvalheim. For many years, he tried to get an important chemometrics conference—Chemometrics in Analytical Chemistry (CAC)—to China. Finally, he succeeded, and in 2015 the conference was organized in Changsha. At that moment, he already was suffering from the disease of which he finally lost the battle. The year after his organization, during the 2016 CAC meeting in Barcelona, Spain, he was awarded the CAC Lifetime Achievement Award for his contributions to chemometrics. Unfortunately, his health prevented him to travel to Europe to personally receive the award.

His legacy includes, besides many good memories to those who have known him, a long list of PhD students and also more than 400 research papers in English. Including his Chinese contributions raises that number to almost 1000 papers. For many years, until the very end of his life, he also served as editor of the Chemometrics and Intelligent Laboratory Systems *journal.*

Dedicating this book to his remembrance hopefully will trigger some readers to shortly think back, from time to time, about this remarkable scientist.

— Łukasz Komsta, Yvan Vander Heyden, and Joseph Sherma

Contents

SECTION I Method Development and Optimization

SECTION II Univariate Analysis

SECTION III Data Preprocessing and Unsupervised Analysis

SECTION IV   Classification, Discrimination, and Calibration

SECTION V   Retention Modeling

SECTION VI Application Overviews

SECTION VII Miscellaneous

Preface

Recalling memories of the near past (1980s), when an average home computer was equipped with 64 or 128 kB of memory together with a CPU clocked with several MHz, one can conclude that 30 years was sufficient time to make the *science fiction* come true. Nowadays, an average smartphone has better computational abilities than many scientific supercomputers in the past. Adding to this fact the development of high-quality open-source software, an average data-handling scientist does not have to bear many significant costs to perform *serious* analysis of huge datasets.

Chemometrics, the science related to the systematic optimization of methods and to the analysis of the gathered data, is at the fingertips of every chromatographer and is not an alien discipline, made by specialists for highly experienced users. It is a great practical and it *helps* in chromatographic experimenting, from a proper design of the research projects (reducing the effort and the costs) to the extraction of hidden information in large datasets collected from modern equipment with hyphenated detectors.

The proper use of chemometrics requires the understanding of how these methods work. However, there is no need to dig deeply into the mathematical background, as contemporary chromatographer do not need to write their own software routines. Compare the situation with using a calculator—computing a square root does not require the knowledge of how the calculator computes the value, but one must understand what the square root is and how to use the obtained value in the context of a given problem. In an analogous way, the researcher cannot treat chemometrics as a "black box" approach (because this often leads to misuse and errors), but the algorithmic details of the software can be neglected if the cognitive value of the results is known.

It can be honestly stated that many chromatographers still have some fear about chemometrics and cannot come to terms with the fact that it became the de facto standard in modern analytical chemistry. Although several great textbooks on the topic exist, they may be quite huge, they focus on other domains of application, and they often terrify potential users, who have to read about several methods (including those applied rarely in chromatography) to make potential connections between pure theory and chromatographic practice.

These facts were the main impulse to start this book project. Our main goal was to create a book dedicated to chromatographers, covering the most important chemometric methods that are now also important from a chromatographer's point of view, explained in strict connection with chromatographic practice. Although this book is written by specialists, effort was made to explain chemometrics from the basics, as chromatographers are generally not trained chemometricians.

We hope that the text will be useful for readers that like to use chemometrics in chromatographic analysis. The book may also be interesting for scientists with some chemometric experience who want to reread about known methods, but now occasionally seen from a different point of view. The theory of chemometric methods is explained in relation to chromatography, but the knowledge gathered after reading this book may be applied in other disciplines of analytical chemistry, as well as in other applications. Depending on the context, it can be considered as a manual, a reference book, or a teaching source.

We thank Barbara Knott, Senior Editor—Chemistry, CRC Press/Taylor & Francis Group, for her support of our book proposal and S. Valantina Jessie, Project Manager, for her help in all aspects of our subsequent editorial work. We also thank the chapter authors for great cooperation and their exceptionally valuable contributions.

MATLAB® is a registered trademark of The MathWorks, Inc. For product information, please contact:

The MathWorks, Inc.
3 Apple Hill Drive
Natick, MA, 01760-2098 USA
Tel: 508-647-7000
Fax: 508-647-7001
E-mail: info@mathworks.com
Web: www.mathworks.com

Contributors

Tomasz Bączek
Department of Pharmaceutical Chemistry
Medical University of Gdańsk
Gdańsk, Poland

Juan José Baeza-Baeza
Department of Analytical Chemistry
University of Valencia
Burjassot, Spain

Elaheh Talebanpour Bayat
Department of Chemistry
Shiraz University
Shiraz, Iran

Alessandra Biancolillo
Department of Chemistry
Sapienza University of Rome
Rome, Italy

Lionel Blanchet
Department of Pharmacology and Toxicology
Maastricht University Medical Center
Maastricht, the Netherlands

and

Thayer School of Engineering
Dartmouth College
Hanover, New Hampshire

Katarzyna Bober
Department of Analytical Chemistry
School of Pharmacy with the Division of
    Laboratory Medicine
Medical University of Silesia
Sosnowiec, Poland

Lutgarde Buydens
Department of Analytical Chemistry
Institute for Molecules and Materials
Radboud University
Nijmegen, the Netherlands

Krzesimir Ciura
Department of Physical Chemistry
Medical University of Gdańsk
Gdańsk, Poland

Michał Daszykowski
Institute of Chemistry
University of Silesia in Katowice
Katowice, Poland

Sven Declerck
Department of Analytical Chemistry, Applied
    Chemometrics and Molecular Modelling
Vrije Universiteit Brussel
Brussels, Belgium

Erdal Dinç
Department of Analytical Chemistry
Faculty of Pharmacy
Ankara University
Ankara, Turkey

Małgorzata Dołowy
Department of Analytical Chemistry
School of Pharmacy with the Division of
    Laboratory Medicine
Medical University of Silesia
Sosnowiec, Poland

Armando C. Duarte
Department of Chemistry
and
Centre for Environmental and Marine Studies
University of Aveiro
Aveiro, Portugal

Regina M.B.O. Duarte
Department of Chemistry
and
Centre for Environmental and Marine Studies
University of Aveiro
Aveiro, Portugal

Pietro Franceschi
Computational Biology Unit
Research and Innovation Centre
Edmund Mach Foundation of
    San Michele all'Adige
Trento, Italy

Charlene Muscat Galea
Department of Analytical Chemistry, Applied
  Chemometrics and Molecular Modelling
Vrije Universiteit Brussel
Brussels, Belgium

María Celia García-Álvarez-Coque
Department of Analytical Chemistry
University of Valencia
Burjassot, Spain

Mohammad Goodarzi
Department of Biochemistry
University of Texas Southwestern
  Medical Center
Dallas, Texas

Bahram Hemmateenejad
Department of Chemistry
Shiraz University
Shiraz, Iran

David Brynn Hibbert
School of Chemistry
UNSW Sydney
Sydney, New South Wales, Australia

David Jenkins
Product Quality and Compliance
Durham, North Carolina

Anna de Juan
Department of Analytical Chemistry and
  Chemical Engineering
University of Barcelona
Barcelona, Spain

Eliangiringa Kaale
School of Pharmacy
Muhimbili University of Health and
  Allied Sciences
Dar es Salaam, Tanzania

Piotr Kawczak
Department of Pharmaceutical Chemistry
Medical University of Gdańsk
Gdańsk, Poland

Łukasz Komsta
Department of Medicinal Chemistry
Medical University of Lublin
Lublin, Poland

Yi-Zeng Liang
College of Chemistry and Chemical
  Engineering
Central South University
Changsha, People's Republic of China

Hong-Mei Lu
College of Chemistry and Chemical
  Engineering
Central South University
Changsha, People's Republic of China

Debby Mangelings
Department of Analytical Chemistry, Applied
  Chemometrics and Molecular Modelling
Vrije Universiteit Brussel
Brussels, Belgium

Federico Marini
Department of Chemistry
Sapienza University of Rome
Rome, Italy

Sílvia Mas
Signal and Information Processing for Sensing
  Systems Group
Institute for Bioengineering of Catalonia
Barcelona, Spain

João T.V. Matos
Department of Chemistry
and
Centre for Environmental and Marine Studies
University of Aveiro
Aveiro, Portugal

Nabiollah Mobaraki
Department of Chemistry
Shiraz University
Shiraz, Iran

Joanna Nowakowska
Department of Physical Chemistry
Medical University of Gdańsk
Gdańsk, Poland

Geert Postma
Department of Analytical Chemistry
Institute for Molecules and Materials
Radboud University
Nijmegen, the Netherlands

Alina Pyka-Pająk
Department of Analytical Chemistry
School of Pharmacy with the Division of
    Laboratory Medicine
Medical University of Silesia
Sosnowiec, Poland

Elmira Rafatmah
Department of Chemistry
Shiraz University
Shiraz, Iran

Samantha Riccadonna
Computational Biology Unit
Research and Innovation Centre
Edmund Mach Foundation of
    San Michele all'Adige
Trento, Italy

Ines Salsinha
Department of Analytical Chemistry and
    Pharmaceutical Technology
Vrije Universiteit Brussel
Brussels, Belgium

Frederik-Jan van Schooten
Department of Pharmacology and Toxicology
Maastricht University Medical Center
Maastricht, the Netherlands

Danstan Hipolite Shewiyo
Department of Analytical Chemistry, Applied
    Chemometrics and Molecular Modelling
Vrije Universiteit Brussel
Brussels, Belgium

Zahra Shojaeifard
Department of Chemistry
Shiraz University
Shiraz, Iran

Ivana Stanimirova
Institute of Chemistry
University of Silesia in Katowice
Katowice, Poland

José Ramón Torres-Lapasió
Department of Analytical Chemistry
University of Valencia
Burjassot, Spain

Yvan Vander Heyden
Department of Analytical Chemistry, Applied
    Chemometrics and Molecular Modelling
Vrije Universiteit Brussel
Brussels, Belgium

Johan Viaene
Department of Analytical Chemistry, Applied
    Chemometrics and Molecular Modelling
Vrije Universiteit Brussel
Brussels, Belgium

Saeed Yousefinejad
Chemistry Department
Shiraz University
Shiraz, Iran

Igor Zenkevich
Institute for Chemistry
St. Petersburg State University
St. Petersburg, Russia

Zhi-Min Zhang
College of Chemistry and Chemical Engineering
Central South University
Changsha, People's Republic of China

Section I

Method Development and Optimization

1 Experimental Design in Chromatographic Method Development and Validation

Łukasz Komsta and Yvan Vander Heyden

CONTENTS

1.1 INTRODUCTION

Chromatography is an experimental separation science. Its progress, both in the context of theory development and finding new applications, is based on performing experiments and making appropriate conclusions from their results. As experiments are the main body of chromatographic research, they must be planned to *contain* the information desired to form a conclusion. Otherwise, the conclusion can be accidentally based on irrelevant information, noise, or some technical artifacts. Moreover, the information hidden inside results must be of good quality.

In most cases, chromatographic experiments are designed to estimate the influence of a *factor* (temperature, mobile phase parameter, flow, etc.) onto the chromatographic process. A second task is to optimize the chromatographic process—to find the best value (level) of a number of factors obtaining the best *response*. To draw the conclusions, experiments should be performed on several levels of a particular factor. For the influence testing (screening), two levels are generally sufficient. At least three levels are required to estimate the relationship between the response and the factors, called the *response surface*, which usually is modeled using a quadratic polynomial model.

For many years, chromatographic experiments were performed investigating one factor in one series of experiments (univariate approach). All other factors are then kept at a constant level. This approach is called OVAT (One Variable At a Time). However, modern method development applies a multivariate approach with experimental designs, which contain experiments at various levels of all factors, spanning the whole (in some cases, only the applicable) multivariate experimental space. An experimental design is a table with a predefined number of experiments, varying a number of factors simultaneously at a given number of levels.

Although the OVAT version of experimental setup seems to be an easier way, univariate investigations should preferably be avoided. The main reasons are as follows:

1. Changing one factor, simultaneously keeping all other factors constant, does not take into account *interactions* between factors. An interaction occurs when the effect of one factor depends on the level of another. The chromatographer should be aware not only of the occurrence of interactions that are obvious (or that *should* be studied) but also of interactions that are really *unknown* and may lead to surprising results. Only the full factorial multivariate design gives adequate and deep understanding of a process. However, most often this design requires too many experiments to be feasible. On the other hand, the design applied in practice does not allow studying individual interactions due to their confounding pattern.

2. Sequential optimization (e.g., optimizing pH, then modifier concentration at chosen pH, then flow at chosen concentration/pH) does not always lead the chromatographer to the best possible results, as it is a *local* approach and the final result depends strictly on previous results and the initial choices. There is no guarantee that there is no combination of factor values giving better results. There may be many *acceptable* combinations of factors, but finding really the *best* one (with multivariate approach) can substantially reduce cost, time, and amount of reagents. Note that the global optimum may also be found by updating the sequential or OVAT procedure. This means that after sequential optimization of the pH, modifier concentration, and flow, one reevaluates the pH starting at the best conditions found, then again concentration, and so on. These reevaluation steps allow finding the global optimum but are usually not performed (too many experiments required).

3. In extreme cases, changing a factor without changing the others will not lead to any performance change (very specific interactions). Only the simultaneous change of several factors will improve separation. In this case, OVAT for the considered factors leads to the wrong assumption that overall performance cannot be improved, unless the updated procedure mentioned earlier is applied. For instance in chiral separations, the pH change of the mobile phase can have almost no effect with low concentration of a modifier, where its effect is strong at higher concentrations.

4. In the majority of cases, a multivariate approach requires *less* experiments and time (sic!) than the univariate sequential.

5. Only a proper experimental design provides full information about the relation between the factors and the response, that is, about the response surface. OVAT experiments, even when sequentially repeated as discussed earlier, and with a higher number, provide less suitable information about this response surface.

6. The univariate approach relies on some a priori knowledge about the process and therefore is *biased*. Very often, this knowledge forces the chromatographer to start from the cheapest experiments, which could be a wrong starting point, and leads to a *local* optimal place in the experimental domain instead of the global. However, as discussed in item 2, this may be overcome by reevaluation of the studied factors.

7. Although it is obvious that some factors always significantly change the chromatographic performance, the screening step is in some cases recommended, as making wrong subjective decisions leads to a so-called *sentimental* design. One should get some distance on the "principle of effect sparsity" (only a small number of possible factors are significant), as the performance of a chromatographic system can be affected by around 50 factors and they can interact [1]. However, if the most important factors are known, the screening step

may be cancelled. For instance, to optimize a reversed-phase liquid chromatographic (LC) method to separate pharmaceutical compounds (and their impurities) one knows that the stationary phase type, the mobile-phase pH, and the mobile-phase composition will mostly affect the selectivity of the separations.

8. The proper multivariate experimental model (yielding the response surface) can predict the performance (response) inside the entire experimental domain sufficiently well, for example, with *at least* similar uncertainty as a real experiment. However, compared to multivariate calibration problems, the quality of the predictions in method optimization is less crucial. The goal is to properly indicate the region with the best responses, a somewhat less good prediction of the response at these conditions is not catastrophic. Therefore, less effort is put in finding optimal models and usually a quadratic polynomial model performs sufficiently well.

9. One experimental design is usually sufficient to find a suitable factor combination. When this occasionally is not the case, then another reformulated design is sufficient to achieve this goal.

This chapter is designed as an introductory tutorial. It lacks detailed mathematical background, as ready-to-use procedures are available in many statistical and numerical software packages, including R and MATLAB®. Moreover, the mathematical issues are widely covered in cited references. The experimental design topic is covered by many papers and for detailed information the reader is referred to [2–6] for recent reviews on the topic. Summaries of applications in analytical chemistry [7], capillary electrophoresis [8,9], factorial-based approaches in chromatography [10], chromatography in food and environmental analysis [11], chiral separations [2], and chromatography–mass spectrometry [12] can easily be found as well.

1.2   INITIAL CONSIDERATIONS

It should be emphasized that there is no *automatical* (sometimes even called "automagical") way to optimize a chromatographic method, as the experimental design is only a *tool* to achieve the goal. Regardless of ready-to-use procedures, the final decisions taken can depend on many details. The analyst must ensure what the real objective of the planned study is and must also be aware how many experiments can be done. The limits are set by time, amounts of reagents, and also financial issues. Moreover, some combinations of factor levels can be impossible to perform and such experiment must be omitted or some restricted design (domain) has to be used. Thus, there should be a careful choice of the ranges of the investigated factors, both in the case of screening and optimization. A strong effort must be put in the choice of the *response(s)*. In chromatography, it can be retention time, total time of analysis, resolution of critical peak pair (or the selectivity parameter), recovery (in robustness), or any other chromatographic response function (CRF). However, it should be emphasized that the main response in separation optimization should always be related to the separation of compounds, especially to the critical peak pair separations.

When multiple responses have to be taken into account, they can be combined (after an initial transformation) into one response that is maximized. The most frequently used method to combine several responses into one response function is the Derringer desirability function [13–16]. However, other possibilities, among which the Pareto optimality approach, are also available [17].

In a full development of a chromatographic method, one uses experimental designs in three possible steps:

1. Screening [1,18–21]—estimation of the impact of several (as many as possible) factors; selection of important factors.
2. Optimization [2,11,14,16,22–25]—obtaining a dependence between a limited number of factors (chosen at first step) and the response(s). Response surface methodology is used to find the best chromatographic conditions.

3. Robustness testing [26–34]—part of method validation; estimation of the impact of several factors on the method performance. This step is analogous to (1), but it is done in a narrow range of the factors and we want to prove *insignificance* of factors instead of finding the significance, as the method should be insensitive to these changes. The main response of interest is related to the content determination (as always in method validation), while in steps 1 and 2 it is separation-related.

After choosing and generation of a design, the so-called *design matrix* is obtained. It is a matrix (table) with a number of rows equal to the experiments and of columns equal to the number of factors. It contains some values, most often −1, 0, and 1 (or −1 and 1), representing the levels of the investigated factors. Occasionally, it can contain more levels. The design matrix is used to obtain the *model matrix*, that is, the design matrix augmented with columns containing higher degrees (most often quadratic values) of the original factor levels, as well as their products (interactions). This model matrix (represented as X) is used to *fit* a response surface or a linear model through the results (represented as y). The experiments are performed according to the design matrix, obtaining the response(s) (y value) for each row (experiment) of this matrix. It is recommended to perform experiments in *random order*, to minimize the influence of time drifts, occasionally giving false significances. However, time drifts ideally should not occur in well-designed experiments (but often cannot be avoided) and randomization of the experiments is not a sufficient way to deal with them. The time drifts in the chromatograms are occasionally caused by worn-out equipment (e.g., lamp in the detector), and their cause preferably should be eliminated before execution of the experiments. However, in practice, chromatographic experimental design experiments are always subjected to drift effects, that is, due to column aging. Often the influence is negligible and can be ignored, but when this is not the case the response is corrected for the observed drift. If not, given estimated effects are affected by the drift.

1.3   SCREENING

The first (and often introductory) type of research is to find significant factors (having a significant effect on the response), which is called screening. Screening occasionally can also be used to evaluate some interactions between studied factors. The screening step is generally done with many factors, a screening design is applied, and the results are used to make a choice of the most relevant factors for optimization. It can combine quantitative and qualitative factors in one approach, opposite to optimization designs.

Let us assume one wants to estimate whether a change in pH significantly affects the retention of a substance. To achieve this goal, one must obtain replicate chromatograms at two distinct pH values. The replicates allow estimating the intra-pH variability (at one pH), representing the experimental error, which is compared statistically to the inter-pH variability (between pH values), representing the effect of the pH, in a *t*-test. If the inter-pH variability is significantly larger than the experimental error, one accepts that the considered pH change has a significant effect on the retention. In an experimental design approach, several factors and interactions are tested in several *t*-tests. An alternative approach applies a multiway ANOVA table in which an equal number of *F*-tests are performed to evaluate the same factor and interaction effects.

As already mentioned, one usually evaluates more than one factor in a real-life situation. Let us consider that also the significance of the influence of the temperature changes is investigated. Two experiments (with replicates) are not enough in this situation. If both pH and temperature are changed simultaneously (considering two experiments: low pH and low temperature, then high pH and high temperature) and the response changes, it is only known whether *at least one* factor (temperature, pH) is responsible. Two experiments will not allow estimating the effect of the two factors individually. It is called a *supersaturated* design. Such an approach is used only exceptionally, for instance, for very expensive experiments. It provides an answer whether *any* factor is significant. However, in rare cases this approach can lead to wrong assumptions, that is, when interaction between both significant factors makes the estimated factor effect very small. Then, we obtain a *false negative* result, when estimated changes caused by several (significant) factors are masked by their interactions.

Let us increase the number of experiments to three, obtaining the following plan: low pH and low temperature, low pH and high temperature, and high pH and low temperature. Such a design would allow investigating the two effects individually (from comparing experiments 1 and 2, then experiments 1 and 3), but these three experiments do not provide any information about the interaction of the factors. To be able to estimate the interaction effect, adding a fourth experimental combination to the design is necessary: high pH and high temperature. This setup is called a full factorial design for two factors.

Adding a third factor to this project increases the full design to eight experiments, the double of the previous design: each factor is four times at the low level of the third factor and four times at the high level.

1.3.1 FULL FACTORIAL DESIGNS

The last designs considered above are examples of two-level full factorial 2^k designs, where k is the number of factors included. It is the best possible and most informative approach in a screening. Evaluating k factors in the design requires 2^k experiments, as each factor is tested at two levels and all possible combinations are performed. For full factorial designs, the number of required experiments increases exponentially with the number of factors. Therefore, this type of design can only be used in the case of a relatively small number of possible factors, when one is not strongly limited by the number of possible experiments.

1.4 FRACTIONAL FACTORIAL DESIGNS

Fractional factorial designs contain only a fraction of the experiments existing in the full version. They are noted as $2^{(k-v)}$, where k is the number of factors and $k - v$ is some smaller power of two, representing the number of selected experiments (their fraction is $1/2^v$ of the full design).

The experiments are carefully selected to include as much information about the process as possible. An example can be seen in Figure 1.1, where half of the cube vertices are chosen to form a tetrahedron. The consequence is that one loses information when executing a fractional factorial design. In this example, the interactions cannot be estimated anymore, separate from the main factor effects. In this design, each two-factor interaction is estimated together with a main effect. It is said that these effects are *confounded*, that is, they cannot be estimated individually anymore. The three-factor interaction that can be considered in this case is confounded with the intercept estimation.

In any fractional design, some interactions cannot be estimated anymore independently from either main factor effects or other interactions (the values in the design matrix of factor/interactions that cannot be estimated independently are always equal), which is called confounding as mentioned earlier. Although the "principle of effect hierarchy" establishes that the main effects should generally be more important than the interactions, this is not always fulfilled. Two-factor interactions can have a considerably high effect. The fractional design is a compromise and allows estimating of main effects and some two-factor interaction effects, but higher-order interaction effects cannot be revealed. However, the latter usually are anyway negligible. It also should be noted that confounding occurs in *all* fractional designs and no effect is estimated unconfounded.

To describe the efficiency of a fractional design, the *resolution* parameter is used. It is represented by a roman numeral: values III–V (3–5) are most often met in practice. Resolution III means that the design can estimate main effects, but these effects may be confounded with two-factor interactions (at least one main effect is). Resolution IV means that main effects can be estimated and they are not confounded by two-factor interactions but by three-factor interactions. Two-factor interaction effects can be then also estimated, but they are confounded with other two-factor

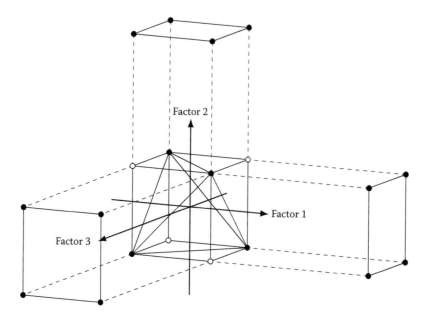

FIGURE 1.1 Three-dimensional representation of a $2^{(3-1)}$ fractional factorial design. From eight full factorial experiments, represented as the vertices of a cube, half of them are chosen in a way to be viewed as a square when projected onto two dimensions. To fulfill this requirement, the remaining points must be arranged as a tetrahedron.

interactions. Resolution V uncovers main effects, two-factor interaction effects, and also gives possibility of computing three-factor interaction effects, but these confound with two-factor interactions (and vice versa).

There is no easy rule to remember what efficiency can be obtained with a particular number of experiments and factors, but the (most often used) combinations can be summarized in the following way:

1. Four experiments give full resolution for 2 factors.
2. Eight experiments give full resolution for 3 factors, IV for 4 factors, and III for 5–7 factors.
3. Sixteen experiments give full resolution for 4 factors, V for 5 factors, IV for 6–8 factors, and III for 9–15 factors.
4. Thirty-two experiments give full resolution for 5 factors, VI for 6 factors, and IV for 7–15 factors.

Table 1.1 contains fractional designs for 8, 16, and 32 experiments. In each case, the three or four smallest numbers of possible factors are shown.

1.4.1 Plackett–Burman Designs

An alternative for the fractional factorial designs are the Plackett–Burman designs [20,35–37]. It can be constructed for a number of experiments n, which is a multiple of four (starting from eight, the most useful example, i.e., with 12 experiments is shown in Table 1.2), and it can estimate up to $n-1$ main effects (the resolution is always III). To be more precise, this design always estimates $n-1$ factor effects, but when the real factor number is smaller, the remaining factors are treated as dummy variables (nothing changes when such factor is either at low or high level; their effects should be insignificant and are used to estimate the experimental error).

TABLE 1.1
Fractional Factorial Screening Designs for 8, 16, and 32 Experiments

This table presents fractional factorial design matrices (values of +1/−1) arranged for four, five, six, and seven factors across 8, 16, and 32 experiments.

Four Factors (2^{4-1}): factors F1, F2, F3, F4; No. 1–8

Five Factors (2^{5-2}): factors F1, F2, F3, F4, F5; No. 1–8

Six Factors (2^{6-3}): factors F1, F2, F3, F4, F5, F6; No. 1–8

Seven Factors (2^{7-4}): factors F1, F2, F3, F4, F5, F6, F7; No. 1–8

Five Factors (2^{5-1}): factors F1, F2, F3, F4, F5; No. 1–16

Six Factors (2^{6-2}): factors F1, F2, F3, F4, F5, F6; No. 1–16

Seven Factors (2^{7-3}): factors F1, F2, F3, F4, F5, F6, F7; No. 1–16

(Continued)

TABLE 1.1 (Continued)
Fractional Factorial Screening Designs for 8, 16, and 32 Experiments

No.	Six Factors (2⁶⁻¹)						Seven Factors (2⁷⁻²)							Eight Factors (2⁸⁻³)							
	F1	F2	F3	F4	F5	F6	F1	F2	F3	F4	F5	F6	F7	F1	F2	F3	F4	F5	F6	F7	F8
1																					
2																					
3																					
4																					
5																					
6																					
7																					
8																					
9																					
10																					
11																					
12																					
13																					
14																					
15																					
16																					
17																					
18																					
19																					
20																					
21																					
22																					
23																					
24																					
25																					
26																					
27																					
28																					
29																					
30																					
31																					
32																					

Note: Each design is presented with three or four possible numbers of screening factors.

TABLE 1.2
Plackett–Burman Design with 12 Experiments (Maximal Number of Factors Is Equal to 11)

No.	F1	F2	F3	F4	F5	F6	F7	F8	F9	F10	F11
1	−1	−1	−1	1	−1	1	1	−1	1	1	1
2	1	1	1	−1	−1	−1	1	−1	1	1	−1
3	1	−1	1	1	1	−1	−1	−1	1	−1	1
4	1	−1	−1	−1	1	−1	1	1	−1	1	1
5	−1	1	−1	1	1	−1	1	1	1	−1	−1
6	−1	1	1	−1	1	1	1	−1	−1	−1	1
7	−1	1	1	1	−1	−1	−1	1	−1	1	1
8	1	−1	1	1	−1	1	1	1	−1	−1	−1
9	−1	−1	1	−1	1	1	−1	1	1	1	−1
10	1	1	−1	−1	−1	1	−1	1	1	−1	1
11	−1	−1	−1	−1	−1	−1	−1	−1	−1	−1	−1
12	1	1	−1	1	1	1	−1	−1	−1	1	−1

1.4.2 SUPERSATURATED DESIGNS

There is a possibility to construct designs with fewer experiments than factors. Such designs are called supersaturated. However, they are very rarely used in chromatography, as their application is usually limited to situations with very high numbers of factors. Moreover, there is no easy possibility to estimate the main factor effects unconfounded from each other [35,38–42].

1.4.3 REFLECTED DESIGNS

There also can be a need to perform screening at three levels when changes in the ranges (−1, 0) and (0, 1) can be different. Such approach requires performing a chosen two-level screening approach twice, with (0, 1) as levels in one part, then (−1, 0) in the other. There is one common experiment with all factors set to zero [4].

1.4.4 SIGNIFICANCE OF EFFECTS

Usually, the significance of the estimated effects is determined. This is usually performed by means of t-tests (occasionally an ANOVA table; see earlier discussion). The experimental error on an effect can be estimated either from replicated experiments, a priori considered negligible effects (interaction effects or dummy effects), or a posteriori considered negligible effects [3,4,33].

1.5 OPTIMIZATION

When a (quantitative) factor is investigated at two distinct levels, it is only possible to estimate whether the factor has any effect (or to estimate the dependence by a *linear* function). As in most cases, the dependences are curvilinear (and we search for the extremum), the designs used to model the dependences and to find the optimal chromatographic conditions must have at least three levels of each factor. Therefore, full factorial designs for two and three factors require $3^2 = 9$ and $3^3 = 27$ experiments, respectively. The multivariate curve of response, called a *response surface*, is then fitted and the optimal conditions are obtained from it by *response surface methodology*. The "cube" (i.e., the experimental domain) of the full factorial 3^3 design is presented in Figure 1.2, with two example designs with a reduced number of experiments. The latter two are discussed in more detail further.

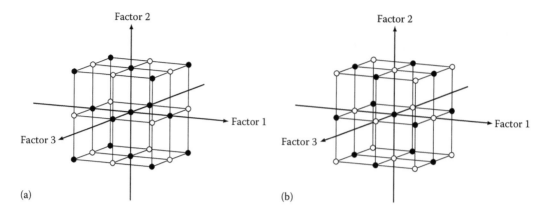

FIGURE 1.2 Designs constructed by selection of points from the full factorial 3^3 design (27 experiments): face-centered central composite design (a) and Box–Behnken design (b). The selected experiments are presented as black-filled points.

In many cases, the range of the factors that is needed to perform the optimization is unknown. This range can be estimated by preliminary experiments, analogous to the screening step. When more factors are considered, a full design is not feasible, as, for example, for 5 factors it requires $3^5 = 243$ experiments. Several compromise designs allow performing well-balanced studies with fewer experiments. They are also often used for two or three factors. The most often applied are central composite designs (CCD, most often applied), Box–Behnken designs (BBD), and Doehlert designs (DD).

1.6 CENTRAL COMPOSITE DESIGNS

CCD (Figure 1.2a) for k factors are extended two-level factorial designs [4–6,43]. They consist of 2^k points of a full factorial two-level design (with levels of factors coded as +1 and −1). In special cases (rarely), a fractional variant of the two-level design can be used. This design is extended by adding axial points lying on the coordinate axes (star design). For each factor, two axial points are added, representing the $+\alpha$ and $-\alpha$ levels for this factor, while remaining factors are set to zero. The design is completed with a center point. To allow the estimation of pure experimental error, several replicates of the center point are performed. Mathematically, these replicates preserve also orthogonality of the whole design.

Several values of the α parameter are regularly applied. When α is equal to one, the design is called a face-centered CCD (FCCD, Figure 1.2a). Increasing this parameter leads to circumscribed CCD (CCCD, Figure 1.3). There are three common choices of $\alpha > 1$: *spherical* CCD, when $\alpha = k^{1/2}$ (all points lie in equal distance from the origin); $\alpha = 2^{k/4}$, when the design is *rotatable* (an equal prediction variance is located at a fixed distance from the origin); and *orthogonal*, when $\alpha = \{[(2^k + 2k + n_0)^{1/2} - 2^{k/2}]^2 \cdot 2^{k/4}\}^{1/4}$ with n_0 the number of center point replicates (approximately, $\alpha = 1.826, 2.191,$ and 2.494 for $k = 3, 4,$ and 5, respectively). The design becomes both orthogonal and rotatable when $n_0 \approx 4.2^{k/2} + 4 - 2k$. The number of experiments in a CCD is $2^k + 2k + C$, where C is the number of replicated center points. Table 1.3 contains CCD for three and four factors.

1.6.1 BOX–BEHNKEN DESIGN

The Box–Behnken class of designs preserves rotatability (Figure 1.4d, Table 1.4) and is an interesting alternative to CCD, as they are generally more efficient (in the sense of the number of model coefficients vs. the number of experiments) [44,45]. They require $2k(k-1) + C$ experiments, where

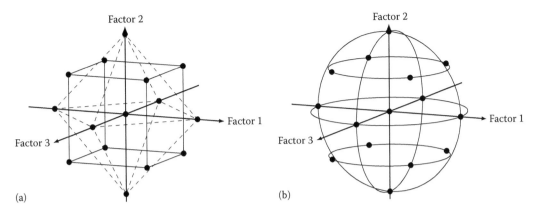

FIGURE 1.3 Central composite design with $\alpha > 1$: representation as a two-factorial cube (2^3 design) extended with six points on the axes and a center point (a) and representation of its sphericity (b).

TABLE 1.3
Central Composite Design for Three and Four Factors

Experiment	Three Factors			Four Factors			
	F1	F2	F3	F1	F2	F3	F4
1	1	1	−1	−1	−1	1	−1
2	1	−1	1	1	1	1	1
3	0	0	0	−1	1	1	1
4	−1	−1	1	0	0	0	0
5	1	−1	−1	1	−1	1	1
6	−1	1	−1	1	−1	1	−1
7	−1	−1	−1	1	1	1	−1
8	−1	1	1	−1	−1	1	1
9	1	1	1	1	−1	−1	1
10	0	0	−α	−1	1	−1	1
11	0	α	0	−1	1	−1	−1
12	α	0	0	−1	−1	−1	−1
13	0	0	α	1	−1	−1	−1
14	0	−α	0	1	1	−1	−1
15	−α	0	0	−1	−1	−1	1
16				−1	1	1	−1
17				1	1	−1	1
18				0	0	α	0
19				−α	0	0	0
20				0	0	−α	0
21				α	0	0	0
22				0	−α	0	0
23				0	0	0	α
24				0	A	0	0
25				0	0	0	−α

Note: See text for possible α values. Replicates of center point are not included.

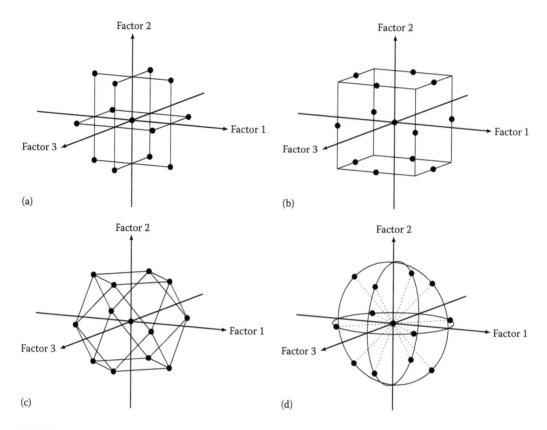

FIGURE 1.4 Geometrical interpretation of Box–Behnken design: as three intersected two-dimensional designs (a), as points on edge middles of a cube (b), as a cuboctahedron with equal edges (c), and as points on a sphere (equal distances to the origin) (d).

C is, as above, the number of center point replicates. They are based on using only three distinct levels (−1, 0, 1). They can be considered as three-level designs but incomplete (Figure 1.4b). The incompleteness is based on interlocking several lower-level designs in an orthogonal manner (Figure 1.4a).

The BBD does not contain any point with all factors set to extreme levels, which can be both advantageous and disadvantageous. Therefore, it is preferred in cases where extreme areas of the experimental domain cannot be investigated. However, it should be noted that the optima are most frequently predicted in an extreme area of the experimental domain.

1.6.2 Doehlert Design

Doehlert designs [46,47] are spherical and every point has the same distance from the center point (Figures 1.5 and 1.6, Table 1.5). The points of the design are also equally spread (uniform space-filling) and they form triangles or hypertriangles in the experimental domain. The design requires $k^2 + k + C$ experiments to be performed. The factors differ in the number of distinct investigated levels (3 for the first factor, 5 for the second, and 7 for all subsequent). These properties can be advantageous in many cases, if the factors are properly assigned. The design is also useful when there is a need to shift to some direction in the experimental domain. Then, only some additional experiments are necessary (Figure 1.5 contains a two-dimensional example).

TABLE 1.4

Box–Behnken Design for Three and Four Factors

Experiment	Three Factors			Four Factors			
	F1	F2	F3	F1	F2	F3	F4
1	0	0	0	0	0	0	0
2	0	1	−1	1	−1	0	0
3	1	−1	0	0	0	1	1
4	1	1	0	1	1	0	0
5	0	1	1	−1	1	0	0
6	−1	0	1	0	0	−1	1
7	0	−1	1	−1	−1	0	0
8	1	0	−1	0	0	1	−1
9	−1	−1	0	0	0	−1	−1
10	0	−1	−1	−1	0	0	−1
11	−1	1	0	−1	0	0	1
12	−1	0	−1	1	0	0	−1
13	1	0	1	0	−1	−1	0
14				1	0	0	1
15				0	1	1	0
16				0	1	−1	0
17				0	−1	1	0
18				0	−1	0	1
19				0	1	0	−1
20				1	0	1	0
21				0	1	0	1
22				−1	0	1	0
23				1	0	−1	0
24				0	−1	0	−1
25				−1	0	−1	0

Note: Replicates of center point are not included.

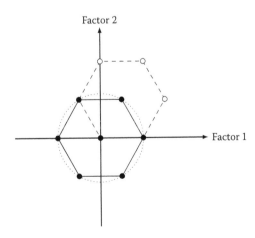

FIGURE 1.5 Two-factor Doehlert design with a dotted circle showing constant distance of each point to the origin. Shifting this design to higher levels of both factors (or in another direction in the experimental domain) requires only three additional experiments (white points and dashed line).

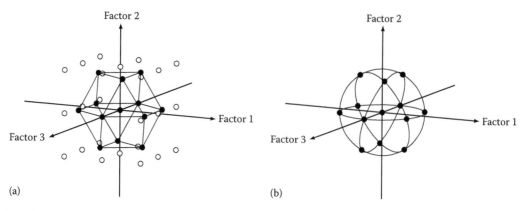

FIGURE 1.6 Three-factor Doehlert design placed inside a grid of a 3^3 full factorial design, forming a cuboctahedron with constant edge lengths (a), while the sphericity (points lie on a unit sphere with distances to the origin always equal to one) is seen in (b).

1.7 MIXTURE DESIGNS

During the optimization of the mobile-phase composition (modifier composition), the fractions of mixture constituents are optimized. Classical designs no longer make sense because mixture variables are not independent, while in the classical designs this is a requirement for the factors included. For example, 0.5 mL acetonitrile (low level) mixed with 0.5 mL water (low level) makes the same mixture as 1 mL (high level) and 1 mL (high level). The solvents cannot be independently changed in the full range, as the sum of their fractions must be equal to unity or 100%. In a mixture of p compounds only $p - 1$ are independent. Consequently, in a classical screening, $p - 1$ mixture factors can be included together with other chromatographic factors (pH, column temperature, etc.). The pth mixture factor is then used to bring the remaining percentage up to 100% or the sum of fractions to 1. This rule applies also when the mobile-phase mixture factors undergo *robustness* testing in a screening design.

When only mixture factors are to be optimized, a so-called mixture design is applied. A typical mixture-design approach in chromatography is performed when optimizing the modifier composition in the mobile phase applying Snyder's solvent triangle [48]. In a mixture design, the number of degrees of freedom is limited and one works basically with the ratios of solvents. A three-component mixture can be represented as a point inside an equilateral triangle [49]. A point lying in a vertex corresponds to a pure solvent. Points lying on the edges are two-component mixtures. Any other point represents a three-component mixture, with equal proportions between constituents in the centroid of the triangle (Figure 1.7). For a four-component phase, the experimental domain is a tetrahedron (Figure 1.8), for example, a mobile phase with three modifiers and water. The triangular domain to optimize the modifier composition is obtained by making an intersection with a plane. For instance, the triangle can represent isoeluotropic conditions (constant solvent strength throughout the triangle). When optimizing a mobile phase, often only the edges of the triangle are examined (mobile phase with two modifiers) since usually in the triangle no better separations are obtained.

This can be generalized for any larger number of mixture factors in multidimensional space, but it cannot be visualized anymore. Generally, a structure with $n + 1$ vertices in n-dimensional space is called a *simplex*. An experimental design for a mixture problem consists of points filling uniformly this simplex, which represents the experimental domain (Figure 1.7).

As the percentages of mixture constituents are not orthogonal (they are intercorrelated), some authors recommend the use of partial least squares (PLS) regression with a proper validation to fit the model (see [49] as starting point for further reading).* However, most frequently, the quadratic polynomial model, discussed in the section on model building, is used.

---

* See also Chapter 14 for PLS description.

TABLE 1.5

Doehlert Designs for Two to Five Factors

Exp.	Two Factors		Three Factors			Four Factors				Five Factors				
	F1	F2	F1	F2	F3	F1	F2	F3	F4	F1	F2	F3	F4	F5
1	0.0000	0.0000	0.0000	0.0000	0.0000	0.0000	0.0000	0.0000	0.0000	0.0000	0.0000	0.0000	0.0000	0.0000
2	1.0000	0.0000	1.0000	0.0000	0.0000	1.0000	0.0000	0.0000	0.0000	1.0000	0.0000	0.0000	0.0000	0.0000
3	-1.0000	0.0000	-1.0000	0.0000	0.0000	-1.0000	0.0000	0.0000	0.0000	-1.0000	0.0000	0.0000	0.0000	0.0000
4	0.5000	0.8660	0.5000	0.8660	0.0000	0.5000	0.8660	0.0000	0.0000	0.5000	0.8660	0.0000	0.0000	0.0000
5	-0.5000	-0.8660	-0.5000	-0.8660	0.0000	-0.5000	-0.8660	0.0000	0.0000	-0.5000	-0.8660	0.0000	0.0000	0.0000
6	0.5000	-0.8660	0.5000	-0.8660	0.0000	0.5000	-0.8660	0.0000	0.0000	0.5000	-0.8660	0.0000	0.0000	0.0000
7	-0.5000	0.8660	-0.5000	0.8660	0.0000	-0.5000	0.8660	0.0000	0.0000	-0.5000	0.8660	0.0000	0.0000	0.0000
8			0.5000	0.2887	0.8165	0.5000	0.2887	0.8165	0.0000	0.5000	0.2887	0.8165	0.0000	0.0000
9			-0.5000	-0.2887	-0.8165	-0.5000	-0.2887	-0.8165	0.0000	-0.5000	-0.2887	-0.8165	0.0000	0.0000
10			0.5000	-0.2887	-0.8165	0.5000	-0.2887	-0.8165	0.0000	0.5000	-0.2887	-0.8165	0.0000	0.0000
11			0.0000	0.5774	-0.8165	0.0000	0.5774	-0.8165	0.0000	0.0000	0.5774	-0.8165	0.0000	0.0000
12			-0.5000	0.2887	0.8165	-0.5000	0.2887	0.8165	0.0000	-0.5000	0.2887	0.8165	0.0000	0.0000
13			0.0000	-0.5774	0.8165	0.0000	-0.5774	0.8165	0.0000	0.0000	-0.5774	0.8165	0.0000	0.0000
14						0.5000	0.2887	0.2041	0.7906	0.5000	0.2887	0.2041	0.7906	0.0000
15						-0.5000	-0.2887	-0.2041	-0.7906	-0.5000	-0.2887	-0.2041	-0.7906	0.0000
16						0.5000	-0.2887	-0.2041	-0.7906	0.5000	-0.2887	-0.2041	-0.7906	0.0000
17						0.0000	0.5774	-0.2041	-0.7906	0.0000	0.5774	-0.2041	-0.7906	0.0000
18						0.0000	0.0000	0.6124	-0.7906	0.0000	0.0000	0.6124	-0.7906	0.0000
19						-0.5000	0.2887	0.2041	0.7906	-0.5000	0.2887	0.2041	0.7906	0.0000
20						0.0000	-0.5774	0.2041	0.7906	0.0000	-0.5774	0.2041	0.7906	0.0000
21						0.0000	0.0000	-0.6124	0.7906	0.0000	0.0000	-0.6124	0.7906	0.0000
22										0.5000	0.2887	0.2041	0.1581	0.7746
23										-0.5000	-0.2887	-0.2041	-0.1581	-0.7746
24										0.5000	-0.2887	-0.2041	-0.1581	-0.7746
25										0.0000	0.5774	-0.2041	-0.1581	-0.7746
26										0.0000	0.0000	0.6124	-0.1581	-0.7746
27										0.0000	0.0000	0.0000	0.6325	-0.7746
28										-0.5000	0.2887	0.2041	0.1581	0.7746
29										0.0000	-0.5774	0.2041	0.1581	0.7746
30										0.0000	0.0000	-0.6124	0.1581	0.7746
31										0.0000	0.0000	0.0000	-0.6325	0.7746

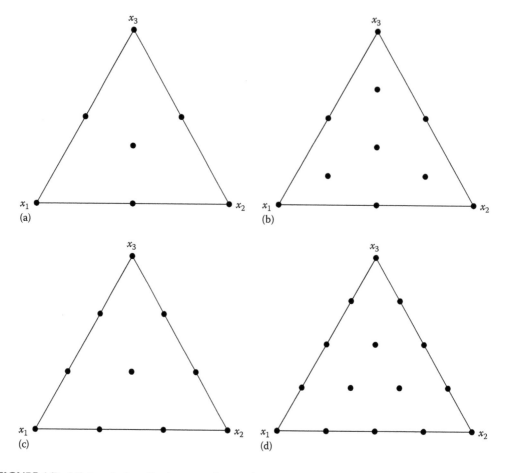

FIGURE 1.7   Mixture designs for three constituents (x_1, x_2, and x_3): simplex centroid design (a), augmented simplex centroid design (b), and simplex lattice designs with three (c) and four (d) levels.

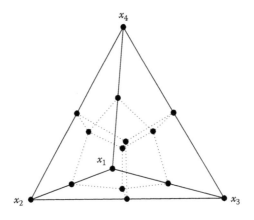

FIGURE 1.8   Simplex centroid design for mixture of four constituents.

The mixture designs are divided into simplex centroid and simplex lattice designs. The simplex centroid design (SCD) for n solvents (examples in Figures 1.7 and 1.8) consists of $2^n - 1$ points. These are all n pure solvents, all binary 50%/50% combinations, all ternary 33.3%/33.3%/33.3% combinations, all quaternary (25%/25%/25%/25%)... (ending with a combination of all solvents in equal proportion 100%/n). The simplex lattice design (SLD, Figure 1.7) requires $(n + k - 1)!/$ $(k!(n - 1)!)$ points, where k is the number of levels of each factor and n is the number of factors.

1.8 RESTRICTED DESIGNS

There are cases when some combinations of the factor levels are impossible (e.g., two solvents do not mix in given proportions, device exceeds the maximal pressure). This can occur for both mixture and orthogonal factors and their combinations. The above means that the resulting experimental domain is restricted and not necessarily symmetric anymore. Then, there is a need to generate custom design matrices, filling the *possible* (restricted) experimental domain uniformly. All approaches are characterized by the generation of a uniform grid of possible experiments from which then a *representative* subset is chosen. This can be done by optimizing the position of the experimental points in the restricted multivariate space to maximize a given criterion. Several criteria exist. However, the most often used concept in this context is called D-optimality, leading to a D-optimal design [50–52]. The generation of the custom designs can still be done in various other ways. The most often used approaches are the Fedorov algorithm [50] and the Kennard and Stone algorithm [53]. The same rules apply for a restricted mixture design [49]. Specialized software is used and the design matrix is always custom generated, depending on particular restrictions.

1.9 BUILDING A MODEL

By *building a model* we assume making a regression, modeling the obtained responses (see Figure 1.9) as a function of the factors (design matrix). The model matrix is obtained from the design matrix by augmenting the matrix with interaction and higher-order terms.

In general, the following equation is fitted, modeling a response as a function of the factors:

$$y = \beta_0 + \sum_{i=1}^{k} \beta_i x_i + \sum_{i=1}^{k} \beta_i x_{ii}^2 + \sum_{1 \le i \le j}^{k} \beta_i x_i x_j + \varepsilon$$

where the first term is the intercept, the second represents the linear effects, the third are the quadratic effects, and the fourth are two-factor interactions. The last term (ε) indicates the error of the model, while β_i represents the model coefficients.

For mixture problems (mixture designs), the above equation, for three factors, becomes

$$y = \beta_1 x_1 + \beta_2 x_2 + \beta_3 x_3 + \beta_{12} x_1 x_2 + \beta_{13} x_1 x_3 + \beta_{23} x_2 x_3 + \beta_{123} x_1 x_2 x_3 + \varepsilon$$

This model is equivalent to the above but rewritten for a mixture situation. The general model can be simplified because for mixtures the following restriction is valid: $x_1 + x_2 + x_3 = 1$.

The linear models (with optional interactions) are fitted during screening, or alternatively factor and interaction effects are calculated. During optimization, a quadratic model is fitted to obtain a curvilinear response surface and find the optimum (minimum, maximum, or specific value of the response to be optimized; Figure 1.9). The coefficients of the model can be useful in interpreting *how strong* the effect of a particular factor to the response is, and whether the interactions are significant. The significance of quadratic terms has also a diagnostic value as it can be perceived as an indication of curvature (curvature can be also detected with a lack-of-fit test).

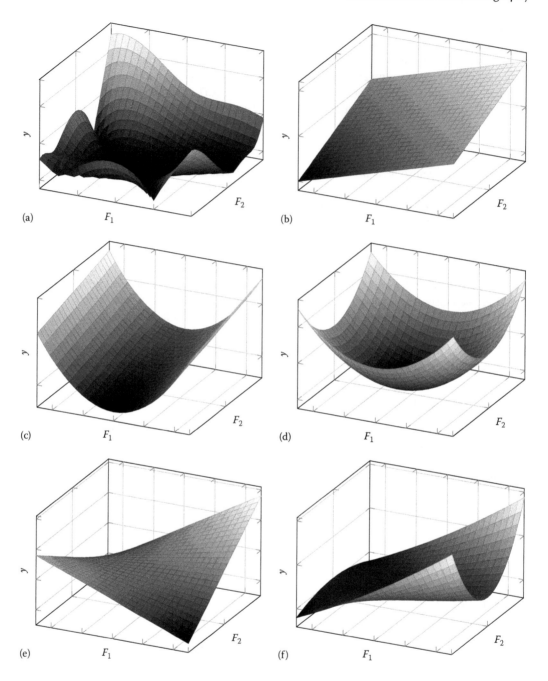

FIGURE 1.9 Examples of response surfaces of two factors: (a) complex response with many interactions; (b) linear response; (c) curvilinear response for one factor, without interactions; (d) curvilinear response for both factors, without interactions; (e) linear response with interaction; and (f) curvilinear response with interaction.

The quality of the model can be validated by leave-one-out cross-validation.* This technique can detect overfitting and when that occurs the complexity of the model should be decreased (e.g., by deleting interaction terms or quadratic terms). For the review of other significance-estimating techniques, the reader is referred to [54].

---

* Theory of model validation can be found in Chapter 14.

It must be emphasized that the design model has a number of *degrees of freedom* equal to the difference between the number of experiments and the number of coefficients. If this number is low, the significances of coefficients and their standard error cannot properly be estimated (or, if it is zero, it cannot be estimated at all, unless one has error estimates obtained externally).

However, one should make a difference between modeling in multivariate calibration and response surface modeling in method optimization. In the first situation, the model should accurately predict a response for future new samples (the predictive ability). Thus, much effort is spent in developing the best model with good predictive properties. In method optimization, one wants to determine the part of the experimental domain where the response is suitable. Therefore, the accuracy of the predictions is less critical. This is also the reason why the above-mentioned model is applied. Usually, it fits the measured results suitably well to be able to determine the suitable regions in experimental domain properly.

In chromatography, when optimizing the separation of a mixture, the response of interest is related to the separation, more specifically to the separation of the worst-separated pair of peaks. This can be quantified by, for instance, minimal resolution R_{Smin} or the minimal separation factor, α_{min}. However, these responses cannot be modeled as such since in the different grid points of the experimental domain (i.e., at the different possible experiments) they are determined by different peak pairs, resulting in an irregular response surface that cannot be modeled. Modeling the R_S or α between consecutive peak pairs is also problematic. Suppose the separation of two peaks A and B. Regardless of the sequence, R_S is positive and $\alpha \geq 1$. This means, for instance, that identical separations of $A - B$ and $B - A$ both get the same R_S value, for example, $R_S = 3$. Thus, different separations result in the same response value (idem for α). However, when this change is due to a given factor, it means that somewhere at an intermediate level R_S went through zero (α through 1). Such behavior cannot be modeled by the above equation unless one would distinguish between both situations by changing the sign of the resolution, for example, $R_S(A - B) = 3$ and $R_S(B - A) = -3$.

Moreover, when taking the above into account, one still has to model all possible α or R_S responses, since one does not know a priori which peak pairs will be relevant at a given grid point. For instance, when separating 6 compounds, at each grid point 5 peak pairs are relevant, while $5 + 4 + 3 + 2 + 1 = 15$ pairs can be considered.

Therefore, R_S or α should not be modeled unless no change in peak sequence occurs—in this case one should model the retention time or the retention factor. Thus, in the example, six retention parameters are modeled (in isocratic elution also their widths). Then at each grid point, the retention factor is predicted and this allows calculating R_S or α for the consecutive peaks (the relevant peak pairs). This allows determining the minimal resolution at each grid point as well as the conditions where the minimal resolution is maximal, that is, the optimum.

REFERENCES

1. Araujo, P.W. and G. Brereton. 1996. Experimental design. I. Screening, *Trends in Analytical Chemistry*, 15(1): 26–31.
2. Dejaegher, B., D. Mangelings, and Y. Vander Heyden. 2013. Experimental design methodologies in the optimization of chiral CE or CEC separations: An overview, *Methods in Molecular Biology*, 970: 409–427.
3. Dejaegher, B. and Y. Vander Heyden. 2011. Experimental designs and their recent advances in set-up, data interpretation, and analytical applications, *Journal of Pharmaceutical and Biomedical Analysis*, 56(2): 141–158.
4. Dejaegher, B. and Y. Vander Heyden. 2009. The use of experimental design in separation science, *Acta Chromatographica*, 21(2): 161–201.
5. Hibbert, D.B. 2012. Experimental design in chromatography: A tutorial review, *Journal of Chromatography B: Analytical Technologies in the Biomedical and Life Sciences*, 910: 2–13.
6. Leardi, R. 2009. Experimental design in chemistry: A tutorial, *Analytica Chimica Acta*, 652(1–2): 161–172.

7. Ebrahimi-Najafabadi, H., R. Leardi, and M. Jalali-Heravi. 2014. Experimental design in analytical chemistry—Part I: Theory, *Journal of AOAC International*, 97(1): 3–11.
8. Altria, K.D., B.J. Clark, S.D. Filbey, M.A. Kelly, and D.R. Rudd. 1995. Application of chemometric experimental designs in capillary electrophoresis: A review, *Electrophoresis*, 16(11): 2143–2148.
9. Hanrahan, G., R. Montes, and F.A. Gomez. 2008. Chemometric experimental design based optimization techniques in capillary electrophoresis: A critical review of modern applications, *Analytical and Bioanalytical Chemistry*, 390(1): 169–179.
10. Jančic Stojanovic, B. 2013. Factorial-based designs in liquid chromatography, *Chromatographia*, 76(5–6): 227–240.
11. Bianchi, F. and M. Careri. 2008. Experimental design techniques for optimization of analytical methods. Part I: Separation and sample preparation techniques, *Current Analytical Chemistry*, 4(1): 55–74.
12. Riter, L.S., O. Vitek, K.M. Gooding, B.D. Hodge, and R.K. Julian Jr. 2005. Statistical design of experiments as a tool in mass spectrometry, *Journal of Mass Spectrometry*, 40(5): 565–579.
13. Bourguignon, B. and D.L. Massart. 1991. Simultaneous optimization of several chromatographic performance goals using Derringer's desirability function, *Journal of Chromatography A*, 586(1): 11–20.
14. Derringer, G. and R. Suich. 1980. Simultaneous optimization of several response variables, *Journal of Quality Technology*, 12(4): 214–219.
15. Nemutlu, E., S. Kir, D. Katlan, and M.S. Beksaç. 2009. Simultaneous multiresponse optimization of an HPLC method to separate seven cephalosporins in plasma and amniotic fluid: Application to validation and quantification of cefepime, cefixime and cefoperazone, *Talanta*, 80(1): 117–126.
16. Vera Candioti, L., M.M. De Zan, M.S. Cámara, and H.C. Goicoechea. 2014. Experimental design and multiple response optimization. Using the desirability function in analytical methods development, *Talanta*, 124: 123–138.
17. Keller, H.R., D.L. Massart, and J.P. Brans. 1991. Multicriteria decision making: A case study, *Chemometrics and Intelligent Laboratory Systems*, 11(2): 175–189.
18. Dejaegher, B., A. Durand, and Y. Vander Heyden. 2009. Identification of significant effects from an experimental screening design in the absence of effect sparsity, *Journal of Chromatography B: Analytical Technologies in the Biomedical and Life Sciences*, 877(23): 2252–2261.
19. Dejaegher, B., J. Smeyers-Verbeke, and Y. Vander Heyden. 2005. The variance of screening and supersaturated design results as a measure for method robustness, *Analytica Chimica Acta*, 544(1–2): 268–279.
20. Li, W. and H.T. Rasmussen. 2003. Strategy for developing and optimizing liquid chromatography methods in pharmaceutical development using computer-assisted screening and Plackett-Burman experimental design, *Journal of Chromatography A*, 1016(2): 165–180.
21. Vander Heyden, Y., M.S. Khots, and D.L. Massart. 1993. Three-level screening designs for the optimisation or the ruggedness testing of analytical procedures, *Analytica Chimica Acta*, 276(1): 189–195.
22. de Almeida, A.A. and I.S. Scarminio. 2007. Statistical mixture design optimization of extraction media and mobile phase compositions for the characterization of green tea, *Journal of Separation Science*, 30(3): 414–420.
23. Araujo, P.W. and R.G. Brereton. 1996. Experimental design II. Optimization, *Trends in Analytical Chemistry*, 15(2): 63–70.
24. Bezerra, M.A., R.E. Santelli, E.P. Oliveira, L.S.A. Villar, and L.A. Escaleira. 2008. Response surface methodology (RSM) as a tool for optimization in analytical chemistry, *Talanta*, 76(5): 965–977.
25. Lundstedt, T., E. Seifert, L. Abramo, B. Thelin, A. Nyström, J. Pettersen, and R. Bergman. 1998. Experimental design and optimization, *Chemometrics and Intelligent Laboratory Systems*, 42(1–2): 3–40.
26. Araujo, P.W. and R.G. Brereton. 1996. Experimental design III. Quantification, *Trends in Analytical Chemistry*, 15(3): 156–163.
27. Dejaegher, B. and Y. Vander Heyden. 2007. Ruggedness and robustness testing, *Journal of Chromatography A*, 1158(1–2): 138–157.
28. Dejaegher, B. and Y. Vander Heyden. 2008. Robustness tests of CE methods, *Capillary Electrophoresis Methods for Pharmaceutical Analysis*, (eds. M. Jimidar and S. Ahuja). Elsevier, Amsterdam, the Netherlands, pp. 185–224.
29. Hartmann, C., J. Smeyers-Verbeke, D.L. Massart, and R.D. McDowall. 1998. Validation of bioanalytical chromatographic methods, *Journal of Pharmaceutical and Biomedical Analysis*, 17(2): 193–218.
30. Vander Heyden, Y., A. Bourgeois, and D.L. Massart. 1997. Influence of the sequence of experiments in a ruggedness test when drift occurs, *Analytica Chimica Acta*, 347(3): 369–384.

31. Vander Heyden, Y., K. Luypaert, C. Hartmann, D.L. Massart, J.B. Hoogmartens, and J.C. De Beer. 1995. Ruggedness tests on the high-performance liquid chromatography assay of the United States Pharmacopeia XXII for tetracycline hydrochloride. A comparison of experimental designs and statistical interpretations, *Analytica Chimica Acta*, 312(3): 245–262.
32. Vander Heyden, Y. and D.L. Massart. 1996. Review of the use of robustness and ruggedness in analytical chemistry, *Robustness of Analytical Methods and Pharmaceutical Technological Products*, (eds. M.M.W.B. Hendriks, J.H. de Boer, and A.K. Smilde), Elsevier, Amsterdam, the Netherlands, pp. 79–147.
33. Vander Heyden, Y., A. Nijhuis, J. Smeyers-Verbeke, B.G.M. Vandeginste, and D.L. Massart. 2001. Guidance for robustness/ruggedness tests in method validation, *Journal of Pharmaceutical and Biomedical Analysis*, 24(5–6): 723–753.
34. Vander Heyden, Y., F. Questier, and L. Massart. 1998. Ruggedness testing of chromatographic methods: Selection of factors and levels, *Journal of Pharmaceutical and Biomedical Analysis*, 18(1–2): 43–56.
35. Dejaegher, B., X. Capron, and Y. Vander Heyden. 2007. Fixing effects and adding rows (FEAR) method to estimate factor effects in supersaturated designs constructed from Plackett-Burman designs, *Chemometrics and Intelligent Laboratory Systems*, 85(2): 220–231.
36. Dejaegher, B., M. Dumarey, X. Capron, M.S. Bloomfield, and Y. Vander Heyden. 2007. Comparison of Plackett-Burman and supersaturated designs in robustness testing, *Analytica Chimica Acta*, 595(1–2): 59–71.
37. Plackett, R.L. and J.P. Burman. 1946. The design of optimum multifactorial experiments, *Biometrika*, 33(4): 305–325.
38. Dejaegher, B., X. Capron, and Y. Vander Heyden. 2007. Generalized FEAR method to estimate factor effects in two-level supersaturated designs, *Journal of Chemometrics*, 21(7–9): 303–323.
39. Dejaegher, B. and Y. Vander Heyden. 2008. Supersaturated designs: Set-ups, data interpretation, and analytical applications, *Analytical and Bioanalytical Chemistry*, 390(5): 1227–1240.
40. Heyden, Y.V., S. Kuttatharmmakul, J. Smeyers-Verbeke, and D.L. Massart. 2000. Supersaturated designs for robustness testing, *Analytical Chemistry*, 72(13): 2869–2874.
41. Lin, D.K.J. 1995. Generating systematic supersaturated designs, *Technometrics*, 37(2): 213–225.
42. Yamada, S. and D.K.J. Lin. 1999. Three-level supersaturated designs, *Statistics and Probability Letters*, 45(1): 31–39.
43. Ebrahimi-Najafabadi, H., R. Leardi, and M. Jalali-Heravi. 2014. Experimental design in analytical chemistry—Part II: Applications, *Journal of AOAC International*, 97(1): 12–18.
44. Box, G.E.P. and D.W. Behnken. 1960. Simplex-sum designs: A class of second order rotatable designs derivable from those of first order, *The Annals of Mathematical Statistics*, 31(4): 838–864.
45. Ferreira, S.L.C., R.E. Bruns, H.S. Ferreira, G.D. Matos, J.M. David, G.C. Brandão, E.G.P. da Silva et al. 2007. Box-Behnken design: An alternative for the optimization of analytical methods, *Analytica Chimica Acta*, 597(2): 179–186.
46. Doehlert, D.H. 1970. Uniform shell designs, *Applied Statistics*, 19: 231–239.
47. Ferreira, S.L.C., W.N.L. Dos Santos, C.M. Quintella, B.B. Neto, and J.M. Bosque-Sendra. 2004. Doehlert matrix: A chemometric tool for analytical chemistry—Review, *Talanta*, 63(4): 1061–1067.
48. Johnson, A.R. and M.F. Vitha. 2011. Chromatographic selectivity triangles, *Journal of Chromatography A*, 1218(4): 556–586.
49. Eriksson, L., E. Johansson, and C. Wikstrom. 1998. Mixture design—Design generation, PLS analysis, and model usage, *Chemometrics and Intelligent Laboratory Systems*, 43(1–2): 1–24.
50. de Aguiar, P.F., B. Bourguignon, M.S. Khots, D.L. Massart, and R. Phan-Than-Luu. 1995. D-optimal designs, *Chemometrics and Intelligent Laboratory Systems*, 30(2): 199–210.
51. Goupy, J. 1996. Unconventional experimental designs theory and application, *Chemometrics and Intelligent Laboratory Systems*, 33(1): 3–16.
52. Jones, B., D.K.J. Lin, and C.J. Nachtsheim. 2008. Bayesian D-optimal supersaturated designs, *Journal of Statistical Planning and Inference*, 138(1): 86–92.
53. Kennard, R.W. and L.A. Stone. 1969. Computer aided design of experiments, *Technometrics*, 11(1): 137–148.
54. Dejaegher, B., X. Capron, J. Smeyers-Verbeke, and Y.V. Heyden. 2006. Randomization tests to identify significant effects in experimental designs for robustness testing, *Analytica Chimica Acta*, 564(2): 184–200.

2 Chromatographic Response Functions

Regina M.B.O. Duarte, João T.V. Matos,
and Armando C. Duarte

CONTENTS

2.1 INTRODUCTION

Investigating chemical structures and quantifying analytes in complex mixtures has historically relied on the use of a chromatographic separation protocol followed by a detection method (e.g., diode array, fluorescence, evaporative light scattering, and mass spectrometry), whose choice is dependent on the problem being solved, as well as on the level of selectivity required. The importance of chromatography has increased as the science evolved, playing an important role not only in target analysis of well-known compounds (e.g., pharmaceutical drugs; food components such as carbohydrates, proteins, lipids, and artificial additives; polymers; organic pollutants; and natural bioactive compounds) [1,2] but also in more explorative applications (i.e., nontarget analysis) such as life sciences (e.g., proteomics and metabolomics) [3] and in the analysis of complex environmental samples [4,5].

Regardless of the field of application, the use of either one- (1D) or two-dimensional (2D) chromatography is still a very complex and challenging task, particularly when the main goal encompasses the comprehensive screening of all the compounds in a sample. As in any successful analytical method, the optimization of the chromatographic separation is crucial for achieving the desired purity of the target analyte(s). In determining the conditions for achieving the best chromatographic performance, in terms of the degree of separation, number of peaks, analysis time, and other chromatographic features (e.g., working conditions [flow rate and temperature] and elution mode [isocratic or gradient]), the analyst must exploit the possibilities of the chromatographic system. Although time-consuming, the conventional "trial and error" approach or "one variable at a time" is still very often used for the experimental optimization [6], which means that the best chromatographic separation depends on the expertise and intuition of the chromatographer [7]. However, this type of optimization strategy does not foresee the possibility of interaction between the experimental variables, being also expensive and unfeasible for solving complex mixtures.

Nowadays, it is of common agreement that experiment-based interpretative methodologies, supported by mathematical models, are the most efficient tools for finding the optimal conditions in chromatography,* particularly in those related to the routine analyses of selected individual compound(s) [6,8]. These methodologies include two main steps: (1) buildup of a retention model of the compounds to be separated (this requires running a limited number of experiments in order to fit equations or train algorithms that will allow the prediction of retention) [6,9], and (2) estimative of the separation quality, and therefore of the optimum conditions, through computer simulation [6]. The selection of a proper mathematical equation for mimicking the retention behavior of the solutes is an important step in those interpretative optimization procedures [10]. This means that the analyst must have an in-depth knowledge of the chromatographic system, including the stationary phase characteristics, eluents composition, flow and temperature effects, analyte–eluent interactions, and characteristics of the target analyte(s). Also core to all these strategies is finding a numerical criterion able to describe the overall separation quality of a chromatogram. The degree of separation between peaks (i.e., elementary and global resolution) is one of the most important requirements that define the quality of a chromatogram [6]. Depending on the purpose of analysis, other secondary requirements can also be considered, such as the number of peaks, total analysis time, shape of the peaks, and the uniformity of peak separation [6,8,11]. In real practice of chromatographic optimization strategies, including the computer-assisted ones, these elementary criteria are combined to construct a numerical criterion that allows describing the overall quality of separation in a single measurable value, which is used to classify each chromatogram in search of the optimal solution. In the literature, this global quality criterion is generally referred to as chromatographic response function (CRF) [7,12–15]. In simple separation problems, where the compounds of interest are known and for which standards are available, the optimization of the experimental conditions with the assistance of CRFs is straightforward, since it is relatively easy to quantify and rank the chromatograms with all chromatographic peaks resolved in a short time of analysis. However, when dealing with complex separation problems, namely, with samples containing unknown compounds or without standards available, the choice of a proper CRF becomes particularly daunting. Yet, this challenge must be overcome if sophisticated chromatographic separation approaches are to be optimized efficiently for addressing such complex mixtures.

Many different CRFs have been proposed in the literature, and a number of studies have been devoted to discuss which CRF is the most suitable for solving separation problems of diverse complexity [6,8,11,12,14–23]. This chapter aims at building upon these earlier works, and it presents a critical assessment of the different optimization criteria currently available for applying in 1D and 2D chromatography. Firstly, several chromatographic elementary criteria and their capabilities are evaluated in detail. Secondly, the most relevant CRFs introduced to date are critically compared in terms of their application and their usefulness. The final section of this review addresses the challenges ahead when using these chromatographic descriptors as tools for assessing chemical complexity through 1D and 2D chromatography.

2.2 ASSESSING THE QUALITY OF CHROMATOGRAMS IN ONE-DIMENSIONAL CHROMATOGRAPHY

2.2.1 Approaches to Measure the Elementary Resolution

In any chromatographic separation problem, the analyst faces one simple, yet complex, question: How does one control and assess the resolution obtained in a chromatographic separation? The key equation (Equation 2.1) for assessing the success of a 1D chromatographic separation indicates that

---

* See Chapter 1 for more information.

resolution $(R_{i,i+1})$ can be affected by three important parameters: efficiency $(\sqrt{N}/2)$, capacity factor $(\bar{k}_{i,i+1}/(1+\bar{k}_{i,i+1}))$, and selectivity factor $((\alpha_{i,i+1}-1)/(\alpha_{i,i+1}+1))$ [24]:

$$R_{i,i+1} = \left(\frac{\sqrt{N}}{2}\right) \times \left[\frac{\bar{k}_{i,i+1}}{(1+\bar{k}_{i,i+1})}\right] \times \left[\frac{(\alpha_{i,i+1}-1)}{(\alpha_{i,i+1}+1)}\right] \tag{2.1}$$

where
  N is the number of theoretical plates
  $\bar{k}_{i,i+1}$ is the average capacity factor $(\bar{k}_{i,i+1}=(k_i+k_{i+1})/2)$
  $\alpha_{i,i+1}$ is the selectivity factor (i and $i+1$ represent two consecutive chromatographic peaks)

The efficiency is a measure of the dispersion of the solute band as it travels through the column, thus reflecting the column performance. In practice, the column efficiency is usually expressed in terms of N, which may be defined as follows (Equation 2.2 [24]):

$$N = 5.54\left(\frac{t_{R,i}}{w_{1/2,i}}\right)^2 \approx 16\left(\frac{t_{R,i}}{w_{B,i}}\right)^2 \tag{2.2}$$

where
  $w_{1/2,i}$ and $w_{B,i}$ are the peak width at half height and the peak width at the baseline, respectively
  $t_{R,i}$ is the time needed for the solute band to elute from the column [24]

By definition, the capacity factor k_i of a solute is equal to the ratio of the quantity of solute in the stationary phase to the quantity in the mobile phase [24]. In practice, the capacity factor k_i is a means of measuring the retention of the solute on the chromatographic column, being easily determined from the obtained chromatographic profile using Equation 2.3 [24]:

$$k_i = \frac{(t_{R,i}-t_0)}{t_0} \tag{2.3}$$

where t_0 is the time that a mobile-phase molecule will spend in the column (also known as the hold-up time, mobile-phase time, or unretained time). The selectivity factor $(\alpha_{i,i+1})$ is the chromatographic parameter that is most directly related to the ability of the chromatographic system to distinguish between sample components. It is usually measured as a ratio of the capacity factors of two consecutive chromatographic peaks (k_i and k_{i+1}) as in Equation 2.4 [24]:

$$\alpha_{i,i+1} = \frac{k_{i+1}}{k_i} \tag{2.4}$$

By definition, $\alpha_{i,i+1}$ is always larger than unity, and it could be considered the simplest measurement for assessing and scoring the separation between each pair of consecutive peaks. However, because peak widths are not considered in this equation, the selectivity factor is only suited for comparing chromatograms where the peaks are relatively narrow with regard to the distance between peaks [6]. In chromatographic practice, the most popular criterion to measure and assess the resolution obtained in a chromatographic separation was defined by the International Union of Pure and Applied Chemistry (IUPAC) [25] (Equation 2.5), and it relates the distance between the retention

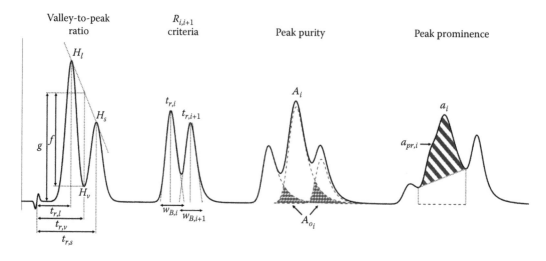

FIGURE 2.1 Chromatograms illustrating the parameters for the estimative of the quality of separation between overlapped 1D peaks using the IUPAC resolution measurement (Equation 2.5), the valley-to-peak ratio (Equation 2.8), the peak purity concept (Equation 2.9), and the peak prominence ratio ($pr_i = a_{pr,i}/a_i$ in Equation 2.23).

time of two consecutive peaks and their width at baseline (normally found by drawing the tangent along the slope of the peak, as depicted in Figure 2.1):

$$R_{i,i+1} = 2(t_{R,i+1} - t_{R,i})/(w_{B,i} + w_{B,i+1}) \qquad (2.5)$$

where

$t_{R,i}$ and $t_{R,i+1}$ are the retention times of two consecutive peaks, i and $i + 1$

$w_{B,i}$ and $w_{B,i+1}$ are the peak widths at baseline of those same peaks

This mathematical definition of chromatographic resolution can be easily applied for Gaussian-shaped individual peaks; however, it is not generally applicable for overlapping and asymmetric peaks, which constitute the most common chromatographic profile of complex mixtures. For such nonideal situations, Schoenmakers et al. [26] developed modifications of the $R_{i,i+1}$ parameter (Equations 2.6 and 2.7) that consider peak widths, asymmetries, and height ratios. In this definition, and when dealing with the separation of a pair of peaks, two values of the effective resolution exist: the first one, $^iR_{i,i+1}$, describes the extent to which peak i is separated from the next peak ($i + 1$) and the second value, $^{i+1}R_{i,i+1}$, reflects the extent to which peak $i + 1$ is separated from the previous one (i). Generally, for optimization purposes, the lowest value characterizing a particular separation (i.e., $^iR_{i,i+1}$ or $^{i+1}R_{i,i+1}$) is kept [27]:

$$^iR_{i,i+1} = \frac{\left[(t_{R,i+1} - t_{R,i})(1 + A_{S,i})(1 + A_{S,i+1})\sqrt{N_i N_{i+1}}\right]}{\left[4A_{S,i}t_{R,i}(1 + A_{S,i+1})\sqrt{N_{i+1}} + 4t_{R,i+1}(1 + A_{S,i})\sqrt{N_i}\sqrt{1 + 0.5\ln(H_{i+1}/H_i)}\right]} \qquad (2.6)$$

$$^{i+1}R_{i,i+1} = \frac{\left[(t_{R,i+1} - t_{R,i})(1 + A_{S,i})(1 + A_{S,i+1})\sqrt{N_i N_{i+1}}\right]}{\left[4A_{S,i}t_{R,i}(1 + A_{S,i+1})\sqrt{N_{i+1}} \times \sqrt{1 + 0.5\ln(H_{i+1}/H_i)} + 4t_{R,i+1}(1 + A_{S,i})\sqrt{N_i}\right]} \qquad (2.7)$$

where

$A_{S,i}$ and $A_{S,i+1}$ are the peak asymmetry factors
H_i and H_{i+1} are the peak heights
N_i and N_{i+1} are the number of theoretical plates of two consecutive peaks i and $i + 1$

In 1972, Carle [28] introduced a good alternative for determining the resolution between unresolved peaks whose heights differ greatly. The mathematical equation was derived using Kaiser's definition for valley-to-peak ratio, $\theta = f/g$, which is a function of peak overlap, as shown in Figure 2.1. The term f in the Kaiser's definition represents the distance between the valley separating the two peaks and a line joining the apexes of the peaks, whereas g represents the distance from the baseline to the line joining both peaks' maximum. As a ratio, this peak-to-valley criterion becomes a normalized measurement that varies between 0, when the peaks are fully overlapped, and 1, when the peaks are completely resolved at the baseline. This boundary condition constitutes an advantage when dealing with chromatographic separation of highly complex systems, since it does not require the prior estimation of an optimum and/or minimum acceptable resolution. This fact contrasts with the $R_{i,i+1}$ criterion that can, in principle, increase from 0.5 to infinity. As shown in Figure 2.1, for Gaussian peaks, the estimate of θ can be easily accomplished, since the construction lines for the determination of f and g can generally be accurately drawn. However, when interpreting real, complex chromatograms, where the sizes of the adjacent peaks are grossly different and disproportionate, the estimative of θ is not straightforward. In such cases, the line joining the peak apexes and the perpendicular through the valley becomes parallel to each other. This makes the intersection point of these lines difficult to locate, and therefore difficult to estimate both f and g in the valley-to-peak ratio. To overcome this problem, Carle [28] suggested the replacement of f and g by their geometrical equivalents. Equation 2.8 was then proposed to estimate the resolution ($\theta_{s,l}$) for adjacent peaks of highly unequal area and also for overlapping and asymmetric peaks. This method has no restrictions on the peaks' quality and only requires the definition of the peaks and valley heights and their respective retention times:

$$\theta_{s,l} = 1 - \left[\frac{\left(H_v \times \left|t_{R,l} - t_{R,s}\right|\right)}{\left(\left|t_{R,v} - t_{R,s}\right| \times \left(H_l - H_{ls}\right) + H_s \times \left|t_{R,l} - t_{R,s}\right|\right)} \right] \tag{2.8}$$

where

$t_{R,l}$, $t_{R,s}$, and $t_{R,v}$ are the retention times of the large and small peaks and their valleys, respectively
H_l, H_s, and H_v are the peak heights of the large and small peaks and their valleys, respectively (all depicted in Figure 2.1)

Another criterion for assessing elementary resolution of a 1D chromatographic separation is the peak purity [6,29], represented by Equation 2.9 and schematically shown in Figure 2.1. This measurement quantifies the peak area percentage free of interference, being the complementary fraction of the ratio between the total area of the peak of interest (A_i) and the area that is overlapped by other peaks (A_{O_i}, overlapped fraction) [29]:

$$P_i = 1 - \left(\frac{A_{O_i}}{A_i} \right) \tag{2.9}$$

The peak purity criterion is a normalized measurement that varies between 0 for a fully overlapped peak and 1 for a "pure" peak. The fractional peak overlap (FO), which is a similar measurement, was proposed by Schoenmakers [24], but often neglected by analysts, since it assumes Gaussian shape peaks and it requires knowledge of the exact peak positions and peak widths in the chromatogram.

In this sense, the peak purity criterion seems to be more advantageous, since it depends on relative peak areas and can be employed for addressing non-Gaussian peaks. Nevertheless, its main disadvantage lies in the need for a model of the chromatogram in order to identify the total and the overlapping areas. However, this limitation has been increasingly suppressed, due to the progress in the techniques of deconvolution of the chromatographic peaks and the increasingly faster computer-assisted calculations [6].

When assessing the quality of the whole chromatographic separation, the elementary criteria $R_{i,i+1}$, $\theta_{s,l}$, P_i, and FO are too simple to account for an overall criteria function for the quality of separation. It becomes necessary then to transform each of these individual criteria into a single descriptor of the separation level produced by the whole chromatogram (i.e., global resolution). Several global separation criteria have been already reviewed [6,24], and the most used are the worst elementary resolution (i.e., the least-resolved peak or peak pair), the summation or the product of the values of one of the elementary criteria previously defined for all pairs of peaks (or individual peaks), and some other more complex, such as the normalized $R_{i,i+1}$ product, which has already been used in target analysis [21,27]. Nevertheless, the analyst should be aware that combining the individual resolutions to yield a representative value of the global resolution in a chromatogram has its own limitations, which could hamper the optimization process. For instance, the product of an elementary criterion will be zero if any single pair of peaks is completely unresolved, even when practically all other peaks are resolved at the baseline [24]. In general, the scores given by the product of elementary criteria tend to be dominated by the separation level reached by the poorly resolved peaks. In this sense, the sum of the individual elementary criteria could be more suited to rectify such inconsistency. Nevertheless, even this alternative has its own weaknesses; for example, the sum of $R_{i,i+1}$ is not a useful criterion for assessing and scoring the quality of a chromatogram, since its value is determined largely by the largest values of $R_{i,i+1}$, that is, by the pairs of peaks that are the least relevant for the optimization of a separation [24]. The sum of $\theta_{s,l}$ or P_i gives a better representation of the actual separation achieved on a given column, since there is a limit to the contribution of fully resolved pairs of peaks [24].

In real optimization processes, however, more advanced criteria functions have been developed to quantify the quality of the chromatograms in order to determine the best separation conditions. These approaches consider that a chromatogram should be classified taking into account not only the "resolution criterion" but a relationship between other secondary requirements, such as the number of peaks and the time of analysis. As a consequence, several CRFs have been introduced aiming at relating the various criteria to quantitatively evaluate and score the quality of chromatograms in any particular experimental work. Section 2.2.2 will address the most relevant CRFs introduced to date for use in chromatographic optimization processes.

2.2.2   EVALUATION OF WHOLE CHROMATOGRAM USING CHROMATOGRAPHIC RESPONSE FUNCTIONS

For any optimization process to succeed, its goals must be defined unambiguously. As mentioned earlier, the quality of the chromatographic performance is best described when a mathematical function is used for classifying and translating each chromatogram into a measurable value, according to some specific elementary criteria. Besides the quality of separation, the number of peaks and the time spent in the analysis are typically the most important secondary requirements during a chromatographic optimization procedure. Regarding the number of peaks, the difficulty in applying this criterion is not on the counting of the actual number of peaks existing in the chromatogram, but in the procedure to correctly identify each peak. This detection becomes even more challenging for nontarget analysis of complex mixtures, since the amount of compounds is unknown and the obtained chromatograms exhibit a large number of overlapped peaks. In 1D chromatography, the peak finding process is usually attended by using derivatives of the chromatographic signals, but it can also be done by using peak filters, and wavelet transforms [30]. After flagging the number of peaks, this criterion may be used under two ways: (1) to maximize the function when the number of peaks is unknown and (2) to compare the number of peaks found in the sample with a known value for the same sample.

The other important secondary criterion is the time spent in analysis, which is normally associated with the minimization of the retention time of the last eluted peak. It is usually translated into a relationship between the total time of the analysis and the optimal analysis time previously defined. As recently reviewed by Matos et al. [19], this relationship can be expressed either by the absolute value of the difference [16] or the ratio [11] between the two above-mentioned times. Divjak et al. [31] proposed a more elaborated arrangement, represented by Equation 2.10, where a sigmoidal transformation compresses the time spent in the analysis between a value close to 1, for very short times, and a value close to 0, for long times:

$$g = \frac{1}{\left[1 + \exp\left(b_2 \times t_L + b_3\right)\right]} \tag{2.10}$$

where
    b_2 and b_3 are sigmoidal transformation parameters
    t_L is the retention time of the last peak in the chromatogram

Presetting a maximum acceptable time for the analysis, as suggested by Berridge [16] and Morris et al. [11], becomes problematic in the nontarget analysis of complex unknown samples. In order to overcome this situation, Duarte and Duarte [8] developed a function, described by Equation 2.11, that does not require any preset value and it varies between 0, for a time of analysis equal to the elution time corresponding to the extra-column volume, and a value close to 1, for a long time of analysis:

$$f(t) = \frac{\left(t_L - t_0\right)}{t_L} \tag{2.11}$$

where t_0 and t_L are the retention times of the extra-column volume and last eluted peak, respectively. A limitation of Equation 2.11 is that when t_L is more than 10 times higher than t_0, the result of $f(t)$ will be approximately 1, making it impossible to differentiate among chromatograms with very high times of analysis. In order to overcome this problem, Matos et al. [23] proposed a logarithmic expansion of the Duarte and Duarte [8] function, as described by Equation 2.12:

$$f(t)_w = N \times \frac{\log\left\{\left[1 - \left(\left(t_L - t_0\right)/t_L\right)\right] \times 100\right\}}{2} \tag{2.12}$$

where N is the total number of chromatographic peaks (resolvable and overlapped) detected in the chromatogram and the remaining parameters are the same as in Equation 2.11. Unlike the initial proposal of Duarte and Duarte [8], this time function maximizes the value obtained for shorter time of analysis, and that value is multiplied by N in order to assure the same relative weight of the other chromatographic elementary criteria [23]. This new criterion reaches a maximum (equal to N) when $t_L \approx t_0$, and zero when t_L is 100 times higher than t_0. According to Matos et al. [23], the logarithm nature of Equation 2.12 ensures that the results provide a better differentiation between chromatograms with small time of analysis, still being able to achieve some degree of differentiation between chromatograms with large time of analysis.

Once established the terms for measuring and classifying the separation between N analytes and the analysis time, several approaches have been proposed to appraise the overall quality of the chromatographic separation. One of the first CRFs was proposed by Watson and Carr [32], Equation 2.13, and this function can be divided into two terms [32–34]. In the first term, the global separation is estimated as the sum of a logarithmic ratio between the degree of separation of each peak pair using the Kaiser's definition (θ_i) and a value that the analyst defined as the desired peak separation (θ_{opt}). According to this definition, all pairs of peaks that meet or exceed the value set for θ_{opt} are given 0 weight and therefore do not contribute to the CRF. On the other hand, only those pairs of peaks whose peak separation (θ_i) does not meet the desired peak separation (θ_{opt}) are used in the estimative of the CRF; in such situations, the value of the logarithmic ratio will decrease as the peaks become increasingly more overlapped. The second term is the time of analysis criterion, where the total time of analysis (t_L) is subtracted from an ideal time of analysis (t_M) defined by the analyst. Again, if t_L is equal to or less than t_M, the result of this term is considered to be 0. In these cases, this second term does not contribute to the final value of the CRF; otherwise, it will decrease even more. This criterion is further controlled by a weight (identified as α) provided by the analyst for regulating its importance in the final result of the CRF:

$$CRF_{Watson} = \sum_{i=1}^{N-1} \ln\left(\theta_i/\theta_{opt}\right) + \alpha \times \left(t_M - t_L\right) \qquad (2.13)$$

As recently discussed by Matos et al. [19], the CRF_{Watson} does not include the number of peaks as a criterion for assessing the overall quality of the separation and, therefore, it can only be used in samples in which the maximum number of peaks is already known. Likewise, the assumption of an optimal time of analysis is an additional limitation for the application of such CRF to the analysis of complex unknown mixtures. Furthermore, the use of a single θ_{opt} value for all pairs of peaks will end up associating the same weight for every peak in the search of the optimal chromatographic conditions. Nevertheless, it should be possible to set different values of θ_{opt} for each pair of peaks, in order to lower the weight associated to peaks with less interest and to increase the weight associated to the target peaks, which will contribute more to the final CRF result [19].

The CRF suggested by Berridge [16], Equation 2.14, is probably the most known CRF that focuses on the well-resolved peaks [35–41]. This function can be divided into four factors: (1) the sum of the IUPAC resolution (R_i, in practice limited to a maximum value of 2) for all pairs of peaks, (2) the number of peaks (N), (3) a time parameter for assessing the time spent in the analysis, and (4) a time parameter related to the elution of the first chromatographic peak. Both the time of analysis and the elution time of the first peak are weighted against values taken by the analyst as acceptable. This function also uses arbitrary weights (α, β, γ), which allow the analyst to define the importance of three of the factors against the sum of R_i:

$$CRF_{Berridge} = \sum_{i=1}^{N-1} R_i + N^\alpha - \beta\left|t_M - t_L\right| - \gamma\left|t_0 - t_1\right| \qquad (2.14)$$

where t_M, t_0, t_L, and t_1 are the maximal acceptable analysis time, the minimal retention time for the first peak, and the retention times for the last and first peaks, respectively. The use of a global criterion such as the sum of R_i makes the $CRF_{Berridge}$ ideal for nontarget analysis, although it does not ensure a uniform separation of all pairs of peaks. Also, the N factor allows the $CRF_{Berridge}$ to be used in the optimization of the separation of unknown samples, where N may not be constant throughout the optimization process. However, the $CRF_{Berridge}$ approach requires the specification of both t_M and t_0, which can be a very challenging task for unknown samples.

The CRF described by Equation 2.15 is the chromatography resolution statistic (CRS) proposed by Schlabach and Excoffier [42]. This function can be divided into three different factors. The first factor allows the optimization to proceed until a minimum of 0, when the IUPAC resolution R_i reaches an optimal resolution (R_{opt}) predefined by the analyst. The second factor uses an average resolution value ($\bar{R}$), and attempts to ensure a uniform separation along the chromatogram. The third factor (t_L/N) is a multiplicative factor that increases the value of CRS when the time of analysis (t_L) increases and the number of peaks (N) decreases [11]:

$$CRS = \left\{ \sum_{i=1}^{N-1} \left[\frac{\left(R_i - R_{opt}\right)^2}{\left(R_i \times \left(R_i - R_{min}\right)^2\right)} \right] + \sum_{i=1}^{N-1} \left(\frac{R_i^2}{a\bar{R}^2} \right) \right\} \times \left(\frac{t_L}{N} \right) \qquad (2.15)$$

where
R_{min} is the minimum acceptable IUPAC resolution between adjacent pairs of peaks
a is the number of resolution elements

The CRS can be used to optimize the separation of unknown samples, since it includes N as one of the criteria. However, when dealing with such types of samples, the possible occurrence of false-positive results is very likely, since similar values for the t_L/N ratio can be obtained for different values of t_L and N. Therefore, the CRS seems to be more adequate for optimizing the chromatographic separation of the so-called representative samples that may well contain an unknown, but predictable, number of compounds [19]. Nevertheless, the CRS can also be employed in nontarget analysis, as long as R_{min} and R_{opt} are kept constant for all pairs of peaks. However, as in the case of the CRF_{Watson}, if the predefined values of R_{min} and R_{opt} are changed accordingly to the analyst goals in a particular separation problem, then the CRS may be also used in target analysis.

A hierarchical chromatography response function (HCRF), which is described by Equation 2.16, has been developed by Peichang and Hongxin [43], and further used by Chen et al. [44] in the optimization of the chromatographic separation of 11 alkaloids, as well as by Ji et al. [45] and Huang and Zhang [46] to optimize the chromatographic fingerprinting of herbal extracts. The HCRF can be divided into three factors: (1) the number of peaks (N) existing in the chromatogram, (2) the IUPAC resolution (R_i) of the least separated pair of peaks, and (3) the total time of analysis (t_L), which is subtracted by 100 (assuming that the total time of analysis does not exceed this value). All factors are multiplied by a different weight to ensure a hierarchy, being the number of peaks considered as the best criterion, followed by the resolution, and finally the time of analysis:

$$HCRF = 100,000 \times N + 10,000 \times R_i + \left(100 - t_L\right) \qquad (2.16)$$

The extremely high relevance given to the total number of peaks makes the HCRF especially suited for nontarget chromatographic analysis of samples in which the number of peaks can change during the optimization process. On the other hand, the use of the least separated pair of peaks as a measure of resolution makes the optimization biased toward the "target" established by this pair of peaks. Thus, if this critical pair of peaks becomes completely resolved at the baseline, then all peaks are uniformly separated. Otherwise, if this separation does not occur, then the optimization may lead to loss of resolution in the other pairs of peaks [19].

Morris et al. [11] suggested a chromatographic exponential function (CEF), Equation 2.17, where the optimal conditions are associated to a minimum value defined by two factors: (1) in the first factor, the quality of separation is assessed by a sum of an exponential value obtained by the difference between the IUPAC resolution (R_i) and a preset optimum resolution value (R_{opt}); and (2) in the second factor, a chromatogram is penalized when the time of analysis (t_L) is greater than a preset maximum acceptable time (t_M). This function does not take into account the number of peaks and, therefore, it can only be used in samples in which the total number of peaks appearing in the chromatogram is known and constant. Moreover, the first term associated with the quality of separation ensures that all pairs of peaks have the same impact in the CEF, as long as the value of R_{opt} is kept constant [19]:

$$CEF = \left[\left(\sum_{i=1}^{N-1} \left(1 - e^{\alpha\left(R_{opt} - R_i\right)} \right)^2 \right) + 1 \right] \times \left(1 + t_L / t_M \right) \tag{2.17}$$

Divjak et al. [31] used a sigmoidal transformation to obtain a normalized CRF, shown in Equation 2.18. The CRF^{exp} varies between 0 (unacceptable) and 1 (optimal) based on two factors [31,47–49]: (1) a separation quality criterion obtained by the product of the results of the sigmoidal transformation of the IUPAC resolution (R_i) of each pair of peaks and (2) a time of analysis criterion (g) obtained by the sigmoidal transformation of the total time of analysis as defined in Equation 2.10. The CRF^{exp} classifies all pairs of peak in the same way, and it becomes suitable for the chromatographic optimization of samples in which the number of peaks does not change [19]:

$$CRF^{exp} = \left\{ \prod_{i=1}^{N-1} \left[1 / \left(1 + \exp\left(b_0 \times R_i + b_1 \right) \right) \right] \right\}^{1/(N-1)} \times g \tag{2.18}$$

The DCRF suggested by Matos et al. [23], Equation 2.19, is a generalization of a first CRF developed by Duarte and Duarte [4,8], being well suited for describing and ranking the separation of peak pairs of highly unequal area, and also for overlapped and asymmetrical peaks. This function is oriented to maximize the result according to three factors: (1) the number of peaks (N); (2) the sum of the degree of separation of each pair of peaks ($\Sigma\theta_i$), estimated through the peak-to-valley ratio in Equation 2.8; and (3) the time-saving criterion, determined by means of Equation 2.12. These three factors have the same relative weight and, therefore, the same relative impact on the final value of the $DCRF$:

$$DCRF_f = \alpha \times \left(N \right) + \beta \times \left(\sum_{i=1}^{N-1} \theta_i \right) + \gamma \times \left(f(t) \right)_w, \quad \text{where } \alpha + \beta + \gamma = 1 \tag{2.19}$$

The number of combinations of values of α, β, and γ in the $DCRF_f$ can be almost unlimited, being associated with the relevance of each criterion for solving the analytical problem in hand. In this regard, Matos and coworkers [23] presented a range of different possible scenarios associated with different combinations of α, β, and γ, describing different cases that are most likely to occur in the application of $DCRF_f$ in chromatography. Using 15 different optimization scenarios and simulated chromatograms, the robustness of this CRF was successfully evaluated and proved to be an adequate response to each criterion and easily adaptable to different combinations of α, β, and γ. Readers are encouraged to consult the work of Matos et al. [23] to obtain additional information on this topic.

In 2011, Jančić-Stojanović and coworkers [18] suggested a new CRF (NCRF), depicted by Equation 2.20, which attempts to find the minimum value corresponding to the optimal conditions. This function is defined by two factors: (1) the quality of separation, which is measured by a ratio between the sum of the degree of separation of each peak pair (estimated by Equation 2.8) and the number of expected pairs of peaks (n), and (2) the ratio between the time of analysis and a preset maximum acceptable time, in a similar fashion to the CEF of Morris et al. [11]. This function also uses two arbitrary weights (α and β), which allow prioritizing the importance of each factor in the analysis:

$$NCRF = \left[\alpha \left(1 - \left(\sum_{i=1}^{n-1} \theta_{s,i} / n - 1 \right) \right) + 1 \right] \times \left(1 + \left(t_L / t_M \right)^{\beta} \right)$$ (2.20)

By taking into account the number of expected peaks, the *NCRF* becomes only usable for optimizing the chromatographic separation of samples in which the total number of peaks is already known. In this regard, the NCRF was developed for describing the chromatographic separation of only five compounds. It should also be mentioned that the NCRF classifies all pairs of peaks in the same way, in a nontarget fashion. Unfortunately, the authors neither present nor discuss how their function behaves against other CRF when dealing with the chromatographic separation of complex unknown samples [19].

In an alternative strategy, Ortín and coworkers [50] suggested the use of a peak count (*PC*) based function for addressing analytical problems with insufficient chromatographic resolution, but still of interest for finding the conditions offering the maximal possible separation. The strategy applies a so-called "fractional peak count" (*fPC*) function (Equation 2.21) based on the number of "well resolved" peaks, which are those exceeding a threshold of elementary peak purity (given by Equation 2.9):

$$fPC = PC + f$$ (2.21)

The integer part of Equation 2.21 indicates the number of compounds that exceeds the threshold, whereas the fractional part (*f*) quantifies the global resolution, taken as the geometric mean of elementary peak purities of the peaks that exceed the established threshold (limited to $0 \leq f < 1$). According to Ortín and coworkers [50], the *fPC* function is able to discriminate among conditions that resolve the same number of peaks in low-resolution situations, being oriented to quantify the success in the separation, and not its failure. Although this strategy could be applied to resolve complex samples, it requires a previous knowledge of the identity of each peak, and therefore, the integer part of Equation 2.21 corresponds to visible peaks of individual compounds. Besides, the computation of peak purities requires previous modeling of the chromatographic behavior using standards. These two requirements cannot be fulfilled when dealing with unknown samples, for which standards are not available and where the detected peaks in the chromatogram can have one or more compounds associated.

In a following study [22], the *fPC* concept has been updated to a so-called "limiting peak count with fractional term (*fLPC*)" function (Equation 2.22):

$$fLPC = LPC + f$$ (2.22)

This case considers that the threshold can be an absolute value (minimal peak purity beyond which the peak is considered "well resolved"), or a relative value (fraction or percentage of peak purity with regard to the limiting value for each compound). *PC* and *LPC* are the number of peaks whose elementary peak purity exceeds the absolute or relative threshold, respectively. As before, the integer part of Equation 2.22 indicates the number of compounds that exceed the threshold, and the fractional part qualifies the peak resolution, here taken as the product of elementary purities of the resolved peaks. The elementary purities are to be normalized inside a range limited by the established threshold (minimal value), and the limiting elementary peak purity (maximal value). The product of the normalized peak purities avoids bias of the f term toward small values for low thresholds. Again, $fLPC$ function suffers from the same limitations as fPC when dealing with unknown complex samples.

Recently, Alvarez-Segura and coworkers [15] suggested a new CRF, named "peak prominence," which quantifies the protruding part of each peak in a chromatogram with regard to the valleys that delimit it, or the baseline (depicted in Figure 2.1). The global separation (R) is evaluated according to Equation 2.23, on the basis of the prominence ratio ($pr_i = a_{pr,i}/a_i$):

$$R = \left(\prod_{i=1}^{n_{pr}} a_{pr,i}/a_i \right)^{1/n_{pr}} \tag{2.23}$$

where

$a_{pr,i}$ is the area (or height) of the protruding part of the peak
a_i represents its total area (or height)
n_{pr} is the number of detected peaks exceeding a preestablished threshold

According to the authors, the most important features of the prominence ratio include the following: its applicability to real-world chromatograms; it does not require standards; it attends to all visible peaks, each one corresponding to one or more underlying compounds; it is a normalized measurement; and it scores individual peaks, which facilitates the combination of the elementary resolution and peak counting. Therefore, in the proposed strategy, the primary criterion is the peak count (i.e., the number of peaks exceeding a threshold), followed by the resolution (i.e., the visibility of the protruding fractions) only when the number of significant peaks coincides [15]. The selection of the threshold is most probably one of the most import requirements for the success of this strategy, particularly when dealing with unknown samples, since this parameter is important to discriminate among the significant peaks and those associated with impurities, baseline perturbations, or noise. As the choice of an unsuitable threshold may lead to wrong decisions, Alvarez-Segura et al. [15] suggested that this choice should take into account the sample complexity: for complex samples, the threshold should be established as an absolute area, whereas for samples with only some unknown minor peaks, the decision should be based on the resolution (secondary criterion) restricted only to the main peaks. Being a normalized measurement, the global separation (R) varies between 0 (when at least one peak is fully overlapped) and 1 (when all peaks are baseline resolved), which means that R tends to be dominated by the peaks exhibiting low resolution (critical peak), being insensitive to the resolution reached by the remaining peaks [15]. This is probably one of the most important shortcomings of using this CRF to find the best separation conditions in the nontarget analysis of complex unknown samples.

In 2014, Tyteca and Desmet [14] reported the only large-scale *in silico* comparison study made to date regarding a set of different CRFs (some of them also reported in this review), focusing on their ability to guide search-based optimizations of chromatographic separations, and for cases where the number of compounds in the sample is unknown. The authors concluded that CRFs of the type $CRF = n_{obs} + NIP$, wherein n_{obs} is the number of observed peaks and NIP is a normalized noninteger part ($0 < NIP < 1$), perform significantly better than CRFs whose values do not increase monotonically with n_{obs} or are based on Snyder resolution R_s (Equation 2.1). However, f/g-based CRFs lose their advantage as soon as the noise level becomes important. The authors also concluded that the performance of the best CRFs depends strongly on the separation efficiency N (Equation 2.2) of the chromatographic column with which the search for the optimum separation conditions is conducted. For obtaining additional information on this comparative study, readers are encouraged to consult the work of Tyteca and Desmet [14].

2.3 ASSESSING THE QUALITY OF CHROMATOGRAMS IN TWO-DIMENSIONAL CHROMATOGRAPHY

2.3.1 EVALUATION OF 2D PEAK SEPARATION

Comprehensive 2D chromatography, either gas or liquid, has become an attractive technique for the separation of very complex multicomponent samples in many fields of modern analytical chemistry, including environmental analysis, food analysis, analysis of plant extracts, pharmaceutical products, and biological material. The groundbreaking aspect of this chromatographic technique is that the entire sample is resolved on two distinct columns (or separation mechanisms) of complementary selectivity, which results in enhanced peak capacities, particularly when compared to those attained in 1D chromatography. Details on the fundamentals and data handling in comprehensive 2D chromatography, either gas or liquid, and their applicability to various complex matrices have been described in depth in several reviews, and readers are encouraged to consult the works of François et al. [51], Malerod et al. [52], Meinert and Meierhenrich [53], Jandera [54], and Matos et al. [55], and references therein.

The optimization strategy of a 2D chromatographic method requires studying the numerous physical (e.g., column dimensions, particle sizes, flow rates, modulation time) and chemical parameters (e.g., mobile-phase composition) that affect the separation power of the whole methodology. Therefore, the first- and second-dimension separations have been largely optimized in an independent fashion, although there are some gradient programs that have focused on the optimization of both dimensions [56,57]. In terms of mathematical approaches for handling and processing 2D chromatographic data, there are still some shortcomings that have been reviewed by Matos et al. [55], whereas in terms of CRFs for 2D chromatography, only those of Duarte et al. [17] and Matos et al. [23] have been proposed to date. The main issue in the development of CRFs for 2D chromatography is that it has been hindered by the challenge of translating the concepts and data processing tools used in CRFs of 1D chromatography into models for 2D chromatography [55].

In terms of individual criteria for measuring the quality of separation in 2D chromatography, most developments assume that the peaks have a Gaussian shape. On the basis of this assumption, in 1997, Schure [58] presented a compilation of resolution metrics, of which this review highlights those of Equations 2.24 and 2.25:

$$Rs(\theta) = \left[\frac{\left(\delta_t \sqrt{\gamma^2 \sin^2 \theta + \cos^2 \theta} \right)}{4\sigma_x} \right] \quad (2.24)$$

where
δ_t are the deviations between peaks using the Euclidean distance
σ_x are the standard deviations of the x peak zone
γ is the ratio between the x and y standard deviations
θ is the angle between the line that links both peaks and a parallel line to the x-axis (depicted in Figure 2.2a)

$$Rs = \sqrt{Rs_x^2 + Rs_y^2} = \sqrt{\left(\delta_x^2/16\sigma_x^2 \right) + \left(\delta_y^2/16\sigma_y^2 \right)} \quad (2.25)$$

where
σ_x and σ_y are the standard deviations of peaks x and y, respectively
δ_x and δ_y are calculated as $\left(\bar{t}_{2,x} - \bar{t}_{1,x} \right)$ and $\left(\bar{t}_{2,y} - \bar{t}_{1,y} \right)$, respectively (illustrated in Figure 2.2a)

In 2007, Peters and coworkers [59] innovated on this field by suggesting the use of a resolution measure (*Rs*, Equation 2.26) based on the Kaiser's valley-to-peak ratio ($V = f/g$) and in the saddle point

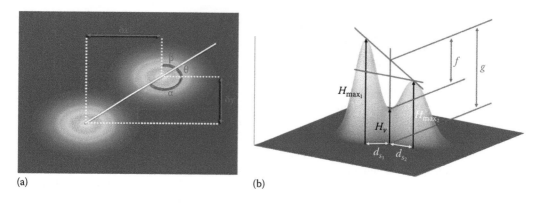

(a)   (b)

FIGURE 2.2 Schematic representation of simulated 2D chromatographic peaks, illustrating the parameters for the estimative of the quality of separation using (a) the resolution metrics in Equations 2.24 and 2.25 and (b) the valley-to-peak ratio between two overlapping 2D peaks by means of Equations 2.26 through 2.28.

concept (i.e., minimal point located in the shortest trajectory line between the two maximal points of Gaussian shape peaks, shown in Figure 2.2b):

$$Rs = \sqrt{-(1/2)\ln\left[(1-V)/2\right]}$$
(2.26)

where g and f of the valley-to-peak ratio (V) are defined by Equations 2.27 and 2.28, respectively:

$$g = \frac{\left(H_{\max_1} \times d_{s_2} + H_{\max_2} \times d_{s_1}\right)}{\left(d_{s_1} + d_{s_2}\right)}$$
(2.27)

$$f = g - H_v$$
(2.28)

where d_{s_1} and d_{s_2} are the distances between the retention time of each peak and that of the saddle point, while $H_{\max_1}$ and $H_{\max_2}$ represent their respective heights, and H_v is the height of the saddle point (all depicted in Figure 2.2b). The approach of Peters et al. [59] has to be commented on the fact that it focuses on the estimate of the quality of separation between pairs of neighboring peaks, which have been previously settled through a peak vicinity algorithm. However, in 2D chromatography, it is very likely that each individual 2D peak has more than only two neighboring peaks, thus making the estimative of the quality of separation through the algorithm of Peters et al. [59] an extremely complicated task. Furthermore, the saddle point concept and, therefore, the algorithm of Peters et al. [59] may fail when applied to a 2D chromatogram containing a large amount of peaks with a poor resolution degree, since it is very likely that no saddle point exists between the overlapping peaks. In order to overcome these difficulties, Duarte et al. [17] suggested the application of another concept for assessing the quality of separation: the 2D peak purity. This concept is inspired in the peak purity concept from 1D chromatography (Equation 2.9), but in 2D chromatography the computation of the peak areas (both total and overlapping) is replaced by the calculation of the 2D peak volumes (Equation 2.29):

$$P_i = 1 - \left(\frac{V_{O_i}}{V_i}\right)$$
(2.29)

where
   V_{O_i} is the volume of the overlapped region of a peak i
   V_i is the volume of peak i free from interferences

Additional details on the mathematical approach for the calculation of both V_{Oi} and V_i can be found in the study of Duarte and coworkers [17]. The methodology for assessing the elementary criterion for 2D peak separation, depicted by Equation 2.29, has the advantage of dealing with each peak at a time, without the need for identifying the multiple vicinities of each 2D peak as suggested by Peters et al. [59]. In fact, the estimation of the quality index of separation is now oriented to each individual 2D peak, instead of each 2D peak pair. However, its application does assume a model of the 2D chromatogram in order to allow the computation of both V_{Oi} and V_i in Equation 2.29. The application of such a model in 2D chromatography is not trivial and it is certainly time-consuming and daunting. Yet, the prospect of employing such a mathematical concept in the evaluation of 2D peak separation provides an extraordinary range of possibilities for solving separation problems using 2D chromatography.

2.3.2 ESTIMATING THE QUALITY INDEX OF SEPARATION IN 2D CHROMATOGRAPHY

Matos and coworkers [23] suggested a flexible CRF, identified as $DCRF_{f,2D}$, and unique in the best of the authors' knowledge, for the estimation of the global quality index of separation in nontarget 2D chromatographic analysis of complex samples. This is a generalization of the DCRF suggested by Duarte and Duarte [8] and of the $DCRF_{2D}$ suggested by Duarte et al. [17], and it includes the degree of purity of each 2D peak, the number of detected peaks, and the time of analysis as shown in Equation 2.30:

$$DCRF_{f,2D} = \alpha \times (N_{2D}) + \beta \times \left(\sum_{i=1}^{N_{2D}} P_{i,2D} \right) + \gamma \times \left(f(t)_{w,2D} \right) \qquad (2.30)$$

where
    α, β, and γ are the weights associated with each criterion (matching the condition $\alpha + \beta + \gamma = 1$)
    $P_{i,2D}$ is the individual purity of each 2D peak (Equation 2.29)
    N_{2D} is the number of 2D peaks appearing in the chromatogram
    $f(t)_{w,2D}$ is a function of time spent in the analysis calculated through Equation 2.31 [23]:

$$f(t)_{w,2D} = N_{2D} \times \log\left[\left(1 - f(t)_{2D}\right) \times 100 \right] / 2 \qquad (2.31)$$

where
    N_{2D} is again the number of peaks detected in the chromatogram
    $f(t)_{2D}$ is the result of Equation 2.32:

$$f(t)_{2D} = \left(\left(\sqrt{t_{L,1D}^2 + t_{L,2D}^2} - \sqrt{t_{0,1D}^2 + t_{0,2D}^2} \right) / \sqrt{t_{L,1D}^2 + t_{L,2D}^2} \right) \times \left[\left(\varnothing - \arctan\left(t_{L,2D} / t_{L,1D}\right) \right) / \varnothing \right] \qquad (2.32)$$

where $t_{L,1D}$ and $t_{L,2D}$, and $t_{0,1D}$ and $t_{0,2D}$ are the elution times of the last eluted 2D peak and the elution times corresponding to the extra-column volumes of the first and second dimensions, respectively, and ϕ is $\pi/2$ or $90°$, depending on whether the arctangent calculation is performed in radians or degrees. Matos and coworkers [23] also tested their $DCRF_{f,2D}$ using 2D simulated chromatograms describing 15 different optimization scenarios. The authors concluded that the $DCRF_{f,2D}$ is an adequate response to each criterion and easily adaptable to different combinations of weights.

2.4 CONCLUSIONS AND FUTURE PERSPECTIVES

Chromatographic fingerprinting of samples with widely differing, yet predictable, composition, or with unknown compounds, requires chromatograms as informative as possible, with an enhanced peak capacity. The development of a chromatographic method to achieve such level of information comprises the search for the optimal separation conditions, where a compromise often has to be found between conflicting goals, such as maximizing the separation while minimizing the analysis time. Several mathematical equations for assessing and ranking the quality of separation—a branch of chromatographic optimization processes—have been proposed to assist the analyst in the automated search of the optimum conditions. The earliest mathematical approaches were mainly focused on peak pair resolution, and the best separation conditions were searched by maximizing this resolution descriptor. However, in real optimization problems, the analyst has the need to include other secondary requirements in a single measurement to appraise the overall separation quality. As a consequence, an array of different CRFs have been proposed in the literature, simultaneously considering the quality of chromatographic separation, the number of peaks, and the time spent in the analysis. The majority of the CRFs addressed in this review were developed for the nontarget analysis of samples, either with a predictable number of compounds or with unknown compounds for which no standards are available. Additionally, only a few CRFs were proposed for the target analysis of the so-called representative samples [19]—a situation that can be explained by the existence of several software packages commercially available (e.g., DRYLAB, PREOPT-W, OSIRIS, MICHROM, and CHROMSWORD, as described by García-Álvarez-Coque et al. [6]) that enable modeling the retention behavior of the compounds of interest, thus facilitating the development of the chromatographic methods. Employing a CRF for optimizing the separation of fractions for which the chromatographic behavior of all the compounds cannot be predictable is obviously more difficult than using a CRF that focuses on a single parameter (i.e., peak resolution). Nevertheless, the requests of modern analytical chemistry for dealing with samples of increasing chemical complexity under unstructured optimization processes require the use of an objective measure that qualifies the separation degree without the need of a priori chromatographic information (e.g., desired peak resolution, number of peaks, and time constraints).

It must be emphasized, however, that the use of CRFs to find the optimum conditions for a given separation problem is not a consensual topic. The use of subjective thresholds and arbitrary weighting coefficients to balance the elementary criteria against each other and, simultaneously, define the search aims is considered to be the main weaknesses of employing a CRF in search of the best separation conditions. As such, the use of CRFs that have neither restrictions on the peak quality nor requirements for the prior definition of an optimum and/or minimum acceptable resolution is an added value for estimating the quality index of separation in both target and nontarget analysis of unknown samples. In this regard, CRFs based on valley-to-peak ratio and peak prominence (definitions, see Equations 2.19 and 2.23, respectively) seem to solve the problem of assessing the quality of separation between pair of peaks of highly unequal area, and also for overlapping and asymmetric peaks, being therefore highly desirable for dealing with target and/or nontarget analysis of complex unknown samples.

Future research should focus on the development of CRFs for use in optimization strategies of 2D chromatography. The only CRF known to date (Equation 2.30) is especially devoted to tackle the nontarget analysis of unknown samples, and it introduces the "2D peak purity" concept for assessing the quality of 2D chromatographic separation. However, the estimative of 2D peak purity requires the mathematical modeling of the 2D chromatogram, which constitutes its main limitation since this is a very complex and time-consuming task. Future research should also focus on the development of analytical expertise for handling the 2D chromatographic data in order to promote the routine application of these techniques to the target and nontarget analysis of complex samples. Future research should also take into account the whole analytical information from the multichannel detectors currently used in chromatography, both in 1D and 2D chromatography, which are usually ignored in chromatographic optimization processes. In this regard, Torres-Lapasió et al. [60] recently proposed two-way chromatographic objective functions for use in 1D chromatography coupled to diode array detection that incorporate both time and spectral information, based on the peak purity (peak fraction free of overlapping) and the multivariate selectivity (figure of merit derived from the net analyte signal) concepts. These functions are sensitive to situations where the components that co-elute in a mixture show some spectral differences [60]. Indeed, with the advent of new mathematical approaches especially devoted to an efficient utilization of the large amounts of data obtained from both 1D and 2D chromatographic analyses, the years ahead promise to gain new insights on the use of CRFs as a tool for the optimization of the chromatographic separation of complex unknown samples.

ACKNOWLEDGMENTS

Thanks are due for the financial support to CESAM (UID/AMB/50017/2013), to FCT/MEC through the European Social Fund (ESF) and "Programa Operacional Potencial Humano – POPH", and the co-funding by the FEDER, within the PT2020 Partnership Agreement and Compete 2020. FCT/MEC is also acknowledged for an Investigator FCT Contract (Regina M.B.O. Duarte, IF/00798/2015) and a PhD grant (João T.V. Matos, SFRH/BD/84247/2012).

REFERENCES

1. S. Fanali, P.R. Haddad, C.F. Poole, P. Schoenmakers, D. Lloyd, *Liquid Chromatography: Applications*, Elsevier, Waltham, MA, 2013.
2. O. Potterat, M. Hamburger, Concepts and technologies for tracking bioactive compounds in natural product extracts: Generation of libraries, and hyphenation of analytical processes with bioassays, *Natural Product Reports*. 30 (2013) 546–564. doi:10.1039/c3np20094a.
3. K.K. Unger, R. Ditz, E. Machtejevas, R. Skudas, Liquid chromatography—Its development and key role in life science applications, *Angewandte Chemie* (International Ed. in English). 49 (2010) 2300–2312. doi:10.1002/anie.200906976.
4. R.M.B.O. Duarte, A.C. Duarte, Optimizing size-exclusion chromatographic conditions using a composite objective function and chemometric tools: Application to natural organic matter profiling, *Analytica Chimica Acta*. 688 (2011) 90–98. doi:10.1016/j.aca.2010.12.031.
5. G.C. Woods, M.J. Simpson, P.J. Koerner, A. Napoli, A.J. Simpson, HILIC-NMR: Toward the identification of individual molecular components in dissolved organic matter, *Environmental Science & Technology*. 45 (2011) 3880–3886. doi:10.1021/es103425s.
6. M.C. García-Álvarez-Coque, J.R. Torres-Lapasió, J.J. Baeza-Baeza, Models and objective functions for the optimisation of selectivity in reversed-phase liquid chromatography, *Analytica Chimica Acta*. 579 (2006) 125–145. doi:10.1016/j.aca.2006.07.028.
7. R. Cela, C.G. Barroso, J.A. Pérez-Bustamante, Objective functions in experimental and simulated chromatographic optimization, *Journal of Chromatography A*. 485 (1989) 477–500. doi:10.1016/S0021-9673(01)89157-3.
8. R.M.B.O. Duarte, A.C. Duarte, A new chromatographic response function for use in size-exclusion chromatography optimization strategies: Application to complex organic mixtures, *Journal of Chromatography A*. 1217 (2010) 7556–7563. doi:10.1016/j.chroma.2010.10.021.
9. R. Cela, J. Martínez, C. González-Barreiro, M. Lores, Multi-objective optimisation using evolutionary algorithms: Its application to HPLC separations, *Chemometrics and Intelligent Laboratory Systems*. 69 (2003) 137–156. doi:10.1016/j.chemolab.2003.07.001.
10. P. Nikitas, A. Pappa-Louisi, Retention models for isocratic and gradient elution in reversed-phase liquid chromatography, *Journal of Chromatography A*. 1216 (2009) 1737–1755. doi:10.1016/j.chroma.2008.09.051.
11. V.M. Morris, J.G. Hughes, P.J. Marriott, Examination of a new chromatographic function, based on an exponential resolution term, for use in optimization strategies: Application to capillary gas chromatography separation of phenols, *Journal of Chromatography A*. 755 (1996) 235–243. doi:10.1016/S0021-9673(96)00600-0.

12. A.M. Siouffi, R. Phan-Tan-Luu, Optimization methods in chromatography and capillary electrophoresis, *Journal of Chromatography A*. 892 (2000) 75–106. doi:10.1016/S0021-9673(00)00247-8.

13. E.J. Klein, S.L. Rivera, A review of criteria functions and response surface methodology for the optimization of analytical scale HPLC separations, *Journal of Liquid Chromatography & Related Technologies*. 23 (2000) 2097–2121. doi:10.1081/JLC-100100475.

14. E. Tyteca, G. Desmet, A universal comparison study of chromatographic response functions, *Journal of Chromatography A*. 1361 (2014) 178–190. doi:10.1016/j.chroma.2014.08.014.

15. T. Alvarez-Segura, A. Gómez-Díaz, C. Ortiz-Bolsico, J.R. Torres-Lapasió, M.C. García-Alvarez-Coque, A chromatographic objective function to characterise chromatograms with unknown compounds or without standards available, *Journal of Chromatography A*. 1409 (2015) 79–88. doi:10.1016/j.chroma.2015.07.022.

16. J. Berridge, Unattended optimisation of reversed-phase high-performance liquid chromatographic separations using the modified simplex algorithm, *Journal of Chromatography A*. 244 (1982) 1–14. doi:10.1016/S0021-9673(00)80117-X.

17. R.M.B.O. Duarte, J.T.V. Matos, A.C. Duarte, A new chromatographic response function for assessing the separation quality in comprehensive two-dimensional liquid chromatography, *Journal of Chromatography A*. 1225 (2012) 121–131. doi:10.1016/j.chroma.2011.12.082.

18. B. Jančić-Stojanović, T. Rakić, N. Kostić, A. Vemić, A. Malenović, D. Ivanović, M. Medenica, Advancement in optimization tactic achieved by newly developed chromatographic response function: Application to LC separation of raloxifene and its impurities, *Talanta*. 85 (2011) 1453–1460. doi:10.1016/j.talanta.2011.06.029.

19. J.T.V. Matos, R.M.B.O. Duarte, A.C. Duarte, Chromatographic response functions in 1D and 2D chromatography as tools for assessing chemical complexity, *TrAC Trends in Analytical Chemistry*. 45 (2013) 14–23. doi:10.1016/j.trac.2012.12.013.

20. W. Nowik, S. Héron, M. Bonose, A. Tchapla, Separation system suitability (3S): A new criterion of chromatogram classification in HPLC based on cross-evaluation of separation capacity/peak symmetry and its application to complex mixtures of anthraquinones, *The Analyst*. 138 (2013) 5801–5810. doi:10.1039/c3an00745f.

21. P.F. Vanbel, P.J. Schoenmakers, Selection of adequate optimization criteria in chromatographic separations, *Analytical and Bioanalytical Chemistry*. 394 (2009) 1283–1289. doi:10.1007/s00216-009-2709-9.

22. A. Ortín, J.R. Torres-Lapasió, M.C. García-Álvarez-Coque, A complementary mobile phase approach based on the peak count concept oriented to the full resolution of complex mixtures, *Journal of Chromatography A*. 1218 (2011) 5829–5836. doi:10.1016/j.chroma.2011.06.087.

23. J.T.V. Matos, R.M.B.O. Duarte, A.C. Duarte, A generalization of a chromatographic response function for application in non-target one- and two-dimensional chromatography of complex samples, *Journal of Chromatography A*. 1263 (2012) 141–150. doi:10.1016/j.chroma.2012.09.037.

24. P.J. Schoenmakers, *Optimization of Chromatographic Selectivity: A Guide to Method Development*, Elsevier, Amsterdam, the Netherlands, 1986.

25. L.S. Ettre, Nomenclature for chromatography (IUPAC Recommendations 1993), *Pure and Applied Chemistry*. 65 (1993) 819–872. doi:10.1351/pac199365040819.

26. P.J. Schoenmakers, J.K. Strasters, Á. Bartha, Correction of the resolution function for non-ideal peaks, *Journal of Chromatography A*. 458 (1988) 355–370. doi:10.1016/S0021-9673(00)90578-8.

27. P.F. Vanbel, B.L. Tilquin, P.J. Schoenmakers, Criteria for optimizing the separation of target analytes in complex chromatograms, *Chemometrics and Intelligent Laboratory Systems*. 35 (1996) 67–86. doi:10.1016/S0169-7439(96)00046-9.

28. G.C. Carle, Determination of chromatographic resolution for peaks of vast concentration differences, *Analytical Chemistry*. 44 (1972) 1905–1906.

29. G. Vivó-Truyols, J.R. Torres-Lapasió, M.C. García-Alvarez-Coque, Complementary mobile-phase optimisation for resolution enhancement in high-performance liquid chromatography, *Journal of Chromatography A*. 876 (2000) 17–35. doi:10.1016/S0021-9673(00)00188-6.

30. G. Vivó-Truyols, Bayesian approach for peak detection in two-dimensional chromatography, *Analytical Chemistry*. 84 (2012) 2622–2630. doi:10.1021/ac202124t.

31. B. Divjak, M. Moder, J. Zupan, Chemometrics approach to the optimization of ion chromatographic analysis of transition metal cations for routine work, *Analytica Chimica Acta*. 358 (1998) 305–315. doi:10.1016/S0003-2670(97)00644-2.

32. M.W. Watson, P.W. Carr, Simplex algorithm for the optimization of gradient elution high-performance liquid chromatography, *Analytical Chemistry*. 51 (1979) 1835–1842.

33. J.L. Glajch, J.J. Kirkland, K.M. Squire, J.M. Minor, Optimization of solvent strength and selectivity for reversed-phase liquid chromatography using an interactive mixture-design statistical technique, *Journal of Chromatography A*. 199 (1980) 57–79. doi:10.1016/S0021-9673(01)91361-5.

34. D.M. Fast, P.H. Culbreth, E.J. Sampson, Multivariate and univariate optimization studies of liquid-chromatographic separation of steroid mixtures, *Clinical Chemistry*. 28 (1982) 444–448.

35. J.C. Berridge, Automated multiparameter optimisation of high-performance liquid chromatographic separations using the sequential simplex procedure, *The Analyst*. 109 (1984) 291. doi:10.1039/an9840900291.

36. A.G. Wright, A.F. Fell, J.C. Berridge, Sequential simplex optimization and multichannel detection in HPLC: Application to method development, *Chromatographia*. 24 (1987) 533–540. doi:10.1007/BF02688540.

37. S.M. Sultan, A.H. El-Mubarak, High performance liquid chromatographic method for the separation and quantification of some psychotherapeutic benzodiazepines optimized by the modified simplex procedure, *Talanta*. 43 (1996) 569–576. doi:10.1016/0039-9140(95)01772-0.

38. R. Karnka, M. Rayanakorn, S. Watanesk, Y. Vaneesorn, Optimization of high-performance liquid chromatographic parameters for the determination of capsaicinoid compounds using the simplex method, *Analytical Sciences: The International Journal of the Japan Society for Analytical Chemistry*. 18 (2002) 661–665. doi:10.2116/analsci.18.661.

39. S. Srijaranai, R. Burakham, R.L. Deming, T. Khammeng, Simplex optimization of ion-pair reversed-phase high performance liquid chromatographic analysis of some heavy metals, *Talanta*. 56 (2002) 655–661. doi:10.1016/S0039-9140(01)00634-8.

40. N. Kuppithayanant, M. Rayanakorn, S. Wongpornchai, T. Prapamontol, R.L. Deming, Enhanced sensitivity and selectivity in the detection of polycyclic aromatic hydrocarbons using sequential simplex optimization, the addition of an organic modifier and wavelength programming, *Talanta*. 61 (2003) 879–888. doi:10.1016/S0039-9140(03)00374-6.

41. K. Mitrowska, U. Vincent, C. von Holst, Separation and quantification of 15 carotenoids by reversed phase high performance liquid chromatography coupled to diode array detection with isosbestic wavelength approach, *Journal of Chromatography A*. 1233 (2012) 44–53. doi:10.1016/j.chroma.2012.01.089.

42. T.D. Schlabach, J.L. Excoffier, Multi-variate ranking function for optimizing separations, *Journal of Chromatography A*. 439 (1988) 173–184. doi:10.1016/S0021-9673(01)83832-2.

43. L. Peichang, H. Hongxin, An intelligent search method for HPLC optimization, *Journal of Chromatographic Science*. 27 (1989) 690–697. doi:10.1093/chromsci/27.12.690.

44. X.G. Chen, X. Li, L. Kong, J.Y. Ni, R.H. Zhao, H.F. Zou, Application of uniform design and genetic algorithm in optimization of reversed-phase chromatographic separation, *Chemometrics and Intelligent Laboratory Systems*. 67 (2003) 157–166. doi:10.1016/S0169-7439(03)00091-1.

45. Y.-B. Ji, Q.-S. Xu, Y.-Z. Hu, Y. Vander Heyden, Development, optimization and validation of a fingerprint of *Ginkgo biloba* extracts by high-performance liquid chromatography, *Journal of Chromatography A*. 1066 (2005) 97–104. doi:10.1016/j.chroma.2005.01.035.

46. T. Huang, X. Zhang, Optimization of conditions for capillary electrophoresis of winged euonymus by a variable dimension expansion-selection method, *Journal of Liquid Chromatography & Related Technologies*. 29 (2006) 1515–1524. doi:10.1080/10826070600674984.

47. P. Ebrahimi, M.R. Hadjmohammadi, Optimization of the separation of coumarins in mixed micellar liquid chromatography using Derringer's desirability function, *Journal of Chemometrics*. 21 (2007) 35–42. doi:10.1002/cem.1032.

48. J. Felhofer, G. Hanrahan, C.D. García, Univariate and multivariate optimization of the separation conditions for the analysis of five bisphenols by micellar electrokinetic chromatography, *Talanta*. 77 (2009) 1172–1178. doi:10.1016/j.talanta.2008.08.016.

49. M. Hadjmohammadi, V. Sharifi, Simultaneous optimization of the resolution and analysis time of flavonoids in reverse phase liquid chromatography using Derringer's desirability function, *Journal of Chromatography B: Analytical Technologies in the Biomedical and Life Sciences*. 880 (2012) 34–41. doi:10.1016/j.jchromb.2011.11.012.

50. A. Ortín, J.R. Torres-Lapasió, M.C. García-Álvarez-Coque, Finding the best separation in situations of extremely low chromatographic resolution, *Journal of Chromatography A*. 1218 (2011) 2240–2251. doi:10.1016/j.chroma.2011.02.022.

51. I. François, K. Sandra, P. Sandra, Comprehensive liquid chromatography: Fundamental aspects and practical considerations—A review, *Analytica Chimica Acta*. 641 (2009) 14–31. doi:10.1016/j.aca.2009.03.041.

52. H. Malerod, E. Lundanes, T. Greibrokk, Recent advances in on-line multidimensional liquid chromatography, *Analytical Methods*. 2 (2010) 110. doi:10.1039/b9ay00194h.

53. C. Meinert, U.J. Meierhenrich, A new dimension in separation science: Comprehensive two-dimensional gas chromatography, *Angewandte Chemie—International Edition*. 51 (2012) 10460–10470. doi:10.1002/anie.201200842.

54. P. Jandera, Comprehensive two-dimensional liquid chromatography—Practical impacts of theoretical considerations. A review, *Central European Journal of Chemistry*. 10 (2012) 844–875. doi:10.2478/s11532-012-0036-z.

55. J.T.V. Matos, R.M.B.O. Duarte, A.C. Duarte, Trends in data processing of comprehensive two-dimensional chromatography: State of the art, *Journal of Chromatography B*. 910 (2012) 31–45. doi:10.1016/j.jchromb.2012.06.039.

56. P. Česla, T. Hájek, P. Jandera, Optimization of two-dimensional gradient liquid chromatography separations, *Journal of Chromatography A*. 1216 (2009) 3443–3457. doi:10.1016/j.chroma.2008.08.111.

57. B.W.J. Pirok, S. Pous-Torres, C. Ortiz-Bolsico, G. Vivó-Truyols, P.J. Schoenmakers, Program for the interpretive optimization of two-dimensional resolution, *Journal of Chromatography A*. 1450 (2016) 29–37. doi:10.1016/j.chroma.2016.04.061.

58. M.R. Schure, Quantification of resolution for two-dimensional separations, *Journal of Microcolumn Separations*. 9 (1997) 169–176. doi:10.1002/(SICI)1520-667X(1997)9:3<169::AID-MCS5>3.0.CO;2-#.

59. S. Peters, G. Vivó-Truyols, P.J. Marriott, P.J. Schoenmakers, Development of a resolution metric for comprehensive two-dimensional chromatography, *Journal of Chromatography A*. 1146 (2007) 232–241. doi:10.1016/j.chroma.2007.01.109.

60. J.R. Torres-Lapasió, M.C. García-Alvarez-Coque, E. Bosch, M. Rosés, Considerations on the modelling and optimisation of resolution of ionisable compounds in extended pH-range columns, *Journal of Chromatography A*. 1089 (2005) 170–186. doi:10.1016/j.chroma.2005.06.085.

3 Chemometric Strategies to Characterize, Classify, and Rank Chromatographic Stationary Phases and Systems

Charlene Muscat Galea, Debby Mangelings,
and Yvan Vander Heyden

CONTENTS

3.1 INTRODUCTION

A stationary phase in any given chromatographic system can be viewed as the core of the separation process. This is mainly because elution or retention of substances of an injected mixture depends on the properties of the stationary phase, in relation to those of the mobile phase. Stationary-phase properties can be studied in different ways. Characterizing stationary phases by retention profiling (Section 3.2.2.2) requires the mobile phase to be kept constant. When working with different mobile-phase compositions, it is no longer correct to say that one is characterizing stationary phases, but rather the characterization of chromatographic systems is taking place. Determining the physicochemical properties of stationary phases, on the other hand, requires the use of different mobile phases (Section 3.2.2.1).

This characterization could help obtaining a quantitative analysis on features that will affect the selectivity of the separation process [1]. The quantified properties of the stationary phase can then be used to develop column rankings or classifications [2]. This may be useful to choose, on one hand, columns of similar selectivity, for instance, when one column needs to be replaced by another, or in the context of compendial analyses (selection of columns with similar selectivity), or to choose, on the other hand, dissimilar columns, for instance, to be applied during method development. Characterization may also inspire the manufacturers to design and produce columns having different properties that are not yet available on the market. Finally, it will save a lot of time and money, when one could choose in a systematic way based on its characteristics rather than simply based on trial and error a column for a specific separation or application. Much effort has been invested by several research groups to characterize chromatographic columns. This chapter will give a brief overview of different methods used to characterize stationary phases and will go over different chemometric techniques that are used to analyze the chemical data obtained from the different characterization methods.

3.2 STATIONARY-PHASE CHARACTERIZATION METHODS

Hundreds of high-performance liquid chromatography (HPLC) columns have been tested and characterized in the literature. Over the years, several methods for the characterization of stationary-phase chemistry have been established. Common approaches include thermodynamic methods [3–6], spectroscopic techniques [7–10], and chromatographic tests [11–17], while chemometric methods have been used to evaluate or visualize the data generated. The chromatographic tests can be further divided into the determination of physicochemical properties, retention profiling, and model-based tests. Similarity indices were defined to combine and compare test results and retention profiles and to classify columns as similar or dissimilar [18,19]. The information on the characterization of stationary phases has been organized into databases, which helps prospective users to select columns needed for their specific purposes.

3.2.1 Nonchromatographic Tests

Thermogravimetric analysis can be used to determine the surface coverage of chemically modified silica because the weight loss perceived between 200°C and 600°C is associated with the loss of the organic groups attached to the surface [3]. On the other hand, elemental analysis allows the determination of the percentages of elements, such as carbon, hydrogen, and nitrogen, in the stationary phase. Such analysis can give a better idea of the stationary-phase surface coverage.

Other nonchromatographic approaches for phase characterization are thermodynamic methods. Van't Hoff plots, that is, plots of the chromatographic retention factors against the reciprocal values of a sufficient range of absolute temperatures (ln k vs. 1/T), can be used to estimate the enthalpy and entropy contributions to chromatographic retention or selectivity [4–6]. In addition, spectroscopic methods, such as infrared (IR), Raman, fluorescence (FL), and nuclear magnetic resonance (NMR), have been used to investigate the structure, conformation, and dynamics of different stationary phases [7–10]. The drawback of the above approaches is that they are destructive for the stationary phase.

3.2.2 Chromatographic Tests

3.2.2.1 Physicochemical Properties of Stationary Phases

Stationary-phase properties, like efficiency, hydrophobicity, silanol activity, ion-exchange capacity, steric selectivity, and amount of metal impurities, are quantified. Retention/separation information of given solutes, which are known to reflect given stationary-phase properties, are determined. Several properties are measured to globally characterize a phase—since one particular property is insufficient to reflect the different physicochemical interactions involved in chromatographic retention [11–13]. More details with examples on studies determining such properties to characterize stationary phases will be discussed further in the chapter.

3.2.2.2 Retention Profiling

A generic set of substances is injected (individually) on different columns. This set should be representative for the future use of the columns. For instance, to test phases to be used in drug impurity profiling, the compounds applied should differ in structure (functional groups and ring structures) and molecular weight and originate from different pharmacological and chemical classes [15,20]. Their composition thus may be different depending on the future use of the separations to be developed on the considered phases. The retention times or retention factors of all compounds on the tested phases are recorded to obtain a retention profile for each phase, which will serve to characterize the phases. This approach has, for instance, been used to characterize C18 phases in HPLC [14], as well as diverse columns in SFC [15]. More details will be discussed further in the chapter.

3.2.2.3 Model-Based Stationary-Phase Characterization

This characterization is based on building a specific retention model [2,11,13,16]. The hydrophobic subtraction model defined by Snyder et al. [16], or the linear solvation energy relationship from Taft et al. [17] adapted by Abraham et al. [21], are such models. Experimental retention data of a test set of compounds, for example, log k or log (k/k_{ref}), with k_{ref} being the retention factor of a nonpolar reference solute, are modeled as a function of a number of their molecular descriptors. The model coefficients may provide information on the properties of the stationary phase, for instance, its hydrophobicity, hydrogen-bonding acidity and basicity, and its steric resistance to insertion of bulky molecules into the stationary phase.

3.3 CHEMOMETRIC METHODS

In order to gain maximal information from the multivariate character of the chromatographic test data, chemometric tools can be applied. They enable the handling of a large amount of output variables, resulting in an easily interpretable result for the complete set of chromatographic systems analyzed. Before starting the chemometric treatment, the stationary-phase data needs to be organized in a matrix, where each row represents one system profile and each column a test compound or a chromatographic parameter. Then several chemometric approaches, such as exploratory data analysis or classification methods, can be applied. These methods will group chromatographic systems with similar properties and thus help classifying stationary phases and selecting either similar or dissimilar phases.

3.3.1 Data Pretreatment

In general, the signals in chromatography may vary from one analysis to another. Data pretreatment procedures are often employed to correct for these nonchemical variations in chromatographic data. Differences in overall abundance between chromatograms measured on the same instrument can result from variation in sample preparation, sample injection, chromatographic conditions, and instrument response. In order to correct this problem, normalization procedures are commonly applied [22]. Other sources of variation that lead to changes in the shape of the baseline-noise signal are the injection solvent and oven temperature. Differences in stationary-phase age and wear can also lead to rise and variation in the signal over time. It is often necessary to apply baseline-correction procedures to minimize this variation [23]. Noise, the high-frequency fluctuation in the signal, is another source of variation, which is hard to identify visually when there is a high signal-to-noise ratio. Noise is often the result of instrumental and electronic variation and can be corrected using a smoothing procedure [24]. Peaks in the chromatogram can elute at slightly different retention times, due to instrumental variation in flow rates, column degradation, mobile-phase composition variation, and manual injection procedures. These retention time variabilities can be corrected by applying a peak alignment algorithm [25]. The above pretreatments are usually applied when the chromatogram is applied as a multivariate profile. They are less frequently required here since stationary-phase characterization is based on information obtained from given compounds, for instance, retention data and peak shape information. The information of entire chromatograms as characteristic profiles is not used then.

In addition to the above pretreatment methods, data transformation or scaling may be required. Autoscaling and data centering are the most commonly used transformations. Column centering results in a scale shift in the data matrix, because for each variable (e.g., a property measurement of a chromatographic stationary phase), the column mean is subtracted from each of its measurements. As a result, all variables have zero average. With autoscaling, this difference is divided by the standard deviation of the considered variable giving rise to variables with average 0 and standard deviation 1, which are thus independent of the unit of measurement, which have equal range and therefore equal weight or importance [26].

3.3.2   DATA HANDLING ON CHROMATOGRAPHIC TEST DATA

3.3.2.1   Exploratory Data Analysis

Exploratory data analysis is an approach used to summarize the data's main characteristics, primarily by visualization methods. In this section of the chapter, a number of techniques commonly applied to chromatographic system or column characterization will be overviewed. Research examples showing how these techniques were applied on different data sets will also be discussed.

3.3.2.1.1   Principal Component Analysis

Principal component analysis (PCA) is a variable reduction technique that helps visualizing data of multivariate nature in a low-dimensional space. Principal components (PCs) are latent variables, defined to describe the maximal variance in the data. The PCs are oriented so that the first (PC1) describes as much as possible the original variance between the objects, the second (PC2) is orthogonal to the first and describes as much as possible of the remaining variation, and so on and so forth.* Projection of an object on a given PC provides its score on the considered PC. Plotting the scores on, for instance, PC1 and PC2 provides a PC1–PC2 plot and shows the relation between the objects (groups or clusters, outlying objects). The scores on a given PC are linear combinations of the original variables and the weights in these combinations are called the loadings. Thus, how important an original variable is for a given PC is described by its loadings, one on each PC. Plotting the loadings in loading plots will allow understanding the relative importance of the original variables, as well as whether correlation is present between the variables.

Numerous groups have used PCA to characterize and classify chromatographic columns. Euerby et al. [27] characterized 135 stationary phases by 6 chromatographic parameters. These were the retention factor for pentylbenzene (k_{PB}), hydrophobicity or hydrophobic selectivity (α_{CH2}), shape selectivity ($\alpha_{T/O}$), hydrogen-bonding capacity ($\alpha_{C/P}$), total ion-exchange capacity ($\alpha_{B/P}$ pH 7.6), and acidic ion-exchange capacity ($\alpha_{B/P}$ pH 2.7), where α represents a selectivity factor or relative retention. The stationary phases were then roughly grouped into three main categories (Figure 3.1a), with high silanol activity and high retentivity (A), with lower silanol activity and high retentivity (B), and polar-embedded phases with high shape selectivity character (C). The corresponding loading plot (Figure 3.1b) allows understanding which of the original variables are most important or if any variables are correlated. Combining the information of both score and loading plots indicates which variables are responsible for given stationary phases to be situated at given positions in the score plot. Studying subgroups of phases, for example, C18 phases based on nonacidic silica or polar-embedded C18 phases, using PCA, will reveal differences between these more similar phases. In addition, the (Euclidean) distance in the six-dimensional variable space between a phase of interest and the other columns in the database can be calculated. Larger distances would signify more dissimilarity in terms of chromatographic properties and such columns will be located at distinct positions on the PC score plot.

---

* See also Chapter 8.

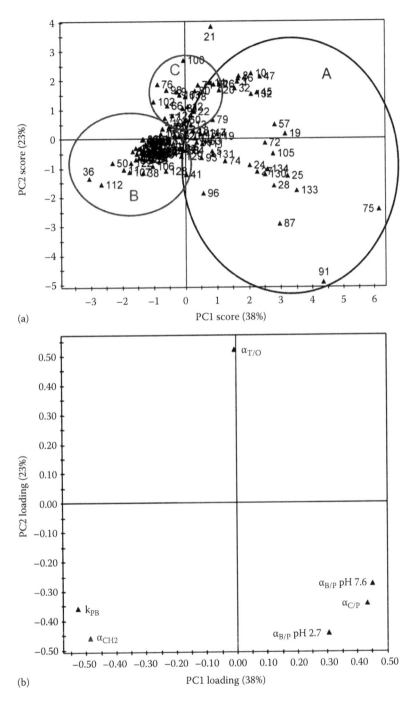

FIGURE 3.1 (a) PC1–PC2 score plot for the stationary phases evaluated by Euerby et al. (b) Corresponding loading plot. (Reprinted from *J. Chromatogr. A*, 994, Euerby, M.R., Petersson, P., Campbell, W., and Roe, W., Chromatographic classification and comparison of commercially available reversed-phase liquid chromatographic columns containing phenyl moieties using principal component analysis, 13–36, Copyright 2003, with permission from Elsevier.)

Forlay-Frick et al. [28] subjected two parameters, column efficiency and the symmetry factors, on different chromatographic columns to PCA. The different responses were analyzed and measured for three compounds, benzoic acid, N,N-dimethyl-aniline, and vancomycin, individually and both responses seemed necessary for adequate characterization and classification. Systems having a similar plate number and symmetry-factor values were grouped. If a system needed to be replaced for a given separation, another having comparable values for both responses would be chosen. However, care has to be exercised when selecting columns for compounds having high masses and multifunctional groups. In such cases, more complex interactions will be present. Therefore, even though simple test compounds suggest good performance, the stationary phase might not be the most suitable for the separation of more complicated compounds.

Van Gyseghem et al. [14,29–32] used 68 pharmaceutical compounds to characterize and classify chromatographic systems based on their retention profiles. PCA was then used as a visualization aid to appreciate the dissimilarity of different phases or systems within the context of drug impurity profiling by HPLC. Similarly, pharmaceutical drugs were also used to characterize stationary phases, in SFC [15]. Once again, retention profiles of the drugs were used to classify the phases by PCA in groups according to similarities.

Visky et al. [13] determined 24 test parameters using different chromatographic methods on 69 stationary phases, with the aim to identify a minimal number of parameters required for a good stationary-phase characterization. The 24 parameters were classified in seven "property" groups: efficiency, hydrophobicity, silanol activity, two groups for metal impurities, steric selectivity, and other properties. PCA was performed on the chromatographic data from the 69 phases, and this resulted in a score plot showing the columns grouped according to similar properties and a loading plot showing the chromatographic test parameters grouped according to the chromatographic properties they measure. The seven chromatographic property groups could be distinguished from the loading plot. The study concluded that four parameters, $k'_{amylbenzene}$, $rk'_{o-terphenyl/triphenylene}$, $rk'_{benzylamine/phenol}$ pH 2.7, and $k'_{2,2-dipyridyl}$, representing hydrophobicity, steric selectivity, ion exchange at low pH, and metal impurities, respectively, are sufficient to characterize and classify the phases. The column classification based on these four parameters was then verified by correlating it with the selectivity obtained for real separations [33].

3.3.2.1.2 Hierarchical Cluster Analysis

In cluster analysis, objects are grouped based on their similarities for a set of variables.* The result is visualized as a dendrogram (Figure 3.2b). In hierarchical cluster analysis, the technique can be agglomerative, that is, objects are sequentially merged and smaller clusters are linked to make up larger ones, or divisive, when clusters are sequentially split until they contain only single objects. The latter methods are less frequently applied, and therefore will not be discussed further in this chapter.

The clustering will be done based on a similarity measure and a linkage method. Objects are iteratively linked, starting from the most similar, until all objects are linked. Different similarity parameters, either correlation or distance based, such as the correlation coefficient or the Euclidean distance, are used to identify similar stationary phases. Other distance parameters exist, for instance, the Manhattan distance, Spearman dissimilarity, Chi-square distance, which however are less widely used. The dendrograms are then created using linkage methods, such as single linkage, complete linkage, average linkage, inner squared distance linkage (Ward's method), and weighted-average linkage, in order to link the different objects/clusters. Different combinations of similarity parameters and linkage methods will lead to different dendrograms and thus different visualizations of the data set. The analyst then has to judge which combination is leading to the most meaningful result.

For instance, retention data of 137 organic compounds in gas chromatography, obtained on one C78 standard alkane phase and nine phases with stand-alone polar interactive groups, has been

---

* See also Chapter 8.

subjected to HCA by Dallos et al. [34]. The resulting clustering reveals similarities between given stationary phases. The differences are caused by the polar groups added onto the C78 alkane, and the molecular interactions they bring about. The results obtained were also confirmed by PCA and correlation analysis.

At RPLC conditions, seven basic compounds were analyzed on base-deactivated supports with mobile phases at pH 7.0 and 3.0 to allow the evaluation of the masking capacity of silanol groups and the hydrophobic properties of the selected supports [35,36]. The retention and asymmetry factor profiles were analyzed by PCA and HCA to discriminate between the different supports (Figure 3.2).

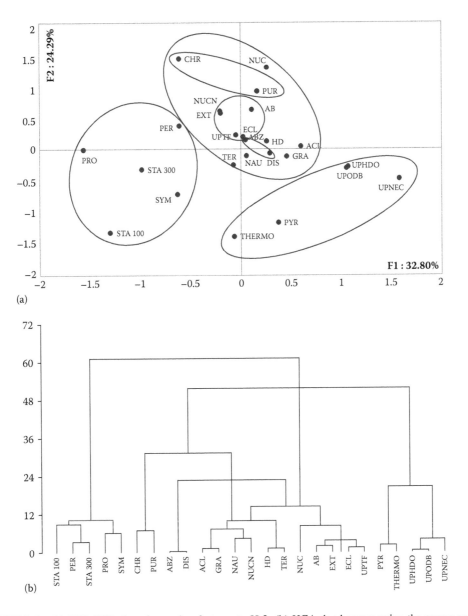

FIGURE 3.2   (a) PC1–PC2 plot of retention factors at pH 3; (b) HCA dendrogram using the scores on first six principal components at pH 3. (Reprinted from *J. Pharm. Biomed. Anal.*, 43, Stella, C., Rudaz, S., Gauvrit, J.-Y., Lantéri, P., Huteau, A., Tchapla, A., and Veuthey, J.-L., Characterization and comparison of the chromatographic performance of different types of reversed-phase stationary phases, 89–98, Copyright 2007, with permission from Elsevier.)

Groups of phases having similar masking capacities, or hydrophobic properties, are linked in the HCA. They are also located in close proximity on the PC score plots.

HCA was also used in column characterization in the hydrophilic interaction LC mode (HILIC) [37]. The selectivity of different stationary phases for peptides was studied from a number of responses, which included plate number, asymmetry factor, limit of detection (LOD), geometric mean resolution, resolution product, time-corrected resolution product, peak capacity, and chromatographic response function. The columns were clustered into three main groups with similar chemistries showing different chromatographic properties for peptides.

Other studies [31,38] compared different similarity measures and linkage techniques with the aim of identifying the best linkage technique to select similar or dissimilar chromatographic systems. The weighted- and unweighted-average linkage, single and complete linkage, centroid, and Ward's methods using correlation coefficients or the Euclidean distance as similarity measures, calculated on retention profiles, were evaluated in a study by Van Gyseghem et al. [31]. Color maps (see further) for the similarity parameters were then drawn to confirm the groups observed in the HCA at a given dissimilarity level. Selection of dissimilar systems can be done by choosing out one system from each group. However, not all similarity measures and linkage techniques provided useful information. With single-linkage clustering, for instance, the groups were much less pronounced than with weighted- and unweighted-average linkage using r-values as similarity measure. The Euclidean distance applied with the different linkage methods resulted in different stationary-phase classifications. In addition, no method leads to the optimal identification of similar and orthogonal relationships.

In another study evaluating chiral polysaccharide-based systems in SFC, both r-values and the Euclidean distances based on enantioresolutions of 29 racemates were used in combination with weighted pair-group method with averaging (WPGMA) and complete linkage to identify dissimilar and complementary chiral systems to be used for screening purposes [38]. The dendrograms based on $1 - |r|$ as dissimilarity criterion allowed the selection of the most dissimilar systems. However, the identification of complementary systems, that is, systems that when used allow the separation of the highest number of racemates (highest cumulative success rate), is not possible by dendrograms.

3.3.2.1.3 Color Maps

Correlation coefficient or Euclidean distance matrices determined above for HCA, for instance, can be transformed into color maps that represent the degree of correlation or the distance between the retention profiles or chromatographic parameters by a color. The sequence of the systems in the color map may be varied to reflect different aspects of the data. For example, De Klerck et al. [38] chose to rank the systems according to the chiral selector and sugar backbone of the stationary phase. Different colors were seen within each group of similar phases, meaning that despite having phases with an identical chiral selector, different enantioselectivity was observed. Van Gyseghem et al. [30], on the other hand, ranked the systems in the color map, initially as the sequence of systems seen in a dendrogram (Figure 3.3a), and later based on increasing dissimilarities in the dendrogram (Figure 3.3b). Defining the sequence of systems in the color maps based on given criteria helps promoting the observation of given clustering, if occurring. In another study, Dumarey et al. [39] ranked systems according to the selection sequence given by the Kennard and Stone algorithm [31,40,41], autoassociative multivariate regression trees (AAMRT) [42,43], and the generalized pairwise correlation method (GPCM) [44] (see next sections). These methods were used to rank the systems studied starting from the most dissimilar systems and going to the least dissimilar. Differences between the different color maps can be seen, meaning that the techniques show different aspects, but are not equally effective in selecting the most dissimilar systems.

Ordering points to identify the clustering structure (OPTICS) color maps [45] are a density-based method that is able to establish a unique sequence of the data. OPTICS color maps use as much dimensions as needed to explain 99.0%–99.5% of the variance in the data, as opposed to PCA

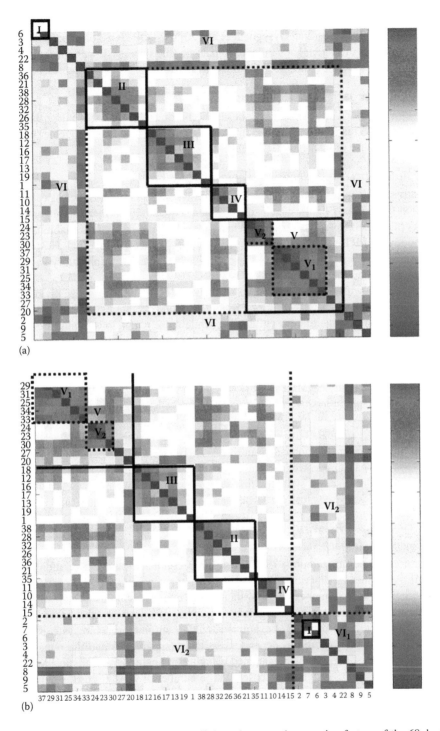

FIGURE 3.3 Color maps of the correlation coefficients between the retention factors of the 68 drugs, measured on the 38 systems, (a) sequence of systems as seen in HCA dendrogram, and (b) sequence of systems based on increasing dissimilarities seen in HCA dendrogram. Brown, high correlation; blue, low correlation. (Reprinted from *J. Chromatogr. A*, 1026, Van Gyseghem, E., Crosiers, I., Gourvénec, S., Massart, D.L., Vander Heyden, Y., Determining orthogonal and similar chromatographic systems from the injection of mixtures in liquid chromatography–diode array detection and the interpretation of correlation coefficients colour maps, 117–128, Copyright 2004, with permission from Elsevier.)

score plots, where only two or three dimensions can be represented simultaneously, but which often represents less variability [46]. Van Gyseghem et al. [20] used OPTICS color maps in an attempt to define classes of either orthogonal or similar chromatographic systems. It was concluded that the use of dendrograms was superior to the OPTICS maps to select the sequence of preference.

3.3.2.2 Other Approaches Distinguishing Dissimilar Phases

In addition to exploratory data analysis, other chemometric techniques that allow the grouping of phases are available. This section will give an overview of these techniques together with examples of studies that made use of them.

3.3.2.2.1 *Autoassociative Multivariate Regression Trees*

Autoassociative multivariate regression trees (AAMRT) is a method based on classical classification and regression trees (CART) [42], which split the data sequentially in more homogeneous or purer groups concerning their response variable. When more than one response variable is available, an extended mode of CART, multivariate regression trees (MRT), can be applied. However, when no response variables are present, an AAMRT approach, which uses the explanatory variables also as response variables, may be used.

An AAMRT starts with a parent node containing all objects. This node is then divided by a binary split based on the value of one explanatory variable. The objects that have a higher value than this split variable are then classified in one child node, while those with a lower value are classified in a second. All measured values of each explanatory variable are considered as possible split values at a given level and the split, for which the highest reduction in impurity is achieved, is selected [42,47]. The splitting process is continued considering each child node as a new parent node. Compared to other clustering methods, AAMRT has the advantage that it allows indicating which variable is responsible for the distinction between two groups, therefore leading to an improved interpretability [42,48].

Dumarey et al. [39] compared the performance of a number of chemometric techniques that can be used in dissimilar chromatographic system selection. The Kennard and Stone algorithm, AAMRT, and the generalized pairwise correlation method (GPCM; see further) performed best in selecting the most dissimilar systems, since the first selected systems have the lowest mutual average r. For WPGMA and OPA, these values are higher and therefore these techniques are less preferred.

3.3.2.2.2 *Generalized Pairwise Correlation Method (GPCM) with McNemar's Statistical Test*

GPCM is a nonparametric technique that takes into account the pairwise relations between the systems [44]. A number of superiority is determined comparing all possible independent variable pairs and is defined as the number of wins, that is, the number of times the considered independent variable was found superior. Analogously, a number of inferiority, that is, of losses, can be determined. Van Gyseghem et al. [31] examined ranking chromatographic systems with GPCM. Each system was once considered as dependent variable (supervisor) and the remaining systems were then ranked. The superiority was evaluated using the nonparametric statistical McNemar's test [49], and the ranking of the systems was based on the number of wins minus the number of losses. Van Gyseghem et al. [31] compared three approaches, the Kennard and Stone algorithm, AAMRT, and GPCM method with McNemar's test, as ranking and selection techniques for dissimilar chromatographic systems. The three methods resulted in similar subset selections; however, all three failed to identify groups of similar systems, making the drawing of a color map, where results are ranked according to the results of these methods, less relevant.

3.3.2.2.3 *Pareto-Optimality Method*

The Pareto-optimality concept is a method that allows obtaining a compromise between different criteria, occasionally in mutual conflict [50,51]. Acceptable conditions for different responses

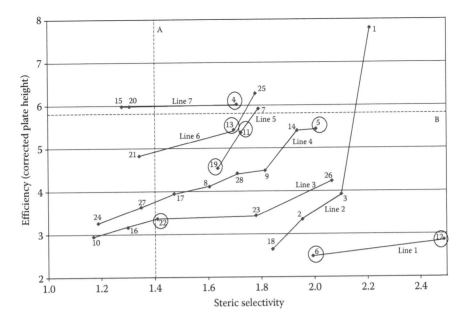

FIGURE 3.4   The efficiency (corrected plate height) vs. the steric selectivity for 28 columns; lines 1–7, sequential Pareto-optimal compromises for the two responses; the circles indicate eight dissimilar columns, earlier selected based on experience. Dotted lines A and B, arbitrarily chosen desirability thresholds. (Reprinted from *J. Chromatogr. A*, 1042, Van Gyseghem, E., Jimidar, M., Sneyers, R., Redlich, D., Verhoeven, E., Massart, D.L., and Vander Heyden, Y., Selection of reversed-phase liquid chromatographic columns with diverse selectivity towards the potential separation of impurities in drugs, 69–80, Copyright 2004, with permission from Elsevier.)

measured on a given stationary phase may be equally desirable, and a compromise has to be found. One can then apply multicriteria decision-making methods, like Pareto-optimality, to find such compromise. An object is Pareto-optimal when there is no other object that gives a better result for one criterion without having an inferior result for the other. The Pareto-optimality concept can be applied to make either a pairwise or a simultaneous evaluation of the responses. In a pairwise comparison, two responses are plotted on Cartesian axes. When all responses are considered simultaneously, Pareto-optimal objects are calculated, but visualization is no longer possible.

In a selection process for RPLC phases to be used as starting points for drug impurity profiling, the group of Vander Heyden considered efficiency and separation capacity or steric selectivity as being of major importance (Figure 3.4) [32]. The Pareto-optimality concept was not used in its most narrow sense (only border points, i.e., line 1 selected), but a series of sequential "Pareto-optimal" lines were selected. These different lines can be considered as sequential Pareto-optimal points after elimination of previously selected lines, that is, after removing the first set of Pareto-optimal points (line 1), line 2 connects these that now become Pareto-optimal, etc. This approach was evaluated since the aim of the study was to select a set of columns, and application of the concept in its most narrow sense would lead to a much too limited selection.

3.3.3   STATIONARY-PHASE RANKING

Stationary phases can also be compared and ranked using chromatographic test results. The following are three examples of characterization and ranking methods applied to stationary phases.

Applying the parameter (F) proposed by Hoogmartens' group [52], columns are ranked relative to the reference column comparing the retention factor of amylbenzene (k_{amb}), the relative retention factor benzylamine/phenol at pH 2.7 ($rk_{ba/pH\,2.7}$), the relative retention factor triphenylene/o-terphenyl ($rk_{tri/ter}$), and the retention factor of 2,2-dipyridyl ($k_{2,2'-d}$). These responses represent hydrophobicity, silanol activity, steric selectivity, and silanol activity plus metal impurity, respectively. The F-value for a column i is calculated according to Equation 3.1 and represents the square of the Euclidean distance with the reference column. A small F-value implies that column i is similar to the reference column (ref) in test parameters, and therefore will be highly ranked on a similarity scale when several columns are compared to the reference column and thus is expected to have a similar enantioselectivity as the reference column:

$$F = \left(k_{amb,ref} - k_{amb,i}\right)^2 + \left(rk_{ba/pH2.7,ref} - rk_{ba/pH2.7,i}\right)^2 + \left(rk_{tri/ter,ref} - rk_{tri/ter,i}\right)^2 + \left(k_{2,2'-d,ref} - k_{2,2'-d,i}\right)^2$$

(3.1)

A similar approach, but based on six responses, was proposed by Euerby et al. [27]. The responses included in this characterization are the amount of alkyl chains, hydrophobic selectivity, steric selectivity, hydrogen-bonding capacity, and the ion-exchange capacities at pH 2.7 and at pH 7.6. Columns are compared using the column difference factor (CDF), which is the Euclidean distance between a tested column and a reference column:

$$CDF = \left[\left(xn_{t1} - xn_1\right)^2 - \left(xn_{t2} - xn_2\right)^2 - \left(xn_{t3} - xn_3\right)^2 - \left(xn_{t4} - xn_4\right)^2 - \left(xn_{t5} - xn_5\right)^2 - \left(xn_{t6} - xn_6\right)^2\right]^{1/2}$$

(3.2)

where xn_{ti} and xn_i refer to the six responses measured for the test and the reference column, respectively. In Equations 3.1 and 3.2, the responses are autoscaled.

Thirdly, a distance parameter Fs (Equation 3.3), based on a weighted Euclidean distance, is calculated to compare two columns, where one column acts as the reference column [53]. The comparisons of multiparameter tests by Euclidean distances have an inherent limitation. Changes in the magnitude of the different parameters can result in a similar Euclidean distance for column types with different combinations of characteristic parameters, not necessarily having the same selectivity differences. The Fs measure is based on five selectivity parameters of the Snyder–Dolan model [16], representing the hydrophobicity (H), steric resistance (S*), hydrogen-bond acidity (A) and basicity (B), and cation-exchange activity (C). The Fs value can be regarded as the distance between two columns in a five-dimensional space, where the weighting factors (12.5, 100, etc.) take into account the effect of each parameter on the relative retention of different compounds in an average sample. An Fs value below three indicates that the two compared columns are similar and most probably will show comparable selectivities:

$$Fs = \left[\left[12.5\left(H_2 - H_1\right)^2\right] + \left[100\left(S_2^* - S_1^*\right)^2\right] + \left[30\left(A_2 - A_1\right)^2\right] + \left[143\left(B_2 - B_1\right)^2\right] + \left[83\left(C_2 - C_1\right)^2\right]\right]^{1/2}$$

(3.3)

Van Gyseghem et al. [32] used Derringer's desirability functions d on the measured properties (steric selectivity, efficiency, silanol activity, ion-exchange capacity, and H-bonding capacity) to rank stationary phases. Measured responses are transformed by a desirability function, to a desirability scale ranging between 0 and 1, with 0 for a completely undesirable and 1 for a completely desirable situation. The transformed responses are then combined to give an overall desirability, which is calculated as the geometric mean, D, of the individual desirability values [54,55]. The more the individual properties are favorable, the higher the overall desirability, that is, the closer it is to 1.

3.3.4   CLASSIFICATION METHODS

Exploratory chemometric techniques, such as PCA and HCA, have also been successful tools for the classification of chromatographic systems. These techniques were applied on different data sets by different research groups, all with the objective of classifying chromatographic systems, for instance, on data about physicochemical properties of phases [27,28], model-based column characterization [56,57], and retention profiling [15,20].

Two chemometric tools closely related to PCA, correspondence factor analysis (CFA) and spectral mapping analysis (SMA), have also been used for chromatographic system classification. Walczak et al. [58] presented differences between stationary-phase packing using CFA by which a set of factors affecting selectivity in the chromatographic systems were extracted. These factors were then transformed into hydrophobic and nonhydrophobic properties, which enabled the classification of the stationary phases. It was concluded that the stationary-phase characteristics affecting solute selectivity are the carbon loading, the nature of the organic ligand, and the accessibility of the silanol groups. Hamoir et al. [59] used SMA to classify 16 stationary phases based on the retention factors of benzene derivatives.

3.4   SUMMARY

This chapter has given an overview of the most common methods to characterize chromatographic systems as well as of the chemometric techniques used to analyze the data generated. Exploratory data analysis techniques allow visualization, therefore facilitating understanding of the clustering and grouping of the chromatographic systems. Even though chemometric techniques help observing the clustering of chromatographic systems based on similar properties, the chromatographers' knowledge is still required to make practically useful selections. Still, chemometric techniques are invaluable tools in enabling the chromatographer to select either similar or dissimilar systems to fulfill one's needs.

REFERENCES

1. Neue UD. Stationary phase characterization and method development. *J Sep Sci* 2007; 30:1611–1627.
2. Claessens H. Trends and progress in the characterization of stationary phases for reversed-phase liquid chromatography. *Trends Anal Chem* 2001; 20:563–583.
3. Auler LMLA, Silva CR, Collins KE, Collins CH. New stationary phase for anion-exchange chromatography. *J Chromatogr A* 2005; 1073:147–153.
4. Ranatunga RPJ, Carr PW. A study of the enthalpy and entropy contributions of the stationary phase in reversed-phase liquid chromatography. *Anal Chem* 2000; 72:5679–5692.
5. Philipsen HJA, Claessens HA, Lind H, Klumperman B, German AL. Study on the retention behaviour of low-molar-mass polystyrenes and polyesters in reversed-phase liquid chromatography by evaluation of thermodynamic parameters. *J Chromatogr A* 1997; 790:101–116.
6. McCalley DV. Effect of temperature and flow-rate on analysis of basic compounds in high-performance liquid chromatography using a reversed-phase column. *J Chromatogr A* 2000; 902:311–321.
7. Kohler J, Chase DB, Farlee RD, Vega AJ, Kirkland JJ. Comprehensive characterization of some silica-based stationary phases for high-performance liquid chromatography. *J Chromatogr* 1986; 352:275–305.
8. Ho M, Cai M, Pemberton JE. Characterization of octadecylsilane stationary phases on commercially available silica-based packing materials by Raman spectroscopy. *Anal Chem* 1997; 69:2613–2616.
9. Ducey MW, Orendorff CJ, Pemberton JE, Sander LC. Structure-function relationships in high-density octadecylsilane stationary phases by Raman spectroscopy. 1. Effects of temperature, surface coverage, and preparation procedure. *Anal Chem* 2002; 74:5576–5584.
10. Albert K, Bayer E. Review—Characterization spectroscopy. *J Chromatogr* 1991; 544:345–370.
11. Cruz E, Euerby MR, Johnson CM, Hackett CA. Chromatographic classification of commercially available reverse-phase HPLC columns. *Chromatographia* 1997; 44:151–161.

12. Visky D, Vander Heyden Y, Iványi T, Baten P, De Beer J, Kovács Z, Noszál B, Roets E, Massart DL, Hoogmartens J. Characterisation of reversed-phase liquid chromatographic columns by chromatographic tests. Evaluation of 36 test parameters: Repeatability, reproducibility and correlation. *J Chromatogr A* 2002; 977:39–58.

13. Visky D, Vander Heyden Y, Iványi T, Baten P, De Beer J, Kovács Z, Noszál B et al. Characterisation of reversed-phase liquid chromatographic columns by chromatographic tests. Rational column classification by a minimal number of column test parameters. *J Chromatogr A* 2003; 1012:11–29.

14. Van Gyseghem E, Jimidar M, Sneyers R, De Smet M, Verhoeven E, Vander Heyden Y. Stationary phases in the screening of drug/impurity profiles and in their separation method development: Identification of columns with different and similar selectivities. *J Pharm Biomed Anal* 2006; 41:751–760.

15. Galea C, Mangelings D, Vander Heyden Y. Method development for impurity profiling in SFC: The selection of a dissimilar set of stationary phases. *J Pharm Biomed Anal* 2015; 111:333–343.

16. Snyder LR, Dolan JW, Carr PW. The hydrophobic-subtraction model of reversed-phase column selectivity. *J Chromatogr A* 2004; 1060:77–116.

17. Taft RW, Abboud M, Kamlet MI, Abraham MH. Linear solvation energy relations. *J Solution Chem* 1985; 14:153–186.

18. Wilson NS, Gilroy J, Dolan JW, Snyder LR. Column selectivity in reversed-phase liquid chromatography. *J Chromatogr A* 2004; 1026:91–100.

19. Dolan JW, Snyder LR. Selecting an "orthogonal" column during high-performance liquid chromatographic method development for samples that may contain non-ionized solutes. *J Chromatogr A* 2009; 1216:3467–3472.

20. Van Gyseghem E, Van Hemelryck S, Daszykowski M, Questier F, Massart DL, Vander Heyden Y. Determining orthogonal chromatographic systems prior to the development of methods to characterise impurities in drug substances. *J Chromatogr A* 2003; 988:77–93.

21. Abraham MH, Ibrahim A, Zissimos AM. Determination of sets of solute descriptors from chromatographic measurements. *J Chromatogr A* 2004; 1037:29–47.

22. Wu Y, Li L. Sample normalization methods in quantitative metabolomics. *J Chromatogr A* 2015; 1430:80–95.

23. Samanipour S, Dimitriou-Christidis P, Gros J, Grange A, Arey JS. Analyte quantification with comprehensive two-dimensional gas chromatography: Assessment of methods for baseline correction, peak delineation, and matrix effect elimination for real samples. *J Chromatogr A* 2015; 1375:123–139.

24. McIlroy JW, Smith RW, McGuffin VL. Assessing the effect of data pretreatment procedures for principal components analysis of chromatographic data. *Forensic Sci Int* 2015; 257:1–12.

25. Niu W, Knight E, Xia Q, McGarvey BD. Comparative evaluation of eight software programs for alignment of gas chromatography-mass spectrometry chromatograms in metabolomics experiments. *J Chromatogr A* 2014; 1374:199–206.

26. Vandeginste BGM, Massart DL, Buydens LMC, de Jong S, Lewi PJ, Smeyers-Verbeke J. Handbook of chemometrics and qualitmetrics: Part B. In: Vandeginste BGM, Rutan SC (eds.), *Data Handling in Science and Technology*. Amsterdam, the Netherlands: Elsevier; 1998; 118–123.

27. Euerby MR, Petersson P, Campbell W, Roe W. Chromatographic classification and comparison of commercially available reversed-phase liquid chromatographic columns containing phenyl moieties using principal component analysis. *J Chromatogr A* 2003; 994:13–36.

28. Forlay-Frick P, Fekete J, Héberger K. Classification and replacement test of HPLC systems using principal component analysis. *Anal Chim Acta* 2005; 536:71–81.

29. Van Gyseghem E, Jimidar M, Sneyers R, Redlich D, Verhoeven E, Massart DL, Vander Heyden Y. Orthogonality and similarity within silica-based reversed-phased chromatographic systems. *J Chromatogr A* 2005; 1074:117–131.

30. Van Gyseghem E, Crosiers I, Gourvénec S, Massart DL, Vander Heyden Y. Determining orthogonal and similar chromatographic systems from the injection of mixtures in liquid chromatography–diode array detection and the interpretation of correlation coefficients colour maps. *J Chromatogr A* 2004; 1026:117–128.

31. Van Gyseghem E, Dejaegher B, Put R, Forlay-Frick P, Elkihel A, Daszykowski M, Héberger K, Massart DL, Vander Heyden Y. Evaluation of chemometric techniques to select orthogonal chromatographic systems. *J Pharm Biomed Anal* 2006; 41:141–151.

32. Van Gyseghem E, Jimidar M, Sneyers R, Redlich D, Verhoeven E, Massart DL, Vander Heyden Y. Selection of reversed-phase liquid chromatographic columns with diverse selectivity towards the potential separation of impurities in drugs. *J Chromatogr A* 2004; 1042:69–80.

33. Dehouck P, Visky D, Vander Heyden Y, Adams E, Kovács Z, Noszál B, Massart DL, Hoogmartens J. Characterisation of reversed-phase liquid-chromatographic columns by chromatographic tests. *J Chromatogr A* 2004; 1025:189–200.
34. Dallos A, Ngo HS, Kresz R, Héberger K. Cluster and principal component analysis for Kováts' retention indices on apolar and polar stationary phases in gas chromatography. *J Chromatogr A* 2008; 1177:175–182.
35. Stella C, Rudaz S, Gauvrit J-Y, Lantéri P, Huteau A, Tchapla A, Veuthey J-L. Characterization and comparison of the chromatographic performance of different types of reversed-phase stationary phases. *J Pharm Biomed Anal* 2007; 43:89–98.
36. Stella C, Seuret P, Rudaz S, Tchapla A, Guavrit J-Y, Lantéri P, Veuthey J-L. Simplification of a chromatographic test methodology for evaluation of base deactivated supports. *Chromatographia* 2002; 56:665–671.
37. Van Dorpe S, Vergote V, Pezeshki A, Burvenich C, Peremans K, De Spiegeleer B. Hydrophilic interaction LC of peptides: Columns comparison and clustering. *J Sep Sci* 2010; 33:728–739.
38. De Klerck K, Vander Heyden Y, Mangelings D. Exploratory data analysis as a tool for similarity assessment and clustering of chiral polysaccharide-based systems used to separate pharmaceuticals in supercritical fluid chromatography. *J Chromatogr A* 2014; 1326:110–124.
39. Dumarey M, Put R, Van Gyseghem E, Vander Heyden Y. Dissimilar or orthogonal reversed-phase chromatographic systems: A comparison of selection techniques. *Anal Chim Acta* 2008; 609:223–234.
40. Daszykowski M, Walczak B, Massart DL. Representative subset selection. *Anal Chim Acta* 2002; 468:91–103.
41. Galvão RKH, Araujo MCU, José GE, Pontes MJC, Silva EC, Saldanha TCB. A method for calibration and validation subset partitioning. *Talanta* 2005; 67:736–740.
42. Questier F, Put R, Coomans D, Walczak B, Vander Heyden Y. The use of CART and multivariate regression trees for supervised and unsupervised feature selection. *Chemom Intell Lab Syst* 2005; 76:45–54.
43. Smyth C, Coomans D, Everingham Y, Hancock T. Auto-associative multivariate regression trees for cluster analysis. *Chemom Intell Lab Syst* 2006; 80:120–129.
44. Forlay-Frick P, Van Gyseghem E, Heberger K, Vander Heyden Y. Selection of orthogonal chromatographic systems based on parametric and non-parametric statistical tests. *Anal Chim Acta* 2005; 539:1–10.
45. Ankerst M, Breunig MM, Kriegel H, Sander J. OPTICS: Ordering points to identify the clustering structure. *Proceedings of the ACM SIGMOD 1999 International Conference of Data*, Philadelphia, PA. 1999, pp. 49–60.
46. Daszykowski M, Walczak B, Massart DL. Looking for natural patterns in analytical data. 2. Tracing local density with OPTICS. *J Chem Inf Comput Sci* 2002; 42:500–507.
47. Put R, Van Gyseghem E, Coomans D, Vander Heyden Y. Selection of orthogonal reversed-phase HPLC systems by univariate and auto-associative multivariate regression trees. *J Chromatogr A* 2005; 1096:187–198.
48. Dumarey M, van Nederkassel AM, Stanimirova I, Daszykowski M, Bensaid F, Lees M, Martin GJ, Desmurs JR, Smeyers-Verbeke J, Vander Heyden Y. Recognizing paracetamol formulations with the same synthesis pathway based on their trace-enriched chromatographic impurity profiles. *Anal Chim Acta* 2009; 655:43–51.
49. Massart DL, Vandeginste BGM, Buydens LMC, De Jong S, Lewi PJ. *Handbook of Chemometrics and Qualimetrics: Part A*. Amsterdam, the Netherlands: Elsevier; 1997.
50. Keller HR, Massatt DL. Program for pareto-optimality in multicriteria problems. *Trends Anal Chem* 1990; 9:251–253.
51. Keller HR, Massart DL. Multicriteria decision making: A case study. *Chemom Intell Lab Syst* 1991; 11:175–189.
52. Visky D, Haghedooren E, Dehouck P, Kovács Z, Kóczián K, Noszál B, Hoogmartens J, Adams E. Facilitated column selection in pharmaceutical analyses using a simple column classification system. *J Chromatogr A* 2006; 1101:103–114.
53. Dolan JW, Maule A, Bingley D, Wrisley L, Chan CC, Angod M, Lunte C et al. Choosing an equivalent replacement column for a reversed-phase liquid chromatographic assay procedure. *J Chromatogr A* 2004; 1057:59–74.
54. Jimidar M, Bourguignon B, Massart DL. Application of Derringer's desirability function for the selection of optimum separation conditions in capillary zone electrophoresis. *J Chromatogr A* 1996; 740:109–117.
55. Derringer G, Suich RJ. Simultaneous optimization of several response variables. *J Qual Technol* 1980; 12:214–219.

56. West C, Lesellier E. Chemometric methods to classify stationary phases for achiral packed column supercritical fluid chromatography. *J Chemom* 2012; 26:52–65.
57. Bączek T, Kaliszan R, Novotná K, Jandera P. Comparative characteristics of HPLC columns based on quantitative structure–retention relationships (QSRR) and hydrophobic-subtraction model. *J Chromatogr A* 2005; 1075:109–115.
58. Walczak B, Morin-Allory L, Lafosse M, Dreux M, Chretien JR. Factor analysis and experiment design in high-performance liquid chromatography. VII. Classification of 23 reversed-phase high-performance liquid chromatographic packings and identification of factors governing selectivity. *J Chromatogr* 1987; 395:183–202.
59. Hamoir T, Cuesta Sanchez PA, Bourguignon B, Massart DL. Spectral mapping analysis: A method for the characterization of stationary phases. *J Chromatogr Sci* 1994; 32:488–498.

Section II

Univariate Analysis

4 Chromatographic Applications of Genetic Algorithms

Mohammad Goodarzi and Yvan Vander Heyden

CONTENTS

4.1 INTRODUCTION

In general, optimizations are very important in both an industrial and a scientific context, since they can significantly improve any process/system not only quantitatively but also qualitatively. Contrarily, if one does not optimize the system under investigation, such a system may drastically limit the yield of an experiment or process/system (e.g., increase in cost, time, and/or needed materials, while resulting in a low yield of products). Optimization is not limited to chemical or biological problems but affects a large range of daily issues. As an example of common practical problems, we can mention what happens if we do not optimize the train scheduling, traffic, or telecommunications. Life cannot be imagined without optimization because one is dealing daily with many processes that must be optimized to reach given goals [1]. In chromatography, the separation between given or all compounds of a mixture needs to be optimized, for instance, in drug impurity profiling or chromatographic fingerprinting.

What makes a process more complicated and more difficult to optimize is the number of factors involved. The higher the number, the more complicated the optimization becomes. As an example to master the problem, imagine a cook who wants to make a pastry. How many factors are involved in making a delicious Belgian waffle? There are many factors that, if they are not optimized, cannot result into an optimized recipe for such a delicious product. For instance, one should know how many people to make waffles for, how much flour is needed, how many eggs, how to mix them, how long and at what temperature they should be baked. If one wants to have all parameters optimized, then an experimental design approach should best be taken into account. Similar to our cooking example, in a chemical process, normally several factors have to be set at proper levels to find optimal conditions. In chromatography also several factors are involved in finding a suitable separation. For instance, the type of the stationary phase, the pH of the mobile-phase buffer, the organic modifier amount and composition, the column temperature, the gradient slope, and the amount of additives all can affect the separation to a major or minor degree.

Once several factors (variables) are involved in our process and some have so-called interaction effects, then setting the proper conditions is not that straightforward anymore. To solve such a problem, normally optimization techniques called "Design of Experiments (DOE)" techniques are used to find both the factor effects (applying screening designs) and to set the proper conditions for a given process or separation (response surface designs).*

Apart from DOE, three types of optimization strategies can be considered [2,3]:

1. Parameter optimization (numerical problems). In this case, an algorithm tries to find an optimal setting for a particular model. As an example, we can refer to curve fitting where an algorithm is used to analyze the unresolved (overlapping) peaks by first estimating, for instance, peak width, height, and position, and then optimizing the shape of individual peaks [4–6].
2. Variable selection (subset selection problems). An algorithm is used to find a small representative and informative set of variables from a large pool.
3. Sequence optimization (e.g., traveling salesman problem, where an algorithm tries to find the shortest path between a given list of cities). This is rarely used in the chemistry domain. It is worth mentioning that GA as an optimization algorithm can be used in all three categories.

Note that optimization strategies in chemometrics are categorized into only two types: one is parameter optimization and the other is variable selection as described above. The concept of the traveling salesman is a variable selection problem with a particularity that in variable selection several subset solutions may exist, while in the traveling salesman one solution is optimal. In the salesman problem, one has a list of cities with the distances between them and tries to find the shortest path. In variable selection, a list of variables and, for instance, a similarity value (between Y property to be modeled and X variables) is available and one tries to find a small subset of X with the highest relationship to Y [7]. Several variable selection algorithms/strategies exist that have been introduced or applied in diverse domains [8]. As a chromatographic example of parameter optimization, one can refer to alignment strategies. For instance, the correlation optimized warping (COW) method tries to correct the chromatographic shifts by piecewise linear stretching and compressing of a given chromatogram [9]. This algorithm uses a parameter (i.e., correlation between the sample profile and a reference profile) to align the profile of a given sample. The parameter (e.g., correlation coefficient) can be assigned to an optimization technique, like GA, to be maximized/optimized leading to, for instance, the best match between the shifted sample and a reference profile. Furthermore, we can use a variable selection–elimination–reduction prior to modeling a given activity (e.g., antioxidant, cytotoxic, etc.) as a function of fingerprints within herbal medicine analysis [10].

In order to optimize a system under study, the optimization task can be defined as follows:

- Define an objective function (OF) for the optimization problem with a scalar output showing the quantitative performance (e.g., it can be system cost, yield, or profit). As an example, in a partial least squares modeling, the number of latent variables that are incorporated into the model should be optimized (minimized). In this case, the OF may be the root mean squared error of cross validation [11].
- Build a predictive model (e.g., basically a least squares solution is used, such as multiple linear regression) that shows the behavior of a process or system.
- Find the most important and effective variables (features), which can be interpreted as degrees of freedom for a system.

The above three steps are general tasks to be solved for any optimization question. As an example, we refer to a study where the goal was to find antioxidant compounds in different herbal species [10]. After obtaining the chromatographic profiles (a matrix X, in this case each profile belongs

---

* See Chapter 1.

to a sample measured for 60 min leading to a matrix with a dimension of n × m, with n being the number of samples and m the number of measured properties, i.e., 60 (min) × 60 (min) × p (measurements/s)) and also experimentally measuring the antioxidant activity of each sample by a standard protocol, for instance, the DPPH method (resulting in a vector $\mathbf{Y}$), one can build a PLS model between $\mathbf{X}$ and $\mathbf{Y}$. In order to find the best latent variables, for instance, a parameter optimization technique as mentioned above can be used (in [10] a cross-validation strategy was used*). After the PLS model is built, the model performance is evaluated by, for instance, the root mean squared error of cross validation or of prediction as internal or external validations, respectively. Finally, the retention times of the most important compounds, which have an effect on the total antioxidant activity, can be found by, for instance, estimating the regression coefficients of the built PLS model. These regression coefficients, or any other technique used as a variable selection method, can be used during the PLS modeling in data to reduce the dimension (m) in $\mathbf{X}$, which usually results in simpler PLS models.

Many optimization algorithms are available for different optimization purposes, such as simulated annealing (SA) [12], genetic algorithms (GAs) [13], and Tabu Search (TS) [14], as well as swarm intelligence methodologies, such as particle swarm optimization (PSO) and ant colony optimization (ACO) [15], all using the potential that computers offer. These algorithms are used for different types of optimization problems, such as parameter optimization or feature selection, depending on the problem and the way they are coded.

Chromatographic techniques have become one of the most advanced analytical tools commonly applied in various fields of chemistry. By the growing development of data handling tools, the chromatographic technique became very accurate, reliable, and user-friendly. Chemometric tools can offer a lot to chromatographers, for example, by resolving overlapping chromatographic peaks, by helping to model retention as a function of structural properties (quantitative structure–retention relationships), by selecting column properties for higher sensitivity or a proper selectivity, by data processing, such as signal enhancement, background elimination, variable selection, and peak alignment [16,17]. In this chapter, we describe in more detail one of the most applied optimization techniques, called genetic algorithm.

4.2 GENETIC ALGORITHM

4.2.1 DEFINITION

Similar to many other techniques, genetic algorithms (GAs) are inspired by evolutionary sciences, specifically by exploiting Charles Darwin's theory, called "natural selection." GA was introduced by John Holland in the 1970s [18]. Later, the advantages of this new method were revealed to the world by David Goldberg, who was one of Holland's students. He used the algorithm in his dissertation to solve a problem involving the control of gas-pipeline transmission [19]. Since then, the method quickly found its place as a very popular heuristic algorithm in different fields and has been successfully applied to solve various optimization problems. The reason for such popularity is its power to solve complex problems. For instance, some advantages over traditional methods can be mentioned: it can solve (non)linear problems, continuous or discrete or even with random noise. Moreover, it is easy to parallelize the algorithm because it uses multiple offspring in a population that can explore the search space in many directions. This algorithm has also a few disadvantages, like population size or setting a proper rate of mutation and crossover (see further), where an improper value may lead the GA to local minima.

Questions always asked by chemists are as follows: What is GA, how can I use it in my research, and how can I understand it? In this chapter, not all questions will be addressed, but we try to explain what genetic algorithm is and show how it was used in chromatography.

---

* For the idea of model validation, see Chapter 14.

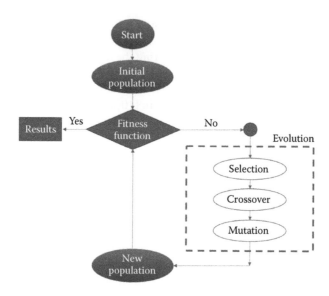

FIGURE 4.1 A general flowchart of genetic algorithms. Yes: the best solution based on the fitness function is reached (or the algorithm was not able to find a solution). No: the algorithm is still performing evolution and a decision was not reached. (Adapted from Goodarzi, M. and dos Santos Coelho, L., *Anal. Chim. Acta*, 852, 20, 2014.)

As mentioned above, this algorithm belongs to the family of evolutionary algorithms, which are categorized into four major types: genetic algorithms (GAs), genetic programming (GP) [20], evolution strategies (ES) [21], and evolutionary programming (EP) [22]. All are based on a population of individuals [23].

The procedure of a genetic algorithm (Figure 4.1) starts with an initial set called "initial population" that are random candidate solutions. As in reality, each individual in the population is called "chromosome" and consists of various segments (called "genes"). Each gene represents the value of a decision variable (called "phenotype"). Phenotypic transformation is a function that converts phenotypes into genotypes (chromosome is also called genotype). From one generation to the next, the chromosomes are evolved while being compared against each other based on a user-defined fitness function. Note that in each generation, new chromosomes are made, called "offspring," by the following steps:

- By mating two chromosomes from the current generation using a crossover method
- By modifying the chromosome using a mutation method

In each new generation, those chromosomes that have a bad or low fitness (user defined) are replaced by the obtained offspring [24].

Assume a QSRR case study where one tries to build a relationship between molecular descriptors and the corresponding retention times for a set of compounds.* In this case, each individual of a population is a chromosome of binary values where each variable (descriptor) is a gene and a set of them forms a chromosome. At first, each chromosome is randomly composed with the values 0 and 1 (binary values). This indicates that the population of the first generation is randomly selected. The state of each variable is represented by the value of 1 (descriptor is selected) or the value of zero (descriptor is not selected). Afterward, a crossover and mutations operators with a (user-defined)

---

* For the introduction about the QSRR methodology, see Chapter 17.

probability can be applied on the first selected variables to generate the next offspring. The main core of a GA is depicted as a flowchart in Figure 4.1 [25]. The various stages of GA are elaborated in the following sections.

4.2.2 PRINCIPLE

In order to start a GA, first an initial population of chromosomes, which is a set of possible solutions to any optimization problem, must be created. Normally, a population is randomly generated and its size can vary from a few to thousands of individuals. Various encoding strategies for a GA exist that depend on the problem. Such encoding is used to transfer the problem solution into chromosomes. As an example, the binary, permutation, value, or tree encodings can be mentioned [26]:

- Binary encoding: In binary encoding, every chromosome is a string of bits (binary, 0 or 1) (example Figure 4.2a). This encoding (where the data is converted into a binary string) is the most common, especially in chemometrics.
- Permutation encoding: In permutation encoding, every chromosome is a string of numbers that represent a position in a sequence. This encoding is suitable for ordering or queuing problems (e.g., it is used for solving the previously mentioned salesman challenge) [27]. Figure 4.2b illustrates such encoding.
- Value encoding: In this encoding, every chromosome is also a sequence of some values. Values can be anything connected to the problem, such as (real) numbers or characters. An example can be found in Figure 4.2c.
- Tree encoding: In the tree encoding, every chromosome is a tree of some objects, such as functions or commands, and this type of encoding is mainly used for genetic programming [27]. Figure 4.2d shows an example of tree encoding.

In this chapter, we are mainly dealing with chemometrics applied to chromatographic problems and usually a binary encoding is then used for GA programming. After encoding, which represents the population generation, the chromosomes will be used in the data handling.

Let's start with a spectroscopic example to explain the principle. Assume that NIR spectra of serum samples were measured and the goal is to find the wavelengths that are important or can influence the prediction of the glucose content in those samples [28]. In case we want to use GA as feature selection technique to find the best wavelengths from the spectra (for instance, select few from hundred or thousand wavelengths). Each chromosome consists of a user-defined number of genes, which is a randomly selected subset from the data set and each gene represents a wavelength.

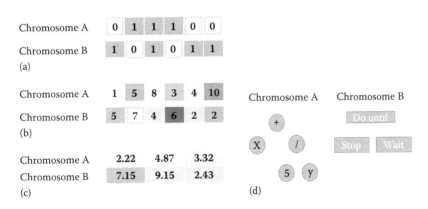

FIGURE 4.2 Examples of different encodings: two possible chromosomes encoded for the same problem are shown. (a) Binary encoding. (b) Permutation encoding. (c) Value encoding. (d) Tree encoding.

If only one chromosome (a subset) randomly is generated in the binary encoding way with a size of 5, something like 01001 would be obtained. This means that the first, third, and fourth wavelengths are not selected, while the second and the fifth are. Note that for a variable selection problem, one can also use a permutation encoding (e.g., a chromosome of size 5 could look like 5 8 9 1 6, which means wavelengths 5, 8, 9, 1, and 6 should be evaluated). However as stated above, a binary encoding is the best and simplest used in chemometrics to solve variable selection problems.

As mentioned above, GA can also be applied on chromatographic data. For instance, one can use GA to find the most active compounds from various species of herbal medicines. In this case, it is used after measuring the chromatographic profiles (fingerprints) and the considered activity of several samples. GA can also be coupled with PLS to find the most significant retention times from the chromatographic profiles of all samples, which influence the prediction performance of a model built for a given activity. Those selected retention times can then be further assigned to biologically active compounds.

After encoding, a set of chromosomes or candidate solutions must somehow be selected from the initial population (selection). The selected set should be the best (this is based on the fitness function) for further operations. This step is mainly performed to evaluate the quality of the chromosomes and whether they meet the expectations or not. In this step, a fitness function is used to evaluate each chromosome in the population. Assume one wants to select the best subset from a vast amount of information as, for instance, described above. The best set is the one that leads to a model (e.g., from PLS) having the highest correlation coefficient between the experimental value of the property to be modeled and the set of predicted property values. For each chromosome, the correlation coefficient (R) resulting from the model built can be calculated, and if they meet the expectations, then the best subset is selected. If not, the GA procedure should be involved in the next step, called evolution. As an example, if the desired R is set to be 0.99 then it can be seen if any chromosome (subset) leads to such R. If yes, then the algorithm finishes searching. If not, two possibilities exist: either the procedure goes many times to the second stage (evolution) and when no better solution is found it brings the algorithm to stop or the algorithm continues searching. In the latter case, all chromosomes with, for instance, R below 0.5 are discarded. By doing so, the population has been shrunk and only the best chromosomes, which can be used for the evolution stage, are kept. Figure 4.1 shows the step as "selection" in the evolution stage of a GA. Note that it is unlikely to find the best set of chromosomes, which address a given problem, from the initial population. Normally, the best sets of chromosomes are kept for the evolution step and the others discarded. Sometimes a suitable solution (chromosome) is never found from a given data set. For such situations, an iteration point is set and an ultimatum is given to the algorithm: if a suitable solution is not found when the ultimatum is reached, the algorithm stops.

Once the evaluation of the initial population is done, a new set of chromosomes is created from the current population. This step selects the best parents from the available population (best set of chromosomes) for mating, which is also called the "breeding population." There are many different types of selection methods, such as roulette wheel selection, rank selection, steady-state selection, and tournament selection [29]. A roulette wheel selection is often performed because of its simplicity [29]. Therefore, this technique is explained a bit more in detail. To understand the procedure, consider a roulette wheel with a handle as shown in Figure 4.3. The wheel has N segments (each segment is a chromosome or subset). Now a random number between 0 and 1 is generated and the wheel is rotated clockwise with the rotation proportional to the random number times 360°. By spinning the wheel, a chromosome can be selected using the arrow at the starting location. This selected chromosome will go to the next step of evolution and the rest is discarded. In Figure 4.3, it is seen that, for the six chromosomes, C5 is selected by this selection process. One of the biggest disadvantages of this type of selection process is that it is biased toward chromosomes that have a higher fitness value (in our example, higher R values). It can easily be seen that the chromosomes with higher R (correlation coefficient) take more space in our wheel and thus have a larger probability of selection.

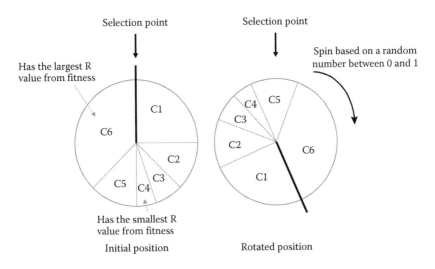

FIGURE 4.3   Roulette wheel selection. C represents a chromosome (subset).

Now the selected chromosome is used for crossover (Figure 4.1), which is the main genetic operator where two chromosomes (parents) are mating resulting in two new chromosomes, called offspring or the next generation. Obviously, this step greatly influences the GA performance. Normally, its rate is defined as the ratio of the number of offspring produced in each generation to the population size. Many attempts have been made to discover an appropriate crossover operator for GAs, called, for instance, single point, double point, uniform, and arithmetic [30,31]. The most used crossover techniques are single points, double points, and uniform points (Figure 4.4). In the single-point method, the genes (each variable of a chromosome) from the two parents (chromosomes) are split at the same size but at random points. The first part of parent A is swapped with the second of the chromosome called parent B, resulting in a new offspring. Moreover, in double-point crossover, the parents are split with the same size but at two random points where two fractions of parent A and the middle part of parent B make the offspring. In Figure 4.4, we tried to illustrate the crossover by both graphic and binary values. Single- and double-point methods are basically the same, with the difference in the fact that the offspring produced based on the double-point

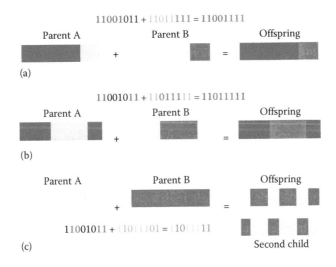

FIGURE 4.4   Three crossover methods: (a) single points, (b) double points, and (c) uniform points.

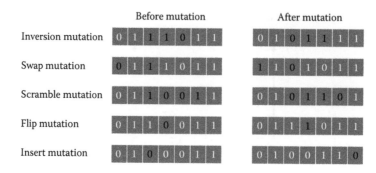

FIGURE 4.5 Examples of mutation methods.

method is less likely to be the same as the original parents. Uniform points is another method used for crossover. In this method, there are several points coming from one parent and several from the other. The second child is also made, generally symmetrical, but both may be independently constructed. In the example in Figure 4.4, four genes are coming from parent A and four from the second parent. In the second child, the exact same splits are made, but the parents are swapped.

After crossover, we must pass the offspring to mutate (mutation), which produces spontaneous random changes to generate new chromosomes. There are various types of mutation methods described, such as inversion, swap, random, and insert mutations. However, not all are applicable to chemometrics [32]. Figure 4.5 illustrates different types of mutation. To perform the inversion, two random points (i.e., positions 3 and 5 in our example) are selected and their genes inverted (1 to 0 and 0 to 1). Swap mutation occurs when two genes are randomly chosen (i.e., genes 1 and 3 in our example) and their positions swapped. In scramble mutation, two random positions are chosen and the genes located between them (i.e., genes 3 through 6 in our example) are scrambled. In flip mutation, a gene is randomly flipped from 0 to 1 or 1 to 0 (i.e., gene 4 in our example). Insert mutation is when one gene is randomly selected (i.e., gene 3 in our example) and is displaced (to gene 7) and inserted into the chromosome. After having performed the mutation, the new chromosomes are ready to be evaluated by the fitness function. If they meet the expectation, then the algorithm stops, otherwise it continues until either the best subset is selected or the algorithm reaches the maximum allowed number of iterations.

Some important points should be taken into account when using GA to solve problems. The population size is very important: when few chromosomes are available, then the chance to perform a crossover is very low (the offspring will be more likely similar to their parents) leading to an incomplete search. When many chromosomes are present in the initial population, the algorithm may take a long time to converge. Furthermore, if the probability of crossover is set to 100% then all offspring is generated based on a crossover method, while a crossover rate of 0% results in an offspring being a copy of their parents without any change. Most commonly, a crossover rate between 50% and 75% is used.

4.3 OPTIMIZATION OF CHROMATOGRAPHIC SYSTEMS AND THEIR ANALYSIS

In the following examples, GAs have been used to optimize different aspects of chromatographic systems. Usually, a single chromatographic column is selected and the HPLC parameters are optimized to solve a given separation problem. In [33], a multicolumn approach using serially coupled columns of different length and selectivity, under isocratic or gradient elution, has been developed for complex mixtures. In a next study [34], an efficient way to predict multilinear gradients accurately for a given mixture under study was discussed. GA was used to reduce computation time when optimizing multilinear gradients for a variety of solutes. The chromosomes consisted of different column combinations and the parameters used for coding were analysis time and solvent content for each node (column), resulting in 8 bits, which were being coded. The column sequence was included as permutation index. Note that for the fitness function, to evaluate the quality of each individual, the global resolution was considered.

In [35], an immune algorithm combined with GA was used to find compounds from overlapping peaks. The method takes an overlapping chromatogram as its input and then processes the chromatogram to find the number of components. This is done by iteratively subtracting the estimated peaks from the input until the residual remains constant.

A genetic algorithm was employed to optimize the buffer system (concentrations of borate, phosphate, and sodium dodecyl sulfate) of micellar electrokinetic capillary chromatography (MEKC) to separate the active components in a Chinese herbal medicine [36]. A hybrid GA was used to optimize the mobile-phase composition in liquid chromatography when using two or more mobile phases with complementary behavior. Each of these mobile phases leads to the optimal separation of certain compounds in a mixture, while the others can remain overlapped [37]. The optimization is based on a local search method, which alternates two combinational search spaces, one defined by combinations of solutes and the other by combinations of mobile phases.

Modeling of response surfaces in LC is another topic that may benefit from GA. GA was used to model retention surfaces by determining the coefficients of the theoretical equation adopted to describe the surfaces (i.e., surfaces describing the combined effect of pH and modifier concentration on the retention factor k of a solute) [38,39]. As another example, GA was used to find the best size of the packing material, the total number of columns, and the total feed concentration in a preparative chromatographic separation with a focus on maximizing simultaneously the purity of the extract and the productivity of the unit [40]. GA was also used to optimize the parameters affecting the separation selectivity and efficiency, such as surfactant concentration, percent of cosurfactant and percent of organic oily solvent, temperature, and pH in a microemulsion liquid chromatographic method for fat-soluble vitamins [41].

4.4   VARIABLE SELECTION

One of the most critical steps in multivariate calibration applications, including those in separation science, is variable selection. The reason is that chromatographic techniques generate chromatograms with hundreds or thousands of variables, of which not all are informative (e.g., baseline sections), but make it more difficult to model and interpret. Variable selection results in simpler models and an improvement in model performance. Moreover, a model with fewer variables can be transferred more readily [8]. In this chapter, GA as a variable selection technique in chromatography is divided into two groups, one when it is used for model building applied on chromatographic data and the second when it is used to build quantitative structure–retention relationship (QSRR) models.

4.4.1   Variable Selection in Modeling

In only very few studies, researchers have tried to use GA for selecting relevant variables from a chromatogram. For instance, it was used to reduce the number of data points from 6376 in the original chromatogram to only 37 for the classification of Chilean wines using HPLC-DAD data [42]. In another study, GA was used as variable selection methodology in the determination of pesticides, such as carbaryl, methyl thiophanate, simazine, dimethoate, and the metabolite phthalimide, in wine samples using HPLC-DAD data [43]. It was also used for the identification and qualification of the adulteration of olive oils with other vegetable oils based on the triacylglycerol profile measured with GC-MS [44]. A model based on GA-PLS led to a better model than a PLS model built without GA as variable selection technique. As another example, we can mention the use of GA-optimized support vector machines as a classification tool employed to classify 64 samples from 7 Citrus herb species based on 10 potential marker compounds derived from liquid chromatography combined with quadrupole time-of-flight mass spectrometry (LC–QTOF-MS) [45]. The obtained classifier showed good prediction accuracy. Thus, GA as variable selection approach on HPLC data leads to better and simpler classification and calibration models.

4.4.2   VARIABLE SELECTION IN QSRR

In a QSRR model, a retention parameter (t_R, k, log k) is modeled as a function of structural information parameters contained in molecular descriptors. The QSRR model could be built based on a limited number of molecular properties as, for instance, in the "linear free-energy relationship (LFER)," which is mainly focusing on solvatochromic or hydrogen-binding descriptors and links them to the retention of a set of molecules [46]. Another approach is to build the QSRR model using a number of so-called theoretical molecular descriptors. However, these latter are often very difficult to interpret from a physical chemical point of view. Nowadays, more than 6000 theoretical descriptors for a given molecule can be calculated. One of the problems in building a QSRR is to find "relevant" descriptors from the large pool. Variable selection or feature selection, on which detailed information can be found in Reference 8, is often performed or even required. However, the main reason for a variable selection is to find an as small as possible set of descriptors that can be linked to the retention, resulting in a model able to predict the retention for a set of new molecules, which are not yet experimentally measured. There are many techniques, including linear (e.g., MLR, PLS) and nonlinear (e.g., SVM, ANN) [47–58], which have been used for model building. Multiple linear regression (MLR) is often used as a modeling technique in QSRR (Table 4.1). Here variable selection is obligatory because the number of descriptors is usually much higher than the number of compounds in the calibration set. In QSRR, the first step usually is to delete the (nearly) constant descriptors because they are not informative. Secondly, highly correlated descriptors are identified, because MLR cannot deal with them. Usually, the one best correlated to the retention is maintained

TABLE 4.1

Some Case Studies Applying QSRR with GA as Variable Selection Technique

Number and Type of Solutes	Variable Selection Approaches	Modeling Approaches	Ref.
83 diverse drugs	Genetic algorithms, stepwise regression, relief method, ant colony optimization	MLR, SVM	[47]
83 basic drugs	UVE, GA	PLS, MLR, CART	[48]
67 hydantoins	GA, mutual information	PLS, SVM	[49]
9 arylpropionic acids	GA	MLR, ANN	[50]
69 peptides	GA, stepwise regression	MLR, PLS	[51]
368 pesticides in animal tissues	GA	PLS, ANN, SVM	[52]
47 acylcarnitines	GA	MLR, PLS	[53]
86 drug compounds from London 2012 Olympic and Paralympic Games	GA, UVE, CARS, IRIV, VISSA	PLS	[54]
116 essential oils	GA	MLR, PLS, KPLS, ANN	[55]
282 amino acids and carboxylic acids	GA	MLR, ANN and SVM	[56]
209 polychlorinated biphenyls	GA	MLR, PLS, ANN	[57]
72 oligonucleotides	GA	MLR	[58]

Notes: UVE, uninformative variable elimination; GA, genetic algorithm; CARS, competitive adaptive reweighted sampling; IRIV, iteratively retaining informative variables; VISSA, variable iterative space shrinkage approach; MLR, multiple linear regression; SVM, support vector machines; CART, classification and regression trees; PLS, partial least squares; ANN, artificial neural network; KPLS, kernel PLS.

and the others discarded. Then, from the remaining pool, the descriptors to be included in the model are selected. There are many different variable selection techniques, which are used not only for MLR but in all types of modeling [8]. Genetic algorithm is one of the known techniques in this field. A number of studies in QSRR that used genetic algorithm are shown in Table 4.1.

4.5 CONCLUSION

Genetic algorithms have been used as chemometric tools for both parameter optimization and variable selection. In chromatography, the parameter optimization has been used to optimize different aspects of chromatographic systems. As variable selection tool, GAs were applied on one hand in multivariate classification and calibration studies, where chromatograms are linked to given sample information, and to select molecular descriptors for QSRR on the other.

REFERENCES

1. C. Blum, X. Li, Swarm intelligence in optimization, in: C. Blum and D. Merkle (Eds.), *Swarm Intelligence*, Springer, Berlin, Germany, 2008: pp. 43–85.
2. D.L. Massart, B. Vandeginste, L.M.C. Buydens, S. de Jong, P.J. Lewi, J. Smeyers-Verbeke, *Handbook of Chemometrics and Qualimetrics, Part A*, Elsevier, Amsterdam, the Netherlands, 1997.
3. J.A. Hageman, Optimal optimisation in chemometrics, PhD dissertation, Radboud University, Nijmegen, the Netherlands, 2004.
4. E. Reh, Peak-shape analysis for unresolved peaks in chromatography: Comparison of algorithms, *TrAC Trends in Analytical Chemistry*. 14 (1995) 1–5.
5. K.J. Goodman, J.T. Brenna, Curve fitting for restoration of accuracy for overlapping peaks in gas chromatography/combustion isotope ratio mass spectrometry, *Analytical Chemistry*. 66 (1994) 1294–1301.
6. B.G.M. Vandeginste, L. De Galan, Critical evaluation of curve fitting in infrared spectrometry, *Analytical Chemistry*. 47 (1975) 2124–2132.
7. G. Laporte, S. Martello, The selective travelling salesman problem, *Discrete Applied Mathematics*. 26 (1990) 193–207.
8. M. Goodarzi, B. Dejaegher, Y. Vander Heyden, Feature selection methods in QSAR studies, *Journal of AOAC International*. 95 (2012) 636–651.
9. N.-P. Vest Nielsen, J.M. Carstensen, J. Smedsgaard, Aligning of single and multiple wavelength chromatographic profiles for chemometric data analysis using correlation optimised warping, *Journal of Chromatography A*. 805 (1998) 17–35.
10. S. Thiangthum, B. Dejaegher, M. Goodarzi, C. Tistaert, A.Y. Gordien, N.N. Hoai, M. Chau Van, J. Quetin-Leclercq, L. Suntornsuk, Y. Vander Heyden, Potentially antioxidant compounds indicated from *Mallotus* and *Phyllanthus* species fingerprints, *Journal of Chromatography B*. 910 (2012) 114–121.
11. M. Goodarzi, M.P. Freitas, R. Jensen, Ant colony optimization as a feature selection method in the QSAR modeling of anti-HIV-1 activities of 3-(3,5-dimethylbenzyl) uracil derivatives using MLR, PLS and SVM regressions, *Chemometrics and Intelligent Laboratory Systems*. 98 (2009) 123–129.
12. M.A. Curtis, Optimization by simulated annealing theory and chemometric applications, *Journal of Chemical Education*. 71 (1994) 775–779.
13. R. Leardi, R. Boggia, M. Terrile, Genetic algorithms as a strategy for feature selection, *Journal of Chemometrics*. 6 (1992) 267–281.
14. K. Baumann, H. Albert, M. von Korff, A systematic evaluation of the benefits and hazards of variable selection in latent variable regression. Part I. Search algorithm, theory and simulations, *Journal of Chemometrics*. 16 (2002) 339–350.
15. M. Goodarzi, Swarm intelligence for chemometrics, *NIR News*. 26 (2015) 7–11.
16. A.C. Duarte, S. Capelo, Application of chemometrics in separation science, *Journal of Liquid Chromatography & Related Technologies*. 29 (2006) 1143–1176.
17. J.R. Chrétien, Chemometrics in chromatography, *TrAC Trends in Analytical Chemistry*. 6 (1987) 275–278.
18. J.H. Holland, *Adaptation in Natural and Artificial Systems*, University of Michigan Press, Ann Arbor, MI, 1975.
19. D.E. Golberg, *Genetic Algorithms in Search, Optimization, and Machine Learning*, Addison Wesley, Reading, MA, 1989.

20. R.A. Davis, A.J. Charlton, S. Oehlschlager, J.C. Wilson, Novel feature selection method for genetic programming using metabolomic ^{1}H NMR data, *Chemometrics and Intelligent Laboratory Systems*. 81 (2006) 50–59.

21. A.H. Kashan, A.A. Akbari, B. Ostadi, Grouping evolution strategies: An effective approach for grouping problems, *Applied Mathematical Modelling*. 39 (2015) 2703–2720.

22. L. Kota, K. Jarmai, Mathematical modeling of multiple tour multiple traveling salesman problem using evolutionary programming, *Applied Mathematical Modelling*. 39 (2015) 3410–3433.

23. T. Bhoskar, O.K. Kulkarni, N.K. Kulkarni, S.L. Patekar, G.M. Kakandikar, V.M. Nandedkar, Genetic algorithm and its applications to mechanical engineering: A review, *Materials Today: Proceedings*. 2 (2015) 2624–2630.

24. T.Y. Park, G.F. Froment, A hybrid genetic algorithm for the estimation of parameters in detailed kinetic models, *Computers & Chemical Engineering*. 22 (1998) S103–S110.

25. M. Goodarzi, L. dos Santos Coelho, Firefly as a novel swarm intelligence variable selection method in spectroscopy, *Analytica Chimica Acta*. 852 (2014) 20–27.

26. R.B. Boozarjomehry, M. Masoori, Which method is better for the kinetic modeling: Decimal encoded or Binary Genetic Algorithm? *Chemical Engineering Journal*. 130 (2007) 29–37.

27. L.H. Randy, H.S. Ellen, *Practical Genetic Algorithms*, John Wiley & Sons, New York, 2004.

28. M. Goodarzi, W. Saeys, Selection of the most informative near infrared spectroscopy wavebands for continuous glucose monitoring in human serum, *Talanta*. 146 (2016) 155–165.

29. R.L. Houpt, S.E. Houpt, *Practical Genetic Algorithms*, John Wiley & Sons, New York, 1998.

30. S. Yuan, B. Skinner, S. Huang, D. Liu, A new crossover approach for solving the multiple travelling salesmen problem using genetic algorithms, *European Journal of Operational Research*. 228 (2013) 72–82.

31. T. Kellegöz, B. Toklu, J. Wilson, Comparing efficiencies of genetic crossover operators for one machine total weighted tardiness problem, *Applied Mathematics and Computation*. 199 (2008) 590–598.

32. S.N. Sivanandam, S.N. Deepa, *Introduction to Genetic Algorithms*, Springer, Heidelberg, Germany, 2007.

33. C. Ortiz-Bolsico, J.R. Torres-Lapasió, M.C. García-Alvarez-Coque, Simultaneous optimization of mobile phase composition, column nature and length to analyse complex samples using serially coupled columns, *Journal of Chromatography A*. 1317 (2013) 39–48.

34. C. Ortiz-Bolsico, J.R. Torres-Lapasió, M.C. García-Alvarez-Coque, Optimisation of gradient elution with serially-coupled columns. Part II: Multi-linear gradients, *Journal of Chromatography A*. 1373 (2014) 51–60.

35. X. Shao, Z. Chen, X. Lin, Resolution of multicomponent overlapping chromatogram using an immune algorithm and genetic algorithm, *Chemometrics and Intelligent Laboratory Systems*. 50 (2000) 91–99.

36. K. Yu, Z. Lin, Y. Cheng, Optimization of the buffer system of micellar electrokinetic capillary chromatography for the separation of the active components in Chinese medicine "SHUANGDAN" granule by genetic algorithm, *Analytica Chimica Acta*. 562 (2006) 66–72.

37. G. Vivo-Truyols, J.R. Torres-Lapasió, M.C. Garcia-Alvarez-Coque, A hybrid genetic algorithm with local search: I. Discrete variables: Optimisation of complementary mobile phases, *Chemometrics and Intelligent Laboratory Systems*. 59 (2001) 89–106.

38. P. Nikitas, A. Pappa-Louisi, A. Papageorgiou, A. Zitrou, On the use of genetic algorithms for response surface modeling in high-performance liquid chromatography and their combination with the Microsoft Solver, *Journal of Chromatography A*. 942 (2002) 93–105.

39. K. Treier, P. Lester, J. Hubbuch, Application of genetic algorithms and response surface analysis for the optimization of batch chromatographic systems, *Biochemical Engineering Journal*. 63 (2012) 66–75.

40. Z. Zhang, M. Mazzotti, M. Morbidelli, Multiobjective optimization of simulated moving bed and Varicol processes using a genetic algorithm, *Journal of Chromatography A*. 989 (2003) 95–108.

41. F. Momenbeik, M. Roosta, A.A. Nikoukar, Simultaneous microemulsion liquid chromatographic analysis of fat-soluble vitamins in pharmaceutical formulations: Optimization using genetic algorithm, *Journal of Chromatography A*. 1217 (2010) 3770–3773.

42. N.H. Beltrán, M.A. Duarte-Mermoud, S.A. Salah, M.A. Bustos, A.I. Peña-Neira, E.A. Loyola, J.W. Jalocha, Feature selection algorithms using Chilean wine chromatograms as examples, *Journal of Food Engineering*. 67 (2005) 483–490.

43. R.L. Carneiro, J.W.B. Braga, C.B.G. Bottoli, R.J. Poppi, Application of genetic algorithm for selection of variables for the BLLS method applied to determination of pesticides and metabolites in wine, *Analytica Chimica Acta*. 595 (2007) 51–58.

44. C. Ruiz-Samblás, F. Marini, L. Cuadros-Rodríguez, A. González-Casado, Quantification of blending of olive oils and edible vegetable oils by triacylglycerol fingerprint gas chromatography and chemometric tools, *Journal of Chromatography B*. 910 (2012) 71–77.

45. L. Duan, L. Guo, K. Liu, E.H. Liu, P. Li, Characterization and classification of seven Citrus herbs by liquid chromatography–quadrupole time-of-flight mass spectrometry and genetic algorithm optimized support vector machines, *Journal of Chromatography A*. 1339 (2014) 118–127.

46. M.H. Abraham, Scales of solute hydrogen-bonding: Their construction and application to physicochemical and biochemical processes, *Chemical Society Reviews*. 22 (1993) 73–83.

47. M. Goodarzi, R. Jensen, Y. Vander Heyden, QSRR modeling for diverse drugs using different feature selection methods coupled with linear and nonlinear regressions, *Journal of Chromatography B*. 910 (2012) 84–94.

48. T. Hancock, R. Put, D. Coomans, Y. Vander Heyden, Y. Everingham, A performance comparison of modern statistical techniques for molecular descriptor selection and retention prediction in chromatographic QSRR studies, *Chemometrics and Intelligent Laboratory Systems*. 76 (2005) 185–196.

49. S. Caetano, C. Krier, M. Verleysen, Y. Vander Heyden, Modelling the quality of enantiomeric separations using Mutual Information as an alternative variable selection technique, *Analytica Chimica Acta*. 602 (2007) 37–46.

50. G. Carlucci, A.A. D'Archivio, M.A. Maggi, P. Mazzeo, F. Ruggieri, Investigation of retention behaviour of non-steroidal anti-inflammatory drugs in high-performance liquid chromatography by using quantitative structure–retention relationships, *Analytica Chimica Acta*. 601 (2007) 68–76.

51. K. Bodzioch, A. Durand, R. Kaliszan, T. Baczek, Y. Vander Heyden, Advanced QSRR modeling of peptides behavior in RPLC, *Talanta*. 81 (2010) 1711–1718.

52. Z. Dashtbozorgi, H. Golmohammadi, E. Konoz, Support vector regression based QSPR for the prediction of retention time of pesticide residues in gas chromatography–mass spectroscopy, *Microchemical Journal*. 106 (2013) 51–60.

53. A.A. D'Archivio, M.A. Maggi, F. Ruggieri, Modelling of UPLC behaviour of acylcarnitines by quantitative structure–retention relationships, *Journal of Pharmaceutical and Biomedical Analysis*. 96 (2014) 224–230.

54. M. Talebi, G. Schuster, R.A. Shellie, R. Szucs, P.R. Haddad, Performance comparison of partial least squares-related variable selection methods for quantitative structure retention relationships modelling of retention times in reversed-phase liquid chromatography, *Journal of Chromatography A*. 1424 (2015) 1–32.

55. H. Noorizadeh, A. Farmany, A. Khosravi, Investigation of retention behaviors of essential oils by using QSRR, *Journal of the Chinese Chemical Society*. 57 (2010) 982–991.

56. M.H. Fatemi, M. Elyasi, Prediction of gas chromatographic retention indices of some amino acids and carboxylic acids from their structural descriptors, *Journal of Separation Science*. 34 (2011) 3216–3220.

57. A.A. D'Archivio, A. Incani, F. Ruggieri, Retention modelling of polychlorinated biphenyls in comprehensive two-dimensional gas chromatography, *Analytical and Bioanalytical Chemistry*. 399 (2010) 903–913.

58. B. Lei, S. Li, L. Xi, J. Li, H. Liu, X. Yao, Novel approaches for retention time prediction of oligonucleotides in ion-pair reversed-phase high-performance liquid chromatography, *Journal of Chromatography A*. 1216 (2009) 4434–4439.

5 Statistics in Validation of Quantitative Chromatographic Methods

Eliangiringa Kaale, Danstan Hipolite Shewiyo, and David Jenkins

CONTENTS

5.1 BACKGROUND INTRODUCTION

5.1.1 QUANTITATIVE CHROMATOGRAPHIC METHODS AND VALIDATION REQUIREMENTS

Chromatographic methods permit the scientist to separate closely related components of complex mixtures [1]. In all chromatographic separations, the sample is transported in a mobile phase, which may be a gas [2], a liquid [3–5], or a supercritical fluid [6–8]. This mobile phase is then forced through an immiscible stationary phase, which is fixed in place in a column or on a solid surface. The two phases are chosen so that the components of the sample distribute themselves between the mobile and stationary phase to varying degrees. Chromatographic methods can either be qualitative, in the sense they are used to identify active drug substances, or quantitative to establish the amount of a given active substance [9,10]. Identification tests [11–13] are intended to ensure the identity of an analyte in a sample. This is normally achieved by comparison of a property of the sample (e.g., spectrum) [14,15], chromatographic behavior [14,15], and chemical reactivity to that of a reference standard [16,17]. The quantitative chromatographic methods are also divided into subtypes depending on the purpose [18], namely, assay, impurities and related substances, residual solvents, and dissolution testing. Analytical methods for each of the above tests will require a slightly different validation approach/requirement [18]. Quantitative tests for impurities establish the amount of the impurities in a sample. It is intended to accurately reflect the purity characteristics of the sample, whereas content determination/assay tests are intended to measure the analyte present in a given pharmaceutical dosage form sample. In the official compendia, the assay represents a quantitative measurement of the major component(s) in the drug substance [19].

Validation of analytical methods is defined as a process to demonstrate that it is suitable for its intended purpose [18,20,21]. The following are typical validation characteristics that should be considered:

1. Precision (repeatability and intermediate precision and reproducibility)
2. Accuracy
3. Linearity
4. Robustness
5. Sensitivity (limit of detection and limit of quantitation)

For the drug product, similar validation characteristics also apply when assaying for the active or other selected component(s). The selection of validation parameters largely depends on the purpose of the method. For assay or dissolution tests, the limit of detection and quantification is not a requirement, whereas linearity range is different for these two tests. The same validation characteristics may apply to assays associated with other analytical procedures (e.g., impurity determination).

Requirements to carry out method validation may be specified in guidelines within a particular sector relevant to the method [10,18,19,22]. It is recommended that these requirements are followed to ensure the specific validation terminology, together with the statistics used are interpreted in a manner consistent within the relevant sector. Official application of a method (e.g., in a monograph of the pharmacopoeia) may require characterization using a collaborative study.

5.2 STATISTICAL APPROACHES FOR VALIDATION PARAMETERS

5.2.1 PRECISION

Precision is the degree of scatter between multiple determinations pooled from the same composite materials [18,23,24]. The results of a particular set of measurements will be randomly distributed around the average value. The measure of dispersion of the results gives information about precision. It is usually expressed by statistical parameters that describe the spread of results, typically

TABLE 5.1

Summary of Minimal Statistical Parameters to Assess Precision in Chromatographic Analytical Method Validation

Precision Assessed	Minimal Set Up	Parameters Estimated	Prerequisite Conditions	References
Repeatability	A minimum of six[a] independent replicates; same composite material	Standard deviation and relative standard deviation of results for each composite	Same analyst and equipment, same laboratory, short period of time (1 day)	[18]
Intermediate precision	A minimum of twelve independent replicates; same composite material and protocol	Standard deviation and relative standard deviation of results for each composite material	Different analyst, equipment, days, and/or calibration. Same laboratory	[10,18]
Reproducibility	A minimum of twelve independent laboratories; same composite material and protocol	Repeatability standard deviation from ANOVA results for each material. Between-laboratory standard deviation calculated from ANOVA results. Sum of both variances equals to reproducibility	Different laboratories (including automatically different equipment, operators, and an extended period of time)	[34–36]

[a] In ICH Q2, it is also listed that a minimum of nine replicates can be analyzed that encompasses the range of the method (i.e., three concentrations with three replicates each).

the standard deviation (or relative standard deviation), calculated from results obtained by carrying out replicate measurements on a suitable material, under specified conditions. Deciding on the "specified conditions" is an important aspect of evaluating precision because it determines the type of precision estimate obtained, namely, repeatability, intermediate precision, or reproducibility. Experiments involving replicate analyses should be designed to take into account all variations in operational conditions, which can be expected during routine use of the method [25], preferably using real sample [26,27]. The aim should be to determine typical variability and not minimum variability (ref Table 5.1). In the next subsections, specific statistics applied in each of the precision subtypes will be reviewed.

5.2.1.1 Repeatability

Repeatability is the first level of precision estimation, which can be defined as the degree of scatter between results obtained under the same operating conditions over a short period of time (intra-day precision). The samples and concentrations tested should cover the range of the chromatographic assay test under consideration. In assessing repeatability, the entire method is replicated, including weighing and sample pretreatment steps, using the same chromatographic plate/column, in the same laboratory, with the same analyst, on the same day, using the same reagents, and the same equipment. Thus, independent test results are obtained with the same method on identical test items in the same laboratory by the same operator using the same equipment within short intervals of time [28,29]. To establish whether the method is repeatable or not, proper parameters results must be obtained. Parameters for the precision estimation widely applied are the mean, variance, and relative standard deviation [30–33].

The standard deviation is the most common measure of data dispersion, expressing how widely spread the values in a data set are. If the data points are close to the mean, then the standard deviation is small [33]. Conversely, if many data points are far from the mean, then the standard deviation is large. If all the data values are equal, then the standard deviation is zero.

The mean, standard deviation, and relative standard deviation can be calculated from Equations 5.1 to 5.3, respectively. Let $x_1, x_2, x_3, \ldots, x_n$ be n observations of a random variable x. The mean of x_i is defined by the formula below (Equation 5.1) and is the measure of the centrality of the data:

$$\bar{x} = \frac{1}{n}\sum_{i=1}^{n} x_i = \frac{x_1 + x_2 + \cdots + x_n}{n} \tag{5.1}$$

The statistics most often used to qualify the variability of data are the sample variance (s^2) and the sample standard deviation (s), where s is calculated with the following formula (Equation 5.2):

$$S = \sqrt{\frac{\sum_{i=1}^{n} (x_i - \bar{x})^2}{n-1}} \tag{5.2}$$

with n equaling the number of independent repetitions (i.e., $n \geq 6$ for repeatability, $n \geq 12$ for intermediate precision and reproducibility).

In validation, a normalized index called the relative standard deviation (rsd or % rsd) or the coefficient of variation is often used. It is expressed as a percentage (Equation 5.3) of the standard deviation relative to the average and is useful for comparing the variability between different measurements of varying absolute magnitude:

$$\%rsd = \frac{s \cdot 100\%}{\bar{x}} \tag{5.3}$$

5.2.1.2 Intermediate Precision

This is a second level of precision measurement occasionally also referred to as run-to-run, analyst-to-analyst, or day-to-day precision depending on the factor that is varied. A simplistic approach, varying one factor only (e.g., different days), uses the same statistical approach as described in Section 5.2.1.1, except the number replicated is larger. Interaction effects between different factors are often ignored. However, analyst-to-analyst variation may interact with a day-to-day variation. In this case, the use of proper statistical approach, such as given designs of experiments, can then be used to estimate the interaction effects as well. Detailed explanation of design of experiment is described elsewhere [30,31,37,38].

Intermediate precision refers to situations where precision is estimated under conditions that are more diverse than in repeatability, yet more limited than for reproducibility. It is a within-laboratory estimation of precision where different factors can be varied. Best known is the time-different intermediate precision (day-to-day or between-day precision). Other factors that can be varied are analyst, instrument, and, occasionally, calibration. All situations where up to approximately four factors are varied represent estimates of intermediate precision.

The following section provides an example data set with calculations provided for repeatability and time-intermediate precision estimates (standard deviations). A "lab" is considered a unique combination of laboratory location, analyst, and equipment. For this example, in one laboratory, two replicate analyses are conducted under repeatability conditions for each of seven different days using a method that determines the percent label claim of a finished pharmaceutical product (i.e., the method being validated). Table 5.2 provides the primary data set and other supporting data that should be referenced in the following explanation of the calculations conducted to determine the repeatability standard deviation (S_r), the between-day standard deviation (S_B), and the time-intermediate standard deviation ($S_{I(T)}$).

The data is collected in a manner such that k groups (days indexed as j) each have n_j replicates (here, thus, individually indexed as i within the jth group), where each result is indicated as x_{ij}. The following sums of squares, namely, SS_B (between days), SS_W (within days), and SS_T

TABLE 5.2

Example Data Summary for n Estimations of Repeatability and Time-Intermediate Precision

	Day 1	Day 2	Day 3	Day 4	Day 5	Day 6	Day 7
Replicate 1	101.01	100.23	99.13	101.03	99.65	98.67	99.34
Replicate 2	101.54	100.78	101.32	98.05	99.34	98.74	100.43

	SS	df	MS	F
Between days	8.137	6	1.356	1.221
Within days	7.775	7	1.111	
Total	15.912	13	1.224	

s_r^2	s_r	s_B^2	s_B	$s_{I(T)}^2$	$s_{I(T)}$
1.111	1.054	0.123	0.350	1.233	1.111

(total), are calculated (Equations 5.4 through 5.6), where $\bar{x}_j$ and $\bar{x}$ represent the average of a jth group and the total average, respectively:

$$SS_B = \sum_{j=1}^{k} n_j \left(\bar{x}_j - \bar{x} \right)^2 \tag{5.4}$$

$$SS_W = \sum_{j=1}^{k} \sum_{i=1}^{n} \left(x_{ij} - \bar{x}_j \right)^2 \tag{5.5}$$

$$SS_T = \sum_{j=1}^{k} \sum_{i=1}^{n} \left(x_{ij} - \bar{x} \right)^2 \tag{5.6}$$

The number of degrees of freedom (df) are given below:

$$df_B = k - 1; \quad df_W = n - k; \quad df_T = n - 1 \tag{5.7}$$

Furthermore, the mean squares (and F) are calculated as

$$MS_B = \frac{SS_B}{df_B}; \quad MS_W = \frac{SS_W}{df_W}; \quad MS_T = \frac{SS_T}{df_T} \tag{5.8}$$

The repeatability variance (s_r^2) is equal to MS_W, and the between-day variance (s_B^2) is estimated as shown below:

$$s_B^2 = \frac{MS_B - MS_W}{n_j} \tag{5.9}$$

The time-intermediate variance ($s_{I(T)}^2$) can then be determined with the following relationship:

$$s_{I(T)}^2 = s_r^2 + s_B^2 \tag{5.10}$$

5.2.1.3 Reproducibility

The third level of precision estimation is called reproducibility, which expresses the precision involving by definition collaborative studies applied, determined under diverse conditions [27]. The statistics used in evaluating proficiency data is described in [39] and can involve an originator laboratory preparing a protocol and composite material quantitated as an exact value where one set of composite samples is distributed to multiple sites. The results for parameters assessed are sent back for statistical analysis in a manner similar to the example below.

An example data set is provided in Table 5.3 and will be used to outline the approach from ISO 5725 [29,40] to estimate the repeatability (s_r) and reproducibility (s_R) standard deviations. For this example, assume that the same type of test method is used as in the previous example, where the same definition of a laboratory applies. Data is collected from p different laboratories (indexed as i) from q different levels of samples, indexed as j and at ~80%, 100%, and 120% of label claim, where n replicates are determined within each laboratory/level combination. In this example, eight laboratories are considered. Three concentration levels (80%, 100%, and 120%) are measured. This could be the case when measuring both API material and formulation. The concentration range will depend on future samples that will be measured. For instance, to measure the drug "active" in the blood, other concentration ranges will be tested. Usually, only two replicates per laboratory are measured under repeatability conditions. Limiting replicates to two results in repeatability and between-laboratory variances that are estimated with similar number of degrees of freedom.

The term "cell" represents the given n replicates within a laboratory/level combination. From the data in Table 5.3, cell means ($\bar{x}_{ij}$) and cell standard deviations (s_{ij}) are determined and presented in Tables 5.4 and 5.5, respectively.

An assessment of statistical outliers should be conducted. Two types of outliers can be considered; either a cell mean is outlying (laboratory shows a bias) or a cell standard deviation is outlying (laboratory has a bad repeatability). Graphical (Mandel h and k) and numerical (Grubbs and Cochran) approaches can be applied. The search for outliers involves a comparison of the magnitude of a cell's mean or standard deviation to the other cells. Regardless of the approach, critical values are used at different levels of significance (i.e., 5% or 1%) to determine whether the cell result can be considered a "straggler" (outside of 5% significance limit) or an outlier (outside of 1% significance limit). Generally, outliers are excluded from the data set when estimating repeatability and reproducibility, where stragglers are allowed to remain in the determination. The identification of

TABLE 5.3
Original Data for Reproducibility Example

Laboratory i	Level j		
	1	2	3
1	81.01	100.23	119.13
	81.54	100.78	121.32
2	79.24	102.52	118.48
	78.74	99.75	120.85
3	80.32	100.48	118.18
	79.52	98.92	118.52
4	80.08	102.42	122.49
	80.87	101.71	121.45
5	78.67	99.83	120.27
	81.54	101.02	118.45
6	79.39	102.59	121.87
	79.28	100.21	119.37
7	80.48	101.63	121.42
	81.93	102.83	119.93
8	78.95	100.03	119.04
	80.47	99.08	121.05

TABLE 5.4

Cell Means ($\bar{x}_{ij}$) for Reproducibility Example

Laboratory i	Level j		
	1	2	3
1	81.275	100.505	120.225
2	78.990	101.135	119.665
3	79.920	99.700	118.350
4	80.475	102.065	121.970
5	80.105	100.425	119.360
6	79.335	101.400	120.620
7	81.205	102.230	120.675
8	79.710	99.555	120.045

TABLE 5.5

Cell Standard Deviations (S_{ij}) for Reproducibility Example

Laboratory i	Level j		
	1	2	3
1	0.375	0.389	1.549
2	0.354	1.959	1.676
3	0.566	1.103	0.240
4	0.559	0.502	0.735
5	2.029	0.841	1.287
6	0.078	1.683	1.768
7	1.025	0.849	1.054
8	1.075	0.672	1.421

---

stragglers and outliers is a useful tool either to investigate improvements that may be needed within a given method or to identify further training needs "prior" to the actual collaborative study. ISO 5725 provides tables with critical values, for both the graphical and numerical approaches, to be found as a function of p and n for a given interlaboratory study.

Starting with the graphical (Mandel) approach, the h and k statistics are determined for each cell with the relationships below, where $\bar{\bar{x}}_j$ is the global total mean for the jth group:

$$h_{ij} = \frac{\bar{x}_{ij} - \bar{\bar{x}}_j}{\sqrt{\frac{1}{(p_j - 1)} \sum_{i=1}^{p_j} (\bar{x}_{ij} - \bar{\bar{x}}_j)^2}} \tag{5.11}$$

$$k_{ij} = \frac{s_{ij} \sqrt{p_j}}{\sqrt{\sum s_{ij}^2}} \tag{5.12}$$

Relative to the entire data set, the h statistic is associated with deviations of the average results of a cell, and the k statistic is related in a similar manner with the standard deviation of a cell. Tables 5.6

TABLE 5.6

h Statistics (h_{ij}) for Reproducibility Example

$p = 8, n = 2$

| h Indicator—1% significance | | | (±) 2.06 |
| h Indicator—5% significance | | | (±) 1.75 |

	Level j		
Laboratory i	1	2	3
1	1.395	−0.370	0.105
2	−1.382	0.257	−0.422
3	−0.251	−1.172	−1.659
4	0.423	1.183	**1.746**
5	−0.027	−0.450	−0.709
6	−0.962	0.521	0.476
7	1.310	1.348	0.528
8	−0.507	−1.316	−0.065

TABLE 5.7

k Statistics (k_{ij}) for Reproducibility Example

$p = 8, n = 2$

| k Indicator—1% significance | | | 2.25 |
| k Indicator—5% significance | | | 1.88 |

	Level j		
Laboratory i	1	2	3
1	0.394	0.345	1.183
2	0.372	1.737	1.280
3	0.595	0.978	0.184
4	0.588	0.445	0.562
5	**2.135**	0.746	0.983
6	0.082	1.493	1.350
7	1.079	0.753	0.805
8	1.131	0.596	1.085

and 5.7 (with the adjacent graphical representation in Figures 5.1and 5.2) provide the h and k statistics for each cell and include the critical values for h and k at the various levels of significance.

From this information, no outliers were observed but two potential stragglers may exist (Lab 4, level 3 and Lab 5, level 1—see bold values in Tables 5.6 and 5.7); this will be confirmed further with the numerical approaches (Grubbs and Cochran). The Cochran test statistic (C) is determined with the relationship below, where S_{max} is the maximum cell standard deviation observed within the p laboratories and s_i represents a given cell standard deviations within the p laboratories:

$$C = \frac{S_{max}^2}{\sum_{i=1}^{p} s_i^2}$$

(5.13)

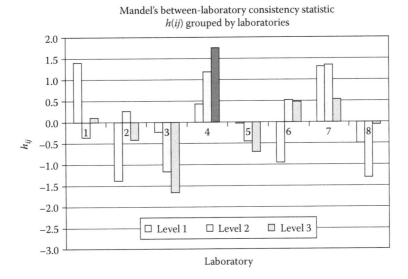

FIGURE 5.1 Graphical representation of h statistics for reproducibility example.

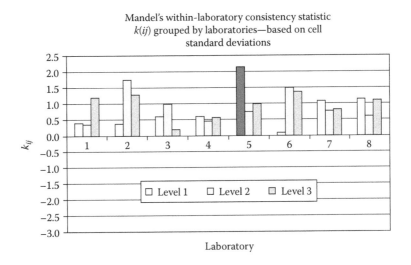

FIGURE 5.2 Graphical representation of k statistics for reproducibility example.

Table 5.8 provides Cochran test statistics for each of the levels, where the sum of variances is the denominator in the above relationship. The results indicate that no stragglers or outliers were observed from this data set based on the level of variance observed amongst the different lab/level combinations.

With respect to cell averages, the Grubbs test statistics are used to quantify the level of potential outliers. The Grubbs statistics for assessing the largest observation (G_p) and the smallest observation (G_1) are provided below:

$$G_p = \frac{\bar{x}_p - \bar{\bar{x}}_j}{\sqrt{\dfrac{1}{p_j - 1} \sum_{i=1}^{p_j} \left(\bar{x}_{ij} - \bar{\bar{x}}_j \right)^2}} \tag{5.14}$$

TABLE 5.8
Cochran's Test Statistics for Reproducibility Example

Critical values; 1%—0.794, 5%—0.680

$p = 8, n = 2$

	Level j		
	1	2	3
Maximum s_{ij}	2.029	1.959	1.768
Sum of variance	7.228	10.168	13.716
Test statistic (C)	0.570	0.377	0.228

No stragglers or outliers identified.

$$G_1 = \frac{\bar{\bar{x}}_j - \bar{\bar{x}}_1}{\sqrt{\frac{1}{p_j-1}\sum_{i=1}^{p_j}\left(\bar{x}_{ij} - \bar{\bar{x}}_j\right)^2}} \tag{5.15}$$

Correspondingly, the Grubbs statistics to access if the two largest ($G_{p-1,p}$) and two smallest ($G_{1,2}$) observations are outliers are provided below:

$$G_{p-1,p} = \frac{\sum_{i=1}^{p-2}\left(\bar{x}_{ij} - \bar{\bar{x}}_{p-1,p}\right)^2}{\sum_{i=1}^{p_j}\left(\bar{x}_{ij} - \bar{\bar{x}}_j\right)^2} \tag{5.16}$$

$$G_{1,2} = \frac{\sum_{i=3}^{p}\left(\bar{x}_{ij} - \bar{\bar{x}}_{1,2}\right)^2}{\sum_{i=1}^{p_j}\left(\bar{x}_{ij} - \bar{\bar{x}}_j\right)^2} \tag{5.17}$$

where $\bar{\bar{x}}_{p-1,p}$ and $\bar{\bar{x}}_{1,2}$ are the averages of the $p - 2$ smallest and the $p - 2$ largest cell means, respectively. From these relationships, the various Grubbs statistics are presented in Table 5.9.

In comparison to the critical values, no stragglers or outliers have been identified and thus all data will be used to determine the repeatability and reproducibility standard deviations.

The repeatability variance (s_{rj}^2) and the between-laboratory (s_{Lj}^2) variances are calculated from the relationships below, respectively:

$$s_{rj}^2 = \frac{\sum_{i=1}^{p}\left(n_{ij} - 1\right)s_{ij}^2}{\sum_{i=1}^{p}\left(n_{ij} - 1\right)} \tag{5.18}$$

$$s_{Lj}^2 = \frac{\left(s_{dj}^2 - s_{rj}^2\right)}{\bar{n}_j} \tag{5.19}$$

TABLE 5.9
Grubbs' Test Statistics for Reproducibility Example

Level	Single Low	Single High	Double Low	Double High	Type of Test
1	−1.382	1.395	0.464	0.302	Grubbs' test statistic
2	−1.316	1.348	0.409	0.388	
3	−1.659	1.746	0.402	0.402	
Stragglers (5%)	Upper 5%—(±) 2.126		Lower 5%—0.1101[a]		Grubbs' critical values
Outliers (1%)	Upper 1%—(±) 2.274		Lower 1%—0.0563[a]		

No stragglers or outliers identified from single or double (largest/smallest).

[a] For the double Grubbs' tests, the test statistic should be lower than the critical value to be significant.

where s_{dj}^2 and $\bar{\bar{n}}_j$ are determined with the following relationships:

$$s_{dj}^2 = \frac{1}{p-1} \sum_{i=1}^{p} n_{ij} \left(\bar{y}_{ij} - \bar{\bar{y}}_j \right)^2 \tag{5.20}$$

$$\bar{\bar{n}}_j = \frac{1}{p-1} \left[\sum_{n=1}^{p} n_{ij} - \frac{\sum_{i=1}^{p} n_{ij}^2}{\sum_{i=1}^{p} n_{ij}} \right] \tag{5.21}$$

The reproducibility variance is calculated as in the following equation:

$$s_{Rj}^2 = s_{rj}^2 + s_{Lj}^2 \tag{5.22}$$

Table 5.10 provides the s_r and s_R at each respective level for the example presented.

ISO 5725 outlines how to determine the potential dependency of the precision relative to the means (m) at the different levels (j). The relation is to be properly modeled by the analyst. For a narrow range, no dependency is expected, and the standard deviation expected can be reported as the average standard deviation. When the concentration range of samples is broader, often a

TABLE 5.10
Standard Deviations (Repeatability/Reproducibility) for Reproducibility Example

	Level j		
	1	2	3
s_r^2	0.904	1.271	1.715
s_L^2	0.225	0.373	0.273
s_R^2	1.129	1.644	1.988
m	80.127	100.877	120.114
s_r	0.951	1.127	1.309
s_R	1.062	1.282	1.410

linear relation between s and m is seen, although some situations (i.e., biological analysis) may require plotting on a log scale to observe linearity.

5.2.2 ACCURACY/TRUENESS

Accuracy and trueness are related terms that are associated with performance of analytical methods. They differ in that the accuracy of an analytical procedure expresses the closeness of agreement between the value that is accepted either as a conventional true value or an accepted reference value and the value found [18,33], whereas trueness is defined as the closeness of agreement between the average result obtained from a large series of test results and the accepted reference value [24,41]. Accuracy includes both bias and precision. In practice, analysts compute the mean of many results found to determine its closeness to the true value, hence performing trueness rather than accuracy. Therefore, in this chapter, we will focus on statistics applicable to trueness rather than the accuracy.

According to EURACHEM guide [24], trueness is quantitatively expressed as bias, which has two components, namely, method bias and laboratory bias. The method bias arises from systematic errors inherent to the method, irrespective of which laboratory uses it, and laboratory bias arises from additional systematic errors specific to the laboratory and its interpretation of the method [24,41]. Laboratory bias is estimated as part of between-laboratory variability during reproducibility testing.

Trueness is quantitatively studied as bias, and the following are equations for statistical evaluation as given by [41,42], where b denotes the bias, $\bar{x}$ mean of results, x_{ref} the reference value, $\bar{xx}'$ mean result of spiked sample, and x_{spike} is the added concentration:

1. Absolute bias

$$b = \bar{x} - x_{ref} \tag{5.23}$$

2. Relative in percent

$$b(\%) = \frac{\bar{x} - x_{ref}}{x_{ref}} \times 100 \tag{5.24}$$

3. Relative spike recovery

$$R'(\%) = \frac{\bar{x}' - \bar{x}}{x_{spike}} \times 100 \tag{5.25}$$

4. Relative recovery

$$R(\%) = \frac{\bar{x}}{x_{ref}} \times 100 \tag{5.26}$$

It is further recommended to perform significance testing when evaluating the bias since it involves comparing the response of method to a reference material with a known assigned value [41,42].

Various experimentation conditions are identified in the literature that affect the way bias is estimated during method validations [41,42], which can be a function of the type of method or the concentration range expected for the samples. These experimental considerations that may be encountered when preparing experimental designs for bias estimation are summarized below and briefly each shall be described:

- Certified reference material is available.
- Limited concentration range and blank material is available (or can be reconstituted).
- Large concentration range and blank material is available (or can be reconstituted).
- Limited concentration range and spiking is possible.
- Large concentration range and spiking is possible.
- Large concentration range and no blank material is available and no spiking possible (comparison with reference method).

5.2.2.1 Certified Reference Materials Available

When CRM is available, it enables a single laboratory to estimate combined (laboratory and method) biases through repeated analysis of the CRM [41,42]. The mean of the results is compared with the known value using equation (Equation 5.24) in order to determine bias, and a lab's bias would be more significant if it approaches 1.5%–2% based on that level being on the order of where "typical" critical h and Grubbs' critical statistical values are at a 5% significance level for under 10 labs.

5.2.2.2 Limited Concentration Range and Blank Material Is Available (or Can Be Reconstituted)

This refers to the situation where the range of concentration to be studied is limited and the blank material without substance of interest is available or can be reconstituted. Typically, such situations are encountered in methods for assay of active drug in pharmaceutical products. ICH guide for validation of analytical procedures [18] recommends trueness study at the range of 80%–120% of the expected drug concentration, commonly at three (3) concentration levels of 80%, 100%, and 120%. During bias estimation, six (6) replicates at each concentration level are proposed though only three replicates are required [18]. Replicates are prepared by adding to the blank known amounts to obtain the concentration levels. In all conditions, regardless of whether six or three replicates have been prepared and analyzed, mean results at each level are compared to the known added values at each level separately using t-test statistic, at $\alpha = 5\%$ significance level. If the $t_{calculated} < t_{critical}$, it is concluded that there is no significant difference between the two values, and, hence, insignificant bias.

5.2.2.3 Large Concentration Range and Blank Material Is Available (or Can Be Reconstituted)

Typical examples of this condition include detection of drug substances or metabolites in biological fluids such as blood, plasma, or urine. Other examples include detection of micronutrients in food or certain substances in environmental samples. In these cases, the sample matrix may be large and the substance of interest may exist in small quantities. Normally, the concentration levels to be studied may cover a large range. In case the blank material is available or can be reconstituted, bias estimation can be carried out by spiking one or few replicates at three concentration levels—one at the lower end close to the quantitation limit, others at the middle and upper level. The responses (y), which is the amount found, are plotted against spiked amounts (x), where regression analysis is conducted. The equation of the line can be written as

$$y = b_0 + xb_1 \tag{5.27}$$

where the following types of errors can exist as a function of different values of b_0 and b_1:

$b_0 = 0$ and $b_1 = 1$: No systematic errors
$b_0 = 0$ and $b_1 \neq 1$: Proportional systematic error
$b_0 \neq 0$ and $b_1 = 1$: Absolute systematic error

This method enables the detection of bias and the type of bias and, hence, makes it easier to investigate its origin and possibly come up with remedial measures.

5.2.2.4 Limited Concentration Range and Spiking Is Possible

This situation is similar to conditions in Section 5.2.2.2, in that spiking of blank material is possible and statistical treatment of data will be the same. It is not recommended in this situation (as also in Section 5.2.2.2) to use regression method for bias estimation as in Section 5.2.2.3 since in such a limited concentration range, the estimation of regression parameters b_0 and b_1 will not be optimal.

5.2.2.5 Large Concentration Range and Spiking Is Possible

Treatment of data as described in Section 5.2.2.3 is applicable to this situation.

5.2.2.6 Large Concentration Range and No Blank Material Is Available and No Spiking Possible (Comparison with Reference Method)

When an alternative or modified method is developed to replace an existing fully validated method, where there is neither blank material nor possibility for spiking, an approach through comparison of two methods is considered. In general, bias estimation of the new method can be estimated by conducting a number of replicate analyses of suitable test materials, covering the desired concentration level using the existing and the new method. It is cautioned to pass the data through homogeneity and normality checks before applying one regular t-test per concentration level and Bonferroni correction to demonstrate if the new method has bias.

5.2.3 LINEARITY OF CALIBRATION LINE

5.2.3.1 Graphical Approaches

A linear calibration line is the preferred mathematical model by analysts due to its simplicity in computations, hence, the need to establish the linear portion of the line whose equation takes the form of

$$y = mx + b \tag{5.28}$$

where
  y is the response
  m is the slope
  x is the concentration
  b is the y intercept

With the insertion of a measured y response, this equation gives the unknown concentration by solving for x, and is critical in the utilization of the method to quantify unknown amounts of analyte based on the method response.

Evaluation of the linearity of the calibration line can initially be studied by plotting responses as a function of concentrations of standard substance [18,24] where they are plotted on y- and x-axis, respectively. This graphical approach involves visual inspection of the plot to assess whether it is linear, before it is then subjected to statistical treatment to determine acceptability of the observed linear relationship within given significance level. Normally, the relationships can graphically be observed as polynomial, exponential, linear, power, etc., as demonstrated in Figure 5.3.

The goodness of fit of the data to the model can be inspected visually using residual plot where residuals values are plotted against x or y values [24,43]. With linear models being the most common with chromatographic methods, residual values are first obtained by computing the difference between the response values during measurement and those responses predicted by the model, at

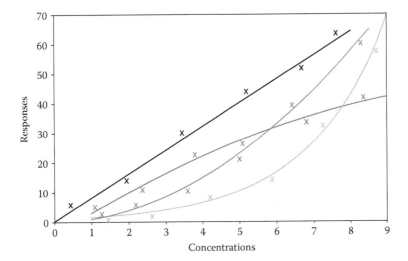

FIGURE 5.3 Showing plots of example responses as functions of concentrations. The relationships are linear (black line), polynomial (red line), power (blue line), and exponential (green line).

each concentration level studied. When residuals are randomly scattered along the horizontal line with the number of positive and negative residues approximately equal, this indicates nonabnormality and, hence, adequacy of the model [43]. Furthermore, if the scatter of the residuals increases along the horizontal line, it indicates homoscedasticity where precision is not constant along the straight line and where a normal distribution of measurement error is lacking at each concentration level. Moreover, residue values showing a U shape indicate lack of fit, where they can best be described by a curve.

5.2.3.2 ANOVA Lack-of-Fit Test

Massart et al. [33] provide a detailed explanation on how the analysis of variance (ANOVA) can be used to detect lack of fit in a regression in order to verify whether the straight line model is adequate. It is important to note that replicate measurements are required in order to estimate variances. With ANOVA, therefore, series of computations are carried out to estimate variations, which then enable to obtain *sum of square due to regression line* and *residual sum of square. Residual sum of square* is divided into a component that measures variations due to pure experimental error uncertainty, called *pure error sum of square* (SS_{PE}), and another component measuring variation of group means about the regression line, called *pure error due to lack of fit* (SS_{LoF}). When these two quantities are divided by their degrees of freedom (df), mean squares (MS) that are required for verifying lack of fit of the model are obtained. MS_{PE} is an estimate of variances on the line due to pure error and MS_{LoF} estimates variances if chosen model is correct and is said to estimate variance plus bias if the model is not adequate [33].

The lack-of-fit test, which is a one-sided test, is performed by comparing the ratio

$$F = \frac{MS_{LoF}}{MS_{PE}} \qquad (5.29)$$

with the F-distribution at $(k-2)$ and $(n-k)$ degrees of freedom. If the ratio is significant at the chosen significance level (MS_{LoF} significantly larger than MS_{PE}), it is concluded that the model (straight line in this case) is inadequate since the variation of the group means about the line cannot be explained in terms of the pure experimental uncertainty. If MS_{LoF} and MS_{PE} are comparable, the model is accepted adequately as a straight line.

5.2.3.3 Significance of Quadratic Term in Quadratic Model

The test of significance of the quadratic term, b_2, may also be used to verify the linearity of calibration line by fitting second degree polynomial to the calibration data [43,44]. The straight line and polynomial models are then compared based on comparison of the mean squares (MS) values of quadratic term (b_2) and residuals, where deviation from linearity is detected when MS are different [44]. In this situation of nonlinearity, F is larger than $F_{critical}$ on the F-distribution table and, hence, b_2 is significant. Similarly, the significance test can be based on t-test where the value

$$t = \frac{b_2}{s_{b2}}$$
(5.30)

Similarly, the significance evaluation can be based on the t-test by calculating t (5.30), where s_{b2} is the standard deviation of b_2. The t value is compared with the tabulated t value for n-3 degrees of freedom at the chosen confidence level (usually 95%). Linearity is proven when b_2 is not significant ($t < t_{tabulated}$).

5.2.3.4 Use of Correlation Coefficient and Quality Coefficient

The correlation coefficient, r, between x and y, evaluates the degree of linear association between these two variables [44], and on many occasions it is interchanged with R^2, a coefficient of determination that is an indicator of the proportion of variability in the response explained by the regression [44]. These two coefficients, especially R^2, are commonly used as indicators of linearity of calibration curves by many authors in the literature. Values close to unit are usually referred to as indicating goodness of fit of data to the straight lines. However, it has been demonstrated that correlation coefficient very close to 1 can also be obtained for a clearly curved relationship [18,44,45]. These coefficients are not useful indicators of linearity, especially when analytical method is being validated. During linearity check in method validation, other means such as visual inspection of calibration plot, residual plot, and statistical treatment of data should be used to determine acceptability of linearity [24,43,44,46]. Nevertheless, calculation of the correlation coefficient is acceptable for a system suitability check if the full method validation has established linearity between the response and the concentration. The check could further consist in a comparison of the correlation coefficient with a default value specified from the method validation results. For instance, one could require that $r > 0.999$. If r is found to be less, this is taken to mean that the calibration line is not good enough. The reason for this can then be ascertained further through visual inspection.

Quality coefficient (g) is also a statistical measure that is used to evaluate the quality of straight line and is computed from percentage deviations calculated from the percentage deviations of the calculated x-values from the ones expected [43]. The better the experimental points fit the line, the smaller the quality coefficient:

$$g = \sqrt{\frac{\sum (\%deviation)^2}{n-1}}$$
(5.31)

where

$$\%deviations = \frac{x_{predicted} - x_{known}}{x_{known}}$$
(5.32)

Another statistical treatment similar to g reported in the literature for the quality coefficient, denoted as QC, is based on percentage deviation of the estimated responses, where assumption for homoscedastic condition is made [43,44,47]. The proposed equation is

$$QC = \sqrt{\frac{\sum \left(\frac{y_i - \hat{y}_i}{\bar{y}} \right)^2}{n-1}} \tag{5.33}$$

where

$\left(y_i - \hat{y}_i \right)$ is the residue

$\bar{y}$ is the mean of all responses measured

If the target value for QC was already set during the full method validation, the goodness of fit of the line to the model is regarded unacceptable if QC value exceeds the target value [43].

5.2.4 ROBUSTNESS

The robustness of an analytical procedure is a measure of its capacity to remain unaffected by small but deliberate variations in method parameters and provides an indication of its reliability during normal usage [18]. Robustness testing results enable conclusions to be made on the influence of factors evaluated on the response functions (e.g., retention times, percentage recovery, and resolution) and hence facilitate, if necessary, inclusion of precautionary statements in the method's procedure [48]. A statistical technique based on Plackett–Burman experimental design can be used to detect factors that have influence or estimate the factor influence [43,48]. Other alternative designs include the saturated fractional factorial and Taguchi-type designs. It can be noted that factors that may influence the method are studied at two different levels depending on the number of experiments desired and dummy factors are included in the design in order to facilitate estimation of experimental errors, which in turn enables statistical interpretation of the factor effects [48]. So, a factor effect is regarded as statistically significant if the t-value calculated for its effect is larger than a critical t-value or if $|E_X|$-value is larger than a critical effect value. The following equations are used:

$$t = \frac{|E_X|}{(SE)_e} \geq t_{critical} \tag{5.34}$$

or

$$|E_X| \geq E_{critical} = t_{critical} \times (SE)_e \tag{5.35}$$

or

$$|\%E_X| \geq \%E_{critical} = \frac{E_{critical}}{Y_n} \times 100 \tag{5.36}$$

with

$$(SE)_e = \sqrt{\frac{\sum E_{dummy:i}^2}{n_{dummy}}} \tag{5.37}$$

where

E_{dummy} is the effect of dummy factor

n_{dummy} is the number of dummies

$t_{critical}$ is the tabulated two-sided t-value with n_{dummy} degrees of freedom at a significant level α

$(SE)_e$ is standard error on E_X

$\%E$ is the normalized effect

Y_n is the value of response when experiment is performed with all factors at nominal condition

5.3 ACCURACY PROFILES

An accuracy profile (Figure 5.4) is a graphical tool that offers the possibility to visually comprehend the ability of an analytical method to fulfill its objectives and to control the risk associated with its routine applications [49]. Based on the concept of total error, analytical methods are validated using accuracy profiles where bias and standard deviation are combined and built with a β-expectation tolerance interval [49–52]. It is also referred to as one stop statistic, which enables to decide on the validity of a method for future routine use in the scope of the intended purpose. In order to construct accuracy profiles, total errors are estimated by incorporating the systematic (absolute bias) and the random errors (intermediate precision standard deviation) so as to reflect on how large measurement errors of a method can be. In this section, a brief account of the accuracy profiles is provided.

5.3.1 ESTIMATION OF PRECISION

Precision, as random error, is estimated at each concentration level that is considered in the validation study where the time-different intermediate precision variance, $s_{I(t),m}^2$, is the sum of repeatability or intra-series variance, $s_{r,m}^2$, and between-series variance, $s_{between,m}^2$, as represented in the following equation:

$$s_{I(t),m}^2 = s_{r,m}^2 + s_{between,m}^2 \qquad (5.38)$$

It is advised that ANOVA is employed in estimation of precision when data sets are balanced [43]; however, when there is imbalance in the data, then the method for restricted maximum likelihood is proposed [51].

5.3.2 ESTIMATION OF TRUENESS

The trueness gives the systematic error of an analytical method and is quantitatively estimated as bias at each concentration level. Bias is computed using equations in Section 5.2.2.

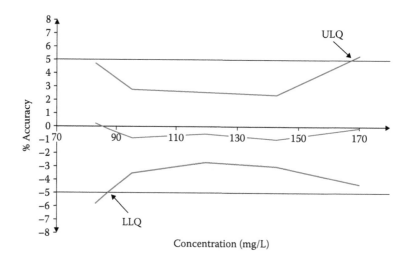

FIGURE 5.4 Example accuracy profile showing accuracy acceptance limits (continuous black lines +5 and −5), upper and lower tolerance limits (red lines), mean recoveries (blue line), and upper (ULQ) and lower (LLQ) limits of quantitation (indicated by arrows).

5.3.3 TOTAL ERROR

Each individual measurement, x_{imn} ($i = 1{:}s$, $m = 1{:}v$, $n = 1{:}w$), is the sum of the true value μ_{imn}, the absolute bias $|bias_m|$, and the intermediate precision standard deviation $s_{I(t),m}$:

$$x_{imn} = \mu_{imn} + |bias_m| + s_{I(t),m} \qquad (5.39)$$

Therefore,

$$x_{imn} - \mu_{imn} = |bias_m| + s_{I(t),m} = \text{Total error or absolute total error} \qquad (5.40)$$

The total error of an analytical method is the sum of the absolute bias ($|bias_m|$) and intermediate precision ($s_{I(t),m}$), and it indicates the ability of the analytical method to produce accurate results. It enables assessment of the quantitative performance of a method. It is used to compute the 95% tolerance interval, which provides guarantee of what results will be produced by the method in future, during routine use of the method [51].

5.3.4 TOLERANCE INTERVAL AND ACCURACY PROFILE

These parameters % $bias_m$, $s_{r,m}^2$, $s_{between,m}^2$, and %$RSD_{I(t),m}$ are used to estimate the expected proportion of observations that falls within the predefined bias acceptance limits ($-\lambda$, $+\lambda$). This expected proportion is calculated using the two-sided 95% β-expectation tolerance interval at each concentration level of the validation standards ($m = 1{:}v$). The lower L_m and the upper U_m tolerance interval limits are calculated as follows:

$$L_m = \%bias_m - Q_{\left(v,\frac{1+\beta}{2}\right)}\sqrt{1 + \frac{1}{swB_m^2}}\,\%RSD_{I(t),m} \qquad (5.41)$$

$$U_m = \%bias_m - Q_{\left(v,\frac{1+\beta}{2}\right)}\sqrt{1 + \frac{1}{swB_m^2}}\,\%RSD_{I(t),m} \qquad (5.42)$$

where

$Q_{\left(v,\frac{1+\beta}{2}\right)} = \beta$ quantile of the Student t distribution with v degrees of freedom, with

$$v = \frac{\left(R_m + 1\right)^2}{\left(\left(R_m + 1/w\right)^2 / (s-1)\right) + \left(\left(1 - 1/w\right)/sw\right)}$$

s is the number of series
w is the number of replicates for validation standards

$$-B_m^2 = \frac{R_m + 1}{wR_m + 1} \quad \text{with} \quad R_m = \frac{s_{r,m}^2}{s_{between,m}^2}$$

Once the two-sided 95% β-expectation tolerance limits at each concentration level of the validation standards, ($m = 1{:}v$), L_m, U_m, $-\lambda$, and $+\lambda$ are then used to construct the accuracy profile [49–51].
The following may be concluded from the accuracy profiles:

- When the 95% β-expectation tolerance interval limits for the considered concentration levels of the validations standards fall within the bias acceptance limits, for example, −5% to 5%, then a guarantee is provided that future results of the method will fall within acceptable bias limits.

- It reflects the adequacy of the response function (calibration equation) used to compute concentrations of the validation standards.
- Since the upper and lower concentrations of the validation standards considered during the validation study define the upper and the lower quantitation limits, when tolerance limits intersect with the bias acceptance limits, the intersection points will define the upper and lower quantitation limits.

5.3.5 Worked Example on Accuracy Profile

A worked example on accuracy profile is presented for demonstration. An analytical method to assay an active ingredient in a tablet dosage form was developed and validated using accuracy profile. The method employed high-performance thin-layer chromatography (HPTLC) technique. Concentration levels of a compound to be quantitated, which were studied, covered 83–170 mg/L. Three independent preparations were made for each concentration level.

A straight line response function was used to back-calculate the validation solutions. Other parameters including precision (intermediate precision), trueness, accuracy, total error, and two-sided 95% tolerance limits were estimated using the respective equations described in Section 5.3. Accuracy (bias) acceptance limits were set at +5 to −5, which is commonly applied for active ingredients in pharmaceutical formulations. An accuracy profile (Figure 5.4) was prepared.

From the figure above, it can be deduced that LLQ and ULQ are 87 and 167 mg/L, respectively, and these limits also define the range of the method since accuracy, precision, and linearity were acceptable. During validation, concentration levels studied were 83–170 mg/L; however, it is demonstrated that future use of the method is guaranteed for concentrations covering 87–167 mg/L, since 95% β-expectation tolerance interval limits for this concentration range fall within the bias acceptance limits, for example, −5% to 5%.

REFERENCES

1. El-Gindy, A., S. Emara, and A. Mostafa, Application and validation of chemometrics-assisted spectrophotometry and liquid chromatography for the simultaneous determination of six-component pharmaceuticals. *J Pharm Biomed Anal*, 2006. **41**(2): 421–430.
2. Delmonte, P., A.R. Kia, Q. Hu, and J.I. Rader, Review of methods for preparation and gas chromatographic separation of trans and cis reference fatty acids. *J AOAC Int*, 2009. **92**(5): 1310–1326.
3. Kaliszan, R., Retention data from affinity high-performance liquid chromatography in view of chemometrics. *J Chromatogr B Biomed Sci Appl*, 1998. **715**(1): 229–244.
4. Gorog, S., The paradigm shifting role of chromatographic methods in pharmaceutical analysis. *J Pharm Biomed Anal*, 2012. **69**: 2–8.
5. Gad, H.A., S.H. El-Ahmady, M.I. Abou-Shoer, and M.M. Al-Azizi, Application of chemometrics in authentication of herbal medicines: A review. *Phytochem Anal*, 2013. **24**(1): 1–24.
6. Sherma, J., New supercritical fluid chromatography instrumentation. *J AOAC Int*, 2009. **92**(2): 67A–75A.
7. Khater, S., M.A. Lozac'h, I. Adam, E. Francotte, and C. West, Comparison of liquid and supercritical fluid chromatography mobile phases for enantioselective separations on polysaccharide stationary phases. *J Chromatogr A*, 2016. **1467**: 463–472.
8. Sherma, J., New gas chromatography, gas chromatography/mass spectrometry, supercritical fluid chromatography/mass spectrometry, and portable mass spectrometry instruments. *J AOAC Int*, 2007. **90**(3): 59A–65A.
9. Le Tarnec, C.L., The European pharmacopoeia: A common European initiative by the council of Europe. *Chimia*, 2004. **58**(11): 798–799.
10. U.S. Pharmacopeia, <1225> Validation of compendial procedures. In: *United States Pharmacopeia*, 40th ed., *National Formulary*, 35th ed., The United States Pharmacopeia Convention, Rockville, MD; United Book Press, Inc., Gwynn Oak, MD, 2017, pp. 1445–1461.
11. Walters, M.J., Liquid chromatographic determination of hydrocortisone in bulk drug substance and tablets: Collaborative study. *J Assoc Off Anal Chem*, 1984. **67**(2): 218–221.

12. FAO, *Guide to Specifications for General Notices, General Analytical Techniques, Identification Tests, Test Solutions and Other Reference Materials*. 1991. Joint FAO/Who Expert Committee on Food Additives, FAO Food Nutrition Paper, Rome, Italy, 5 Revis 2, pp. 1–307.

13. Sherma, J., Analysis of counterfeit drugs by thin layer chromatography. *Acta Chromatogr*, 2007. **19**: 5–20.

14. Henneberg, D., K. Casper, E. Ziegler, and B. Weimann, Computer-aided analysis of organic mass spectra. *Angew Chem Int Ed Engl*, 1972. **11**(5): 357–366.

15. Doorenbos, N.J., L. Milewich, and D.P. Hollis, Spectral properties and reactions of 3-beta-hydroxy-21-formyl-22-oximinopregn-5-en-20-one. *J Org Chem*, 1967. **32**(3): 718–721.

16. Boelens, H.F., P.H. Eilers, and T. Hankemeier, Sign constraints improve the detection of differences between complex spectral data sets: Lc-Ir as an example. *Anal Chem*, 2005. **77**(24): 7998–8007.

17. Sadlej-Sosnowska, N. and A. Ocios, Selectivity of similar compounds identification using Ir spectroscopy: Steroids. *J Pharm Sci*, 2007. **96**(2): 320–329.

18. ICH, Ich Q2 (R1) validation of analytical procedures : Text and methodology. *Int Council Harmon*, 2005. **1994**(November 1996): 17.

19. Europe, Council of Europe, General chapters, in *European Pharmacopoeia 8.0*. 2016. Europe, Council of Europe, Strasbourg, France, p. 1896.

20. Ferenczi-Fodor, K., B. Renger, and Z. Végh, The frustrated reviewer—Recurrent failures in manuscripts describing validation of quantitative Tlc/Hptlc procedures for analysis of pharmaceuticals. *J Planar Chromatogr*, 2010. **23**(3): 173–179.

21. Ferenczi-Fodor, K., Z. Végh, A. Nagy-Turák, B. Renger, and M. Zeller, Validation and quality assurance of planar chromatographic procedures in pharmaceutical analysis. *J AOAC Int*, 2001. **84**(4): 1265–1276.

22. Kaale, E., P. Risha, and T. Layloff, Tlc for pharmaceutical analysis in resource limited countries. *J Chromatogr A*, 2011. **1218**(19): 2732–2736.

23. ISO/IEC Guide 99:2007, I.G. International vocabulary of metrology—Basic and general concepts and associated terms (Vim), Jcgm 200:2012, www.bipm.org. A previous version is published as ISO/IEC Guide 99:2007. ISO, Geneva, Switzerland. JCGM 2007. p. 200.

24. Magnusson, O., *Eurachem Guide: The Fitness for Purpose of Analytical Methods—A Laboratory Guide to Method Validation and Related Topics*. 2014, Eurachem, Olomouc, Czech Republic, pp. 1–70.

25. Hendix, C., What every technologist should know about experiment design. *Chem Technol*, 1979. **9**: 167–174.

26. Jenke, D.R., Chromatographic method validation: A review of current practices and procedures. I. General concepts and guidelines. *J Liq Chromatogr Relat Technol*, 1996. **19**(5): 719–736.

27. Pocklington, W.D., *Guidelines for the Development of Standard Methods by Collaborative Study: The Organisation of Interlaboratory Studies, a Simplified Approach to the Statistical Analysis of Results from Collaborative Studies, and the Drafting of Standardised Methods*. 1990. Laboratory of the Government Chemist, London, U.K.

28. ISO 28-1:1994, 5725-1:1994, I, *Accuracy (Trueness and Precision) of Measurement Methods and Results—Part 1: General Principles and Definitions*. 1994. ISO, Geneva, Switzerland.

29. ISO 29-2:1994, 5725-2:1994, I, *Part 2: Basic Method for the Determination of Repeatability and Reproducibility of a Standard Measurement Method*. 1994. ISO, Geneva, Switzerland.

30. Massart, D.L., *Chemometrics Tutorials: Collected from Chemometrics and Intelligent Laboratory Systems—An International Journal*, Vols. 1–5, 13th ed., R.G. Brereton, D.L. Massart, C.H. Spiegelman, P.K. Hopke, and W. Wegscheider (Eds.). 1990. Elsevier Science, Oxford, U.K.

31. Massart, D.L. and L. Buydens, Chemometrics in pharmaceutical analysis. *J Pharm Biomed Anal*, 1988. **6**(6–8): 535–545.

32. Vandeginste, B.G.M., D.L. Massart, L.M.C. Buydens, S. De Jong, P.J. Lewi, and J. Smeyers-Verbeke, Chapter 36—Multivariate calibration, in *Data Handling in Science and Technology*, B.G.M. Vandeginste and J. Smeyers-Verbeke (Eds.). 1998. Elsevier, Amsterdam, the Netherlands, pp. 349–381.

33. Slutsky, B., Handbook of Chemometrics and Qualimetrics: Part A By D. L. Massart, B. G. M. Vandeginste, L. M. C. Buydens, S. De Jong, P. J. Lewi, and J. Smeyers-Verbeke. Data Handling in Science and Technology Volume 20a. Elsevier: Amsterdam. 1997. Xvii + 867 Pp. Isbn 0-444-89724-0. $293.25. *J Chem Inf Comput Sci*, 1998. **38**(6): 1254.

34. AOAC, Guidelines for collaborative study procedures to validate characteristics of a method of analysis, www.aoac.org. 2002. Accessed on October 12, 2017.

35. INAB, Guide to method validation for quantitative analysis in chemical testing laboratories, INAB guide Ps15, www.inab.ie. 2012. Accessed on October 12, 2017.

36. IUPAC, Guidelines for collaborative study procedures to validate characteristics for a method of analysis. *J Assoc Off Anal Chem*, 1989. **72**: 694–704.

37. DeBeer, J.O., C.V. Vandenbroucke, D.L. Massart, and B.M. DeSpiegeleer, Half-fraction and full factorial designs versus central composite design for retention modelling in reversed-phase ion-pair liquid chromatography. *J Pharm Biomed Anal*, 1996. **14**(5): 525–541.

38. (a) Miller, J.N., and J.C. Miller, Statistics and Chemometrics for Analytical Chemistry, 6th ed. 2010. Pearson Education Limited, Harlow, England; (b) Massart, D.L., B.G.M. Vandeginste, L.M.C. Buydens, S. De Jong, P.J. Lewi, and J. Smeyers-Verbeke, Handbook of Chemometrics and Qualimetrics: Part A, in *Data Handling in Science and Technology*, Vol. 20A. 1997. Elsevier, Amsterdam, the Netherlands.

39. ISO, *ISO 13528:2015—Statistical Methods for Use in Proficiency Testing by Interlaboratory Comparison*. 2015. ISO, Geneva, Switzerland.

40. ISO, *ISO 5725-1:1994—Accuracy (Trueness and Precision) of Measurement Methods and Results—Part 1: General Principles and Definitions*. 1994. ISO, Geneva, Switzerland.

41. Thompson, M., S.L.R. Ellison, and R. Wood, Harmonized guidelines for single-laboratory validation of methods of analysis (IUPAC technical report). *Pure Appl Chem*, 2002. **74**(5): 835–855.

42. Linsinger, T.P.J., Use of recovery and bias information in analytical chemistry and estimation of its uncertainty contribution. *Trends Anal Chem*, 2008. **27**(10): 916–923.

43. Massart, L. and B. Vandeginste, *Handbook of Qualimetrics: Part A*. 1995, Elsevier, Amsterdam, the Netherlands, pp. 519–556.

44. Raposo, F., Evaluation of analytical calibration based on least-squares linear regression for instrumental techniques: A tutorial review. *Trends Anal Chem*, 2016. **77**: 167–185.

45. Shewiyo, D.H., E. Kaale, P.G. Risha, B. Dejaegher, J. Smeyers-Verbeke, and Y. Vander Heyden, Optimization of a reversed-phase-high-performance thin-layer chromatography method for the separation of isoniazid, ethambutol, rifampicin and pyrazinamide in fixed-dose combination antituberculosis tablets. *J Chromatogr A*, 2012. **1260**: 232–238.

46. Shewiyo, D.H., E. Kaale, P.G. Risha, B. Dejaegher, J. Smeyers-Verbeke, and Y. Vander Heyden, Hptlc methods to assay active ingredients in pharmaceutical formulations: A review of the method development and validation steps. *J Pharm Biomed Anal*, 2012. **66**: 11–23.

47. De Beer, J.O., T.R. De Beer, and L. Goeyens, Assessment of quality performance parameters for straight line calibration curves related to the spread of the abscissa values around their mean. *Anal Chim Acta*, 2007. **584**(1): 57–65.

48. Heyden, Y.V., C. Hartmann, D.L. Massart, L. Michel, P. Kiechle, and F. Erni, Ruggedness tests for a high-performance liquid chromatographic assay: Comparison of an evaluation at two and three levels by using two-level Plackett-Burman designs. *Anal Chim Acta*, 1995. **316**(1): 15–26.

49. Hubert, P., J.-J. Nguyen-Huu, B. Boulanger, E. Chapuzet, P. Chiap, N. Cohen, P.-A. Compagnon et al., Harmonization of strategies for the validation of quantitative analytical procedures. A Sfstp proposal—Part I. *J Pharm Biomed Anal*, 2004. **36**(3): 579–586.

50. Hubert, P., J.-J. Nguyen-Huu, B. Boulanger, E. Chapuzet, P. Chiap, N. Cohen, P.-A. Compagnon et al., Harmonization of strategies for the validation of quantitative analytical procedures. A Sfstp proposal—Part II. *J Pharm Biomed Anal*, 2007. **45**(1): 70–81.

51. Hubert, P., J.-J. Nguyen-Huu, B. Boulanger, E. Chapuzet, N. Cohen, P.-A. Compagnon, W. Dewé et al., Harmonization of strategies for the validation of quantitative analytical procedures. A Sfstp proposal—Part III. *J Pharm Biomed Anal*, 2007. **45**(1): 82–96.

52. Hubert, P., J.J. Nguyen-Huu, B. Boulanger, E. Chapuzet, N. Cohen, P.A. Compagnon, W. Dewe et al., Harmonization of strategies for the validation of quantitative analytical procedures. A Sfstp proposal—Part IV. Examples of application. *J Pharm Biomed Anal*, 2008. **48**(3): 760–771.

6 Calibration Curves in Chromatography

Sven Declerck, Johan Viaene, Ines Salsinha, and Yvan Vander Heyden

CONTENTS

6.1 INTRODUCTION

In scientific research or routine applications, quantitative measurements are almost indispensable. When the samples contain several compounds of interest or when the compounds of interest are accompanied by other compounds that possibly can influence their measurement, first a separation should be performed, allowing a selective quantification. Several chromatographic techniques (e.g., liquid, gas, supercritical fluid) can be applied for this purpose. Peak areas or peak heights derived from the chromatograms are used as responses to quantitatively compare the compound(s) of interest with reference standards. A necessary step in the quantification procedure is to determine the relation between the measured response and the analyte's concentration, which is called calibration. It is done by measuring the responses of solutions with known concentrations of the compounds of interest, which finally results in a calibration line or curve. If possible, a linear calibration, that is, a straight-line model, is preferred. The first part of this chapter will discuss the classic least-squares straight-line calibration, considering also common mistakes and misunderstandings. The second part will handle some variations on the classical calibrations, such as one-point calibration, calibration models with heteroscedastic data, calibration using internal standards, calibration with standard addition, non-straight-line calibration, calibration after linearization transformation, inverse calibration, orthogonal calibration, and multivariate calibration.

6.2 STRAIGHT-LINE CALIBRATION

6.2.1 UNIVARIATE LINEAR REGRESSION: ORDINARY LEAST-SQUARES CALIBRATION

The relationship between the concentrations of the calibration standards and the instrumental response is expressed by a mathematical model, which allows predicting the concentrations of new samples [1,2]. Usually, a linear-regression model is used to determine the calibration equation. In addition to the simple linear-regression model $y = a \cdot x + b$, others, such as the exponential ($y = a \cdot b^x$) [3], logarithmic ($y = a + b \cdot \log x$) [4], or power ($y = a \cdot x^b$) [5] models, can be used. In these models, y represents the measured response or dependent variable and x the independent variable. The coefficients of the linear-regression model are determined by the least-squares regression method. This method determines the best fit of the calibration curve through the data, minimizing the sum of the squared residuals. In this approach, only errors on the dependent variable y are assumed to be present, while errors in x, the concentration, are considered negligible. The true model for a linear calibration is given by Equation 6.1, with true regression coefficients β_0 and β_1, the intercept and the slope of the regression line, respectively:

$$y = \beta_0 + \beta_1 x + \varepsilon \qquad (6.1)$$

The true error (ε) is the difference between the observed response and the true value. A working model with estimated coefficients is mostly used because the true model is often unknown. This working model is given by Equation 6.2, with $\hat{y}$ the predicted value for y, and b_0 and b_1 the estimates of β_0 and β_1, respectively:

$$\hat{y} = b_0 + b_1 x + e \qquad (6.2)$$

The residual (e_i) for a given concentration x_i is the difference between the observed response (y_i) and the value ($\hat{y}_i$) predicted by the calibration model.

Ordinary least-squares (OLS) and weighted least-squares (WLS) regression are two modeling techniques that can be used to estimate the coefficients of the linear-regression model [6]. The choice of the best technique is based on (1) the concentration range of interest and (2) the homogeneity of the variances in the considered range.

6.2.2 OLS PREREQUISITES

A number of prerequisites must be fulfilled; else significant errors will be present, such as biased estimators of the slope and intercept of the calibration curve. Homoscedasticity of the errors in y, that is, the residuals, is one of the OLS prerequisites (Figure 6.1a) [2]. This means that the experimental output/signal (y) must have a constant variance. However, when the data are heteroscedastic (Figure 6.1b), the coefficients will be affected disproportionally by the outcomes with the largest

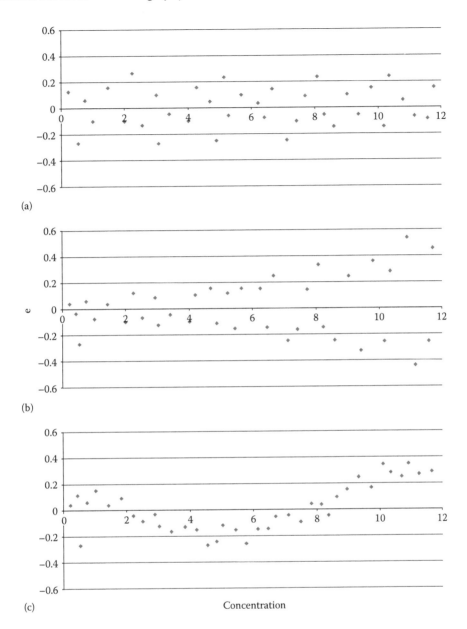

FIGURE 6.1   Residual plots of (a) homo- and (b) heteroscedastic data, and (c) wrong model fitting (lack of fit); e: experimental error, residual.

variances. Mathematical transformations of the data are then possible and common ways to convert heteroscedastic to homoscedastic data are discussed in Section 6.3.2. WLS regression is an alternative to deal with heteroscedastic data. Homoscedasticity must be tested prior to modeling because often heteroscedastic results are obtained [7]. This can, for instance, be performed by means of the Cochran test [1]. Homoscedasticity is influenced by the concentration range of the standards or samples. When it covers more than one order of magnitude, higher variances at higher concentrations will be obtained, resulting in heteroscedastic data. Broad concentration ranges are, for instance, commonly applied in the determination of drug compounds in plasma and biological samples, as well as in pharmacokinetic and pharmacodynamic studies.

6.2.3 VALIDATION OF THE MODEL

The validation of the model is a critical step and a compulsory part in the construction of calibration curves. By validating the model, it is evaluated whether the model adequately describes the relationship between the variables x and y, that is, whether there is no lack of fit. In addition, the normality and constant variance of the residuals must be checked when validating the model.

6.2.3.1 Analysis of the Residuals

A number of prerequisites must be checked when validating the model [1]. The most important prerequisite is the homoscedastic behavior of the variances of y for all calibration standards. This can be tested by the analysis of the residuals, where the residuals can first be plotted for visualization (Figure 6.1), followed by a statistical test. The graphical method is simple, easy, and straightforward and provides a visual evaluation of the pattern of the residuals. Figure 6.1 shows the plots of homo- and heteroscedastic residuals as well as the residual plot when a wrong model has been used (lack of fit). The residuals are often plotted against the concentration of the standards, but other possibilities occur (for instance, the residuals vs. the estimated response of the regression line). However, the interpretation of these residual plots requires some experience and can be indecisive, often also because their number is limited. Therefore, a statistical test for the variances (e.g., the Cochran, Levene's, or Bartlett's test) is beneficial in order to confirm the observed graphical pattern [8]. The differences between the statistical tests are explained in [9].

6.2.3.2 Verifying Linearity

6.2.3.2.1 ANOVA-LOF Test

The second step, when validating the model, is to evaluate the quality of the linear fit. A first test that can be used to test the linearity is the lack-of-fit analysis of variance (ANOVA-LOF) test. It is a statistical test that compares the variability of the measurement at a given concentration (pure error) with the deviation from the calibration model (lack-of-fit error) [1,7,10]. To perform the ANOVA-LOF test, replicate measurements are needed. To avoid underestimation of the pure error, the replicates have to be real replicates. The residual (RES or $e = y_{ij} - \hat{y}_i$), representing the deviation of a measurement y_{ij} at x_i from its predicted value $\hat{y}_i$, is split in the random experimental error or pure error ($PE = y_{ij} - \bar{y}_i$) and the LOF error ($= \bar{y}_i - \hat{y}_i$), that is, the difference between the predicted $\hat{y}_i$ value at x_i and the average of the y values at x_i (Figure 6.2) [1]. Equations 6.3 and 6.4 represent the split but expressed as sums of squares (SS), with n_i the number of replicate measurements at x_i.

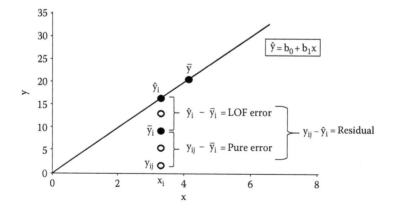

FIGURE 6.2 Splitting the residuals in a pure error and a lack-of-fit compound. $\bar{y}$ represents the global average of all measured responses. (Based on Massart, D.L. et al., *Handbook of Chemometrics and Qualimetrics—Part A*, Elsevier Science, Amsterdam, the Netherlands, 1997.)

The total sum of squares (SS_{TOT}), which is the sum of the SS of the residuals and that of the regression (REG), is given by Equation 6.5. Further information about the SS of the regression can be found in Section 6.2.3.4.

$$SS_{RESIDUAL} = \sum_{ij}\left(y_{ij} - \bar{y}_i\right)^2 + \sum n_i\left(\hat{y}_i - \bar{y}_i\right)^2 \tag{6.3}$$

$$SS_{RESIDUAL} = SS_{PE} + SS_{LOF} \tag{6.4}$$

$$SS_{TOT} = SS_{RES} + SS_{REG} \tag{6.5}$$

The variability of the measurement error is characterized by the SS of the random experimental error (SS_{PE}), while the variability around the regression line is characterized by the SS of the LOF error. As a consequence, the mean square (MS) of the LOF to the MS_{PE} must be nonsignificant when a proper model is used. This can be evaluated by the Fisher–Snedecor test, which is a typical F-test between the two variances (MS values) [11] (see Table 6.1). The model is inadequate if it shows lack of fit, that is, when the calculated F is larger than the critical F-value ($F_{k-2, n-k}$). This means that the variance due to the LOF is larger than the measurement error. Table 6.1 represents a typical table of an ANOVA-LOF test.

It is worth mentioning that the ANOVA-LOF test depends on the precision of the method [12,13]. The test is sensitive for precise data, which reduces the chance to pass the test. Contrary, imprecise data will pass the test without detecting larger deviations from linearity.

6.2.3.2.2 Mandel's Test and Significance of the Quadratic-Term Test

Two alternative tests are possible to validate the model, that is, the Mandel's and the significance of the quadratic-term (SQT) tests. The Mandel's test assumes that relatively large deviations of the responses of the standards from the calibration curve, that is, a straight-line model, are caused by nonlinearity [14]. These deviations can then be reduced using an appropriate regression model, for example, a second-order function. The test consists in comparing the residual standard deviations of both first- and second-order models by means of an F-test. The second-order model or quadratic model provides no better fit compared to the first-order function or straight-line model when F_{crit} is larger than F_{exp} (see Equation 6.6, with n the total number of standards). In other words, the quadratic model is not required to generate the calibration line. Notice that the Mandel's test requires a higher number of calibration standards than a routine calibration [15]. It therefore cannot be used for routine calibrations (e.g., six concentration levels and four replicates of each) as it will not be able to make a proper distinction between the two regression models (straight vs. curved), even when the

TABLE 6.1

Typical ANOVA-LOF Table, with k and n the Numbers of Standards and of Measurements, Respectively

Source of Variance	SS	Degrees of Freedom	MS	F
Residuals	SS_{RES}	$n - 2$	$MS_{RES} = \dfrac{SS_{RES}}{n-2}$	
LOF	SS_{LOF}	$k - 2$	$MS_{LOF} = \dfrac{SS_{LOF}}{k-2}$	$F = \dfrac{MS_{LOF}}{MS_{PE}}$
Pure error	SS_{PE}	$n - k$	$MS_{PE} = \dfrac{SS_{PE}}{n-k}$	

variances of both regression models differ by 25%. IUPAC suggests a Mandel's test with simplified degrees of freedom, but which only can be used when the difference between the variances of both regression models is below 10% [16].

$$F_{exp} = \frac{(n-2)s_{first\ order}^2 - (n-3)s_{second\ order}^2}{s_{second\ order}^2} \tag{6.6}$$

with $s_{first\ order}^2$ and $s_{second\ order}^2$ the variances (s^2) of the first-order (straight) and second-order (curved) regression models, respectively.

The second alternative to validate the model is the SQT test. This test compares a straight-line regression model with a polynomial (second order) by comparing the MS values of the residuals and the quadratic term (QT) [10]. Calibration curves are considered as nonlinear when F_{exp} (see Equation 6.7) exceeds the F_{crit} or when the second-order regression coefficient (b_2) confidence interval does not include zero.

$$F_{exp} = \frac{MS_{QT}}{MS_{RES}} \tag{6.7}$$

6.2.3.3 Quality Coefficient

The quality coefficient (QC) is calculated by Equation 6.8. It is a parameter describing the fit of the measured responses toward a given model. Even though QC is not a parameter indicating linearity, sometimes when QC is below a given limit, for example, $QC \leq 5\%$, one assumes that the used model (for instance, a straight-line model) fits the data sufficiently well and predictions will be acceptable even if the response does not change linearly as a function of the concentration [17]:

$$QC = 100\sqrt{\frac{\sum\left(\frac{y_i - \hat{y}_i}{\hat{y}_i}\right)^2}{n-1}} \tag{6.8}$$

with n the number of data points.

6.2.3.4 Misunderstandings about the Correlation Coefficient
and Coefficient of Determination

The coefficient of determination and the correlation coefficient are two regression parameters that, because of their simplicity, are commonly incorrectly used to indicate linearity. It is also important to notice the difference between these two coefficients.

The correlation coefficient (r) expresses the degree of linear association between two random variables x and y. It can be calculated from the covariance that evaluates also the direction of variance of the variables or, in other words, that measures their joint variation [7]. The correlation coefficient is calculated using Equation 6.9. It ranges between −1 and +1, where $r = -1$ and $r = +1$ correspond to the perfect negative and positive correlation, respectively:

$$r(x,y) = \frac{\sum(x_i - \bar{x})(y_i - \bar{y})}{\sqrt{\sum(x_i - \bar{x})^2 \sum(y_i - \bar{y})^2}} \tag{6.9}$$

with $\bar{x}$ and $\bar{y}$ the averages of all x_i and y_i values, respectively.

When no correlation between x and y exists, the correlation coefficient is zero. Notice that Equation 6.9 does not contain any term related to the model and as such it is impossible to use the correlation coefficient to describe the quality of the model, that is, to decide on linearity.

The coefficient of determination (r^2) contains a model-related factor, that is, SS_{REG} (see Equation 6.10):

$$r^2 = \frac{\sum(y_i - \bar{y})^2 - \sum(y_i - \hat{y})^2}{\sum(y_i - \bar{y})^2} = \frac{\sum(\hat{y}_i - \bar{y})^2}{\sum(y_i - \bar{y})^2} = \frac{SS_{REG}}{SS_{TOT}} \tag{6.10}$$

The coefficient of determination expresses the proportion of the total variation that is explained by the regression (by the model). The better the data points fit the model, the closer r^2 is to one. Besides, from Equation 6.10, it can be calculated from the output of an ANOVA (Table 6.2), which estimates the significance of the regression. Despite the obtained limited information (F-value is usually highly significant), this ANOVA is often reported. The significance of F is due to the minimal contribution of the MS_{RES} compared to the MS_{REG}. Thus, significance of the test for a linear model does not guarantee at all that one is dealing with a straight-line behavior.

It is worth mentioning that there is a misconception about these coefficients. Despite the absence or limited relationship of these coefficients with the linearity of a calibration curve, many scientists use them to report on the linearity of a calibration curve. Especially, the correlation coefficient that is not related to a model is used in this context.

6.2.4 Outliers

A distinction should be made between outliers in the calibration set when standards are injected once and outliers in replicate measurements at a given concentration, since the former will have more influence on the calibration model than the latter. Outliers of replicate measurements can statistically be tested by, for instance, the Grubbs tests, but in this chapter the focus is on the outliers in a calibration set. The easiest way to check for outliers is to plot the residuals. However, in some cases, it is not possible or difficult to conclude whether or not a given measurement is an outlier (because the outlier attracts the model into its direction). In such cases, it is better to perform a statistical test such as the Cook's distance test. This test measures the change in the regression coefficients that occurs if the ith observation is omitted from the data, or, in other words, it measures the effect of deleting a calibration point. When more outliers are present, this test will fail. However, the occurrence of several outliers is also not evident. The more "outliers" there are, the less evident it is that they really are outliers. It usually is rather a situation of bad data (bad precision). Robust regression can be used to eliminate the bias of outlying observations on the model coefficients (see Section 6.7). Even when an outlier is present, it provides a good prediction of the slope and intercept.

TABLE 6.2
ANOVA Table to Estimate the Significance of the Regression

Model	SS	Degrees of Freedom	MS	F
Regression	SS_{REG}	1	$MS_{REG} = \dfrac{SS_{REG}}{1}$	$F_{calc.} = \dfrac{MS_{REG}}{MS_{RES}}$
Residual	SS_{RES}	$n-2$	$MS_{RES} = \dfrac{SS_{RES}}{n-2}$	
Total	SS_{TOT}	$n-1$		

6.2.5   Confidence Intervals of Slope and Intercept

In Equation 6.2, b_0 and b_1 are estimates of β_1 and β_0. The reliability of these estimates is important because otherwise wrong interpolations and estimated concentrations will result. It is therefore necessary to verify the quality of these estimates. They are evaluated by estimating the confidence intervals of the slope and intercept of the calibration line.

6.3   VARIANTS ON CLASSIC CALIBRATION

OLS calibration is widely used and often gives good results, despite some limitations. In this section, a number of variants are discussed that either reduce the workload in routine analysis or solve specific issues where common OLS calibration is suboptimal.

6.3.1   One-Point Calibration and Calibration without Standards

When performing a regular OLS calibration, a number of calibration standards in a given concentration range are analyzed in order to model the measured response as a function of the concentration, that is, to estimate the slope and intercept of the calibration line. However, when no significant intercept is assumed or expected (for instance, from prior method validation), the estimation of the slope of the calibration line is possible based on the measurement of (a number of replicates of) only one calibration standard and the origin (0,0). The International Standards Organization recommends measuring at least two replicates of a standard with a higher concentration than the expected samples [11]. The difference between OLS and one-point calibration is illustrated in Figure 6.3. When using the one-point calibration approach, a linear response–concentration relationship through zero is assumed.

Now, why is the one-point calibration approach not always applied if it reduces the workload? Taking multiple concentrations into account has some clear advantages. Each measurement is affected by an experimental error. Assuming that experimental errors on the measurements are randomly distributed, their influence on the estimations of slope and intercept in a regression model is reduced when several standards are used. Additionally, considering multiple concentrations enables the analyst to check in which concentration range the linear relationship holds and whether outliers are present in the measurements. Although the above aspects are not valid or applicable when performing a one-point calibration, acceptable estimates can be obtained in conditions where the real response–concentration relationship does not deviate too much from a linear and provided that the intercept is close to zero, that is, methods with a good measurement precision. However, when the experimental error on the only measured calibration standard is high, the concentration estimation of a sample may be dramatically affected.

One-point calibrations are often used in the context of quality control. Pharmacopoeial monographs of active pharmaceutical ingredients (API) often include chromatographic tests (usually

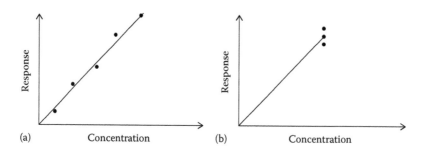

FIGURE 6.3   (a) OLS versus (b) one-point calibration.

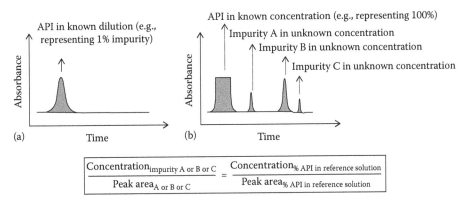

FIGURE 6.4 One-point calibration in the context of a related-substances test. (a) Diluted API sample (reference solution). (b) Test API sample.

high-pressure liquid chromatography with ultraviolet detection at a single wavelength, HPLC-UV) to check whether the related substances in a bulk product sample are within allowable limits. In this situation, the peak area in a chromatogram obtained from a diluted solution of the API with a known and detectable concentration close to the concentration of the impurities (e.g., representing 1% impurity) is used as standard to estimate the concentration–response ratio (Figure 6.4a). The chromatogram of a test solution representing 100% API, where the concentration is chosen to result in quantifiable peaks for the impurities, is considered as well. A chromatogram obtained for such test solution is shown in Figure 6.4b, where besides the major compound, three impurities are observed. Some of these impurities may be known, while others are not. For the known compounds, a correction factor may be applied (if known), in order to take into account the absorption-coefficients ratio of the major compound and the impurities. In the monograph, limit values are set on the contents (% of API) of the impurities, both individually and in total. Using the ratio between the concentration (% of API) and the peak area of the major compound in the reference solution, the concentration of the impurities can be estimated from their peak areas. The calibration is thus performed by means of one calibration standard, that is, the diluted API solution.

The use of one-point calibration in pharmacopoeias is not limited to the evaluation of related substances. In order to evaluate the content of a major active or marker compound in a given herbal formulation (e.g., extract), in certain monographs, a chromatographic analysis is required [18]. A reference standard solution of the marker should then also be analyzed. The content of the marker compound is then finally expressed as a mass percentage of the weighted (dried) sample. On this content, a limit is often set to decide on the conformity of the sample. This approach is, for instance, applied in the *European Pharmacopoeia* monograph on Fleeceflower root (*Polygoni multiflori radix*) [18].

Calibration can even be performed without standards. Examples can be found in several monographs where spectroscopy is used for quantification. The absorption coefficient (which corresponds to the slope of the calibration line) is then given. However, to our knowledge this approach has no applications in chromatography.

6.3.2 HETEROSCEDASTICITY

OLS regression requires a homoscedastic behavior of the responses, meaning that the variance of the measurements is not dependent on the value of the independent variable, that is, the concentration. In that case, a calibration model is developed that gives equal weights to all calibration points [11]. If heteroscedastic behavior is observed, meaning that the variance of the measurements does depend on the concentration level, then OLS regression results in estimators for the slope and intercept that

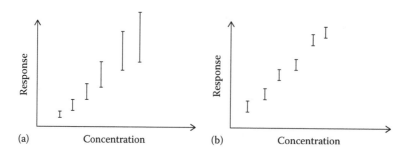

FIGURE 6.5   (a) Heteroscedastic versus (b) homoscedastic data.

are unbiased, but show a large variance [19]. However, the estimator for the error is biased, resulting in unreliable statistical tests (t- and F-tests) and confidence intervals that use this error estimator [11]. Additionally, the coefficient of determination (r^2) is usually overestimated [20].

To detect heteroscedasticity, replicates at each concentration level are measured. The characteristic pattern of heteroscedastic data (Figure 6.5) can be detected by simple visual inspection or also by a statistical test. Testing of unequal variances (s^2) can, for instance, be done using a Cochran C test or a Levene's test [11,21].

If the homoscedasticity requirement is not fulfilled, two approaches can be used. A first option is to mathematically transform the independent variables and the responses, which can result in homoscedastic data. Logarithmic transformations have proved to be useful to correct for heteroscedasticity when the relative standard deviation is rather constant over the concentration range (i.e., when the variance of the y values is proportional to y^2 or the standard deviation is proportional to y). An OLS regression is then applied to the transformed variables. The following model is then built: $\log(y) = b_0 + b_1 \cdot \log(x)$. However, when the variance of the y values is proportional to y, then a square-root transformation is used to create homoscedastic results: $\sqrt{y} = b_0 + b_1 \cdot \sqrt{x}$ [1].

A second approach is to give weights to the calibration results depending on their concentration, which is applied in weighted least-squares (WLS) regression. Since the variances at lower concentrations are smaller, the calibration curve should fit these points best. This is done in practice by giving a higher weight to the lowest concentration levels [11]. The influence on the estimations of slope and intercept is often rather negligible, but the estimator for the error often differs dramatically between OLS and WLS regression results. Different possible weights are reported in different studies. Using a weight of 1/variance is theoretically a good choice. However, in several research papers, different weighting factors are applied, like $1/x$, $1/x^2$, $1/y$, and $1/y^2$, in order to correct for the heteroscedastic behavior [22–24].

6.3.3 Internal Calibration

Several analytical methods contain a number of steps (such as sample preparation, including liquid–liquid extraction, injection in a gas chromatographic system without a loop, derivatization, drying, redissolving) that increase the random error on the results of the analyzed compounds. In such cases, the variance of the response can be large, both for calibration standards and samples. This high variability leads to a poor OLS regression, and eventually to bad quantifications of samples due to errors on the predicted concentrations. The method of internal calibration offers a solution to this problem. The principle is that a compound (internal standard, IS) is added to all calibration and sample solutions to a constant concentration. The addition of the IS needs to be done in the very beginning of the sample pretreatment and at least prior to the steps in the analytical procedure that cause the large variability. The procedure is illustrated in Figure 6.6.

A compound with the potential of being a good IS should fulfill a number of requirements. It should be similar in chemical structure to the compound of interest (COI) in order to ascertain

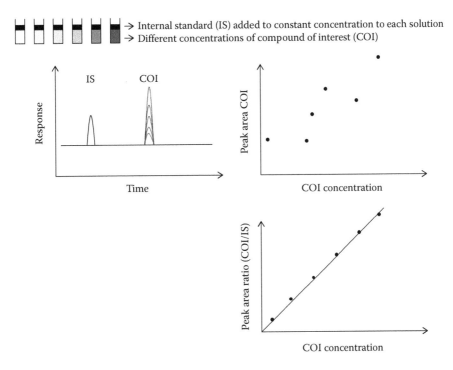

FIGURE 6.6 Principle of internal calibration.

a similar behavior during the entire procedure; its signal should not interfere with that of the COI; its concentration level should be in the order of magnitude of the COI; and its signal should be distinguishable from the COI. If these requirements are fulfilled, the IS and the COI, both undergoing the same procedure steps, should be similarly influenced during sample preparation. Instead of the signal of the compound alone, the signal ratio COI/IS is considered as response. By using the ratio, the analyst corrects for the large random error occurring (Figure 6.6). An OLS regression then is performed, using the concentration of the COI as independent variable and the ratio as dependent. The concentration of an unknown sample is determined by interpolation of its signal ratio on the calibration line.

Liquid chromatography coupled to mass spectrometry quantifications are prone to matrix interferences, in particular when electrospray ionization is used [25]. As a result, a compound with a given concentration in different samples could be detected with a lower or higher signal, depending on the other compounds in the samples, that might influence the COI's signals, even if the interfering compounds are not detectable themselves. To correct for the matrix effect in mass spectrometry, a specific type of IS calibration can be applied using isotopic analogues (e.g., the molecule where a number of hydrogens are exchanged for deuteriums) as IS. Matrix effects are corrected by taking the ratio COI/deuterated analogue as a response, because both the COI and its deuterated analogue are equally influenced by the matrix effects [26].

6.3.4 STANDARD ADDITION CALIBRATION

In some cases, the measurement of a compound can be affected by the sample matrix. The consequence is that the detected signal for a given concentration is different in a solvent or in the sample matrix. Consequently, using classic calibration standards in a solvent will lead to a wrong estimation of the sample concentrations. To solve this problem, an approach can be applied where the

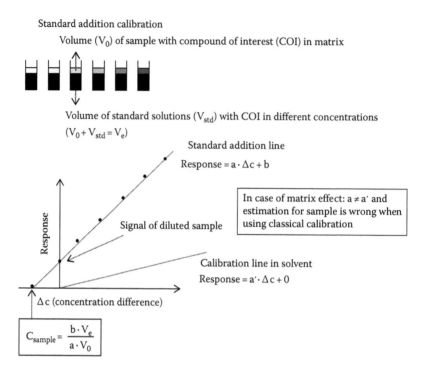

Standard addition calibration

Volume (V_0) of sample with compound of interest (COI) in matrix

Volume of standard solutions (V_{std}) with COI in different concentrations

$(V_0 + V_{std} = V_e)$

Standard addition line

Response = $a \cdot \Delta c + b$

In case of matrix effect: $a \neq a'$ and estimation for sample is wrong when using classical calibration

Signal of diluted sample

Calibration line in solvent

Response = $a' \cdot \Delta c + 0$

Δc (concentration difference)

$$C_{sample} = \frac{b \cdot V_e}{a \cdot V_0}$$

FIGURE 6.7 Principle of standard addition calibration.

calibration standards are prepared in the sample matrix, which is called standard addition calibration. Each standard is thus prepared by taking a fixed volume of the sample, and spiking it with a given volume of standard solutions with different concentrations. The concentration of the sample is in this case estimated by means of extrapolation (Figure 6.7). This is a drawback and as a result, the random errors on the estimated concentrations are expected to be higher. The procedure is illustrated in Figure 6.7.

Application of this methodology is only performed when necessary, that is, to cope with matrix interferences, since for each sample, several solutions need to be measured to determine its concentration [11]. However, occasionally one-point standard addition is also applied.

Standard addition to correct matrix effects is usually applied in spectroscopy, for example, in fluorimetry, where the matrix may cause fluorescence quenching. In classical chromatography, standard addition is usually not performed for matrix effect correction, because at the moment of detection, the COI for both samples and standards is dissolved in the mobile phase. Notice that often in chromatography, the standard is made in the sample matrix (though blank matrix here), for example, to analyze pharmaceuticals in plasma or serum. Often this is the case when internal calibration is needed. Secondly, a standard addition may be required to assay samples for which no blank can be obtained, for example, to assay iron or some other metal ions in blood.

6.4 NON-STRAIGHT CALIBRATION CURVES

6.4.1 QUADRATIC AND POLYNOMIAL MODELS

The application of a straight-line model is often preferred for its mathematical simplicity [27]. However, sometimes it is insufficient to model the data, for example, when curvature is observed. In practice, for wide calibration ranges, linearity is only observed at low concentrations [10]. In order to improve the estimated values based on calibration curves, polynomial functions can be applied.

Quadratic and cubic (i.e., second- and third-order polynomials, respectively) models are represented by Equations 6.11 and 6.12 [10,28]:

$$y = b_0 + b_1 x + b_2 x^2 \tag{6.11}$$

$$y = b_0 + b_1 x + b_2 x^2 + b_3 x^3 \tag{6.12}$$

In practice, quadratic and cubic equations are the main alternatives chosen when the straight-line model does not fit the data appropriately. Even more complex (higher order) models can be applied, but it should be taken into account that polynomial models with many terms can compromise the ability to predict, due to overfitting (see also Chapter 14) [29]. The quadratic model is used when curvature is observed in the calibration curve. A cubic model is occasionally applied to model the retention of an acidic or basic compound as a function of the pH of the mobile-phase buffer. This behavior has an S-like shape that can reasonably be well described by the cubic model. A similar shape may be observed for calibration curves of size exclusion chromatography measurements, where the logarithm of the molecular weight is plotted against the retention time [28].

6.4.2 LINEARIZATION APPROACHES

In several cases, nonlinear models best fit the behavior of the response y (Table 6.3). Estimating response variables (y) for new samples from nonlinear equations becomes more difficult than from a straight-line model. Linearization approaches can alternatively be applied in order to simplify the calculations (see Table 6.3). The data are transformed so that a nonlinear relationship becomes linear.

6.4.3 INVERSE REGRESSION

For a classic calibration, the measured response (y) is in general related to the concentration (x) of a standard through an OLS regression model $y = b_0 + b_1 x$ or $y = f(x)$ (Figure 6.8a). For the standard OLS regression model, the variable x is, in contrast to y, assumed to be free of error [27]. By inverse regression $x = f(y)$ the predicted variable x can be obtained from the measured variable y, which is considered as the stochastic and subject-to-error variable (Figure 6.8b). The used function will specify the x variable as a function of the dependent y variable, resulting in Equation 6.13 [28]:

$$x = b_0' + b_1' y \tag{6.13}$$

For curvilinear models, the inverse regression model facilitates the prediction of x for new samples, where this is somewhat more complex for a classical curvilinear model. The inverse calibration is thus the preferred approach when dealing with several dependent variables, as is the case in multivariate analysis.

TABLE 6.3
Linearization Approaches

Nonlinear Function	Transformation	Resultant Function
Exponential: $y = be^{mx}$	$y' = \ln(y)$	$y' = \ln(y) = \ln(b) + mx$
Power: $y = bx^m$	$y' = \log(y), x' = \log(x)$	$y' = \log(b) + mx'$

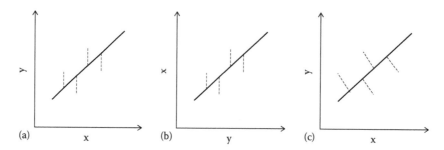

FIGURE 6.8 The ordinary least-squares regression method (a—direct and b—inverse regression) and the orthogonal least-squares regression (c).

6.5 ORTHOGONAL LEAST-SQUARES REGRESSION

The regression coefficients are in general estimated by OLS or WLS because for one variable (usually x) the error is considered negligible. However, when there is also variance in x, that is, x is not "error free," those models are less good. Such situation occurs, for instance, when one compares the results obtained for a series of samples by two measurement methods. The orthogonal least-squares regression can then be considered to tackle this problem. The classical least-squares linear regression of y on x assumes that y is measured subjected to error and that x is fixed or measured without error (Figure 6.8a). Differently from the OLS approach, the orthogonal regression minimizes the orthogonal distance from the observed data points to the regression line, taking into account errors in both variables x and y (Figure 6.8c).

6.6 MULTIVARIATE CALIBRATION

Different from univariate calibration where predictions for new samples are based on one measured variable, which is sufficiently selective for the response of interest, in multivariate calibration more than one variable is used to predict the property of interest—usually represented by $\mathbf{y}$ [30]. The variables providing information on the samples are collected in a matrix—usually represented as $\mathbf{X}$. In multivariate calibration, inverse regression is mostly applied, $\mathbf{y} = f(\mathbf{X})$. Notice that the symbols x or $\mathbf{X}$ and y are switched compared to the previous section (Section 6.4.3).

The most commonly used techniques for multivariate analysis are multiple linear regression (MLR), partial least squares (PLS), and principal component regression (PCR) (see also Chapter 14) [31,32]. The MLR method involves equations as

$$y_i = b_0 + b_1 x_{1i} + b_2 x_{2i} + \cdots + b_n x_{ni} \tag{6.14}$$

where
  y_i corresponds to the modeled and predicted response
  $b_0, b_1, b_2, \ldots, b_n$ are the regression coefficients
  x_n are the variables from the $\mathbf{X}$ matrix used in the modeling

An example of the application of MLR in chromatography is the construction of "quantitative structure–retention relationships" to model the retention of solutes (y_i) as a function of structural information (molecular descriptors) in order to characterize stationary phases (see Chapter 3). The linear solvation energy model, for instance, includes five molecular descriptors (x_n) to model the retention [33,34]. Notice that to perform MLR regression, the number of calibration objects/samples must be larger than the number of predictor variables x_i in the model. However, very often, the $\mathbf{X}$ matrix contains many more variables than the number used in the MLR model. A variable selection or reduction approach is then required to help in selecting the best (the most relevant) descriptors.

In PLS and PCR models, the x_n variables are replaced by so-called latent variables. Latent variables are new axes defined in the domain of the data, according to given criteria. The projections of the samples on these new axes provide coordinates (scores) that are used in the modeling. Models can be represented as $y_i = b_0 + b_1LV_{1i} + b_2LV_{2i} + \cdots + b_nLV_{ni}$, where LV_n represents the scores on a given latent variable. (For more information, see Chapter 14.)

6.7 ROBUST REGRESSION: THE LEAST MEDIAN OF SQUARES METHOD

Robust approaches have gained an increased interest among analytical scientists as these methods are not affected by outliers. Usually, they require higher calculation times. That is why the interest in these methods progressed as the development of computers evolved. One type of robust regression is the least median of squares (LMS) method, which is based on minimizing the median of squared residuals, instead of minimizing the sum of the squared residuals as performed in least-squares regression [35].

To estimate the slope and the intercept, (1) all possible pairs of data points are used to construct a line, (2) relative to each line the squared residuals for all data points are calculated, and (3) the line for which the median of the squared residuals is minimal is retained [17,35]. The median of squares method is a very robust method since its breakdown point, which refers to the fraction of outliers that can be tolerated, is around 50% [11].

6.8 CONCLUSION

Calibration lines are often used in chromatography, where the preference is given to a straight-line OLS model. However, the prerequisites for this model are not always fulfilled and therefore different solutions are available, such as WLS and linearization approaches. In this chapter, we focused, besides the classical straight-line calibration, also on other calibration methods such as one-point calibration, internal calibration, standard addition, non-straight calibration curves, orthogonal least-squares regression, multivariate calibration, and robust regression.

REFERENCES

1. Massart, D.L., Vandeginste, B.G.M.L., Buydens, M.C., de Jong, S., Lewi, P.J. and J. Smeyers-Verbeke. 1997. *Handbook of Chemometrics and Qualimetrics—Part A*. Amsterdam, the Netherlands: Elsevier Science.
2. Montgomery, D.C., Peck, E.A. and G.G. Vining. 2006. *Introduction to Linear Regression Analysis*. Hoboken, NJ: John Wiley & Sons.
3. Xiong, J., Jiang, H.P., Peng, C.Y., Deng, Q.Y., Lan, M.D., Zeng, H., Zheng, F., Feng, Y.Q. and B.F. Yuan. 2015. DNA hydroxymethylation age of human blood determined by capillary hydrophilic-interaction liquid chromatography/mass spectrometry. *Clin. Epigenetics* 7: 72.
4. Stahnke, H., Kittlaus, S., Kempe, G. and L. Alder. 2012. Reduction of matrix effects in liquid chromatography-electrospray ionization-mass spectrometry by dilution of the sample extracts: How much dilution is needed? *Anal. Chem.* 84 (3): 1474–1482.
5. Lynen, F., Roman, S. and P. Sandra. 2006. Universal response in liquid chromatography. *Anal. Chem.* 78 (9): 3186–3192.
6. Vanatta, L.E. and D.E. Coleman. 2007. Calibration, uncertainty, and recovery in the chromatographic sciences. *J. Chromatogr. A* 1158 (1–2): 47–60.
7. Ermer, J. and J.H.M. Miller. 2008. *Method Validation in Pharmaceutical Analysis*. Darmstadt, Germany: Wiley-VCH.
8. National Institute of Standards and Technology. Cochran variance outlier test. Accessed February 15, 2017. http://www.itl.nist.gov/div898/software/dataplot/refman1/auxillar/cochvari.htm.
9. De Souza, S.V.C. and R.G. Junqueira. 2005. A procedure to assess linearity by ordinary least squares method. *Anal. Chim. Acta* 552 (1–2): 23–35.
10. Raposo, F. 2015. Evaluation of analytical calibration based on least-squares linear regression for instrumental techniques: A tutorial review *Trends Anal. Chem.* 77: 167–185.

11. Miller, J.C. and J.N. Miller. 2005. *Statistics and Chemometrics for Analytical Chemistry*. 5th edn. Harlow, U.K.: Pearson Education.
12. Tetrault, G. 1990. Evaluation of assay linearity. *Clin. Chem.* 36 (3): 585–587.
13. Kroll, M.H. and K. Emancipator. 1993. A theoretical evaluation of linearity. *Clin. Chem.* 39 (3): 405–413.
14. Mandel, J. 1964. *The Statistical Analysis of Experimental Data*. New York: Interscience.
15. Andrade, J.M. and M.P. Gomez-Carracedo. 2013. Notes on the use of Mandel's test to check for nonlinearity in laboratory calibrations. *Anal. Methods* (The Royal Society of Chemistry) 5 (5): 1145–1149.
16. Danzer, K. and L.A. Currie. 1998. Guidelines for calibration in analytical chemistry. Part I. Fundamentals and single component calibration (IUPAC Recommendations 1998). *Pure Appl. Chem.* 70 (4): 993–1014.
17. Vankeerberghen, P., Vandenbosch, C., Smeyers-Verbeke, J. and D.L. Massart. 1991. Some robust statistical procedures applied to the analysis of chemical data. *Chemometr. Intell. Lab. Syst.* 12: 3–13.
18. European Directorate for the Quality of Medicines & Healthcare. European Pharmacopoeia. 2011. Accessed January 18, 2017. http://online.pheur.org/EN/entry.htm.
19. Williams, R. Heteroskedasticity. University of Notre Dame, Notre Dame, IN. Accessed January 20, 2017. https://www3.nd.edu/~rwilliam/stats2/l25.pdf.
20. Seddighi, H., Lawler, K.A. and A.V. Katos. 2000. *Econometrics: A Practical Approach*. 1st edn. London, U.K.: Routledge.
21. Bartolucci, A., Singh, K.P. and S. Bae. 2016. *Introduction to Statistical Analysis of Laboratory Data*. Hoboken, NJ: John Wiley & Sons.
22. Emory, J.F., Seserko, L.A. and M.A. Marzinke. 2014. Development and bioanalytical validation of a liquid chromatographic-tandem mass spectrometric (LC-MS/MS) method for the quantification of the CCR5 antagonist maraviroc in human plasma. *Clin. Chim. Acta* 431: 198–205.
23. Hubert, P., Nguyen-Huu, J.J., Boulanger, B., Chapuzet, E., Chiap, P., Cohen, N., Compagnon, P.A. et al. 2004. Harmonization of strategies for the validation of quantitative analytical procedures: A SFSTP proposal—Part I. *J. Pharm. Biomed. Anal.* 36 (3): 579–586.
24. Gu, H., Liu, G., Wang, J. and M.E. Arnold. 2014. Selecting the correct weighting factors for linear and quadratic calibration curves with least-squares regression algorithm in bioanalytical LC-MS/MS assays and impacts of using incorrect weighting factors on curve stability, data quality, and assay perfo. *Am. Chem. Soc.* 86: 8959–8966.
25. Matuszewski, B.K., Constanzer, M.L. and C.M. Chavez-Eng. 2003. Strategies for the assessment of matrix effect in quantitative bioanalytical methods based on HPLC–MS/MS. *Anal. Chem.* 75: 3019–3030.
26. Viaene, J., Goodarzi, M., Dejaegher, B., Tistaert, C., Hoang Le Tuan, A., Nguyen Hoai, N., Chau Van, M., Quetin-Leclercq, J. and Y. Vander Heyden. 2015. Discrimination and classification techniques applied on *Mallotus* and *Phyllanthus* high performance liquid chromatography fingerprints. *Anal. Chim. Acta* 877: 41–50.
27. Ratkowsky, D.A. 1990. *Handbook of Nonlinear Regression Models*. 10th edn. New York: Marcel Dekker.
28. Vander Heyden, Y., Rodríguez Cuesta, M.J. and R. Boqué. 2007. Calibration. *LC-GC Europe* 20 (6): 349–356.
29. Lee, J.W., Devanarayan, V., Barrett, Y.C., Weiner, R., Allinson, J., Fountain, S., Keller, S. et al. 2006. Fit-for-purpose method development and validation for successful biomarker measurement. *Pharm. Res.* 23 (2): 312–328.
30. Forina, M., Lanteri, S. and M. Casale. 2007. Multivariate calibration. *J. Chromatogr. A* 1158: 61–93.
31. Liang, Y. and O.M. Kvalheim. 1996. Robust methods for multivariate analysis—A tutorial review. *Chemometr. Intell. Lab. Syst.* 7439: 1–10.
32. Feudale, R.N., Woody, N.A., Tan, H., Myles, A.J., Brown, S.D. and J. Ferre. 2002. Transfer of multivariate calibration models: A review. *Chemometr. Intell. Lab. Syst.* 64: 181–192.
33. West, C., Khater, S. and E. Lesellier. 2012. Characterization and use of hydrophilic interaction liquid chromatography type stationary phases in supercritical fluid chromatography. *J. Chromatogr. A* 1250: 182–195.
34. West, C. and E. Lesellier. 2013. Effects of mobile phase composition on retention and selectivity in achiral supercritical fluid chromatography. *J. Chromatogr. A* 1302: 152–162.
35. Hartmann, C., Vankeerberghen, P., Smeyers-Verbeke, J. and D.L. Massart. 1997. Robust orthogonal regression for the outlier detection when comparing two series of measurement results. *Anal. Chim. Acta* 344: 17–28.

Section III

Data Preprocessing
and Unsupervised Analysis

7 Introduction to Multivariate Data Treatment

Łukasz Komsta and Yvan Vander Heyden

CONTENTS

7.1 BASIC CONSIDERATIONS

Chromatographic experiments often derive a *multivariate* dataset, which can (and should) be presented and analyzed as a *matrix*. A matrix is a rectangular array of numbers, and it can be easily presented as a table. A fragment of a spreadsheet, filled with numbers, is a good example of a matrix. If one considers the peak areas of 5 compounds and 30 samples are measured, then, for instance, a 30×5 table (matrix) can be built.

Multivariate datasets are obtained for *objects* (e.g., samples) having some *properties* (e.g., peak areas), and for each object all properties are measured. In chemometrics, the standard convention to arrange data in a matrix is that matrix rows correspond to objects and matrix columns to properties. Such multivariate dataset is a *two-way* array with consistency inside the rows and columns, in the sense that a value in ith row and jth column must represent the jth property of ith object.

In general, the choice of what is considered as object is arbitrary and depends on the context of the data analysis. Any matrix can be *transposed* by changing the rows and columns. Then everything reverses—the rows become the properties and the columns become the objects. To understand this concept, several examples are given in the following:

1. 50 compounds were chromatographed in 30 chromatographic systems and the retention times were measured. The matrix can have 50 rows and 30 columns. Each cell will contain the retention time of the corresponding compound in the corresponding system.

By transposing the above matrix, a reverse interpretation can be obtained—30 chromatographic systems are considered as objects, each characterized by 50 properties (the retention times of 50 compounds), that is, by the retention profile of the compounds. This type of matrix is, for instance, created in column characterization or classification.

2. 50 samples were chromatographed in one of the above systems. The areas of 20 main peaks were recorded. One can arrange the results as a matrix with 50 rows and 20 columns, filled with peak areas. Each sample (object) is characterized by 20 properties, representing the chemical composition of a sample. Hence the chemical profile of a sample is then considered.

3. 50 samples were chromatographed in one of the above systems. Chromatograms were recorded during 10 minutes of analysis, with 0.1 second step (6000 response points). The results can be arranged as a matrix with 50 rows and 6000 columns (detector responses at particular time points). Although this dataset is multivariate by definition, it contains random shifts in time, so its real chemical *bilinearity* is corrupted. This means that due to experimental variability, the retention of compounds shows some variability as well (random error of retention times), resulting in the fact that required information of the samples ends up in different matrix columns. Such phenomenon usually is filtered during data preprocessing by warping techniques.*

The above examples can be extended by adding additional information from an extra factor. The results can then be arranged into a *tensor* (a data cube), and *multiway* methods† can be used for its analysis. The multiway methods can be circumvented by *unfolding* the obtained tensor along a chosen dimension, obtaining a matrix (however, this approach performs worse). Three often concerned examples are as follows:

1. 20 compounds were chromatographed with 30 mobile phases, each mobile phase on 10 columns. The resulting retention times can be arranged as a tensor with dimensions 20 × 30 × 10. For typical multivariate treatment, the cube can be unfolded into 20 × 300 matrix (the compounds are the objects), 30 × 200 matrix (the phases are the objects), or 10 × 600 matrix (the chromatographic columns are the objects).

2. 20 compounds were chromatographed on 30 systems. Chromatograms were registered during 20 minutes with 0.1 second step (12,000 time points). For each time point, a DAD spectrum (100 points) was obtained. Then, each chromatogram is a 12,000 × 100 matrix and these matrices can be stacked as a three-way 20 × 12,000 × 100 cube.

3. If the above example is repeated on 10 columns, the data can be arranged as a *four-way* array (20 × 1200 × 100 × 10). This dataset can be considered as a *hypercube* (or a set of cubes).

7.2   BASIC OPERATIONS ON MATRICES

Besides the aforementioned transposition, matrices (denoted by bold uppercase letters in mathematical notation) can undergo many mathematical operations, often used in chemometrics. In chemometric convention, the main dataset is, in most cases, denoted as $\mathbf{X}$ (with dimensions $n \times p$), where the number of rows is denoted as n (number of objects) and the number of columns is denoted as p (properties). The transposed matrix is referred to as $\mathbf{X}^T$ (occasionally as $\mathbf{X}'$) and this matrix has switched dimensions ($p \times n$). Performing transposition again, one can obtain back the original matrix: $(\mathbf{X}^T)^T = \mathbf{X}$. The notation $x_{i,j}$ or $X_{i,j}$ refers to one cell (element, entry) in the ith row and jth column. In this notation, a transposed cell can be noted as $(\mathbf{X}^T)_{i,j} = X_{j,i}$. A special case of a matrix is

---

* See also Chapter 10 for further details on warping and Chapter 19 for the details on fingerprinting.
† See Chapter 23.

a vector (a series of numbers). It can be treated as a matrix with one column or one row, depending on the context. A vector is denoted with a small bold-cased letter, for example, $\mathbf{y}$, with dimensions $(n \times 1)$ for a column vector and $(1 \times p)$ for a row vector.

Addition of matrices is possible only if they have the same size, and the result is then calculated elementwise, that is, $(\mathbf{A} + \mathbf{B})_{i,j} = A_{i,j} + B_{i,j}$. The matrix can be multiplied by a scalar (a number), and this operation multiplies all entries of the matrix by this number.

Multiplication of matrices is more complicated. Two matrices can be multiplied only if the number of columns in the first is the same as the number of rows in the second. If $\mathbf{A}$ is an $n \times p$ matrix and $\mathbf{B}$ is a $p \times m$ matrix, the product $\mathbf{AB}$ is an $n \times m$ matrix. Each element in the resulting product is the *dot product* of the corresponding rows of $\mathbf{A}$ and $\mathbf{B}$: $AB_{i,j} = A_{i,1}B_{1,j} + A_{i,2}B_{2,j} + \cdots + A_{i,p}B_{p,j}$. The matrix multiplication is not commutative: rectangular matrices cannot be reversely multiplied. In the case of non-square matrices, they can be reversely multiplied ($\mathbf{BA}$ is then not possible), while square matrices can, but yield different results ($\mathbf{AB} \neq \mathbf{BA}$). However, if several matrices are multiplicable, the order of multiplication does not make a difference: $(\mathbf{AB})\mathbf{C} = \mathbf{A}(\mathbf{BC})$. The multiplication is also distributive: $(\mathbf{A} + \mathbf{B})\mathbf{C} = \mathbf{AC} + \mathbf{BC}$, or $\mathbf{C}(\mathbf{A} + \mathbf{B}) = \mathbf{CA} + \mathbf{CB}$.

Any matrix can be multiplied by its transpose, and the result is a square matrix called a *cross product*. Two cross products can be obtained from any matrix $\mathbf{X}$ of dimensions $n \times p$: the product $\mathbf{X}^T\mathbf{X}$ of dimensions $p \times p$, called the *outer product*, and the matrix $\mathbf{XX}^T$ of dimensions $n \times n$, called the *inner product*. The cross products have special properties from the point of view of data analysis, as their cell values are proportional to variance (on diagonal) and covariance (off diagonal). The cross product matrix is always a *symmetrical* matrix, that is, it is equal to its own transpose $\mathbf{A} = \mathbf{A}^T$.

7.2.1 Matrix Inversion

The scalar inversion $1/x$ yields a number, which multiplied with the original x leads to unity. This concept can be extended to matrices. The inverse of a square matrix $\mathbf{X}$, denoted as $\mathbf{X}^{-1}$, is also a square matrix. Multiplying the matrix by its inverse results in a matrix analogue to "unity": the *identity* matrix, denoted as $\mathbf{I}$. This matrix is a square *diagonal* matrix, which means that all values besides the diagonal are zeros. All diagonal values of $\mathbf{I}$ are equal to 1. This operation is (exceptionally) commutative $\mathbf{XX}^{-1} = \mathbf{X}^{-1}\mathbf{X} = \mathbf{I}$.

However, not every square matrix is *invertible*. Some of them do not have the inverse, that is, a matrix satisfying the above concept simply does not exist. The noninvertible matrices have a special mathematical property: their *determinant* is equal to zero. The determinant is a number that can be calculated from any square matrix. As the determinant of the transposed matrix is equal to that of the original matrix, the transposal of invertible matrix is also invertible. The inverse of the transpose is the transpose of inverse: $(\mathbf{X}^T)^{-1} = (\mathbf{X}^{-1})^T$. Moreover, the determinant of the inverse is the inverse of determinant: $\det(\mathbf{X}^{-1}) = 1/\det(\mathbf{X})$. That's why there is no inverse of matrix with the determinant equal to zero.

If a matrix is not invertible, one can say that it is *singular* or *degenerate* (on the contrary, invertible matrices are called *nonsingular* and *nondegenerate*). To give a practical chemometric example, the outer cross product (mentioned earlier) of a non-full-rank matrix (introduced later) is always singular. If columns of a matrix are linearly dependent, such matrix is also noninvertible. A matrix can be also *almost* singular: it means that the determinant of such a matrix is very small (close to zero) and the inverse is biased by a huge error raised from the computational (in)accuracy.

7.2.2 Matrix Pseudoinverse

The idea of the inverse can be extended to nonsquare (rectangular) matrices. In this case, not all properties of the inverse matrix can be satisfied, so it is called the *pseudoinverse* and denoted as $\mathbf{X}^+$. A pseudoinverse has the same dimensions as the transpose of the original matrix and satisfies only one of the two above equations: $XX^+ = I$ or $X^+X = I$; the other does not yield the unity matrix. The pseudoinverse can be computed as $X^+ = X^T(XX^T)^{-1}$ if the cross product XX^T is invertible; else other methods exist (e.g., the singular value decomposition discussed later).

7.3 THE GEOMETRICAL POINT OF VIEW

The analysis of multivariate data is based on a geometrical assumption, that the objects can be seen as points in a multivariate space with the dimensionality equal to the number of properties. One can imagine or visualize a dataset with two properties as the points in a two-dimensional planar coordinate system (very often located in the first quarter of this system, when all values are positive). This imagination or visualization can be extended to a 3D space as well as to a space of *any* dimensionality, though in this case it cannot be visualized anymore.

When *signals* or profiles are recorded (chromatograms, DAD spectra, fluorescence spectra), the obtained dataset is a *finite* approximation of a function (a continuous signal is *sampled* with a chosen raster or frequency). Such profile is then also a point in a multivariate space, but of a very high dimensionality. Theoretically, such signal is a very complex mathematical function, being a point in the *infinite-dimensional* space, but we analyze only a finite-dimensional approximation. It is very important that neighboring points have a similar signal, so they are intercorrelated (multicollinear). Therefore, a restricted placement of all signals in the multivariate space occurs. It allows *dimensionality reduction*, which will be explained later.

7.3.1 DISSIMILARITY MEASURES

The multivariate treatment allows defining many measures of similarity (or dissimilarity) between objects. The simplest parameter is the Euclidean distance. The distance between two points in a two-dimensional plane or in a three-dimensional space can be measured. Such distances exist in a space of any dimension, but again they cannot be visualized anymore. The distance between any two objects in the full multivariate space can be computed. The obvious property of such distance is that it is not sensitive to other objects—adding, removing, or changing the position does not affect the considered distance, which is constant.

Each object can be represented not only as a point but also as a vector pointing from the origin to that point. The direction of this vector represents some *proportion* (ratio) of the properties and the relation between them. All samples obtained by dilution or concentration of a raw sample are represented by vectors of the same direction, but different lengths (as the proportions between the constituents are not changed). As a measure of similarity, one can consider the *cosine* of the angle between two vectors. It can be computed regardless of space dimensionality (as the angle is always in the range of $0°–180°$). The angle does not depend on the vector length. Therefore, it can be used to compare composition (mixture properties) of the samples, neglecting the differences being caused by dilution or concentration.

A third measure of similarity between two objects (samples) is the correlation coefficient between their multivariate profiles. The properties of this measure are similar to angular measures, as only the proportions (ratio) of properties are taken into account.

7.3.2 SPANNING AND SUBSPACES

Let us consider the case where two samples are mixed in different ratios. For the full extension of the idea, let us allow that they can be added also in *negative* amounts and concentrations. It is not a silly concept, as in practice the datasets are *centered* and they are situated *relative* to some center point, which becomes their origin. Mixing two solutions, each consisting of two substances *in the same* proportions, we cannot obtain a mixture with different proportions of the constituents. For example, mixing a 10 % solution of two substances (equal concentrations) with 50 % solution of the same substances (also equal concentrations), one cannot obtain a solution with 20 % of the first compound and 30 % of the second. All possible mixtures are located on the straight line passing through the origin in the same direction represented by both vectors. This example is a simple case of a *subspace*. The line is a one-dimensional subspace of the two-dimensional full space, and these two vectors *span* this subspace. The other ratios of the compounds cannot be obtained because the vectors *do not span* the full space.

However, by taking any two vectors of *different* composition (vectors pointing in different directions), any mixture can be obtained by mixing, concentrating, and diluting (again, including negative concentrations and amounts). It can be said that the two vectors of distinct directions in two-dimensional space *span* the whole two-dimensional space.

This important property can be generalized to any dimensionality. In three-dimensional space, an infinite number of one-dimensional subspaces (straight lines passing through origin) and two-dimensional subspaces (planes passing through origin) exist. In higher-dimensional spaces, an infinite number of subspaces of any smaller dimensionality exist. However, having a set of k vectors in an n-dimensional space, the full space cannot be spanned if $k < n$. The maximum possible dimensionality of a subspace is equal, then, to k.

Two solutions of three constituents in different ratios (two vectors with different direction in a three-dimensional space) span only a two-dimensional subspace (plane), thus a mixture outside this plane cannot be obtained. In a very special case, when two vectors are pointing in the same direction, they can span only one-dimensional subspace and all possible mixtures occupy only the straight line.

7.3.3 MATRIX RANK

The *rank* of a matrix is defined as the maximum number of its linearly independent row vectors. The matrix can be of *full rank* (when the rank is equal to the number of columns) and then the vectors of the matrix span the whole space. Full-rank matrices can be *square* (equal number of columns and rows) or *tall* (more rows than columns). If a matrix is *wide* (more columns than rows), the vectors never span the whole space, but they can span *at most* a subspace with dimensionality equal to number of rows. When matrix has a *reduced rank*, all its vectors span only the subspace of this (reduced) dimensionality.

Regardless of the mathematical rank definition, chemometricians often use the concept of *chemical rank* of the matrix. In a high dimensional space, a signal of a sample can be a mixture of a small number of vectors (representing the constituents). Although a set of samples spans a space of more dimensions, all variations in these additional dimensions are the result of measurement errors or some instrumental artifacts. In this case, as the relevant information is placed along other directions than irrelevant information and noise, it can easily be filtered. From a mathematical point of view, the "other directions than a given subspace" is called the *orthogonal complement* of that subspace.

7.3.4 ORTHOGONAL COMPLEMENT

The orthogonal complement of a k-dimensional subspace in an n-dimensional space is also a subspace, with dimensionality $n - k$. For example, on a plane, the orthogonal complement for any linear subspace (a line passing through the origin) is another perpendicular (orthogonal) line passing through the origin (of course, there is the only one perpendicular line possible). In a three-dimensional space, the orthogonal complement of a line is a perpendicular plane and vice versa. This rule can be extended to any dimensionality. If a set of objects occupies a subspace, it is said that there is no information in the subspace being the orthogonal complement. The orthogonal complement to a subspace spanned by a matrix of non-full rank is called the *null space* of this matrix.

7.4   PROJECTION ONTO A SUBSPACE

Consider a set of points in a two-dimensional coordinate system. It is obvious that each point has an x and a y value. The axes of this coordinate system are special cases of one-dimensional sub-spaces (lines passing through the origin, see Figure 7.1). Moving the point to an axis perpendicularly (setting the other coordinate to zero) is a special case of the *orthogonal projection* to this one-dimensional subspace. The distance to this projected point from the origin (in this special case, equal to the coordinate) can be defined as a *score* of this projection. The projection to a subspace is a way of *dimensionality reduction*. One obtains one variable (score of the projection) starting from two variables. The projection is *lossy*, as the information lying on the orthogonal complement to this subspace (in the considered example—the other coordinate) can be perceived as some residual and is lost. Of course, it is possible to perform projection to any axis (any direction, any subspace).

The points can be projected onto an axis rotated by a given angle. Then, another variable is obtained, which is a linear combination of the two original variables. The points projected to the new line occupy this line and all lie inside this subspace. They still have two coordinates, but the coordinates are linearly dependent, as the information on the perpendicular direction was lost. The vectors associated with the projected points span only a one-dimensional subspace, and the matrix of projected points has the rank equal to 1.

The idea can be extended to a three-dimensional subspace (Figure 7.2). In this case, a point can be projected onto one axis and occupy a one-dimensional subspace having one score. The information connected with the distance of the point from this line (and direction of that distance) is then lost, as it is located in the orthogonal complement of this subspace. The matrix of the coordinates of the projected points, although formed of three columns, still has rank equal to one. The other possibility is to project the points onto a plane (a two-dimensional subspace). Then, two scores (the coordinates of the projected points on this plane) are obtained and the information on the orthogonal complement (the distance from that plane) is lost. The matrix of the coordinates of projected points is of rank 2, and the projected objects represent vectors spanning this two-dimensional subspace.

Generally speaking, a cloud of points in an n-dimensional space can be projected onto any k-dimensional subspace (considering $k < n$). This results in "lossy" data *compression* (dimensionality reduction) from n original variables to k scores. The information spanning the $(n - k)$-dimensional orthogonal complement is lost. Figure 7.3 shows the illustration of orthogonal complement in the considered examples.

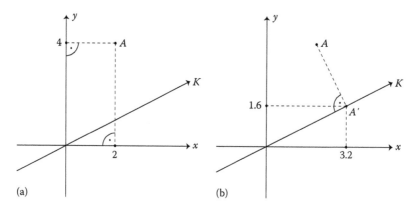

FIGURE 7.1   Two-dimensional example of projection: (a) the original coordinates (2,4) are the scores of projection of the point onto the directions (subspaces) represented by the axes. (b) Projection onto the direction represented by line K changes both x and y coordinates of the projected point.

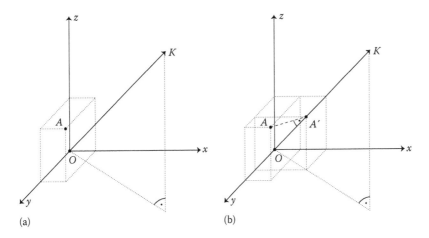

FIGURE 7.2 Example of the projection of three-dimensional point A onto the one-dimensional subspace represented by the line K. (a) Original coordinates of the point, (b) the change of the coordinates during projection (A' is the projected point).

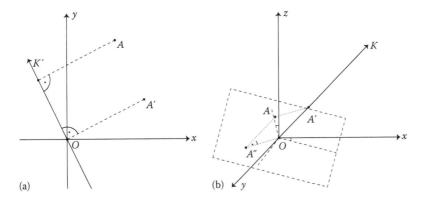

FIGURE 7.3 Orthogonal complement: (a) in a two-dimensional space, the orthogonal complement to line K (see Figure 7.2) is a perpendicular line K'. (b) In a three-dimensional space, the orthogonal complement is an orthogonal plane.

In multivariate data analysis, chemometricians seek for specially positioned subspaces to maximize the compressed information and to minimize the lost information. This can only be done if the data is compressible, and the most popular approach for defining such subspaces is principal component analysis.*

From the mathematical point of view, scores can be obtained by multiplying the data matrix by a matrix containing k orthogonal vectors of unit length. For instance, the vector $\mathbf{v} = (0.707,0.707)$ has the length equal to 1 (its length can be computed using the Pythagorean theorem), and it represents a subspace being a line with the angle equal to $45°$ to both x- and y-axes. Arranging the coordinates of a set of points into $n \times 2$ matrix, then multiplying this matrix by the transposed above vector (i.e., a 2×1 column vector), an $n \times 1$ vector is obtained, containing the scores (seen as the "coordinates" on the rotated axis). To convert these projected points back to the original space (i.e., to obtain coordinates of the projected points), one has to multiply the score vector by the pseudoinverse of the projection vector. It switches the data back to the $n \times 2$ matrix, but then this matrix contains the

---

* See Chapter 8.

two-dimensional coordinates of the projected points. The information passed through a "bottleneck" of a reduced dimension, so the information on the orthogonal complement was lost.

Look at the whole process again: it can be written in matrix notation as $\mathbf{XVV^+}$. When the scores are not needed, the whole process can be achieved in one step, by one multiplication by square *projection matrix* $\mathbf{P = VV^+}$. This matrix is not an *identity* matrix (in this case, $\mathbf{V^+V}$ would be). The matrix P is symmetric and also *idempotent* : $\mathbf{PP = P}$, which means that when making the projection many times the dataset is not further changed.

When making a projection onto a multidimensional subspace, the matrix $\mathbf{V}$ must contain an *orthonormal vector base* of this subspace. It is a set of orthogonal vectors spanning that space, with lengths equal to one. Only in this case the scores can be obtained, and they represent the coordinates in a coordinate system embedded in this subspace. However, if scores are not important, the orthogonality is not needed, because the matrix $\mathbf{P = VV^+}$ is exactly *the same* for any $\mathbf{V}$ spanning this subspace.

7.4.1 Centering as a Projection

A special case of projection is the *centering*, a very important step in the data preprocessing. It can be perceived as the transfer of the origin of the coordinate system to the middle (center) of the data cloud. The centered matrix can be obtained by subtracting the *column mean* from each element of this column. However, the same result is achieved by the projection: multiplying a dataset by the matrix $\mathbf{C = I} - (1/n)\mathbf{O}$, where $\mathbf{I}$ is the identity matrix and $\mathbf{O}$ the square matrix containing only elements equal to one. Centering is therefore a projection to an $(n - 1)$-dimensional subspace and the lost information is located in the one-dimensional orthogonal. In other words, the centering decreases the rank of a dataset by one.

7.4.2 Model and Orthogonal Distances

As it was stated earlier, the original data point can be represented by a vector. The projected data point can also be represented by a vector. The angle between the original and the projected vector is always in the range of 0°–90°. The length of the projected vector, called *model distance*, can be computed by multiplying the length of the original vector by the cosine of this angle, called the *angle between the object and the subspace*. The vector positioned between the projected point and the original point represents the lost information. Its length is proportional to the sine of this angle

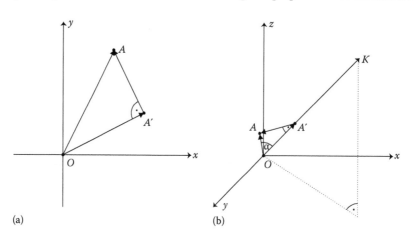

(a)   (b)

FIGURE 7.4 Idea of model distance and orthogonal distance in two (a) and three (b) dimensions. The original vector OA, when projected onto a subspace K, preserves a part of information becoming a vector OA'. The vector $A'A$ represents the orthogonal complement information (orthogonal distance), which is lost during the projection.

and it is called the *orthogonal distance*. If the angle between the object and the subspace is equal to 90°, the projected point is located in the origin and all information is lost (the orthogonal distance is equal to the original object distance). In the opposite situation, when this angle is equal to zero, all information is preserved (the projection does not change the position of a point)—the model distance is equal to the object distance and the orthogonal distance is equal to zero, as no information is lost. Figure 7.4 contains the illustration of the above idea.

7.5   MATRIX FACTORIZATIONS

The factorization of a matrix is its decomposition into a product of several matrices. A matrix can be factorized in an infinite number of ways (similar to the same principle that a number can be the product of many number pairs). However, only some special cases are interesting from the chemometric point of view.

7.5.1   ROTATION AS FACTORIZATION

The matrix $\mathbf{X}$ can be represented as the product $\mathbf{X} = \mathbf{TP}$, where $\mathbf{P}$ is an orthonormal matrix, containing orthogonal vectors of unit length, spanning the whole space. This set of vectors can be seen as a *rotated* coordinate system, while this matrix for unrotated coordinates is equal to identity matrix $\mathbf{I}$. Since the inverse of an orthonormal matrix is equal to its transpose, the coordinates of points in any rotated coordinate system can be obtained by a simple multiplication, $\mathbf{T} = \mathbf{XP}^{-1} = \mathbf{XP}^{\mathrm{T}}$. As the rotated system has the same dimensionality and also *spans* the same dimensionality, $\mathbf{P}^{\mathrm{T}}\mathbf{P} = \mathbf{I}$. The coordinate system can be rotated in an infinite number of ways; however, there is only one way that maximizes the information in subsequent rotated variables (compresses the data). This coordinate system can be derived by *singular value decomposition*, the step being a base of the *principal component analysis** (loadings and scores can be seen as axes and coordinates in the rotated system).

7.5.2   SINGULAR VALUE DECOMPOSITION

Any matrix $\mathbf{X}$ with dimensions $n \times p$ can be presented as the product of three matrices $\mathbf{X} = \mathbf{UDV}^{\mathrm{T}}$, satisfying the following criteria: $\mathbf{U}$ is an $n \times n$ real orthonormal matrix, $\mathbf{D}$ is an $n \times p$ rectangular diagonal matrix (with nonnegative diagonal values), and $\mathbf{V}$ is a $p \times p$ orthogonal matrix. Such decomposition is unique (there is only one set of matrices satisfying these properties), and their finding is called singular value decomposition.

The diagonal entries of the matrix $\mathbf{D}$ are called the *singular values* of $\mathbf{X}$, and they are sorted in descending order. These values are related to the variance of independent (orthogonal) trends in the dataset. If any singular values are equal to zero, the matrix is not of full rank, as it can be decomposed to a smaller number of independent trends than its dimensionality.

The columns of $\mathbf{U}$ and $\mathbf{V}$ are called the *left-singular vectors* and the *right-singular vectors*, respectively. The right-singular vectors represent the coordinate system, rotated in the optimal manner to compress the data: the first k right-singular vectors span a k-dimensional subspace, preserving as much information as possible in k dimensions. The singular value decomposition is thus a special case of the rotation decomposition $\mathbf{X} = \mathbf{TP}$, where $\mathbf{P} = \mathbf{V}$ and $\mathbf{T} = \mathbf{UD}$.

7.5.3   EIGENVALUE DECOMPOSITION

A *diagonalizable* square matrix $\mathbf{X}$ can be decomposed as the unique (one possible solution) product $\mathbf{X} = \mathbf{Q\Lambda Q}^{-1} = \mathbf{Q\Lambda Q}^{\mathrm{T}}$, where $\mathbf{\Lambda}$ is a diagonal matrix containing the *eigenvalues* of the matrix (on its diagonal) and $\mathbf{Q}$ is an orthonormal matrix containing the *eigenvectors* of this matrix.

---

* See Chapter 8.

Not all matrices are diagonalizable, for example, a rotation matrix, mentioned earlier, is not. The eigenvalue decomposition is connected with the singular value decomposition: the singular vectors of a matrix are the same as the eigenvectors of its cross products: $\mathbf{XX^T = UDV^TVDU^T = UD^2U^T}$ (as $\mathbf{V^TV = I}$), $\mathbf{X^TX = VDU^TUDV^T = VD^2V^T}$. From these expressions, one can also derive a conclusion that the singular values of a matrix are the square roots of the eigenvalues of the cross product. The matrix determinant is the product of its eigenvalues, so any eigenvalue equal to zero makes the determinant equal to zero and indicates a singular matrix.

7.6   CONCLUDING REMARKS

The information given in this chapter should be treated as a basis of background knowledge before reading about different chemometric methods in detail, which are provided in the other chapters of the book. However, chronological reading of the chapters is not mandatory and depends on the knowledge of the reader. In many cases, a correct understanding may require switching between chapters to improve the degree of understanding.

8 Introduction to Exploratory and Clustering Techniques

Ivana Stanimirova and Michał Daszykowski

CONTENTS

8.1 INTRODUCTION

Nowadays, studying a given problem or phenomenon is done indirectly and involves the process of collecting a considerable amount of evidence. This evidence is usually gathered through a sampling process or by means of an experiment that is designed in the hope that the analytical data that are obtained will support the exploration of the studied objects and that their analysis will eventually lead to drawing meaningful conclusions. Therefore, from this perspective, the information content and its efficient analysis are crucial.

Among the available instrumental techniques, chromatographic methods have gained considerable attention. They are valued mostly for the possibility to separate the components of a mixture, as well as their ability to detect and quantify. Chromatographic methods have the potential to describe samples in a comprehensive manner. This can be achieved by using different types of detectors, performing separation in diverse chromatographic systems, by taking advantage of orthogonal chromatographic systems, or by means of hyphenation. The analytical data that are obtained in the course of the chromatographic separation process can be presented as a peak table (containing the peak areas or concentrations of the selected chemical components) or as a collection of chromatographic signals. Regardless of the end form, chromatographic data are multivariate and complex. That is why studying the data structure in order to unravel the relations among the samples and/or variables and examining the effect of the variables are challenging tasks. These tasks can be facilitated with the use of different multivariate techniques, which are well suited for data exploration. They are called exploratory techniques and extend the toolbox of available methods of multivariate data analysis considerably [1].

Exploratory techniques assist in uncovering groups of similar objects, local fluctuations of data density, and revealing the presence of atypical objects. Obtaining a simplified data representation, which is much easier to analyze and interpret, is possible either by the construction of low-dimensional data projections or by performing data clustering, that is, by the effective grouping of objects that have similar chemical characteristics. While both types of approaches differ substantially in how useful information can finally be inferred, in fact, they share the same objective. It is important to stress that in order to accomplish the objectives, exploratory techniques do not require additional knowledge about data, unlike supervised techniques, for example, the classification/discrimination methods and regression methods.

In this chapter, we discuss two types of approaches that are widely used in the exploratory analysis of multivariate chromatographic data. Although the catalog of exploratory methods is much larger than the few techniques presented here, our greater attention toward these particular techniques was mostly motivated by their popularity and their wide applicability in separation sciences.

8.2 PREPROCESSING OF MULTIVARIATE DATA

Multivariate data often require preprocessing prior to their further analysis and multivariate modeling. The aim of preprocessing methods is to remove undesired effects from the collected data that could hamper data interpretation [2]. It can be applied to (1) individual samples, (2) individual variables, or (3) simultaneously to all of the data elements. Preprocessing that focuses on transforming individual samples, in general, enhances the signal-to-noise ratio by eliminating background and noise components, correcting the peak shifts that are observed in chromatographic signals [3] or compensating for the effects of sample size using different normalization procedures [4]. Variable scaling accounts for any differences in the observed variance, which involve standardization or autoscaling. Transforming data elements is performed when, for instance, a logarithm or power transformations are believed to diminish heteroscedasticity [5].

The list of different preprocessing methods is very long. In many situations, their use requires experience and intuition, but the impact of the applied transformations on the data analysis and data interpretation can be large because the correlation structure of data is affected. Sometimes, combining certain preprocessing methods may indeed have a positive effect. Therefore, they should be used with caution and will eventually lead to the meaningful interpretation of the results.

8.3 PROJECTION METHODS

The projection methods are based on the concept of low-dimensional projections of multivariate data. Assuming that their construction is effective, multivariate data can be displayed and visually assessed. The notion of "effective" can have different meanings depending on the problem at hand. Therefore, in general, projection methods can be considered as a group of bilinear methods that search in a multivariate data space a few directions that describe possible specific structural features as well as possible [6].

8.3.1 PROJECTION PURSUIT

Projection pursuit is an exploratory technique whose purpose is to identify a few "interesting" directions in the multivariate data space [7]. Depending on the anticipated definition of an "interesting direction," projection pursuit provides different solutions in terms of the information content. For instance, it is possible to look for directions that reveal a low-dimensional data projection, their clustering tendency, or to uncover the presence of outlying samples to the largest possible extent. In the course of projection pursuit, interesting directions are usually found iteratively by maximizing the so-called projection index, for instance, entropy, kurtosis, or the Yenyukov index [8,9].

In general, these indices measure any departures from the normal distribution of a projection by assuming that the projection that follows a normal distribution is the least interesting one.

Projection pursuit can be regarded as a decomposition model that can characterize multivariate data by a set of f score vectors, $\mathbf{T}$, and loading vectors, $\mathbf{P}$:

$$\mathbf{X}_{[m,n]} = \mathbf{T}_{[m,f]}\mathbf{P}^{\mathrm{T}}_{[f,n]} + \mathbf{E}_{[m,n]} \tag{8.1}$$

In the columns of matrix $\mathbf{T}$, there are f score vectors that maximize the value of a certain projection index, whereas in the columns of matrix $\mathbf{P}$, there are f orthonormal loading vectors that define interesting directions in the multivariate data space. Matrix $\mathbf{E}$ contains the residuals corresponding to the part of data variance that were unexplained with the f first components as elements.

8.3.2 PRINCIPAL COMPONENT ANALYSIS

Principal component analysis (PCA) is a projection method that maps samples, which are characterized by a relatively large number of physicochemical parameters, into a low-dimensional space [10]. The low-dimensional space is spanned only by a few new latent variables, which are called principal components. They are constructed to describe as much of the data variance as possible and are linear combinations of the explanatory variables. Bearing these assumptions in mind, the major advantage of PCA is associated with its data dimensionality reduction feature, which opens the possibility of data visualization in the space of the major principal components.

Like the projection pursuit method, PCA is a bilinear model whose objective is to maximize the description of the data variance. It decomposes the multivariate data into a set of f principal components. These are the f vectors of the scores, $\mathbf{T}$, and loadings, $\mathbf{P}$. The columns of matrix $\mathbf{T}$ are constituted by a set of the f orthogonal vectors of the scores, whereas the columns of matrix $\mathbf{P}$ are formed by the f orthonormal vectors of the loadings. Matrix $\mathbf{E}$ contains the residuals that correspond to the part of data variance that was not modeled using the PCA model with the f first principal components as elements.

A property of the mutual orthogonality of principal components is essential because it guarantees the largest possible compression. In this manner, each principal component describes different information.

Scores are the result of the orthogonal projection of the m samples that are described by the n explanatory variables onto the f loading vectors that define the directions in the multivariate data, thus maximizing their variance:

$$\mathbf{T}_{[m,f]} = \mathbf{X}_{[m,n]}\mathbf{P}_{[n,f]} \tag{8.2}$$

On the other hand, f loading vectors can be obtained by projecting n explanatory data variables onto space of f score vectors:

$$\mathbf{P}_{[n,f]} = \mathbf{X}^{\mathrm{T}}_{[n,m]}\mathbf{T}_{[m,f]} \tag{8.3}$$

The PCA model decomposes the information that is contained in the multivariate data and allows the relationships among the samples ($\mathbf{T}$) and the relationships among the explanatory variables ($\mathbf{P}$) to be studied. Therefore, principal components are actively used to visualize and interpret the data structure.

The success and relatively wide popularity of PCA as an exploratory and compression method depends on the level of mutual correlation that is observed among the explanatory variables. Usually, chemical data contain a large number of strongly correlated variables, and thus their compression into a few principal components that summarize the majority of the data variance is straightforward.

It is also interesting to mention that maximizing the data variance is equivalent to minimizing the total sum of squared perpendicular distances that are drawn from the data samples points to the line with the direction that is defined by the corresponding loadings [1]. From the mathematical point of view (rank of data matrix), "wide" data (with a larger number of explanatory variables than the number of samples) always contain a certain number of correlated variables. These can effectively be represented by their linear combinations without a substantial loss of information. In many practical situations, chemical data, including chromatographic data, are composed of a very large number of variables. For instance, the chromatographic fingerprints that are obtained from monochannel detectors are usually constituted from hundreds to dozens of hundreds of variables, depending on the assumed sampling rate and the maximal elution time. When it is feasible, chromatographic data can also be represented by a peak table. In this case, although the number of variables is usually smaller than the number of samples, one can also profit from the presence in multivariate data correlation structure and use the PCA model to summarize and explore the data structure. In many situations, the principal components that are obtained from PCA are considered to be the end product and are used to interpret the structure of the multivariate data. However, due to their properties, the score vectors often serve as input data to other methods that cannot handle correlated variables. For instance, a subset of significant score vectors can serve as a substitute for the original explanatory variables, for instance, in principal component regression or as inputs into a neural network, thereby reducing the number of neurons in the input layer and considerably limiting the complexity and efforts that are necessary to train a network.

8.3.2.1 Visualization of Data Structure: Analysis of Similarities

Visualization of the multivariate data structure is done using the selected pairs of principal components (score and loading vectors) and the construction of their low-dimensional projections. Assuming that the relevant chemical information is distributed along the directions that maximize the description of the data variance, usually a few first principal components are the subjects of interpretation. Owing to the separation of the chemical information that is offered by the PCA model, the analysis of the score and loading projections assists in exploring any similarities observed among the samples and explanatory variables.

The projection of the scores indicates the distribution of the samples in the space of the principal components. The proximity of the samples on the projection, in terms of the Euclidean distances, depends on their chemical similarity, which is manifested by the levels of the chemical components. Therefore, samples that are close to each other on the score projection can be regarded as similar. The projection of the pairs of loading vectors reveals the relationships among the explanatory variables. Assuming that the values of the loadings on certain principal components define the end of a vector with the beginning at the origin of the coordinate system, the angle observed between the pairs of vectors provides information about the strength of their correlation, whereas their orientation provides information about their relation.

With the increasing amount of data variance that is explained by the principal components, the confidence that is placed on the interpretation of the score and loading projections grows substantially. In addition to the analysis of similarities among the samples and among the variables, it is useful to interpret the observed structures on the score projections in terms of their explanatory variables. In other words, the simultaneous analysis of the score and loadings projections, to a certain extent, enables the identification of the explanatory variables that have impact on forming clusters of the similar samples that are revealed on the score projections.

8.3.2.2 Influence of Samples on the PCA Model: The Distance–Distance Plot

Depending on the location of the samples in the space of the principal components, their impact on the PCA model can be different. Samples that are close to the data majority and in the model space are characterized by short Mahalanobis distances and small residuals (orthogonal distances). Therefore, they can be regarded as regular samples that have virtually no influence. Samples that have small residuals and large Mahalanobis distances extend the range of the PCA model and there-fore are called good leverage samples. Regular and good leverage samples have little impact on the PCA model in contrast to high residual and bad leverage samples. High residual samples have short Mahalanobis distances, but they do not fit the PCA model. Bad leverage samples have the largest influence on the construction of a PCA model, since they are far from the data majority and they have large residuals. These four types of samples can be diagnosed using a so-called distance–distance plot, which is constructed based on the parameters of the PCA model that is built with a certain number of principal components.

8.3.3 ROBUST PRINCIPAL COMPONENT ANALYSIS

The construction of principal components is affected by the presence of outlying samples because the PCA model identifies the directions of the largest data variance and the variance is a nonro-bust estimator of scale [11]. Outlying samples, in general, do not follow a similar tendency as that observed for the data majority. Even the presence of a single outlying sample in the data can strongly influence the parameters of a PCA model, and thus it cannot describe the data majority well. The impact of such an atypical observation on the construction of a PCA model can be different depend-ing on its location in the multivariate space. Therefore, in the literature, robust variants of the PCA model were proposed in order to diminish their negative influence [12,13]. These enable the con-struction of robust principal components, that is, that are insensitive to the presence of the outliers in data. This effect can be achieved by (1) identifying the outlying samples followed by constructing a classic PCA model, (2) down weighting the influence of the outlying samples, and (3) optimizing a robust scale as a cost function in robust PCA.

8.4 CLUSTERING METHODS

Projection methods assist in exploring the structure of multivariate data by revealing regions with similar objects on low-dimensional data projections. However, grouping objects is based on a sub-jective judgment—the visual assessment of data projections. Alternatively, the purpose of cluster-ing methods is to group similar objects, usually samples, in an objective manner. As an outcome of a clustering procedure, a membership list of the objects is returned that defines the assignment of each object to a given group. The process of clustering is based on the fundamental assumption that objects within a group are more similar to each other than to any object that is assigned to other groups [14,15].

Depending on problem at hand, data clustering can be oriented toward three major aspects [16]. These are related to (1) selecting representative samples that form a compact data representation, (2) assessing the complexity/diversity that is described by the multivariate data, and (3) identifying the variables that carry similar information.

From the practical point of view, the result of clustering depends on two factors, namely, the definition of a similarity measure and the assumed mechanism of the linkage of the objects. Bearing in mind the subject of clustering (samples, variables) and the type of data being clus-tered, the chemical similarity among objects can be assessed differently. To date, in the literature there are over 60 similarity measures that have been described, among which the Euclidean, Mahalanobis distance, and the correlation coefficient are the most popular [17,18]. Although both the terms "similarity" and "distance" are often used as synonyms of similarity in the chemical sense, their mathematical properties are different. The similarity score, s_{ij}, which is obtained for two objects i and j, is a real number and fulfills three conditions:

1. $s_{ij} = s_{ji}$ $1 \leq i, j \leq m$
2. $0 \leq s_{ij} \leq 1$ $1 \leq i, j \leq m$
3. $s_{ii} = 0$ $1 \leq i \leq m$

Any distance measure between two objects i and j, d_{ij}, satisfies the following properties:

1. $d_{ij} = d_{ji}$ $1 \le i, j \le m$
2. $d_{ij} \ge 0$ $1 \le i, j \le m$
3. $d_{ii} = 0$ $1 \le i \le m$

The Euclidean distance expresses the closeness of two objects in a multivariate space. It is invariant, that is, the distances among the data objects are preserved when the multivariate data are reflected, shifted, or rotated (invariant transformation). However, it is scale-dependent in contrast to the Mahalanobis distance, which takes into account the correlation structure. The Euclidean and Mahalanobis distances between two objects x and y, d_{xy}, which are described by two row vectors with n elements, can be expressed using the general formula:

$$d_{xy} = (\mathbf{x} - \mathbf{y}) \cdot \mathbf{Q} \cdot (\mathbf{x} - \mathbf{y})^{\mathrm{T}} \tag{8.4}$$

where matrix $\mathbf{Q}$ is a weighting matrix and the superscript "T" denotes the operation of vector transposition.

For the Euclidean distance, matrix $\mathbf{Q}$ corresponds to the identity matrix, $\mathbf{I}$, with the nonzero diagonal elements equal to one. In order to remove the effect of scale and correlation among the variables, the weighting matrix corresponds to the inverse of the data variance–covariance that leads to the Mahalanobis distance.

While the Euclidean distance is sensitive to any changes in the chemical content, one can also be interested in evaluating the similarities of the chromatographic profiles taking into account their shapes. This can be done using the Pearson's correlation coefficient, r, which expresses the strength of the linear correlation between two vectors. Since it takes the values from the closed interval from -1 to 1, the result that is often obtained is expressed as $1 - |r|$ in order to fulfill the conditions of a similarity measure.

8.4.1 Nonhierarchical Clustering

Nonhierarchical clustering techniques, which are also known as partitioning methods, partition the available data into a number of exclusive groups that are defined by the user, K [15]. Assuming that the groups of objects are formed by maximizing the similarity measure within groups, a typical objective function involves minimizing the within-group variance criterion:

$$e = \sum_{k=1}^{K} \sum_{i \in C_k}^{m} \left\| \mathbf{x}_i - \boldsymbol{\mu}_k \right\|^2 \tag{8.5}$$

where
$\mathbf{x}$ is an object described by n measurements
$\boldsymbol{\mu}_k$ is the center of the kth group

Alternatively, one can also consider other criteria as potential candidates for the objective function. These are, for instance, minimizing the trace of the "within" sums of squares, its determinant, or maximizing the trace of the between to within sums of squares [1].

K-means is a classic representative of the nonhierarchical clustering techniques. It partitions data objects in the course of an iterative procedure. At the start, one defines the desired number of groups, K, and selects K objects that define the position of the initial cluster centers. Then, the closest objects in terms of the applied similarity measure are assigned to each cluster. New centers are estimated as the average of the coordinates, which are based on the actual members of each group. Once again, the membership list is updated and, when necessary, modified by reassigning the objects to their nearest groups (the closest center). This iterative process is continued until no further changes are observed in the membership list. Once the initial centers of the groups are selected, only a few iterations are usually needed to achieve the convergence of the algorithm.

The definition of the objective function causes the results of the partitioning scheme to be sensitive to different shapes of clusters. In particular, it leads to the satisfactory clustering for compact clusters with many objects that form spherical shapes. Therefore, partitioning methods that minimize the objective function defined earlier cannot detect clusters with arbitrary shapes properly. Likewise, it is also possible that the obtained partition can be a suboptimal result as a consequence of the initial conditions. In order to solve this problem, data clustering can be performed a number of times starting with very different distributions of the initial centers of the clusters. Among several possible data partitions, the optimal one will lead to the minimal value of the cost function.

Although the K-means is the best known partitioning method, there are also different ones that minimize the same cost function. These are, for instance, self-organizing Kohonen's maps, neural gas, or growing neural gas [19].

8.4.2 HIERARCHICAL CLUSTERING

Hierarchical clustering methods reveal the structure of multivariate data in the form of a tree, which is called a dendrogram. It is built step-by-step using either the agglomerative or divisive principle. A tree uncovers the different levels of similarity that are observed among the groups of objects. Assuming the agglomerative approach, one starts with the number of clusters that correspond to the number of objects in the data. In the iterative process, consecutive objects or groups of objects are merged with other ones, taking into account any decrease in the dissimilarity measure and introducing a specific order resembling the hierarchy of the data structures. Along the horizontal axis of a dendrogram, a sequence of objects is presented, whereas the vertical axis provides information about the level of dissimilarity. It is important to stress that the clustering process is irreversible. Objects that were linked at a certain step cannot be relocated to other clusters later on.

While the nonhierarchical methods partition data into a number of groups that have been defined by the user, the hierarchical clustering methods group objects in an automatic manner and their number can be inferred by the careful analysis of a dendrogram. The clustering outcome is guided by the selection of two criteria, namely, the similarity measure and the type of linkage (for instance, single, average, complete, and Ward's linkage). Depending on the linkage criterion that is applied, hierarchical methods can be classified as space contracting, dilating, or conserving [15]. Of all of the types of linkage approaches, the single linkage can be useful when data contain elongated clusters. Due to the chaining effect, the single linkage has the potential to process and identify elongated clusters, and therefore it is better suited for detecting clusters that have a natural shape.

The basic hierarchical clustering of multivariate data is concerned with clustering samples or variables. On the other hand, the exploratory possibilities of the hierarchical methods can be extended considerably. By clustering samples and variables, a specific order of samples and variables is established. These can be used to sort the collection of measurements. The dendrograms that are obtained, which display the hierarchy of the samples and variables, can be augmented with a color map (a heat map). Each pixel of the map has a color that is proportional to a particular measurement value that is observed in a data set. Two-way clustering that is enhanced with a color map supports a detailed interpretation of a data structure through the straightforward identification of the variables with the largest contribution.

8.4.3   Density-Based Clustering

Density-based clustering techniques use the data density concept to define groups of objects. In the data density sense, a cluster is a region in the multivariate space where the data density exceeds a certain threshold value. Therefore, clusters are considered to be high data density regions, and they are separated from each other by zones that have a relatively low data density [20].

The major advantage of the density-based clustering techniques is associated with their ability to cluster groups of objects that form arbitrary shapes. This unique functionality is achieved by a specific approach to processing data objects. Namely, starting with an object that is located in a high data density zone, density-based techniques use the agglomerative scheme to extend the cluster by adding additional object(s) that fulfill the data density condition and that are located within the neighborhood of the objects that have already been assigned to a cluster. Data density can be estimated by counting the number of objects found in the neighborhood of a fixed radius. This concept is used in the DBSCAN (density-based spatial clustering of applications with noise) algorithm [20,21]. Alternatively, one can also consider the adaptive neighborhood size, which is modified with respect to the local fluctuations in data density. Such a property can be obtained by taking the neighborhood radius equal to the Euclidean distance to the kth nearest neighbor, as it is done in the OPTICS (ordering points to identify the clustering structure) algorithm [22,23].

8.4.3.1   Density-Based Spatial Clustering of Applications with Noise

In the DBSCAN method, three types of objects are distinguished. Assuming a certain density threshold, that is, the number of objects in the neighborhood of a given object, k, with a fixed radius, e, one can identify the core, border, and noise objects. A core object contains at least k objects in its neighborhood of a specified radius. A border object has less than k objects in its neighborhood, but at least one of these is recognized as a core object. When any object that is found in the neighborhood is a core object, then the considered object is regarded as noise.

The DBSCAN algorithm processes objects one by one, that is, in a single data scan. It starts with finding any core object, which is regarded as the seed of a cluster. Then, a cluster is expanded step-by-step by incorporating the other core and border objects that are reachable by the density condition. Therefore, a cluster is formed by the core and border objects. The results of the clustering obtained from DBSCAN are in the form of a membership list that is associated with information about the core, border, and noise objects.

When data contain adjacent clusters, the classic DBSCAN algorithm may provide an unstable solution in the region between the clusters. A modified DBSCAN approach was proposed to overcome this problem [24].

8.4.3.2   Ordering Points to Identify the Clustering Structure

The DBSCAN algorithm clusters multivariate data with an assumption that neighborhood of each object has a fixed size [22]. Therefore, one expects that the data density is the same regardless of the region in the data space. It is known that clusters may have different densities and that the data density fluctuates to a great extent even in a single cluster. Bearing this in mind, in the OPTICS method, the size of each object neighborhood is different depending on the local data density [22,23]. This is possible by relating the size of a neighborhood with the distance to the kth nearest neighbor. As a consequence, in dense regions of the data space, the size of the neighborhood will be substantially smaller than in low data density regions. In contrast to DBSCAN, the OPTICS method does not explicitly cluster the available data. Like the hierarchical methods, it reveals their structure using the so-called reachability plot. It represents the reachability distances as consecutive bars. Although the first reachability distance is displayed in the reachability plot as an arbitrary large value, the next one corresponds to the reachability distance to its nearest neighbor. The sequence of the way in which the objects are processed in OPTICS and in the nearest neighbor linkage approach are identical—the order of processing data objects is forced by ranking the similarity measure with respect to an already processed object. Therefore, in OPTICS, the same effect of chaining is achieved. This can uncover zones in the data space that contain clusters of arbitrary shapes. The concept of the reachability distance incorporates information about the local data density scored by the so-called core distance, which expresses the distance to the kth nearest neighbor. The reachability distance of the ith object is defined as the maximal value between the core distance of the nearest object and the Euclidean distance to the nearest object. Such a definition of the reachability distance causes a smoothing effect and therefore the plot of reachability distances is easier to interpret. Information about the regions where similar objects are located can be inferred directly from the reachability plot by analyzing the content of the characteristic valleys. The reachability plot reveals any differences in the local data densities of data. In fact, OPTICS can be interpreted as a generalization of DBSCAN to a situation in which the data are scanned with a different neighborhood radius.

REFERENCES

1. Rencher, A.C. and W.F. Christensen. 2012. *Methods of Multivariate Analysis*. John Wiley & Sons, Hoboken, NJ.
2. Engel, J., J. Gerretzen, E. Szymańska, J.J. Jansen, G. Downey, L. Blanchet, and L.M.C. Buydens. 2013. Breaking with trends in pre-processing? *Trends in Analytical Chemistry*, 50: 96–106.
3. Jellema, R.H. 2009. Variable shift and alignment, in *Comprehensive Chemometrics*. Elsevier, Amsterdam, the Netherlands.
4. Bylesjö, M., O. Cloarec, and M. Rantalainen. 2009. Normalization and closure, in *Comprehensive Chemometrics*. Elsevier, Amsterdam, the Netherlands.
5. Kvalheim, O.M., F. Brakstad, and Y. Liang. 1994. Preprocessing of analytical profiles in the presence of homoscedastic or heteroscedastic noise, *Analytical Chemistry*, 66(1): 43–51.
6. Daszykowski, M. 2007. From projection pursuit to other unsupervised chemometric techniques, *Journal of Chemometrics*, 21(7–9): 270–279.
7. Friedman, J.H. and J.W. Tukey. 1974. A projection pursuit algorithm for exploratory data analysis, *IEEE Transactions on Computers*, C-23(9): 881–890.
8. Daszykowski, M., B. Walczak, and D.L. Massart. 2003. Projection methods in chemistry, *Chemometrics and Intelligent Laboratory Systems*, 65(1): 97–112.
9. Hou, S. and P.D. Wentzell. 2014. Re-centered kurtosis as a projection pursuit index for multivariate data analysis, *Journal of Chemometrics*, 28(5): 370–384.
10. Wold, S., K. Esbensen, and P. Geladi. 1987. Principal component analysis, *Chemometrics and Intelligent Laboratory Systems*, 2(1–3): 37–52.
11. Daszykowski, M., K. Kaczmarek, Y. Vander Heyden, and B. Walczak. 2007. Robust statistics in data analysis—A review: Basic concepts, *Chemometrics and Intelligent Laboratory Systems*, 85(2): 203–219.
12. Stanimirova, I., B. Walczak, D.L. Massart, and V. Simeonov. 2004. A comparison between two robust PCA algorithms, *Chemometrics and Intelligent Laboratory Systems*, 71(1): 83–95.
13. Croux, C. and A. Ruiz-Gazen. 2005. High breakdown estimators for principal components: The projection-pursuit approach revisited, *Journal of Multivariate Analysis*, 95(1): 206–226.
14. Bratchell, N. 1987. Cluster analysis, *Chemometrics and Intelligent Laboratory Systems*, 6(2): 105–125.
15. Massart, D.L. and L. Kaufman. 1989. *The Interpretation of Analytical Chemical Data by the Use of Cluster Analysis*. Krieger Publishing Co., Melbourne, FL.
16. Drab, K. and M. Daszykowski. 2014. Clustering in analytical chemistry, *Journal of AOAC International*, 97(1): 29–38.
17. Todeschini, R., D. Ballabio, and V. Consonni. 2006. Distances and other dissimilarity measures in chemometrics, in *Encyclopedia of Analytical Chemistry*. John Wiley & Sons, New York.
18. De Maesschalck, R., D. Jouan-Rimbaud, and D.L. Massart. 2000. The Mahalanobis distance, *Chemometrics and Intelligent Laboratory Systems*, 50(1): 1–18.
19. Daszykowski, M., B. Walczak, and D.L. Massart. 2002. On the optimal partitioning of data with K-means, growing K-means, neural gas, and growing neural gas, *Journal of Chemical Information and Computer Sciences*, 42(6): 1378–1389.
20. Ester, M., H.-P. Kriegel, J. Sander, and X. Xu. 1996. A density-based algorithm for discovering clusters in large spatial databases with noise. *Proceedings of the Second International Conference on Knowledge Discovery and Data Mining (KDD-96)*, Portland, OR. AAAI Press. pp. 226–231.

21. Daszykowski, M., B. Walczak, and D.L. Massart. 2001. Looking for natural patterns in data: Part 1. Density-based approach, *Chemometrics and Intelligent Laboratory Systems*, 56(2): 83–92.
22. Ankerst, M., M.M. Breunig, H.-P. Kriegel, and J. Sander. 1999. OPTICS: Ordering points to identify the clustering structure, *ACM Sigmod Record*, 28: 49–60.
23. Daszykowski, M., B. Walczak, and D.L. Massart. 2002. Looking for natural patterns in analytical data. 2. Tracing local density with OPTICS, *Journal of Chemical Information and Computer Sciences*, 42(3): 500–507.
24. Tran, T.N., K. Drab, and M. Daszykowski. 2013. Revised DBSCAN algorithm to cluster data with dense adjacent clusters, *Chemometrics and Intelligent Laboratory Systems*, 120: 92–96.

9 Denoising of Signals, Signal Enhancement, and Baseline Correction in Chromatographic Science

Zhi-Min Zhang, Hong-Mei Lu, and Yi-Zeng Liang

CONTENTS

9.1 INTRODUCTION

Chromatographic techniques, particularly hyphenated with mass spectrometry, are essential analytical tools in separation, detection, quantification, and identification of compounds in complex systems [1]. They have been used in various applications, including metabolomics [2–4], proteomics [5,6], foodomics [7–9], fingerprinting analysis of herbal medicines [10–13], environmental chemistry [14–16], and so on. In chromatographic analysis, the detected signals are recorded as chromatograms, which are monitored as a function of time. Chromatograms basically consist of analytical signal of any presented analyte, baseline (systematic behavior not related to analyte), and random noise, which provide extensive amounts of analytical data for complex systems and are time-consuming to be analyzed. When multiple samples are compared and analyzed statistically, one should align peaks representing the same analyte from different samples because of retention time shifts. Therefore, chemometric methods have been used frequently to extract information from chromatograms effectively.

This chapter focuses on chemometric methods for solving problems when handling chromatographic datasets, and they will be elucidated as clearly as possible, from the simplest methods to the more sophisticated ones, in the following sections. We will start with a brief introduction on basic data structure of chromatograms and common artifacts in chromatograms (including noise, baseline, retention time shifts, as well as co-elution). Then we will describe denoising, signal enhancement, and baseline correction methods for individual chromatogram. Peak alignment is critical when analyzing massive chromatograms, so it will be introduced and discussed after methods for individual chromatogram. Finally, advantages and disadvantages of each method will be discussed in the last section.

9.2 CHROMATOGRAPHIC DATASET

A chromatogram is a representation of the separation that has occurred in the chromatographic system, which is the most commonly used representation of chromatographic dataset. X-axis of chromatogram represents retention time and the y-axis represents signal intensity. Ideal chromatogram should only contain well-separated peaks of each compound with fixed retention time, which is impossible to achieve. Commonly, chromatogram basically consists of signal of analyte, baseline, and random noise. Co-elution occurs when compounds elute from a chromatographic column at the same time, and it makes separation and identification difficult. There is always difference in samples due to instrument drift in retention time, which will cause retention time shifts between peaks across different chromatograms. Baseline, noise, co-elution, and retention time shifts are common problems that need to be handled in chromatographic datasets; these problems have been shown in Figure 9.1. In this section, we illustrate these problems in chromatograms and how they can be reduced by preprocessing.

9.2.1 CHROMATOGRAPHIC PEAK

When a sample mixture enters the chromatographic column, it is separated into individual compounds because the affinity between mobile and stationary phase of each compound is different. Each specific analyte is made up of many molecules. The center contains the highest concentration of molecules, while the leading and trailing edges are decreasingly less. When the separated compounds leave the column, they pass immediately into a detector. The electrical signal of the detector responds proportionally to the concentration of a compound, which is converted and recorded as a peak by the computer. Ideal chromatographic peaks can be considered as Gaussian or exponentially modified Gaussian peaks, which can be effectively detected by continuous wavelet transformation with Mexican hat wavelet [17,18].

9.2.2 RANDOM NOISE

A chromatogram from instruments includes chemical information that stems from the analyzed compounds. This part can ideally be described by the model. The other part is the random variation that prevents a perfect representation of the model, which is typically produced by instrumentation deficiencies such as a noise generated by the detector or a noise produced by the amplifier. The noise deteriorates signal-to-noise ratio (SNR) of a chromatogram, which can negatively affect the predictions and the interpretability of multivariate models. Reducing the contribution of noise

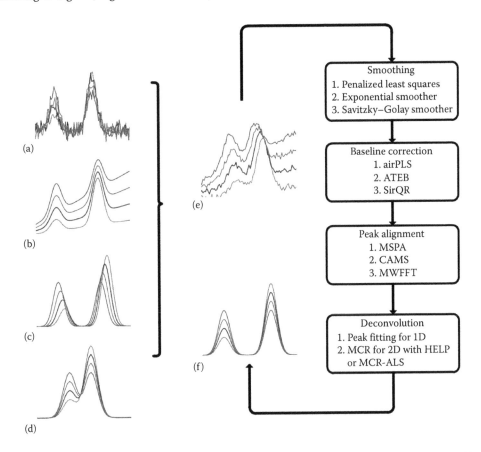

FIGURE 9.1 Common deviations from ideality in chromatograms and achieving acceptable results with chemometric methods. (a) Noise, (b) baseline, (c) time shift, (d) co-elution, (e) combined, and (f) results.

without adversely affecting the underlying signal is an essential part of any strategy for extracting information from chromatogram.

9.2.3 BASELINE

Baseline drift is a steady increasing or decreasing curve independent to the sample, which occurs because of disequilibrium of column, contamination of the system (injector, mobile phase, stationary phase, or detector), stationary phase damage, and so on. It always blurs or even swamps signals and deteriorates analytical results, particularly in multivariate analysis. It is necessary to correct baseline drift to perform further data analysis.

9.2.4 RETENTION TIME SHIFT

Chemical components are separated from each other based on their physical properties in chromatographic instrument. Ideally, each component in different samples should have an equal retention time. But in real analysis, retention time shifts occur frequently due to instrumental drift, poor mixing of mobile phase, stationary phase decomposition, columns changes during usage, interaction between analytes, and so on. So the same peaks in different chromatograms should be aligned at the same mathematical column with peak alignment methods.

9.2.5 Co-elution

Co-elution occurs when compounds cannot be chromatographically separated because of retention times of compounds differ by less than the resolution of the analytical method. If the chromatogram is obtained with hyphenated instrument that follows a bilinear model, then noniterative (HELP) [19,20] or iterative (MCR-ALS) [21,22] multivariate curve resolution methods can be used to obtain both spectra and concentrations of a given co-elution region. Since this chapter focuses on denoising of signals, signal enhancement, and baseline correction in chromatographic science, methods for solving co-elution problem will not be described in detail.

9.3 DENOISING OF CHROMATOGRAMS

Noise in chromatogram significantly reduces both accuracy and precision of the analytical results. Therefore, denoising of signals and signal enhancement are important to increase the SNR of chromatogram and reduce the uncertainty of the analytical result. The most popular denoising method in analytical chemistry is moving window polynomial smoothing, which was regards as a historic collaboration between Savitzky and Golay [23,24]. However, the well-known Savitzky–Golay smoother has several disadvantages. It is relatively complicated for the Savitzky–Golay smoother to handle missing values and boundary of the signal. Furthermore, the Savitzky–Golay smoother cannot control over-smoothness continuously. Therefore, new smoothers should be adapted into chromatography for improvements in flexibility, speed, and ease of usage. Penalized least squares and exponential smoothing for improving the SNR of chromatograms are introduced in the following section.

9.3.1 Penalized Least Squares

Penalized least squares algorithm is a flexible smoothing method published by Whittaker [25]. Then, Silverman [26] developed a new smoothing technique in statistics, which was called the roughness penalty method. Penalized least squares algorithm can be regarded as roughness penalty smooth by least squares, which balanced between fidelity to original data and roughness of fitted data. Liang et al. [27] introduced it into chemistry as a smoothing technique to improve the signal detection and resolution of chemical components with very low concentrations in hyphenated chromatographic two-way data. Recently, Eilers extended its application scopes to general chemical signal smoothing [28].

Assume $\mathbf{x}$ is vector of chromatogram, and $\mathbf{z}$ is the smoothed chromatogram. Lengths of both are m. Fidelity of $\mathbf{z}$ to $\mathbf{x}$ can be expressed as the sum of square errors between them:

$$F = \sum_{i=1}^{m}\left(x_i - z_i\right)^2 \tag{9.1}$$

Roughness of the fitted chromatogram $\mathbf{z}$ can be written as its squared and summed differences:

$$R = \sum_{i=2}^{m}\left(z_i - z_{i-1}\right)^2 = \sum_{i=1}^{m-1}\left(\Delta z_i\right)^2 \tag{9.2}$$

The first differences penalty is adopted to simplify the presentation here. In most cases, the square of second differences penalties can be a natural way to quantify the roughness. The balance of fidelity and smoothness can be then measured as fidelity plus penalties on the roughness, and it can be given by

$$Q = F + \lambda R = \left\|\mathbf{x} - \mathbf{z}\right\|^2 + \lambda\left\|\mathbf{Dz}\right\|^2 \tag{9.3}$$

Here λ can be adjusted by user. Larger λ brings smoother fitted vector. Balance of fidelity and smoothness can be achieved by tuning this parameter. $\mathbf{D}$ is the derivative of the identity matrix such that $\mathbf{Dz} = \Delta\mathbf{z}$. For example, when $m = 5$, $\mathbf{D}$ would be

$$\mathbf{D} = \begin{bmatrix} -1 & 1 & 0 & 0 & 0 \\ 0 & -1 & 1 & 0 & 0 \\ 0 & 0 & -1 & 1 & 0 \\ 0 & 0 & 0 & -1 & 1 \end{bmatrix} \tag{9.4}$$

By finding for the vector of partial derivatives and equating it to 0 ($\partial Q/\partial\mathbf{z}=\mathbf{0}$), we get the linear system of equations that can be easily solved:

$$\left(\mathbf{I}+\lambda\mathbf{D}'\mathbf{D}\right)\mathbf{z} = \mathbf{x} \tag{9.5}$$

Then

$$\mathbf{z} = \left(\mathbf{I}+\lambda\mathbf{D}'\mathbf{D}\right)^{-1}\mathbf{x} \tag{9.6}$$

Equation 9.6 is smoothing method using penalized least squares algorithm.

9.3.2 Two-Sided Exponential Smoothing

Exponential smoothing is a useful technique that can be widely applied to many kinds of analytical signals to produce smoothed data with high SNR. Almost, chromatograms are a sequence of observations with noise. Whereas in the simple moving average, the past observations are weighted equally, exponential smoothing assigns exponentially decreasing weights over increasing variables. The exponential smoothing filter can normalize the weights so that they sum to one, which means that when this filter is applied to a sequence with constant input value $x_i = constant$ (already as smooth as can be), the output values will be the same $y_i = constant$. Meanwhile, Figure 9.2 illustrates

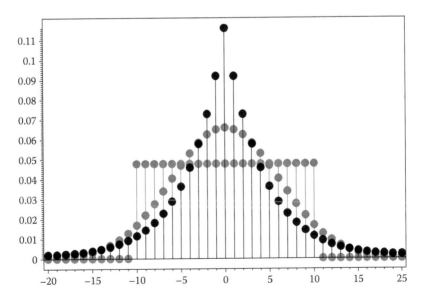

FIGURE 9.2 Coefficients for exponential (black), Gaussian (blue), and boxcar (red) smoothing. For these filters, each output sample is a weighted average of nearby input samples. (Reprinted from *Chemom. Intell. Lab. Syst.*, 139, Liu, X., Zhang, Z., Liang, Y. et al., Baseline correction of high resolution spectral profile data based on exponential smoothing, 97–108. Copyright 2014, with permission from Elsevier.)

weights for the exponential filter algorithm and two alternative smoothing filter algorithms with comparable widths [29].

In practice, the exponential smoothing algorithm was first suggested by Robert Goodell Brown [30], and then expanded by Charles C. Holt [31]. The formulation given in the following, which is the one commonly used, is attributed to Brown and is known as "Brown's simple exponential smoothing." Moreover, it was not only commonly applied to statics data and digital signal processing but also can be used with any discrete set of repeated measurements. The raw data sequence is often represented by $\{x_t\}$, and the output of the exponential smoothing algorithm is commonly written as $\{s_t\}$, which may be regarded as the best estimate of what the next value of x will be. When the sequence of smoothing observations begins at time $t = 0$, we set the starting value to be $s_0 = x_0$. Then, the simplest form of exponential smoothing is given by the formulas

$$s_t = \alpha x_t + (1-\alpha)s_{t-1}, \quad t > 1 \tag{9.7}$$

where α is the smoothing factor, and $0 < \alpha < 1$. In other words, the smoothed signal value s_t is a simple weighted average of the previous observation x_{t-1} and the previous smoothed signal s_{t-1}. The term smoothing factor applied to α here might be something of a misnomer, as larger values of α actually reduce the level of smoothing, and in the limiting case with $\alpha = 1$, the output series is just the same as the original series (with lag of one variable unit). Simple exponential smoothing is easily and broadly applied; simultaneously, it produces a smoothed signal result as soon as two observations are available.

However, when the value of α is close to one, it will have less of a smoothing effect and give greater weight to recent changes in the data, while values of α closer to zero have a greater smoothing effect and are less responsive to recent changes. In order to optimize the estimated process with α and observations s_t, the two-side exponential smoothing algorithm has been introduced into our further smoothing procedure. Similarly, as the earlier description in exponential smoothing algorithm, the raw data sequence of observations is still represented by $\{x_t\}$, beginning at $t = 0$. Meanwhile, S indicates the exponential smoothing values, so as these $S_t^{(1)}$ and $S_t^{(2)}$ represent the first exponential smoothing values and the second exponential smoothing values with the period of t. Briefly, the recursive formula of this two-side exponential smoothing algorithm could be listed as follows:

$$\begin{cases} S_t^{(1)} = \alpha X_t + (1-\alpha)S_{t-1}^{(1)} \\ S_t^{(2)} = \alpha_t^{(1)} + (1-\alpha)S_{t-1}^{(2)} \end{cases} \tag{9.8}$$

where α is also the data smoothing factor, and the value $0 < \alpha < 1$. In addition, unlike the traditional double exponential algorithm, we set the initial values of each-side estimation as $S_t^{(1)} = x_0$, $S_n^{(2)} = S_n^{(1)}$. It means that this algorithm starts with a forward direction in the first smoothing, while it revises the smoothing direction to the backward in the second smoothing process. So it starts with the first smoothing result $S_t^{(1)}$ at $t = n$ in the second smoothing. Note that the formula (9.8) could be simplified to

$$\begin{cases} S_t^{(1)} = \alpha \sum_{i=0}^{t-1} (1-\alpha)^i x_{t-i} + \alpha(1-\alpha)^t x_0 \\ S_t^{(1)} = \alpha \sum_{i=0}^{t-1} (1-\alpha)^i S_{t-i}^{(1)} + \alpha(1-\alpha)^t S_0^{(2)} \end{cases} \tag{9.9}$$

Since the initial value of the second smoothing is

$$S_t^{(2)} = S_{t=n}^{(1)} = \alpha \sum_{i=0}^{n-1} (1-\alpha)^i x_{n-i} + \alpha(1-\alpha)^n x_0 \tag{9.10}$$

combing with the formula (9.9), the two-side exponential smoothing result could be summarized as

$$S_t^{(2)} = \alpha \sum_{i=0}^{t-1}(1-\alpha)^i \left[\alpha \sum_{j=0}^{t-i-1}(1-\alpha)^j x_{t-j} + \alpha(1-\alpha)^{t-i} x_0 \right] + \alpha(1-\alpha)^t \left[\alpha \sum_{i=0}^{n-1}(1-\alpha)^i x_{n-i} + \alpha(1-\alpha)^n x_0 \right]$$

(9.11)

As described in Equation 9.11, the important point is that all of the raw data points and the corresponding weighted values are considered in estimating each smoothing point during the whole process.

Figure 9.3 is a simulated chromatogram smoothed by different smoothers. Penalized least squares can control over-smoothness continuously through adjusting the lambda parameter easily; the

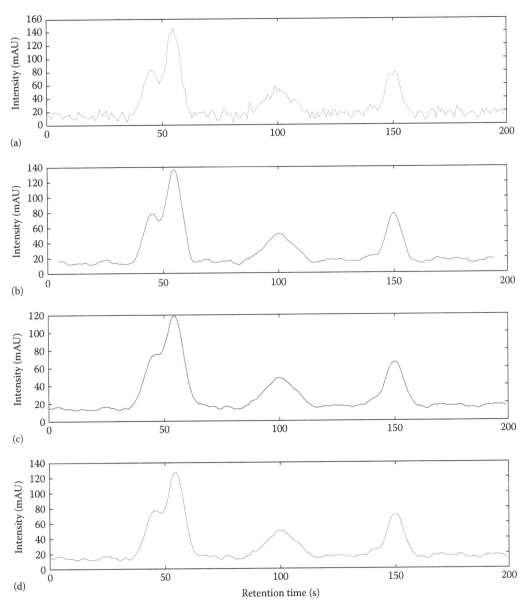

FIGURE 9.3 Smoothing chromatograms with different smoothers. (a) Raw chromatogram (b) smoothed by Savitzky–Golay smoother (c) smoothed by penalized least squares and (d) smoothed by exponential smoother.

smoothed chromatogram can be seen in Figure 9.3c. Two-sided exponential smoother is extremely fast, which can be used for high-resolution dataset. One can see the smoothed chromatogram by this method in Figure 9.3d.

9.4 BASELINE CORRECTION

Simple or modified polynomial fitting [32–34], penalized or weighted least squares [35–46], wavelet [47–53], derivatives [54,55], robust methods [56–59], and morphological operations [60,61] were frequently used for baseline correction in analytical chemistry. However, each of them has some drawbacks in certain aspects: (1) simple manual polynomial fitting is not so effective and its accuracy clearly depends on the user's experience; the modified polynomial fitting methods overcome drawbacks of their predecessor, but their performances are poor in low signal-to-noise and signal-to-background ratio environments; (2) penalized least squares initially proposed for smoothing, which relies on peak detection and is prone to produce negative regions in complex signals; (3) wavelet baseline correction algorithms always suppose that the baseline is well separated in the transformed domain from the signal, but the real-world signals do not agree with this hypothesis; (4) derivatives algorithms change original peak shapes after the correction, which may cause difficulty in the interpretation of the preprocessed spectra; (5) robust local regression requires that the baseline must be smooth and vary slowly, and it also needs specifying the bandwidth and the tuning parameters by user.

In this section, we will focus on relatively new methods including weighted penalized least squares, quantile regression for robust correcting, and iteratively exponential smoothing for baseline correction of chromatograms. They are efficient, flexible, and fast, which are suitable for preprocessing large-scale chromatograms in real-world applications.

9.4.1 WEIGHTED PENALIZED LEAST SQUARES

In order to correct baseline using penalized least squares algorithm, one should introduce weights vector of fidelity, and set to an arbitrary value, say 0, to weights vector at position corresponding to peak segments of $\mathbf{x}$. Fidelity of $\mathbf{z}$ to $\mathbf{x}$ is changed to

$$F = \sum_{i=1}^{m} w_i (x_i - z_i)^2 = (\mathbf{x} - \mathbf{z})' \mathbf{W} (\mathbf{x} - \mathbf{z})$$ (9.12)

where $\mathbf{W}$ is a diagonal matrix with w_i on its diagonal.

Equation 9.12 changes to

$$(\mathbf{W} + \lambda \mathbf{D}'\mathbf{D})\mathbf{z} = \mathbf{W}\mathbf{x}$$ (9.13)

Solve the above linear equations, and the fitted vector can be obtained easily:

$$\mathbf{z} = (\mathbf{W} + \lambda \mathbf{D}'\mathbf{D})^{-1} \mathbf{W}\mathbf{x}$$ (9.14)

Adaptive iteratively reweighted procedure is similar to weighted least squares and iteratively reweighted least squares but uses different ways to calculate the weights and adds a penalty item to control smoothness of fitted baseline. Each step of the proposed adaptive iteratively reweighted procedure involves solving a weighted penalized least squares problem of the following form:

$$Q^t = \sum_{i=1}^{m} w_i^t |x_i - z_i^t|^2 + \lambda \sum_{j=2}^{m} |z_j^t - z_{j-1}^t|^2$$ (9.15)

The weights vector $\mathbf{w}$ is obtained adaptively using an iterative method. One should give an initial value $\mathbf{w}^0 = \mathbf{1}$ at the starting steps. After initialization, $\mathbf{w}$ of each iterative step t can be obtained using the following expressions:

$$w_i^t = \begin{cases} 0 & x_i \geq z_i^{t-1} \\ e^{\frac{t\left(x_i - z_i^{t-1}\right)}{\left\|\mathbf{d}^t\right\|}} & x_i < z_i^{t-1} \end{cases} \tag{9.16}$$

Vector $\mathbf{d}^t$ consists of negative elements of differences between $\mathbf{x}$ and $\mathbf{z}^{t-1}$ in t iteration step.

The fitted value $\mathbf{z}^{t-1}$ in the previous $(t-1)$ iteration is a candidate of baseline. If the value of the ith point is greater than the candidate of baseline, it can be regarded as part of a peak. So its weight is set to zeros to ignore it at the next iteration of fitting. In the airPLS algorithm, the iterative and reweighted methods are used to automatically and gradually eliminate the points of peaks and preserve the baseline points in the weight vector $\mathbf{w}$.

Iteration will either stop the maximal iteration times or the terminative criterion is reached. The termination criterion is defined by

$$\left| \mathbf{d}_t \right| < 0.001 \times \left| \mathbf{x} \right| \tag{9.17}$$

Here, vector $\mathbf{d}^t$ also consists of negative elements of differences between $\mathbf{x}$ and $\mathbf{z}^{t-1}$. The flowchart describing architecture of the proposed algorithm is shown in Figure 9.4.

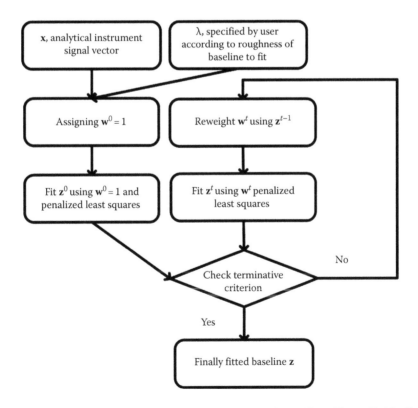

FIGURE 9.4 Flowchart describing framework of the airPLS algorithm. (From Zhang, Z.-M., Chen, S., and Liang, Y.-Z., Baseline correction using adaptive iteratively reweighted penalized least squares, *Analyst*, 135, 1138–1146, 2010. Reproduced by permission of The Royal Society of Chemistry.)

Firstly, simulated dataset has been generated to illustrate the details of airPLS. Simulated data consist of linear or curved baseline, analytical signals, and random noise, which can be mathematically described as follows:

$$s(x) = p(x) + b(x) + n(x) \tag{9.18}$$

where
 $s(x)$ denotes the resulted simulated data
 $p(x)$ the pure analytical signal
 $b(x)$ the linear or curved baseline
 $n(x)$ the random noise

Pure signals are three Gaussian peaks with different intensities (listed in Table 9.1), means, and variances. Curved baseline is a sin curve. Random noise $n(x)$ is generated using the random number generator (the rnorm() function of R language), whose intensity is about 1% of the simulated signals. The corrected results of the airPLS algorithm can be seen in Figure 9.5. Both linear and curved baselines are removed successfully, which has proven the flexibility of the airPLS algorithm. One can also see that both the linear and curved baselines are fitted only in three iterations. It means that the airPLS algorithm converges swiftly.

Chromatograms, analyses of the Red Peony Root using HPLC-DAD, were selected to test proposed algorithm. Eight samples of Red Peony Root were collected from different areas in China, and a standard sample was also bought from the National Institute for Control of Pharmaceutical and Biological Products. The experiments were performed at Chromap Co., Ltd Zhuhai, China. Two UV spectra per second from 200 to 600 nm with a bandwidth of 4 nm resulted in 100 data points in each UV spectrum; then, the "most peaks rich" wavelength 230 nm was selected. The data were transformed into ASCII format using HP chemstations (version A.09.01) for further analysis. The chromatograms could be seen in Figure 9.6a, and one could obviously see that baseline drifts vary from sample to sample. Eight HPLC chromatograms of Red Peony Root were corrected using $\lambda = 30$. Figure 9.6b depicts the corrected chromatograms. As there was a standard chromatogram, principal component analysis (PCA) was applied to the matrix consisting of original, corrected, and standard chromatograms. Then, the scores of the first and the second principal components were plotted in Figure 9.6c to investigate the influences on clustering analysis of the proposed airPLS algorithm. In Figure 9.6c, circle means standard chromatograms; plus signs mean corrected chromatograms; and triangles mean original chromatograms. Since movement trends of points were indicated using arrows in Figure 9.6c, one can obviously observe that corrected chromatograms tend to approach the standard chromatogram after correction. It can demonstrate the validity of the airPLS algorithm. The corrected chromatograms were more compact in pattern space and closer to

TABLE 9.1
Execution Times of Signals with Different Numbers of Variables

Algorithms	Simulated Dataset (900 Variables)	Chromatograms (4000 Variables)	MALDI-Tof-MS (36,802 Variables)	NMR Signal (63,483 Variables)
ALS	0.0608	0.0885	0.5367	1.4529
FABC	0.0284	0.0669	0.3643	1.5160
airPLS	0.0213	0.0460	0.2413	0.5185
ATEB	0.0172	0.0296	0.2283	0.3662

Source: Reprinted from *Chemom. Intell. Lab. Syst.*, 139, Liu, X., Zhang, Z., Liang, Y. et al., Baseline correction of high resolution spectral profile data based on exponential smoothing, 97–108. Copyright 2014, with permission from Elsevier.

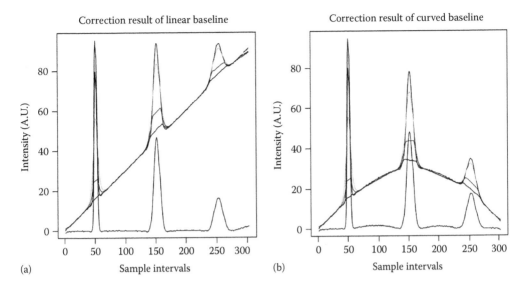

FIGURE 9.5 Correction results of simulated data with different baselines, and the iteration steps are illustrated using gray colors. (a) Linear baseline; (b) curved baseline. (From Zhang, Z.-M., Chen, S., and Liang, Y.-Z., Baseline correction using adaptive iteratively reweighted penalized least squares, *Analyst*, 135, 1138–1146, 2010. Reproduced by permission of The Royal Society of Chemistry.)

the standard chromatogram. The compactness and closeness in principal components pattern space would improve clustering and classification results to some extent.

9.4.2 ROBUST CORRECTION VIA QUANTILE REGRESSION

The objective function of penalized least squares is the sum of squares of errors and differences. The error value of this equation could absolutely be enlarged by square. The minimized Q is not robust, which might not be the ideal result for real-world signals. The solution for this issue is to change the objective function from squares to absolute value:

$$Q_1 = \sum_{i=1}^{m} |y_i - z_i| + \lambda \sum_{i=2}^{m} |z_i - z_{i-1}| \tag{9.19}$$

The sums of squares of residues have been replaced by sums of absolute values, which mean that the L_2 norm had been modified into L_1 norm. Focusing in the right side of (9.19) it could be deemed the functions of y as a single summation of $2m - 1$ absolute values of terms. Then, a novel method could be proposed by combining the ideas from quantile regression with linear programming to fit the dataset.

Koenker and Basset proposed [62] the following problem: a vector y, a regression basis B, and n regression coefficients α. With τ a parameter between 0 and 1, we can minimize the following equation:

$$S(t) = \sum_{i}^{m} \rho_\tau \left(y_i - \sum_{j}^{n} b_{ij}\alpha_j \right) \tag{9.20}$$

Here, $\rho_\tau(\mu)$ is the check function; it is $\tau\mu$ when $\mu > 0$ and instead $(\tau - 1)\mu$ when $\mu \le 0$. In formula (9.20), it will return the weighted absolute values of residuals, such as τ for the positive one otherwise $1 - \tau$ for the negative one. The weights were independent of the sign with $\tau = 0.5$, so solving Equation 9.20 is equivalent to solving (9.19). This idea is the so-called median regression.

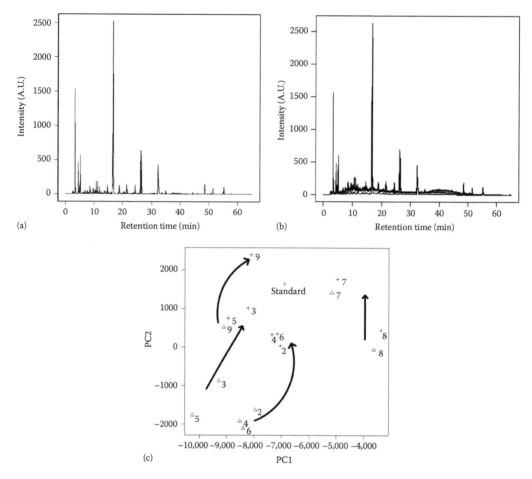

FIGURE 9.6   airPLS baseline correction results of chromatograms of Red Peony Root. (a) Chromatograms of Red Peony Root to correct. (b) Correction results of chromatograms of Red Peony Root. (c) First two principal components of the PCA scores of original, corrected, and standard chromatograms. Circle means standard; plus signs mean corrected; and triangles mean original. Movement trends are marked out with arrows. (From Zhang, Z.-M., Chen, S., and Liang, Y.-Z., Baseline correction using adaptive iteratively reweighted penalized least squares, *Analyst*, 135, 1138–1146, 2010. Reproduced by permission of The Royal Society of Chemistry.)

It was Portnoy and Koenker again who presented a detailed account of quantile regression and also an efficient algorithm based on the interior point method for linear programming [63]. These methods have been implemented in both R and MATLAB® programming languages. However, some modification should be made for signal smoothing and baseline correction:

$$y^* = \begin{pmatrix} y \\ 0 \end{pmatrix} \quad \text{and} \quad B = \begin{pmatrix} I \\ \lambda D \end{pmatrix} \tag{9.21}$$

where
   y is original dataset
   0 is a $m-1$ zeros vector
   I is the $m \times m$ identity matrix
   D is a matrix so that $D_z = \Delta_z$

Thus, $\mathbf{D}$ is an adjusted matrix $m-1 \times m$, whose purpose was to transform z into differences of neighboring elements. For a desired result, y^* and B were the best choose for nest steps.

The key produce of selectively iteratively reweighted quantile regression is similar to the weighted least squares [64], iteratively reweighted least squares [65–67], but using different ways to calculate the weights and utilizing adjustment item to control smoothness of fitted baseline. As is shown

$$S'(t) = \sum_i^m \omega_i \rho_\tau \left(y_i - \sum_j^n b_{ij} \alpha_j \right)$$ (9.22)

where ω_i is the weight vector that selectively acquires the changed values by using iterative method. The initial value of ω_{i0} should be assigned by 10^{-4} at the starting step, which has been determined by several tests and calculations from 0 to 10^{-10}. After initialization, the ω of each iterative step could be acquired using the following expression:

$$w_i^t = \begin{cases} 10^{-10} & d_i \geq d_m \\ e^{\frac{t(x_i - z_i^{t-1})}{|\mathbf{d}^t|}} & d_i < d_m \end{cases}$$ (9.23)

where $d_i = x_i - z_i^{t-1}$ and vector $\mathbf{d}^t$ consists of the elements of the differences between x and z^{t-1}, the value of which was below $d_m = 5.0 \times 10^{-5}$ for a better fitting treating in the iteration step of t. Meanwhile, the value of d_m could be selectively designed by the users for better fitting to the original signal dataset, not just for the default 5.0×10^{-5}. For such datasets with many variables and large orders of magnitude, the corresponding value should be designed bigger than the default one for a better approximation. The fitted value z^{t-1} in the previous $(t-1)$ iteration is a candidate of baseline. If the value of the ith point is greater than the candidate of the baseline, it can be regarded as part of the peak. So the weight of the corresponding ω will be assigned to a tiny value 10^{-10} (cannot be zero in the quantile regression) to almost negligible at the next iteration of fitting. Although the value part of the peak seems to set to 0 to obtain a better fitting baseline, after considering the influence of quantile regression method in whole dataset the corresponding ω should not set to 0 instead of a tiny value. In this SirQR algorithm, in order to achieve automatically and gradually eliminate the points of peaks and preserve the baseline points in the weight vector ω, the iterative and reweight methods were adopted. For iteration procedure, it will reach the destination either in the maximal iteration times or when the termination criterion is arrived in.

Chromatographic analyses were made of tobacco smoke, whose raw tobacco leaves were collected from Yunan province using GCT Premier™ GC-oa-TOF-MS; all the chromatograms with various baselines of GC-ToF-MS analysis of tobacco smoke have been illustrated in Figure 9.7a. The baseline was corrected by the proposed SirQR algorithm as well. Meanwhile, the corrected results were further analyzed by PCA, also comparing with another novel algorithm called morphological weighted penalized least squares (MPLS) in Figure 9.7. One can see clearly the original chromatograms and corrected chromatograms in Figure 9.7a and b, and the results show that the SirQR algorithm is flexible enough to remove the baseline drifts that vary from chromatogram to chromatogram, even in excessive variables. What is more, PCA has been also implemented to assess the validity of the SirQR algorithm proposed. As is commonly known, numerical differentiation could eliminate the tardily shifting baseline; thus, the PCA method has been first executed in the original and has corrected chromatogram signals with first-order numerical differentiation preprocessing. In Figure 9.7c, the triangle, for original, and cross, for corrected, illustrate the good matching in principal component spaces, which suggests that the SirQR algorithm will not eliminate the important information from original chromatograms. Moreover, since all these eight samples are parallel samples, if the effect of baseline could be ignored, they would locate close to each other in principal component spaces. In Figure 9.7d, the triangle represents a chromatogram without any baseline correction by SirQR or by MPLS in principal component spaces; the plus symbols represent the corrected chromatogram signals by proposed SirQR algorithm; then, the diamonds represent the corrected chromatogram signals by MPLS algorithm. The direction in the first principal component mainly conveys the sample difference, as the value reaching 92% of the total variance in first principal component. If we let the scope of these original samples' chromatograms signals be indicated as L in the first principal component direction, it is not difficult to obtain the scope of these corrected sample chromatograms signals in both SirQR method and MPLS method. Comparing with these two distances, SirQR method with $0.410 \times L$ indicate a little better than the MPLS method with $0.430 \times L$. So these values could point out that SirQR algorithm can decrease the variance mostly originating from the baseline as MPLS algorithm does or even better to some extent.

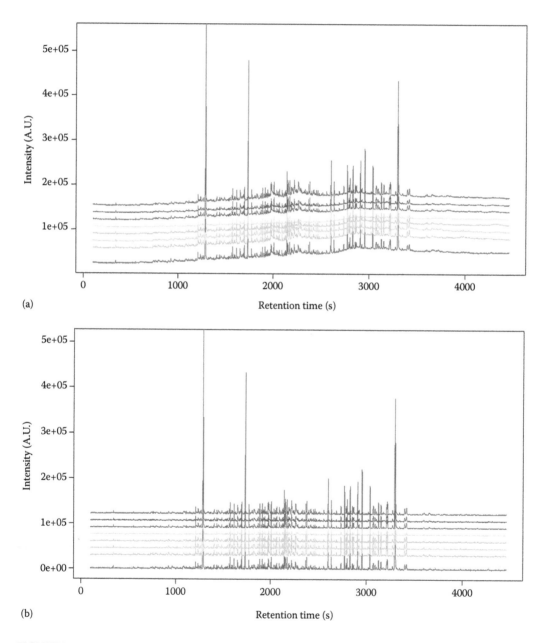

(a)

(b)

FIGURE 9.7 Baseline correction results for the GC-TOF-MS dataset of tobacco smoke by SirQR. (a) Original dataset of eight samples with various baselines. (b) Baselines have been corrected by the SirQR method.

(*Continued*)

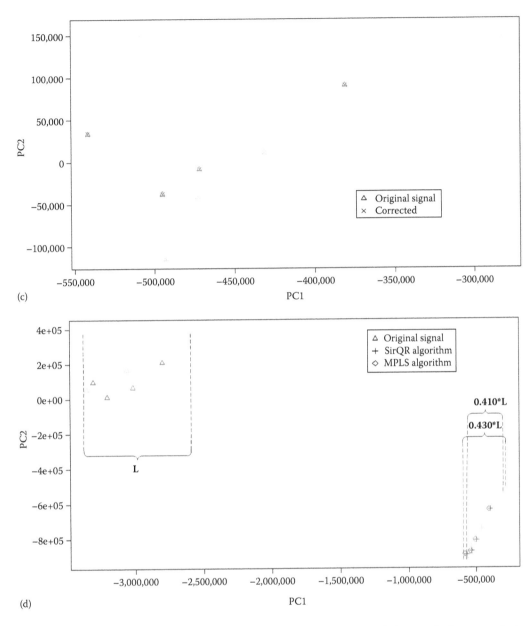

(c)

(d)

FIGURE 9.7 (*Continued*) Baseline correction results for the GC-TOF-MS dataset of tobacco smoke by SirQR. (c) Compare the distribution of samples before and after baseline correction in principal component spaces. (d) First two principal components of the original and corrected chromatograms with first-order numerical differentiation preprocessing and compared with MPLS method. (With kind permission from Springer Science+Business Media: *Anal. Bioanal. Chem.*, Selective iteratively reweighted quantile regression for baseline correction, 406, 2014, 1985–1998, Liu, X., Zhang, Z., Sousa, P.F.M. et al. [67].)

According to the analysis above, one can demonstrate that the larger variation in the first principal component direction of original chromatograms could be owing to varied baseline from chromatogram to chromatogram. This proposed SirQR method can remove this baseline variation amidst a series of chromatograms without missing useful and important information.

The quantile regression method is a clear advantage of our correction algorithm, compared with the ALS correction method, the FABC approach, and airPLS algorithm. As shown in Figure 9.8,

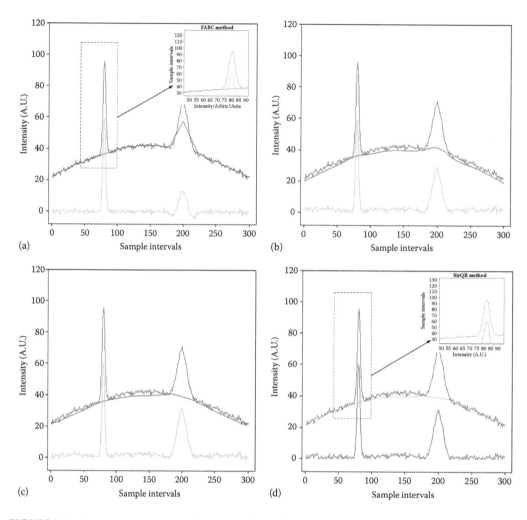

FIGURE 9.8 The robustness testing of four algorithms in high signal-to-noise background. (a) FABC method corrected result and the focusing partial figure. (b) ALS method corrected result. (c) airPLS method corrected result. (d) SirQR method corrected result and the focusing partial figure. (With kind permission from Springer Science+Business Media: *Anal. Bioanal. Chem.*, Selective iteratively reweighted quantile regression for baseline correction, 406, 2014, 1985–1998, Liu, X., Zhang, Z., Sousa, P.F.M. et al.)

these three algorithms have been applied in correcting these data in high signal-to-noise, compared with the proposed SirQR algorithm. One can see clearly that although the FABC method (in Figure 9.8a) could be better in solving the high noise in these methods, it could not correct the wider peak. For another ALS (in Figure 9.8b) and airPLS (in Figure 9.8c) algorithms, the effect of processing noise were not desirable for us, whose fitting line is obviously under the original data. However, the SirQR algorithm could do a better fitting, as shown in Figure 9.8d. Although the corrected result in high noise is not better than the FABC method, this fitting line kept closing to the original data, which was obviously better than the other two methods. In particular, the SirQR algorithm could be flexible to deal with this wide peak notably better than the FABC method. In summary, combing with the two aspects we can say that the SirQR algorithm seems to be a robust correction method. The reason could mostly be owing to L_1 norm, which is the sum of absolute values during the correcting process. These absolute values can not only reach the purpose of fitting the aim line but also, unlike the sums of squares, cannot enlarge the error value. So the robustness of this algorithm could be improved to some extent, obviously more than others in general.

9.4.3 HIGH-SPEED CORRECTION BY AUTOMATIC TWO-SIDE EXPONENTIAL METHOD

Automatic two-side exponential baseline correction basically consists of two steps. First, the original dataset can be smoothed and used to fit the baseline with two-side exponential smoothing algorithm in an iterative fitting process. Subsequently, the corrected final baseline can be automatically determined, when the approximate fitting result reaches the corrected termination in the iterative procedure. It turns the problems of chemical baseline recognition into problems of digital signal processing. Before fitting the baseline, we first assume that there are two types of points in an original dataset: "noise points" and "signal points." A "noise point" is defined as an unprocessed data point when the signal intensity is in the range of $x \leq (\mu - 3\sigma)$; and a "signal point" is as an unprocessed data point when the value of signal intensity is in the range of $x > (\mu - 3\sigma)$. Therefore, after subtracting those "noise points," the processed signals will be smoothed via two-side exponential algorithm. During this process, the smoothing factor α can effectively make this smoothed result close to baseline signals. Moreover, the value of α (smoothing factor) can be flexibly adjusted by user according to different kinds of datasets.

After classifying and smoothing the signal points, the next step is to fit an approximate baseline. Because signal intensities will obscure the position of baseline, it is difficult to estimate the real baseline drift that occurs in correspondence to the signal segments, especially for automatically determining the fitting termination. However, if the peak points (high intensity) and noise points (low and random intensity) can be identified accurately and robustly, a suitable repeatedly iterative process through the two-side exponential smoothing algorithm can well estimate the baseline and preserve the signals. Therefore, immediately following the smoothing process, the second derivative of the smoothed result is taken to confirm the smoothness of the baseline. Then, the sum of absolute derivative value will be taken into fitting termination determination, as follows:

$$d_s = \sum \left| \Delta^2 x_i \right| = \sum \left| \Delta (\Delta x_i) \right| = \sum \left| (x_i - x_{i-1}) - (x_{i-1} - x_{i-2}) \right| = \sum \left| x_i - 2x_{i-1} + x_{i-2} \right| \qquad (9.24)$$

With the fitted curve tending to be a smooth baseline, the smoothness of the fitting baseline will be close to a minimum and stabilized value in a decreasing function. Once the value of the decreasing function tends to be a stabilized value (the difference between d_{s-1} and d_s was below 5.0×10^{-5}), it can be determined that the whole correcting process is reaching the final termination.

One advantage of the ATEB method is its processing speed, which guarantees its performance in large-scale high-resolution signals. Therefore, signals with different numbers of variables were used to test the quickness of the proposed algorithm. In addition, this was compared with three different algorithms at the same time, with the same dataset. The execution times of each algorithm (including ALS [45], FABC [43], airPLS [39], and ATEB [29]), implementing in different numbers of variables, are detailed in Table 9.1. One can notice that the result of the ATEB algorithm's execution time is astonishingly fast.

As represented in Table 9.1, it is evident that the proposed ATEB algorithm is obviously faster than the other, especially in large datasets like NMR signals. Although the subtle advantage of processing speed is difficult to observe with small data in application, when it comes to processing the large dataset with high resolution, the superiority is well reflected immediately. Take, for instance, a fast preprocessing of classification analysis with thousands of different samples including tens of thousands of variables. The relationship between number of variables and execution time was also investigated in detail. It was verified that the relationship between the execution time and the number of variables is exactly linear, which can be observed in Figure 9.9. Obviously, with the increasing of the number of variables, the corresponding execution time will also increase in a linear trend. The exact linear relationship between number of variables and the whole execution time guarantees the good performance of the ATEB algorithm in data even with a larger number of variables. This is mainly attributed to the use of the combination of two-side exponential smoothing algorithm. It can be summarized that the use of the exponential smoothing algorithm enables the application of the

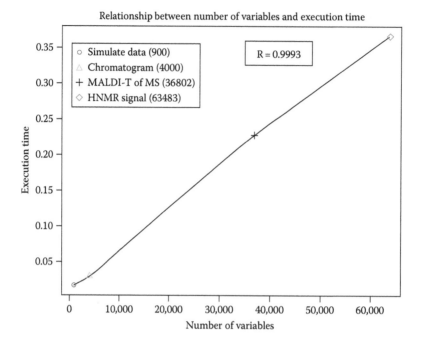

FIGURE 9.9 The relationship between numbers of variables and the total execution time. (Reprinted from *Chemom. Intell. Lab. Syst.*, 139, Liu, X., Zhang, Z., Liang, Y. et al., Baseline correction of high resolution spectral profile data based on exponential smoothing, 97–108. Copyright 2014, with permission from Elsevier.)

ATEB algorithm in more high-throughput domains and effectively meets the needs of data analysis with great speed.

9.5 PEAK ALIGNMENT*

In order to capture differences among samples caused by their composition, the key point of an experiment is to limit the experimental variability as much as possible. However, deviations from normal conditions may appear, causing peak shifts observed among signals. For this reason, the acquired datasets are often too complex to easily extract meaningful information easily. The same peaks in different chromatograms should be aligned at the same mathematical column with peak alignment methods.

Alignment methods that synchronize entire signals usually divide signals into segments, warping these segments by interpolation or transformation to maximize correlation between reference signal and signal to be aligned. The concept of time warp was initially introduced to align retention time shift of chromatograms by Wang and Isenhour [68]. By then, in 1998, two practical alignment methods were introduced, the dynamic time warping (DTW) [69], applied to the analysis and monitoring of batch processes, and the correlation optimized warping (COW) [70], proposed by Nielsen for chromatograms. Both DTW and COW utilize dynamic programming to search all solutions with respect to all possible combinations of parameters, and demonstrated to be effective on chromatograms at that time. But, currently, chromatograms often contain several thousands of data points, and original DTW and COW are no longer suitable for these signals due to the large requirements in both execution time and memory. Also, DTW often "over-warps" signals and introduces artifacts to the aligned profiles, when signals were only recorded using a mono-channel detector [71].

---

* The reader is also referred to Chapter 10 for more information on warping techniques.

Therefore, many heuristic optimization methods, parametric model, and fast correlation algorithms have been applied to accelerate this time-consuming procedure and improve the aligning result. For instance, genetic algorithm [72] and beam search [73] were adopted to align large signals in acceptable time, but it is difficult to optimize the segment size. Eilers proposed a parametric model for the warping function, and presented parametric time warping (PTW) [45], which is fast, stable, and consumes little memory. Pravdova [74], Nederkassel [75], and Jiang [76] compared DTW, COW, PTW, VPdtw, and RAFFT for chromatograms' alignment. Salvador [77] introduced FastDTW, an approximation of DTW that has a linear time and space complexity. Wang [78,79] applied fast Fourier transform (FFT) cross-correlation to estimate shift between segments, which is amazingly fast, and has solved computational inefficiency problems. However, both peak alignment by FFT (PAFFT) and recursive alignment by FFT (RAFFT) move segments by insertion and deletion of data points at the start and end of segments without considering peak information, which may change the shapes of peaks by introducing artifacts and removing peak points [80]. Based on RAFFT and PAFFT, recursive segment-wise peak alignment (RSPA) [81] was proposed by Veskelov to improve the accuracy of alignment using peak position information for recursive segmentation and interval correlation shift (icoshift) [80]. This method can reduce the artifacts by inserting missing values instead of repeating the value on boundary. Variable penalty dynamic time warping was proposed by Clifford [71] to overcome DTW's "over-warps" shortcomings. Recently, Daszykowski [82] proposed an automatic peak alignment method by explicitly modeling the warping function for chromatographic fingerprints. Multiscale peak alignment was developed by Zhang to align peaks against a reference chromatogram from large to small scale gradually, which can not only preserve the shapes of peaks during alignment but also be accelerated by FFT cross-correlation [83]. Moving window fast Fourier transform (MWFFT) [84,85] cross-correlation is introduced to perform nonlinear alignment of high-throughput chromatograms, which can take advantage of elution characteristics of chromatograms will produce local similarity in retention time shifts. Chromatogram alignment via mass spectra (CAMS) [86] is presented here to correct the retention time shifts among chromatograms accurately and rapidly by aligning hyphenated chromatograms against the reference via the correlation of mass spectra.

Among others, fuzzy warping and reduced set mapping often convert signals into peaks' lists, which can speed up alignment by reducing the dimensions of problems dramatically [87–90]. But they align major peaks at the expense of minor peaks, which are harder to detect. Besides, they are prone to misalignment in special peak regions, such as peaks with shoulders, overlapping peaks and peaks' dense region. There are also many mature and competing alignment algorithms or toolbox including alignment algorithms in metabolomics and bioinformatics, including MSFACTs [91], MZmine [92,93], XCMS [94], MetAlign [95,96], and so on.

In the following section, we will focus on MSPA and CAMS methods for massive chromatographic datasets. They have the capacity to preserve the shapes of peaks, perform well with nonlinear retention time shifts, can avoid locally optimal problem, and seem robust and not sensitive to noise and baseline. If hyphenated spectral information is available, CAMS can take full advantage of this information over the entire profile to assure accuracy of alignment.

9.5.1 Multiscale Peak Alignment

At the heart of MSPA is the use of local maximums in FFT cross-correlation as candidate shifts, which can guarantee accuracy and alignment speed. Additionally, it also includes several techniques for peak detection, width estimation, iterative segmentation, and optimal shift determination. Figure 9.10 describes the architecture and overview of the MSPA method.

9.5.1.1 Peak Detection, Width Estimation

Peak detection and width estimation are universal problems in instrument signal analysis, and various criteria have been proposed such as SNR, intensity threshold, slopes of peaks, local maximum, shape

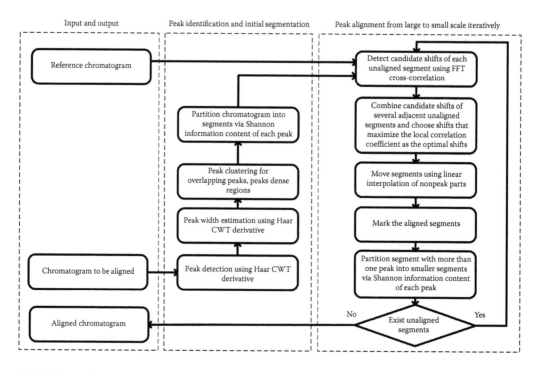

FIGURE 9.10 Flowchart for describing the architecture and overview of the MSPA method. (Reprinted from *J. Chromatogr. A*, 1223, Zhang, Z.-M., Liang, Y.-Z., Lu, H.-M. et al., Multiscale peak alignment for chromatographic datasets, 93–106. Copyright 2012, with permission from Elsevier.)

ratio, ridge lines, and model-based criterion [97]. In this study, a derivative calculation method via CWT was used for peak detection and width estimation, and SNR to eliminate false-positive peaks.

In order to detect peak position and estimate its start and end points, derivative calculations are often applied. However, the simplest numerical differentiation is not very effective for real signal due to the noise increasing drawback, so derivative calculations via Haar CWT were adopted to improve SNR. Wavelet transform is one of the most powerful tools in signal analysis. Wavelet is a series of functions $\psi_{a,b}(t)$, which are derived from $\psi(t)$ by scaling and shifting, according to the equation

$$\psi_{a,b}(t) = \frac{1}{\sqrt{a}}\psi\left(\frac{t-b}{a}\right), \quad a \in R^{+}, \, b \in R \tag{9.25}$$

where
   a is the scale parameter to control scaling
   b the shift parameter to control shifting
   $\psi(t)$ is the mother wavelet

Wavelet transform is defined as the projection of signal onto the wavelet function ψ. Mathematically, this process can be represented as

$$C(a,b) = \langle s(t), \psi_{a,b}(t) \rangle = \int_{-\infty}^{+\infty} s(t)\psi_{a,b}(t)dt \tag{9.26}$$

where
   $s(t)$ is the signal
   C is a 2D matrix of wavelet coefficients

The approximate nth derivative of an analytical signal can be obtained by applying Haar CWT n times to the signal. Haar wavelet is the simplest wavelet function among all the wavelet functions, which can be defined as

$$\psi(t) = \begin{cases} 1 & 0 \leq t < 1/2 \\ -1 & 1/2 \leq t < 1 \\ 0 & \text{other} \end{cases} \qquad (9.27)$$

Peak can be defined as a local maximum of N neighboring points, whose intensity is significantly larger than the noise level. The local maximums can be found from the derivative calculation via Haar CWT of the signal. Then, false-positive peaks are eliminated from whose SNR is lower than a prespecified threshold. Similar to the finding of peak position, the start and end peak points can also be found by this derivative calculation method. For each peak, its start and end points can be obtained by searching the nearest point from its peak position in the vector of detected peak position and peak width. Principles for peak detection and width estimation are concisely illustrated in Figure 9.11. The middle part of Figure 9.11 depicts the derivative calculation via Haar CWT. The top and bottom parts of the figure describe peak detection and width estimation, respectively.

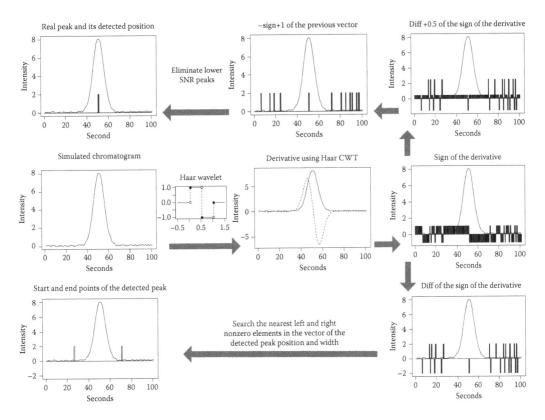

FIGURE 9.11 Principles for peak detection and width estimation using derivative calculation based on continuous wavelet transform with Haar wavelet as the mother wavelet. (Reprinted from *J. Chromatogr. A*, 1223, Zhang, Z.-M., Liang, Y.-Z., Lu, H.-M. et al., Multiscale peak alignment for chromatographic datasets, 93–106. Copyright 2012, with permission from Elsevier.)

9.5.1.2   Segmentation Based on Shannon Information Content

There exist different scale peaks in signal. When correlation-based alignment methods are used, it is intuitive that alignment of large-scale peak is easier than small-scale peak. One could also say that the difficulty level of aligning the specific scale peak has a direct relationship with the uncertainty of this peak in the entire signal profile. In information theory [98], Shannon information content is a good measurement of uncertainty. Consequently, Shannon information content of peaks can be regarded as a good measurement of the difficulty level of specific scale peak. The Shannon information content is defined as

$$\mathbf{h}_i = -\log_2 \mathbf{p}_i \tag{9.28}$$

where
$\mathbf{p}_i$ is the probability of distribution function
$\mathbf{h}_i$ is Shannon information content

A reasonable modification [99] should be employed on the equation above to calculate Shannon information content of peaks in the chromatogram. Firstly, the signal is normalized with its overall peak area equal to one, and then its information content is calculated based on

$$\mathbf{h}_i = -\log_2 \frac{\mathbf{p}_i}{\Sigma \mathbf{p}_i} \tag{9.29}$$

where
$\mathbf{p}_i$ is the area of the ith peak or peak cluster of the real chromatogram
$\Sigma \mathbf{p}_i$ is the overall peak area of the real chromatogram

A small Shannon information content of a peak means a small uncertainty and a large scale and vice versa. It is intuitive that the large-scale peak with small uncertainty should have priority over small-scale peaks during alignment, so peaks were aligned in MSPA against the reference chromatogram from larger to smaller scale, gradually. The iterative segmentation scheme is illustrated in Figure 9.12.

9.5.1.3   Candidate Shifts Detection via FFT Cross-Correlation

The direct evaluation of cross-correlation requires $O(N^2)$ time complexity for a chromatogram of length N, which is time-consuming for chromatograms with several thousands of data points. Fortunately, cross-correlation can be calculated via FFT to achieve a much better performance, which can dramatically reduce time complexity of cross-correlation from $O(N^2)$ to $O(N\log N)$. FFT computes discrete Fourier transform (DFT) and produces the same result as DFT. In order to clarify how to calculate cross-correlation via FFT, a brief introduction about DFT is required. The forward and reverse DFT are defined by the formulas

$$
\begin{aligned}
X_k &= \sum_{n=0}^{N-1} x_n e^{-\frac{2\pi i}{N}kn} \qquad k = 0,\ldots,N-1 \\
x_n &= \frac{1}{N}\sum_{n=0}^{N-1} X_k e^{-\frac{2\pi i}{N}kn} \qquad k = 0,\ldots,N-1
\end{aligned}
\tag{9.30}
$$

where $\mathbf{X}$ is the discrete Fourier transformed data in the wavelength domain. DFT and inverse DFT are often denoted by $\mathbf{X} = \mathcal{F}\{\mathbf{x}\}$ and $\mathbf{x} = \mathcal{F}^{-1}\{\mathbf{X}\}$, respectively.

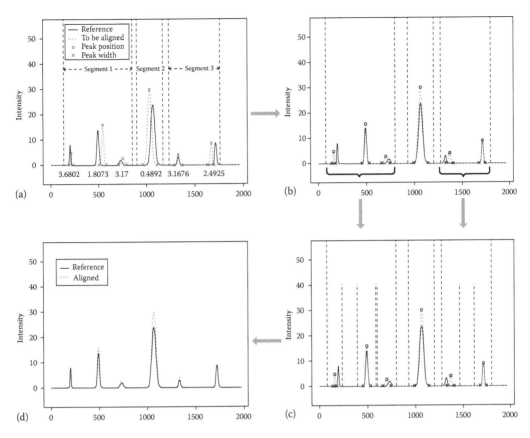

FIGURE 9.12 Scheme of iterative segmentation of chromatogram based on Shannon information content. (a) Peaks, their Shannon information contents and initial segmentation, (b) alignment using FFT cross-correlation and local combination, (c) subdivide segment with more than one peak into smaller segments, and (d) final alignment result. (Reprinted from *J. Chromatogr. A*, 1223, Zhang, Z.-M., Liang, Y.-Z., Lu, H.-M. et al., Multiscale peak alignment for chromatographic datasets, 93–106. Copyright 2012, with permission from Elsevier.)

If **R** and **S** are DFTs of **r** and **s**, then circular convolution theorem and cross-correlation theorem for DFT state

$$\mathbf{c} = \mathcal{F}^{-1}\left\{\mathbf{R}\cdot\mathbf{S}^{*}\right\} \tag{9.31}$$

where
  c is cross-correlation between **r** and **s**
  S* is a complex conjugate of **S**

FFT cross-correlation can only estimate linear shift between signals, but retention time shifts are often nonlinear for chromatograms of real sample. In MSPA, chromatograms to be aligned will be broken into small segments iteratively and FFT cross-correlation will be used to estimate candidate shifts for each segment and align peaks from large scale to small scale, gradually. This strategy can solve the alignment of nonlinear retention time shifting problem by FFT cross-correlation.

Previous alignment methods based on FFT cross-correlation only use the maximum of cross-correlation as the optimal shift. But the maximum of cross-correlation of small segment as the optimal shift may sometimes be the optimal shift locally but not optimal at larger scale or globally. Therefore, all the local maximums of FFT cross-correlation should be detected as the candidate

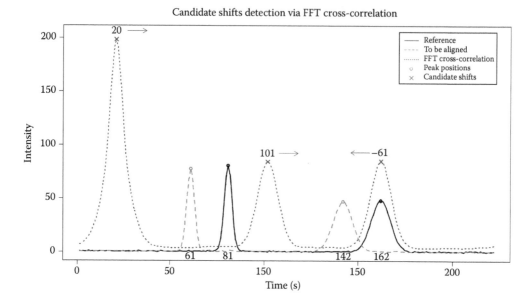

FIGURE 9.13 Candidate shifts' detection of simulated chromatograms by finding the local maximums in FFT cross-correlation. (Reprinted from *J. Chromatogr. A*, 1223, Zhang, Z.-M., Liang, Y.-Z., Lu, H.-M. et al., Multiscale peak alignment for chromatographic datasets, 93–106. Copyright 2012, with permission from Elsevier.)

shifts via CWT derivative calculation. Then, the optimal shift is found by combining the candidate shifts of several adjacent segments to avoid locally optimal problem.

Here is one simple example for candidate shifts' detection of simulated chromatograms. Consider two chromatograms (the reference one is denoted as **r** and test one is **s**) that differ by an unknown shift along the retention time. One can rapidly calculate cross-correlation **c** between **s** and **r** using FFT cross-correlation, and the candidate shifts between **s** and **r** can be found at the local maximums of **c**. Figure 9.13 depicts this procedure visually. The number 20, of candidate shifts in Figure 9.13, means the following: shift the test profile by 20 points, and the maximum cross-correlation between test profile and reference profile will be obtained.

9.5.1.4 Optimal Shift Determination

Sometimes the maximum of cross-correlation of small segment as the optimal shift may be not the optimal shift at larger scale or globally. To avoid this problem, the candidate shifts can be detected using FFT cross-correlation and CWT derivative, according to Section 9.2.3; then, the optimal shift of each segment can be determined by combining the candidate shifts of several adjacent segments to maximize the correlation coefficient between test profile and reference profile.

9.5.1.5 Move Segments

By warping the nonpeak parts to move the segments with peaks using linear interpolation, the detected peaks in each segment can be aligned without altering their shapes. It can reduce the emergence of artifacts by linear interpolation of the nonpeak parts. The linear interpolation of the nonpeak parts can also conserve the information of small peaks as much as possible, which are difficult to detect.

A total of 121 overnight fasting plasma samples were collected from patients at the Xiangya Hospital of Hunan, Changsha city of China. Chromatography and mass detection was performed on Shimadzu GC2010A (Kyoto, Japan) coupled with GCMS-QP2010 mass spectrometer. The MSPA method was run on TIC of free fatty acids in plasma to test its performance on metabolomics dataset. Both unaligned chromatograms (left part) and chromatograms aligned by MSPA (right part) are illustrated in Figure 9.14, and some peaks are zoomed in to demonstrate the performance of MSPA method more clearly. One can see from the zoomed regions of unaligned chromatogram that there exist variations in peak positions. After processing by MSPA, the aligned chromatograms were also amplified at the same peaks' regions to show the aligning results. One can see in detail from the zoomed regions of aligned chromatograms that all peaks were successfully synchronized. Two images were also created to display overall variations in peak positions and alignment results of

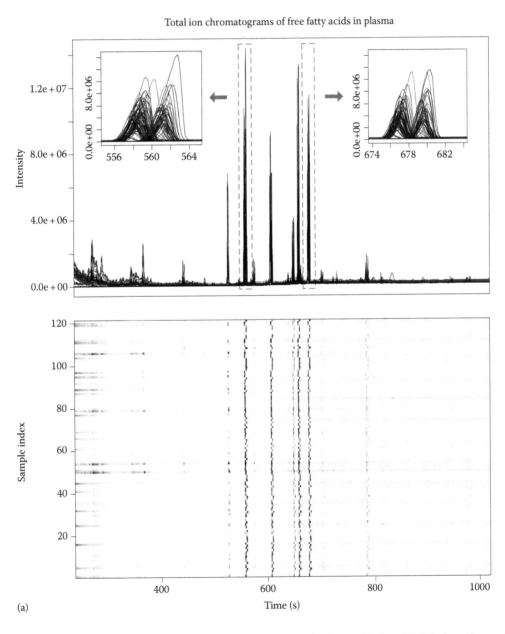

FIGURE 9.14 Total ion chromatograms (TIC) of free fatty acids in plasma, (a) plot of TIC before alignment, (*Continued*)

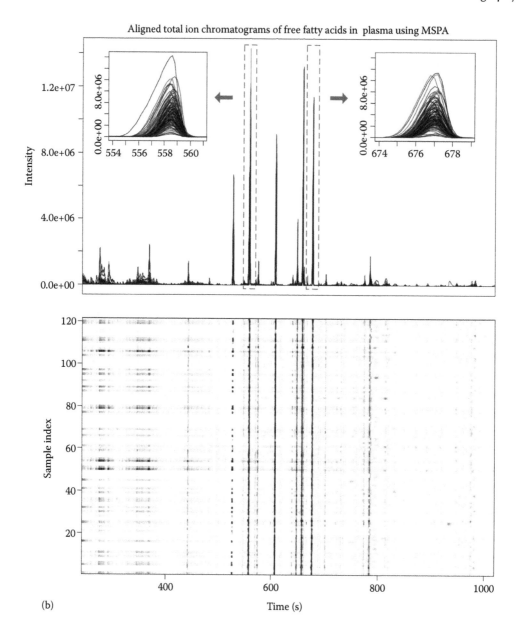

FIGURE 9.14 (*Continued*) Total ion chromatograms (TIC) of free fatty acids in plasma, (b) plot of the aligned TIC using MSPA. (Reprinted from *J. Chromatogr. A*, 1223, Zhang, Z.-M., Liang, Y.-Z., Lu, H.-M. et al., Multiscale peak alignment for chromatographic datasets, 93–106. Copyright 2012, with permission from Elsevier.)

the entire chromatograms in a global manner. It can obviously be seen from the bottom-left image that the lines are "zigzag" and not straight, which means that there are variations in peak positions of unaligned chromatograms from sample to sample. But after aligning them with MSPA, all the "zigzag" lines in the bottom-left image became the straight lines in the bottom-right image, which means that MSPA can effectively eliminate variations in peak positions from sample to sample. By comparing the same zoomed regions of unaligned and aligned by MSPA, it can be seen that all peaks' shapes are intact, which demonstrates that MSPA has the capacity to preserve peaks' shapes during alignment.

9.5.2   CHROMATOGRAM ALIGNMENT VIA MASS SPECTRA

CAMS is the upgraded version of MSPA for hyphenated instruments, and it uses mass spectra to validate candidate shifts. Correlation coefficient between two variables is defined as the covariance of the two variables divided by their standard deviations. The correlation coefficient between two variables is defined as

$$r = \frac{\sum(x-\bar{x})(y-\bar{y})}{\sqrt{\sum(x-\bar{x})^2+(y-\bar{y})^2}} \tag{9.32}$$

where
  x and y are two variables
  $\bar{x}$ and $\bar{y}$ are the means of the two variables

The correlation coefficient ranges from -1 to 1. A value of 1 implies that there exists the linear relationship between x and y perfectly, with all data points lying on a line for which y increases as x increases. A value of -1 implies that all data points lie on a line for which y decreases as x increases. A value of 0 implies that there is no linear correlation between the variables. Correlation coefficient is calculated using mass spectra of candidate points in both the reference and signs to be aligned. However, correlation coefficient between two variables is not robust, so its value can be misleading if outliers are present. In this study, correlation coefficient between two matrices is adopted to this algorithm. Correlation coefficient between A and B matrix is defined as

$$r = \frac{\sum_m\sum_n\left(A_{mn}-\bar{A}\right)\left(B_{mn}-\bar{B}\right)}{\sqrt{\left(\sum_m\sum_n\left(A_{mn}-\bar{A}\right)^2\right)\left(\sum_m\sum_n\left(B_{mn}-\bar{B}\right)^2\right)}} \tag{9.33}$$

where
  A and B are matrices of the same size
  $\bar{A}$ and $\bar{B}$ are the mean of the values in A and B

The row of the matrix is a mass spectra and the size of the matrix is the width of mass chromatograms, and maximum of the correlation coefficient among candidate shifts must be larger than a value (such as 0.9), if $\mathbf{r}$ is lower than the value, which means there exists no peaks can be matched to the detected peaks and the shift of this peak will be assigned as zero.

Here is one simple example for correlation coefficient calculation of simulated chromatograms (the reference is denoted as black and the chromatogram to be aligned is red), which is shown in Figure 9.15. After peak detection and FFT cross-correlation calculation, correlation coefficient of mass spectra will be calculated between peak positions of chromatogram to be aligned and reference chromatogram in each of the candidate shifts, respectively, and the candidate shift that corresponds to the maximums correlation coefficient is optimal peak shift.

Original GC-MS fingerprints of 16 blood serum samples of rats were gathered from Shimadzu GC2010A (Kyoto, Japan) coupled with GCMS-QP2010 mass spectrometer. Figure 9.16a shows the raw chromatograms whose baselines were corrected by airPLS (lambda = 10^4, order = 2) and Figure 9.16b shows the results of the aligned chromatograms by using the CAMS method. A number of peaks were detected by setting the threshold for SNR as 20, and the candidate shifts were calculated by the FFT cross-correlation. By setting the value at 0.9, the optimal shift can be confirmed easily and the chromatograms were well aligned rapidly. Two grayscale images are created to display the detailed view of overall variations in peak positions; the x-axis of the images is retention

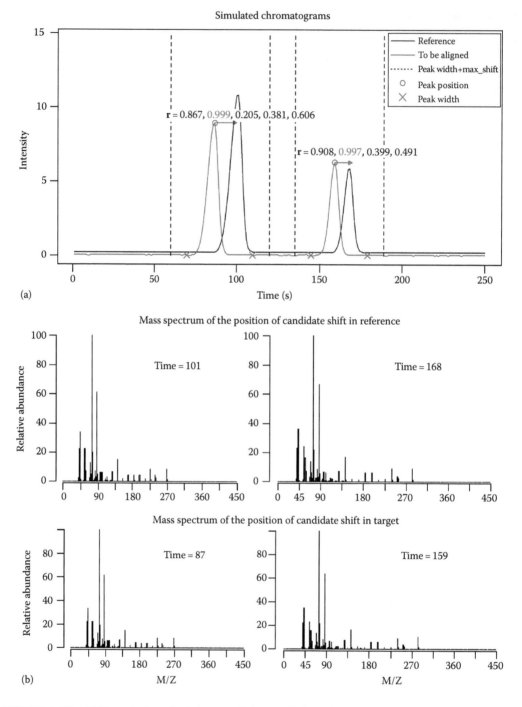

FIGURE 9.15 (a) Example for calculating correlation coefficient of mass chromatograms and finding the local maximums of **r**. The mass spectrum of each peak position in the candidate shift is shown in (b) and the correlation coefficient of mass spectra among the candidate shift in peak one is [0.867, 0.999, 0.205, 0.381, 0.606]; the candidate shift that corresponds to the maximums **r** is the optimal shift. (Reprinted from *J. Chromatogr. A*, 1286, Zheng, Y.-B., Zhang, Z.-M., Liang, Y.-Z. et al., Application of fast Fourier transform cross-correlation and mass spectrometry data for accurate alignment of chromatograms, 175–182. Copyright 2013, with permission from Elsevier.)

time and the y-axis of the images is sample index. Compared to the left grayscale image, it can be seen obviously that the lines of the right grayscale image become straightened, which means all the peaks of the chromatograms were successfully synchronized and aligned. By combining both the detailed views of the zoomed region and the overview of the images, it is easy to draw the conclusion naturally that CAMS can align the fingerprints of GC-MS accurately.

It is a relatively simple task to evaluate the quality of alignment result for real chromatograms by visual comparison, and correlation coefficient is usually used as a basis for evaluation of the alignment quality measure. CAMS is compared with COW (Seg = 80, Slack = 50) and RAFFT (shift = 80) on real chromatograms, and GC-MS fingerprints of 16 blood serum samples above applied here to test the algorithm. The results are shown in Table 9.2. From Table 9.2, the following advantages can be drawn from the above discussion about the CAMS method: (1) average correlation coefficient of CAMS is much better than COW's and RAFFT's; (2) RAFFT achieves greater alignment speed among these methods, and alignment speed of CAMS is also acceptable; (3) CAMS will not change the peak shape of the chromatograms because of peak detection step and the accuracy of the mass spectrogram.

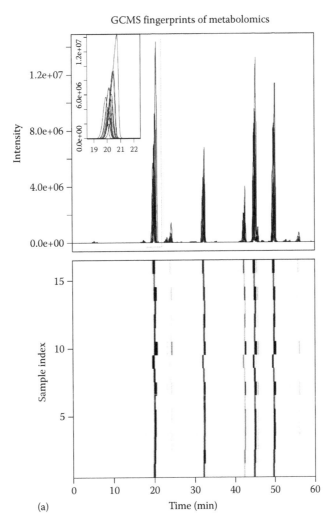

(a)

FIGURE 9.16 (a and b) Total ion chromatograms (TIC) of 16 blood serum samples of rats with left part illustrating TIC before alignment and right part for the aligned TIC using CAMS. *(Continued)*

Aligned GCMS fingerprints of metabolomics using CAMS

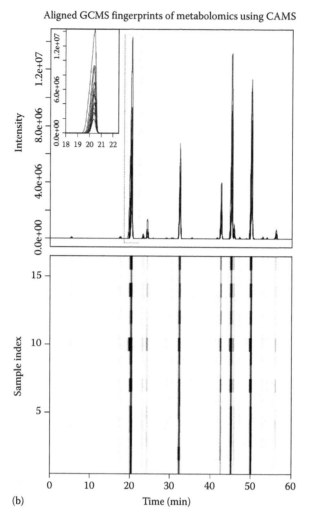

(b)

FIGURE 9.16 (*Continued*)   (a and b) Total ion chromatograms (TIC) of 16 blood serum samples of rats with left part illustrating TIC before alignment and right part for the aligned TIC using CAMS. (Reprinted from *J. Chromatogr. A*, 1286, Zheng, Y.-B., Zhang, Z.-M., Liang, Y.-Z. et al., Application of fast Fourier transform cross-correlation and mass spectrometry data for accurate alignment of chromatograms, 175–182. Copyright 2013, with permission from Elsevier.)

TABLE 9.2
Comparison of Three Alignment Algorithms

Algorithm	Platform	Average r	Computation Time (s)
COW	MATLAB 2011b	0.9172	3.2625 ± 0.0366
RAFFT	MATLAB 2011b	0.9564	0.0038 ± 0.000
CAMS	MATLAB 2011b	0.9586	0.0620 ± 0.0007

Source: Reprinted from *J. Chromatogr. A*, 1286, Zheng, Y.-B., Zhang, Z.-M., Liang, Y.-Z. et al., Application of fast Fourier transform cross-correlation and mass spectrometry data for accurate alignment of chromatograms, 175–182. Copyright 2013, with permission from Elsevier.

9.6 CONCLUSIONS AND PERSPECTIVES

Chromatography with various detectors offers a powerful tool for separating complex compounds to create characteristic chromatograms. The information of complex samples should be mined from chromatograms, which are certainly large datasets. It is not a trivial task, and they should be processed sufficiently and systematically through chemometric methods. Therefore, in this chapter, we have devoted attention to the use of chemometric methods for handling chromatograms. Smoothing, baseline correction, peak alignment, and peak detection are discussed to improve the quality of the generated chromatograms. The researchers can generate chromatograms for complex samples with chromatography and preprocess, analyze, interpret, and extract useful information from these datasets within an acceptable time using chemometric techniques.

REFERENCES

1. J.M. Amigo, T. Skov, R. Bro, ChroMATHography: Solving chromatographic issues with mathematical models and intuitive graphics, *Chem. Rev.* 110 (2010) 4582–4605.
2. W.B. Dunn, D.I. Ellis, Metabolomics: Current analytical platforms and methodologies, *Trends Anal. Chem.* 24 (2005) 285–294.
3. J. Lisec, N. Schauer, J. Kopka et al., Gas chromatography mass spectrometry–based metabolite profiling in plants, *Nat. Protoc.* 1 (2006) 387–396.
4. K. Dettmer, P.A. Aronov, B.D. Hammock, Mass spectrometry-based metabolomics, *Mass Spectrom. Rev.* 26 (2007) 51–78.
5. R. Aebersold, M. Mann, Mass spectrometry-based proteomics, *Nature* 422 (2003) 198–207.
6. N.J. Krogan, G. Cagney, H. Yu et al., Global landscape of protein complexes in the yeast *Saccharomyces cerevisiae*, *Nature* 440 (2006) 637–643.
7. A. Cifuentes, Food analysis and foodomics, *J. Chromatogr. A* 1216 (2009) 7109.
8. M. Herrero, C. Simó, V. García-Cañas et al., Foodomics: MS-based strategies in modern food science and nutrition, *Mass Spectrom. Rev.* 31 (2012) 49–69.
9. M. Castro-Puyana, M. Herrero, Metabolomics approaches based on mass spectrometry for food safety, quality and traceability, *Trends Anal. Chem.* 52 (2013) 74–87.
10. Y.-Z. Liang, P. Xie, K. Chan, Quality control of herbal medicines, *J. Chromatogr. B* 812 (2004) 53–70.
11. P. Xie, S. Chen, Y. Liang et al., Chromatographic fingerprint analysis—A rational approach for quality assessment of traditional Chinese herbal medicine, *J. Chromatogr. A* 1112 (2006) 171–180.
12. Y. Jiang, B. David, P. Tu et al., Recent analytical approaches in quality control of traditional Chinese medicines—A review, *Anal. Chim. Acta* 657 (2010) 9–18.
13. Z. Zhang, Y. Liang, P. Xie et al., Chromatographic fingerprinting and chemometric techniques for quality control of herb medicines, in: J. Poon, S.K. Poon (Eds.), *Data Analytics for Traditional Chinese Medicine Research*, Springer International Publishing, Sydney, Australia, 2014, pp. 133–153.
14. M.C. Bruzzoniti, R.M.D. Carlo, C. Sarzanini, The challenging role of chromatography in environmental problems, *Chromatographia* 73 (2011) 15–28.
15. O. Núñez, H. Gallart-Ayala, C.P.B. Martins et al., New trends in fast liquid chromatography for food and environmental analysis, *J. Chromatogr. A* 1228 (2012) 298–323.
16. H. Shaaban, T. Górecki, Current trends in green liquid chromatography for the analysis of pharmaceutically active compounds in the environmental water compartments, *Talanta* 132 (2015) 739–752.
17. Z.-M. Zhang, X. Tong, Y. Peng et al., Multiscale peak detection in wavelet space, *Analyst* 140 (2015) 7955–7964.
18. P. Du, W.A. Kibbe, S.M. Lin, Improved peak detection in mass spectrum by incorporating continuous wavelet transform-based pattern matching, *Bioinformatics* 22 (2006) 2059–2065.
19. O.M. Kvalheim, Y.Z. Liang, Heuristic evolving latent projections: Resolving two-way multicomponent data. 1. Selectivity, latent-projective graph, datascope, local rank, and unique resolution, *Anal. Chem.* 64 (1992) 936–946.
20. Y.Z. Liang, O.M. Kvalheim, H.R. Keller et al., Heuristic evolving latent projections: Resolving two-way multicomponent data. 2. Detection and resolution of minor constituents, *Anal. Chem.* 64 (1992) 946–953.
21. R. Tauler, Multivariate curve resolution applied to second order data, *Chemom. Intell. Lab. Syst.* 30 (1995) 133–146.

22. R. Tauler, A. Smilde, B. Kowalski, Selectivity, local rank, three-way data analysis and ambiguity in multivariate curve resolution, *J. Chemom.* 9 (1995) 31–58.

23. P.A. Gorry, General least-squares smoothing and differentiation by the convolution (Savitzky-Golay) method, *Anal. Chem.* 62 (1990) 570–573.

24. A. Savitzky, A historic collaboration, *Anal. Chem.* 61 (1989) 921A–923A.

25. E.T. Whittaker, On a new method of graduation, *Proc. Edinb. Math. Soc.* 41 (1922) 63–75.

26. P.J. Green, B.W. Silverman, *Nonparametric Regression and Generalized Linear Models: A roughness Penalty Approach*, 1st edn., Chapman and Hall/CRC, London, U.K./New York, 1993.

27. Y.-Z. Liang, A.K.-M. Leung, F.-T. Chau, A roughness penalty approach and its application to noisy hyphenated chromatographic two-way data, *J. Chemom.* 13 (1999) 511–524.

28. P.H.C. Eilers, A perfect smoother, *Anal. Chem.* 75 (2003) 3631–3636.

29. X. Liu, Z. Zhang, Y. Liang et al., Baseline correction of high resolution spectral profile data based on exponential smoothing, *Chemom. Intell. Lab. Syst.* 139 (2014) 97–108.

30. R.G. Brown, *Smoothing, Forecasting and Prediction of Discrete Time Series*, Courier Corporation, New York, 2004.

31. C.C. Holt, Forecasting seasonals and trends by exponentially weighted moving averages, *Int. J. Forecast.* 20 (2004) 5–10.

32. Y. Liu, X.G. Zhou, Y.D. Yu, A concise iterative method with Bezier technique for baseline construction, *Analyst* 140 (2015) 7984–7996.

33. F. Gan, G. Ruan, J. Mo, Baseline correction by improved iterative polynomial fitting with automatic threshold, *Chemom. Intell. Lab. Syst.* 82 (2006) 59–65.

34. C.A. Lieber, A. Mahadevan-Jansen, Automated method for subtraction of fluorescence from biological Raman spectra, *Appl. Spectrosc.* 57 (2003) 1363–1367.

35. Y. Xie, L. Yang, X. Sun et al., An auto-adaptive background subtraction method for Raman spectra, *Spectrochim. Acta A Mol. Biomol. Spectrosc.* 161 (2016) 58–63.

36. Y.-J. Yu, Q.-L. Xia, S. Wang et al., Chemometric strategy for automatic chromatographic peak detection and background drift correction in chromatographic data, *J. Chromatogr. A* 1359 (2014) 262–270.

37. S. He, W. Zhang, L. Liu et al., Baseline correction for Raman spectra using an improved asymmetric least squares method, *Anal. Methods* 6 (2014) 4402–4407.

38. Z.-M. Zhang, Y.-Z. Liang, Comments on the baseline removal method based on quantile regression and comparison of several methods, *Chromatographia* 75 (2012) 313–314.

39. Z.-M. Zhang, S. Chen, Y.-Z. Liang, Baseline correction using adaptive iteratively reweighted penalized least squares, *Analyst* 135 (2010) 1138–1146.

40. S. Chen, X.-N. Li, Y.-Z. Liang et al., Raman spectroscopy fluorescence background correction and its application in clustering analysis of medicines, *Spectrosc. Spectr. Anal.* 30 (2010) 2157–2160.

41. Z.-M. Zhang, S. Chen, Y.-Z. Liang et al., An intelligent background-correction algorithm for highly fluorescent samples in Raman spectroscopy, *J. Raman Spectrosc.* 41 (2010) 659–669.

42. W. Cheung, Y. Xu, C.L.P. Thomas et al., Discrimination of bacteria using pyrolysis-gas chromatography-differential mobility spectrometry (Py-GC-DMS) and chemometrics, *Analyst* 134 (2009) 557–563.

43. J. Carlos Cobas, M.A. Bernstein, M. Martín-Pastor et al., A new general-purpose fully automatic baseline-correction procedure for 1D and 2D NMR data, *J. Magn. Reson.* 183 (2006) 145–151.

44. V. Mazet, C. Carteret, D. Brie et al., Background removal from spectra by designing and minimising a non-quadratic cost function, *Chemom. Intell. Lab. Syst.* 76 (2005) 121–133.

45. P.H.C. Eilers, Parametric time warping, *Anal. Chem.* 76 (2004) 404–411.

46. H.F.M. Boelens, R.J. Dijkstra, P.H.C. Eilers et al., New background correction method for liquid chromatography with diode array detection, infrared spectroscopic detection and Raman spectroscopic detection, *J. Chromatogr. A* 1057 (2004) 21–30.

47. C.G. Bertinetto, T. Vuorinen, Automatic baseline recognition for the correction of large sets of spectra using continuous wavelet transform and iterative fitting, *Appl. Spectrosc.* 68 (2014) 155–164.

48. Y. Liu, W. Cai, X. Shao, Intelligent background correction using an adaptive lifting wavelet, *Chemom. Intell. Lab. Syst.* 125 (2013) 11–17.

49. Q. Bao, J. Feng, F. Chen et al., A new automatic baseline correction method based on iterative method, *J. Magn. Reson.* 218 (2012) 35–43.

50. Y. Hu, T. Jiang, A. Shen et al., A background elimination method based on wavelet transform for Raman spectra, *Chemom. Intell. Lab. Syst.* 85 (2007) 94–101.

51. X.-G. Shao, A.K.-M. Leung, F.-T. Chau, Wavelet: A new trend in chemistry, *Acc. Chem. Res.* 36 (2003) 276–283.

52. X. Shao, W. Cai, Z. Pan, Wavelet transform and its applications in high performance liquid chromatography (HPLC) analysis, *Chemom. Intell. Lab. Syst.* 45 (1999) 249–256.

53. Z. Pan, X.-G. Shao, H. Zhong et al., Correction of baseline drift in high-performance liquid chromatography by wavelet transform, *Chin. J. Anal. Chem.* 24 (1996) 149–153.

54. M.N. Leger, A.G. Ryder, Comparison of derivative preprocessing and automated polynomial baseline correction method for classification and quantification of narcotics in solid mixtures, *Appl. Spectrosc.* 60 (2006) 182–193.

55. C.D. Brown, L. Vega-Montoto, P.D. Wentzell, Derivative preprocessing and optimal corrections for baseline drift in multivariate calibration, *Appl. Spectrosc.* 54 (2000) 1055–1068.

56. Y. Wu, Q. Gao, Y. Zhang, A robust baseline elimination method based on community information, *Digit. Signal Process.* 40 (2015) 53–62.

57. Ł. Komsta, Comparison of several methods of chromatographic baseline removal with a new approach based on quantile regression, *Chromatographia* 73 (2011) 721–731.

58. K.H. Liland, E.-O. Rukke, E.F. Olsen et al., Customized baseline correction, *Chemom. Intell. Lab. Syst.* 109 (2011) 51–56.

59. A.F. Ruckstuhl, M.P. Jacobson, R.W. Field et al., Baseline subtraction using robust local regression estimation, *J. Quant. Spectrosc. Radiat. Transf.* 68 (2001) 179–193.

60. Z. Li, D.-J. Zhan, J.-J. Wang et al., Morphological weighted penalized least squares for background correction, *Analyst* 138 (2013) 4483–4492.

61. R. Perez-Pueyo, M.J. Soneira, S. Ruiz-Moreno, Morphology-based automated baseline removal for Raman spectra of artistic pigments, *Appl. Spectrosc.* 64 (2010) 595–600.

62. R.W. Koenker, G.W. Bassett, Four (pathological) examples in asymptotic statistics, *Am. Stat.* 38 (1984) 209–212.

63. S. Portnoy, R. Koenker, The Gaussian hare and the Laplacian tortoise: Computability of squared-error versus absolute-error estimators, *Stat. Sci.* 12 (1997) 279–300.

64. P.W. Holland, R.E. Welsch, Robust regression using iteratively reweighted least-squares, *Commun. Stat. Theory Methods* 6 (1977) 813–827.

65. D.B. Rubin, Iteratively reweighted least squares, in: S. Kotz, C.B. Read, N. Balakrishnan, B. Vidakovic (Eds.), *Encyclopedia of the Statistical Science*, John Wiley & Sons, Inc., New York, 2004.

66. P.J. Green, Iteratively reweighted least squares for maximum likelihood estimation, and some robust and resistant alternatives, *J. R. Stat. Soc. Ser. B Methodol.* 46 (1984) 149–192.

67. X. Liu, Z. Zhang, P.F.M. Sousa et al., Selective iteratively reweighted quantile regression for baseline correction, *Anal. Bioanal. Chem.* 406 (2014) 1985–1998.

68. C.P. Wang, T.L. Isenhour, Time-warping algorithm applied to chromatographic peak matching gas chromatography/Fourier transform infrared/mass spectrometry, *Anal. Chem.* 59 (1987) 649–654.

69. A. Kassidas, J.F. MacGregor, P.A. Taylor, Synchronization of batch trajectories using dynamic time warping, *AIChE J.* 44 (1998) 864–875.

70. N.-P.V. Nielsen, J.M. Carstensen, J. Smedsgaard, Aligning of single and multiple wavelength chromatographic profiles for chemometric data analysis using correlation optimised warping, *J. Chromatogr. A* 805 (1998) 17–35.

71. D. Clifford, G. Stone, I. Montoliu et al., Alignment using variable penalty dynamic time warping, *Anal. Chem.* 81 (2009) 1000–1007.

72. J. Forshed, I. Schuppe-Koistinen, S.P. Jacobsson, Peak alignment of NMR signals by means of a genetic algorithm, *Anal. Chim. Acta* 487 (2003) 189–199.

73. G.-C. Lee, D.L. Woodruff, Beam search for peak alignment of NMR signals, *Anal. Chim. Acta* 513 (2004) 413–416.

74. V. Pravdova, B. Walczak, D.L. Massart, A comparison of two algorithms for warping of analytical signals, *Anal. Chim. Acta* 456 (2002) 77–92.

75. A.M. van Nederkassel, M. Daszykowski, P.H.C. Eilers et al., A comparison of three algorithms for chromatograms alignment, *J. Chromatogr. A* 1118 (2006) 199–210.

76. W. Jiang, Z.-M. Zhang, Y. Yun et al., Comparisons of five algorithms for chromatogram alignment, *Chromatographia* 76 (2013) 1067–1078.

77. S. Salvador, P. Chan, Toward accurate dynamic time warping in linear time and space, *Intell. Data Anal.* 11 (2007) 561–580.

78. J.W.H. Wong, C. Durante, H.M. Cartwright, Application of fast Fourier transform cross-correlation for the alignment of large chromatographic and spectral datasets, *Anal. Chem.* 77 (2005) 5655–5661.

79. J.W.H. Wong, G. Cagney, H.M. Cartwright, SpecAlign—Processing and alignment of mass spectra datasets, *Bioinformatics* 21 (2005) 2088–2090.

80. F. Savorani, G. Tomasi, S.B. Engelsen, Icoshift: A versatile tool for the rapid alignment of 1D NMR spectra, *J. Magn. Reson.* 202 (2010) 190–202.
81. K.A. Veselkov, J.C. Lindon, T.M.D. Ebbels et al., Recursive segment-wise peak alignment of biological 1H NMR spectra for improved metabolic biomarker recovery, *Anal. Chem.* 81 (2009) 56–66.
82. M. Daszykowski, Y. Vander Heyden, C. Boucon et al., Automated alignment of one-dimensional chromatographic fingerprints, *J. Chromatogr. A* 1217 (2010) 6127–6133.
83. Z.-M. Zhang, Y.-Z. Liang, H.-M. Lu et al., Multiscale peak alignment for chromatographic datasets, *J. Chromatogr. A* 1223 (2012) 93–106.
84. M. Zhang, M. Wen, Z.-M. Zhang et al., Robust alignment of chromatograms by statistically analyzing the shifts matrix generated by moving window fast Fourier transform cross-correlation, *J. Sep. Sci.* 38 (2015) 965–974.
85. Z. Li, J.-J. Wang, J. Huang et al., Nonlinear alignment of chromatograms by means of moving window fast Fourier transform cross-correlation, *J. Sep. Sci.* 36 (2013) 1677–1684.
86. Y.-B. Zheng, Z.-M. Zhang, Y.-Z. Liang et al., Application of fast Fourier transform cross-correlation and mass spectrometry data for accurate alignment of chromatograms, *J. Chromatogr. A* 1286 (2013) 175–182.
87. Z.-M. Zhang, S. Chen, Y.-Z. Liang, Peak alignment using wavelet pattern matching and differential evolution, *Talanta* 83 (2011) 1108–1117.
88. B. Walczak, W. Wu, Fuzzy warping of chromatograms, *Chemom. Intell. Lab. Syst.* 77 (2005) 173–180.
89. K.J. Johnson, B.W. Wright, K.H. Jarman et al., High-speed peak matching algorithm for retention time alignment of gas chromatographic data for chemometric analysis, *J. Chromatogr. A* 996 (2003) 141–155.
90. R.J.O. Torgrip, M. Åberg, B. Karlberg et al., Peak alignment using reduced set mapping, *J. Chemom.* 17 (2003) 573–582.
91. A.L. Duran, J. Yang, L. Wang et al., Metabolomics spectral formatting, alignment and conversion tools (MSFACTs), *Bioinformatics* 19 (2003) 2283–2293.
92. M. Katajamaa, J. Miettinen, M. Orešič, MZmine: Toolbox for processing and visualization of mass spectrometry based molecular profile data, *Bioinformatics* 22 (2006) 634–636.
93. T. Pluskal, S. Castillo, A. Villar-Briones et al., MZmine 2: Modular framework for processing, visualizing, and analyzing mass spectrometry-based molecular profile data, *BMC Bioinformatics* 11 (2010) 395.
94. C.A. Smith, E.J. Want, G. O'Maille et al., XCMS: Processing mass spectrometry data for metabolite profiling using nonlinear peak alignment, matching, and identification, *Anal. Chem.* 78 (2006) 779–787.
95. A. Lommen, MetAlign: Interface-driven, versatile metabolomics tool for hyphenated full-scan mass spectrometry data preprocessing, *Anal. Chem.* 81 (2009) 3079–3086.
96. A. Lommen, H.J. Kools, MetAlign 3.0: Performance enhancement by efficient use of advances in computer hardware, *Metabolomics* 8 (2012) 719–726.
97. C. Yang, Z. He, W. Yu, Comparison of public peak detection algorithms for MALDI mass spectrometry data analysis, *BMC Bioinformatics* 10 (2009) 4.
98. C.E. Shannon, A mathematical theory of communication, *ACM SIGMOBILE Mob. Comput. Commun. Rev.* 5 (2001) 3–55.
99. F. Gong, Y.-Z. Liang, P.-S. Xie et al., Information theory applied to chromatographic fingerprint of herbal medicine for quality control, *J. Chromatogr. A* 1002 (2003) 25–40.

10 Alignment of One- and Two-Dimensional Chromatographic Signals

Michał Daszykowski

CONTENTS

10.1 INTRODUCTION

Nowadays, chromatographic methods are probably the most appreciated analytical techniques and well suited for the analysis of complex mixtures (samples). Their strongest advantages are primarily related with the possibility of separation of individual mixture components that can be further qualitatively and quantitatively determined. Moreover, the diversity of available chromatographic methods offers the possibility to analyze liquid and gas samples. These also use various types of detectors (single channel, specific, or multichannel detectors) and can easily be combined or extended in order to take advantage of separation carried out in orthogonal chromatographic systems and/or to gather complementary information that is collected from different types of detectors. Such advanced instrumental approaches support the improvement of the resolution power of a method and has the potential to decrease its detection limits.

Owing to the large resolution potential of chromatographic techniques, they are preferred to characterize complex samples or systems that use the so-called one- or two-dimensional chromatographic fingerprints [1] or the fusion of chromatographic data [2]. In general, chromatographic fingerprints are instrumental signals that are recorded for a given set of samples using the same chromatographic conditions whose aim is to reflect their chemical content as well as possible. They are regarded as unique sample fingerprints that help to explore the differences among samples and/ or the various states of a sampled system. A typical example of its application concerns exploring the differences that are observed between healthy and diseased patients, which are manifested in the chemical patterns of sampled biofluids that are characterized by a set of chromatographic fingerprints (e.g., proteomic and metabolomic profiles).

At this point, it is important to recall that the fingerprinting approach (also called the untargeted approach) is primarily focused on finding the differences among samples that are described by their chromatographic fingerprints and can be considered in the analytical workflow as an exploratory step of the analytical process.* Often, this stage of an experiment is facilitated to a great extent by using diverse chemometric techniques that assist in exploring the differences among samples, help in understanding relationships by means of multivariate data modeling, and facilitate the interpretation of the studied multidimensional data using, for instance, variable selection techniques [3]. Once the potential differences among samples are determined, if necessary, further identification and determination of individual mixture components can be carried out. Probably, the biggest advantage of the fingerprinting strategy is associated with eliminating the necessity of using expensive and often unavailable chromatographic standards at an early research stage. Therefore, the differences among studied samples can be examined using the multivariate modeling of chromatographic fingerprints, which are considered to be representations of chromatographic data, without having to construct a peak table.

10.2 ASSUMPTIONS AND PREREQUISITES FOR THE ALIGNMENT PROCESS

The comparative analysis of chromatographic fingerprints implicitly assumes that the chromatographic signals are recorded under the same and stable experimental conditions. Therefore, in a set of collected chromatogram peaks that originate from the same chemical components are expected to elute at the same position along the elution time axis. Unfortunately, in many situations this is not the case. Even slight deviations from the assumed experimental conditions affect the elution process. They are caused by, for instance, the effect of column aging, small fluctuations in the composition of the mobile phase, differences in the mobile phase viscosity, variations in the gradient program, changes of temperature, etc., and result in peak shifting across a collection of chromatographic fingerprints. Depending on the number of influencing external factors and the strength of their impact, the magnitude of peak shifting that is observed can vary substantially in different regions of a signal. When the location of the peaks in a signal is compared with a template fingerprint, the so-called target signal, they may elute faster or have a delay in different signal regions or both situations are possible. The peak shift issue is depicted in Figure 10.1, in which two exemplary chromatographic peaks, one from a target signal (a red line) and one from an aligned signal (a blue line), are translated by two elution time points.

The peak shifts that are observed in a collection of chromatograms are a major obstacle in their comparative analysis using different multivariate chemometric methods. The chemometric analysis of multivariate data usually involves data exploration and/or data modeling and therefore includes the use of a wide range of exploratory techniques (e.g., principal component analysis [4], clustering methods [5,6], etc.) and data modeling techniques such as calibration and discriminant or classification approaches (e.g., partial least squares regression [7], partial least squares discriminant analysis [8,9], SIMCA, one-class classifiers [10], etc.). All of these multivariate methods require the alignment of the corresponding data variables of one-dimensional chromatographic signals as a prerequisite, that is, the appropriate placement of the corresponding variables in the columns of a data matrix [11]. For two one-dimensional chromatographic signals, such as the ones portrayed in Figure 10.1, the corresponding data matrix (containing two rows and twenty columns) does not fulfill this fundamental assumption.

The correspondence of chromatographic data is also necessary when multichannel detectors are used. In this case, the chromatographic fingerprints can be organized as a collection of data arrays

---

* See Chapters 7 and 8.

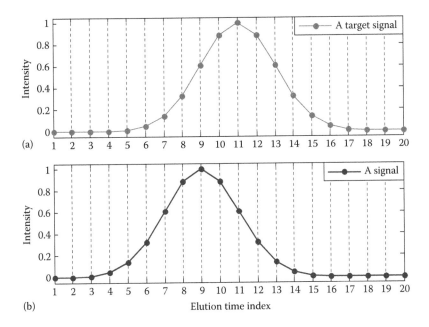

FIGURE 10.1 Two chromatographic peaks from (a) a target signal and (b) an aligned signal.

and each element in a data array represents the signal intensity that is registered at a given elution time and a certain detector channel (e.g., wavelength, m/z). Often, peak shifts are only observed along the chromatographic data dimension (elution time). However, in certain situations, the alignment of peaks may require corrections along two data dimensions or even across both dimensions.

Henceforth, in order to facilitate further discussion and for a better understanding of the peak shifting issue, we will make the following assumptions:

1. A chromatogram is an analytical signal that is sampled uniformly across the time domain, t_i (this implies sampling intervals of equal length, h, and the same sampling frequency for all chromatographic fingerprints, $t_i = ih$).
2. The same separation conditions are used for all of the chromatographic fingerprints.
3. All of the chromatographic fingerprints have the same sampling frequency and are recorded within the same range of the elution time.

10.3 ALIGNMENT OF CHROMATOGRAPHIC SIGNALS

The presence of peak shifts in signals introduces additional variability and thus has a substantial impact on the construction of different chemometric models and their final interpretation. Therefore, signal alignment is a necessary step. There are different correction mechanisms that can be used to compensate for the peaks' shifts in signals. They may involve modifications of the sampling rates, for instance, signal binning, warping (piecewise linear stretching and compression of certain signals sections) [12], and/or the insertion/deletion of certain signal elements [13].

One of the simplest approaches frequently considered for the correction of relatively small peak shifts in spectroscopic signals is known as the so-called bucketing or binning [14]. The reduction of peak shifts is achieved by decreasing a signal's resolution by integrating it over a number of sampling intervals. When peak shifts are relatively large and the peak shifting has an irregular behavior, the binning procedure is insufficient. Unfortunately, this is often the case when chromatographic signals are preprocessed.

Many alignment techniques have been proposed to compensate for the peak shifts in chromatographic signals. They are widely discussed in the literature along with many examples of their successful application. They primarily aim to correct the peak shifts that are observed between a target signal and a collection of chromatograms by finding the optimal transformation of the elution time axis. In the general transformation of m, the consecutive points of the signal being aligned, $x(t_i)$, with respect to a target signal, $y(t_i)$, can be expressed as:

$$y\left(t_i\right) = x\left(w\left(t_i\right)\right) \tag{10.1}$$

where
   $w(\cdot)$ is the alignment function
   $t_i = ih \ (i = 1, 2, ..., m)$
   h is a fixed sampling interval

As a result of this transformation, the peaks that are found in a chromatogram match the corresponding peaks in the target signal. As a matter of fact, there are at least four important technical aspects of signal alignment that may affect the final result. These include (1) preprocessing individual chromatographic signals, (2) selecting a target signal, (3) identifying the corresponding elution time points (peaks), and (4) finding the optimal parameters of the transformation (modeling of the alignment function).

10.3.1 PREPROCESSING OF INDIVIDUAL CHROMATOGRAPHIC SIGNALS

Often, prior to the alignment step, individual chromatographic signals are preprocessed [15]. In general, the aim is to improve their signal-to-noise ratio by suppressing the noise component and eliminating the negative effect of the baseline. This can be achieved using a number of different techniques [1]. Bearing in mind the alignment, presence of baseline seems to be the most crucial. Usually, its shape is different from signal to signal and the influence of the baseline is greater with increasing intensity. Therefore, many alignment algorithms are sensitive to the presence of the baseline.

10.3.2 SELECTION OF A TARGET SIGNAL

The selection of a target signal is as an essential step in the overall alignment workflow [16]. A target signal plays the role of a template for the alignment, and all chromatograms are aligned with respect to the corresponding features that are found in the target signal. Therefore, a good target candidate should be as representative as possible. Of course, there is no clear definition of representativity. However, one expects that a representative signal should, in general, enable an easier alignment process, that is, less demanding with respect to the degree of the necessary peak shift correction. Therefore, it can be regarded as a signal that contains as many of the common chromatographic features that are observed in a collection of chromatographic signals as possible. In many situations, identifying a good target candidate can be a challenging task. This can happen when the chromatograms describe complex mixtures that have relatively large differences in their chemical composition and/or when a set of chromatograms has a structure that is induced either by using the design of the experiments approach (DoE)* [17] or due to other sources of variation, for example, production season, quality grade, and steam pasteurization [18].

In the literature, there are several proposals for how to determine a target signal, but in general, they represent two groups of strategies. The first one assumes the selection of a target signal from

---

* See Chapter 1.

the available group of chromatograms by using a predefined principle, for instance, the largest average correlation coefficient that is observed between a potential candidate and the remaining signals [16]. The other one focuses on the construction of a synthetic target signal, which is artificially generated from the available set of chromatograms, for instance, a mean chromatogram. In certain situations, both target selection concepts may lead to a satisfactory alignment. However, a synthetic target signal (a mean chromatogram, a target that is created by taking the maximal values of the intensities that are observed at each point of the elution time, etc.) usually contains superfluous chromatographic features such as additional peaks and shoulders. Since the number of observed peaks in a synthetic target signal frequently increases compared to collected chromatograms, the relatively large degree of alignment flexibility that is offered by many alignment techniques favors the chance of a misalignment [16].

10.3.3 THE OPTIMAL ELUTION TIME AXIS TRANSFORMATION

Probably, one of the greatest challenges in alignment is connected with finding a suitable transformation of the elution time axis, which will eventually lead to the appropriate peak matching. Unfortunately, the correspondence of chromatographic peaks is *a priori* unknown. Therefore, matching them requires an objective criterion that is capable of scoring the overall quality of the alignment. Comparing a target signal with a transformed signal involves the selection of a similarity measure, for instance, the Euclidean distance or the correlation coefficient. In most applications, the result of peak matching is confirmed using the correlation coefficient score. Therefore, all of the features of a signal, including any local fluctuations of a signal, guide the alignment process and help in determining the optimal transformation parameters. Corrections of the peak shifts in chromatographic signals recorded using monochannel detectors is probably the most sophisticated alignment task because for peak matching, in the most promising situation, the shape of two aligned signals is the objective criterion.

A less popular strategy, but one that is steadily gaining popularity, encourages the use of multichannel detectors, such as diode-array or mass spectrometry detectors, which describe a portion of the eluent by its unique chemical fingerprint—a given spectrum. In such a case, finding the elution time point at which a similar mixture with respect to the chemical composition of the target chromatogram elutes seems to be easier. On the other hand, one has to deal with much larger and far more complex data compared to regular chromatograms because every sample is described by a data matrix that contains multiple chromatograms that are measured at different wavelengths or m/z channels.

As a result of the influence of external factors, peak shifts may have different magnitudes in specific signal regions. Some peaks may elute faster or slower compared to the position of the corresponding peaks in the target signal. Moreover, there are varied tendencies in peak shifting from sample to sample. Therefore, effective compensation for peak shifts usually requires the nonlinear transformation of the elution time axis in order to match the corresponding elution time points in the target signal. A simple linear transformation (translation) of a signal can only help when very simple elution time shifts need to be corrected. The general form of a transformation is either *a priori* assumed, for example, quadratic, or modeled directly by an approximation of the alignment function. In both situations, it is necessary to find the optimal coefficients of the alignment function. This can be done in a number of ways. For instance, in the correlation optimized warping approach, the alignment function is found by using the dynamic programming principle [12,19,20]. In parametric time warping, the coefficients of the alignment function, which has a quadratic form, are estimated in the course of the iterative procedure [21]. In the automated alignment approach, a general form of the alignment function is approximated using a number of basis functions (e.g., Gaussians, splines), whereas their coefficients are found using an optimization procedure [22].

Regardless of the alignment procedure being considered, the aim is to maximize the objective criterion, that is, the given similarity score that will be obtained by comparing the target signal and the aligned signal. Although finding the appropriate correspondence of peaks is very important, it is not the only concern during the alignment process. One also has to pay attention to the possible deformation of the peaks as a consequence of the large degree of flexibility of the alignment algorithms. For this reason, different measures of peak deformations that focus on monitoring the changes in their shape and location have been proposed in the literature [23].

10.3.4   ALIGNMENT OF ONE-DIMENSIONAL CHROMATOGRAPHIC SIGNALS

At this point, we will briefly describe a few alignment approaches in order to illustrate the major differences in how signals alignment is achieved.

10.3.4.1   Correlation Optimized Warping

Correlation optimized warping (COW) is one of the oldest and most popular alignment approaches [12]. The alignment process is realized by means of the piecewise linear stretching and compression of signal sections and comparing them with the corresponding sections in the target signal. At the beginning of the alignment procedure, an aligned signal and the target signal are divided into a number of sections. Then, except for the first and the last sections in a signal, the section being processed is either shortened or lengthened assuming different locations for the start and end points or is left unchanged. The positions of the section borders are permitted to change by a specified number of points, which are defined by the so-called slack parameter. In order to compare a given signal section with its corresponding target section, a linear interpolation is carried out when their lengths are different. In the course of the procedure, all possible combinations of locations of the sections are evaluated. When the optimal locations of the section borders are identified, the linear interpolation of these sections (warping) results in the largest overall correlation coefficient that can be observed between a signal and the target. This principle is referred to as dynamic programming [19].

The quality of alignment that is obtained using the COW method depends on two parameters that are specified by a user—the number of signal sections and the slack parameter, both of which strongly influence the flexibility of the alignment. In general, with a larger number of sections and a higher slack parameter, larger peak shifts can be successfully compensated. On the other hand, their selection should be done with care in order to minimize the risk of misalignment of the peaks. It should be also mentioned that one can obtain virtually the same alignment result for (1) a small number of sections and a relatively large value of the slack parameter and (2) a larger number of sections and a small value of the slack parameter. However, the computational effort is considerably smaller in the second situation because the number of possible sections (due to the different combinations of the ending positions) is substantially smaller. Bearing in mind the properties of the COW algorithm, the largest degree of alignment flexibility can be achieved in the middle region of a signal. Therefore, when a greater degree of alignment flexibility is required close to both of a signal's far sides, one can consider (1) using a larger number of sections, (2) extending a signal using a number of baseline points, or (3) augmenting a signal with its mirror image.

10.3.4.2   Parametric Time Warping

The parametric time warping approach aims to determine the optimal coefficients of the alignment function, $w(t_i)$, which has an *a priori* known form, for instance, the polynomial of the second degree [24]:

$$w(t_i) = \sum_{k=0}^{2} a_k t_i^2 = a_0 + a_1 t_i + a_2 t_i^2 \tag{10.2}$$

The coefficients of the alignment function, a_0, a_1, and a_2, are determined in the course of the iterative procedure that minimizes the sum of squared differences between the target signal and the signal after the transformation of the elution time axis. Bearing this criterion in mind, it is strongly recommended that similar chromatograms be aligned with respect to their amplitude. Moreover, the baseline component, which considerably contributes to the overall signal intensity, can affect the sum of squared differences. Therefore, after correcting the baseline, chromatograms are usually scaled using their average values. It is also advised to consider a smoothing procedure when chromatograms contain very sharp peaks in order to meet the convergence of the algorithm. Although this method is fast, in certain situations, the alignment flexibility that it offers may be insufficient to compensate for large peak shifts.

10.3.4.3 Automated Alignment

The automated alignment approach focuses explicitly on modeling the alignment function shape with a number of the basis functions, for instance, splines [22]:

$$w(t_i) = \sum_{j=1}^{k} a_j b_{ij}(t_i) \qquad (10.3)$$

where
a_j are the coefficients of the k basis functions ($j = 1, 2, ..., k$)
b_{ij} is the ith element of the jth basis function

The major goal of this method is to determine the optimal values of the coefficients for the basis functions that will lead to the optimal alignment result, which is scored with the correlation coefficient that is calculated between the aligned signal and the target signal. Identifying the optimal coefficient values can be done using any optimization procedure.

With an increasing number of the basis functions, a more complex shape of the alignment function can be approximated and thus, in general, a greater degree of alignment flexibility is achieved. Peak positions are corrected locally using the linear interpolation of a signal. The elution time axis of a signal is then transformed with respect to the alignment function. The degree of the peak shifting that is necessary depends on the absolute coefficient value that is associated with the basis function located within the peak region. The sign of the coefficient provides information about the direction of the correction that is required to match the corresponding peak in the target signal.

10.3.5 ALIGNMENT OF TWO-DIMENSIONAL CHROMATOGRAPHIC SIGNALS THAT IS GUIDED BY SPECTRAL INFORMATION

The use of multichannel detectors, for example, a diode-array detector or mass spectrometry, can assist in the reduction of the misalignment of peaks and thus guide the alignment process more efficiently. The correspondence of peaks that are present in the two chromatographic signals being considered can be verified by actively using the collected spectral information. It is assumed that the two peaks in the aligned signal and the target signal should be matched when the corresponding portions of the samples that are eluting within a certain elution time interval have a similar chemical composition. This can be confirmed by comparing the observed spectral profiles and scoring their similarity using a certain similarity measure, for instance, the correlation coefficient or the angle that is observed between two spectra. While the majority of the alignment algorithms identify corresponding peaks based solely on comparing their local shape (or distance), multichannel detectors open new and very attractive possibilities for designing new alignment algorithms that offer a better performance in terms of the alignment improvement. These new features are especially valuable when chromatographic fingerprints that describe very complex mixtures of a natural origin are compared, for example, in metabolomic and proteomic studies. In these cases, the use of advanced hyphenated chromatographic techniques such as gas or liquid chromatography coupled with mass spectrometry is a necessity.

In general, the available alignment algorithms that handle two-dimensional chromatographic signals belong to two categories: (1) peak-based and (2) raw data-based techniques [25]. Peak-based methods require the identification of chromatographic peaks and sometimes even their further deconvolution before the alignment step. Their performance strongly depends on the effectiveness of the appropriate detection of chromatographic peaks or features and, therefore, certain issues including the construction of peak tables and missing information about certain peaks have to be handled [26]. On the other hand, although alignment algorithms that operate on raw chromatographic data do not require peak identification, this feature is obtained at the cost of a considerably larger computational effort.

The landmark selection algorithm, which was proposed by Krebs et al. [27], is one of the early alignment algorithms proposed in the literature that takes advantage of multichannel information. It belongs to the first category proposed in the literature and focuses on the alignment of two-dimensional chromatograms using a set of landmark peaks, which are identified in two chromatograms assuming that the intensities of landmark peaks are above a certain threshold value in the total ion chromatogram (TIC). The landmark peaks that are found (within a certain tolerance of elution time) in both signals are then compared, and their correspondence is confirmed by the angle that is observed between two spectra ($s_{1,i}$ and $s_{2,i}$) observed at two elution time points:

$$\cos(\alpha) = \frac{\sum s_{1,i} s_{2,i}}{\sqrt{\sum s_{1,i}^2 \sum s_{2,i}^2}} \qquad (10.4)$$

Once the landmark peaks of two chromatographic signals are matched, that is, their correspondence is above the assumed $\cos(\alpha)$ level, information about their location along the elution time axis assists in the alignment of a collection of two-dimensional chromatograms. Signal sections of the aligned chromatograms that are located between the neighboring landmark peaks are warped using nonlinear signal interpolation. Similar alignment strategies that incorporate spectral information in order to minimize the misalignment issue are discussed in detail in References 28 and 29.

The raw data-based alignment techniques use the entire set of mass spectra (or their binned form) in order to verify their correspondence along the elution time axis. To reduce the computational effort that is required, the majority of such algorithms consider the elution time shift of a mass spectrum as a whole rather than performing an alignment of individual elution time points. Therefore, the raw data-based techniques identify the position of a spectrum along the elution time axis and the adjusted elution time. In order to perform an alignment, a pairwise similarity function is constructed for the TIC profiles or mass spectra. Locally identified correspondence is further analyzed in order to obtain the optimal alignment. Methods such as ObiWarp [30], signal maps [31], bidirectional best hits peak assignment and cluster extension [25], center-star multiple alignment by pairwise partitioned dynamic time warping [25], etc. are included in this category of alignment techniques. The demand for the effective processing of two-dimensional chromatographic signals strongly encourages the development of new alignment techniques that have an increased functionality or the improvement of the existing techniques.

10.3.6 Alignment of Two-Dimensional Gel Electrophoretic Images

The alignment techniques that have been discussed to this point assume that shifts of the peaks in a signal are observed along one data dimension—the elution time axis and that possible peak shifts in the spectral data dimension can be successfully compensated for with a simple binning procedure. However, in certain situations, the compensation for peak shifts may require a bidirectional correction. Such an alignment strategy has to be considered to diminish the effect of the peak shifts that are observed in a collection of two-dimensional images that are obtained from two-dimensional gel electrophoresis, which is the standard separation technique that is used for the analysis of proteins. Two-dimensional images of samples are obtained by first carrying out separation with respect to an isoelectric point (the first dimension) and then orthogonally using the differences in molecular weights on a polyacrylamide gel that is organized as a two-dimensional sheet (the second dimension) and then staining them using a specific staining method after the separation. The resulting data can be regarded as an image on which each spot is characterized by three coordinates—the location along the first dimension (X), the location along the second dimension (Y), and the signal's intensity. In order to compare different samples, the protein spots that are observed on the electrophoretic images have to be aligned with respect to the location of the corresponding protein spots that are located on a target image. This can be done using an alignment procedure that determines the correspondence of the protein spots according to the fuzzy warping principle [32–34].

At first, a number of the most intense protein spots is identified in the target image as well as a candidate for the alignment, and these spots then serve as specific landmarks for the alignment. In order to ensure the optimal alignment, their number should be relatively large (usually from 200 to 500) and they should cover the image domain as well as possible. The next step consists of finding the correspondence between pairs of protein spots. This is done in the course of an iterative procedure that (1) verifies the correspondence of the spots using the fuzzy matching principle taking into account their proximity on the electrophoretic image and (2) estimates the transformation parameters with respect to actual correspondence of the spots. After the procedure, the one-to-one correspondence of the spots is established in the course of the Sinkhorn iterative standardization procedure [35], which is applied to the rows and columns of the correspondence matrix. This defines the most likely matching peaks as scored by the output of the two-dimensional Gaussian membership function (located at the centers of the spots on the target image). At the stage of identifying the correspondence of the spots, a global transformation of signals is applied (polynomial of the second order) that maximizes the correlation coefficient. Once the correspondence among the protein spots is found, the alignment of the image is further tuned locally using the piecewise linear transformation. A more detailed description of the algorithm can be found in References 32 and 34.

10.4   OTHER ALTERNATIVES TO SIGNAL ALIGNMENT

Another interesting approach that permits the chemometric analysis of two-dimensional chromatographic fingerprints has been proposed by Danielsson et al. [36] and Ullsten et al. [37], who have described a fingerprinting study of urine chromatographic fingerprints that were obtained from capillary electrophoresis using mass spectrometry detection (CE-MS), which were compared without the necessity for their prior alignment. The main idea behind this strategy lies in the construction of a so-called blurred or fuzzy correlation matrix that is transformed into similarity matrices and their later use to compare the two-dimensional fingerprints using, for example, principal component analysis. The concept proposed is very attractive because it allows the troublesome step of aligning the two-dimensional signals to be skipped. Therefore, it offers a considerable gain of time and can easily be applied in online settings.

In Reference 38, another principle was exploited in order to compare two-dimensional signals without their prior alignment. The data dimension along which the peak shifts are observed can be removed by representing each of the two-dimensional signals as a so-called Gram matrix. A Gram matrix is the product of the data matrix and its transposed variant, which is constructed either as $\mathbf{XX}^T$ or $\mathbf{X}^T\mathbf{X}$. In the case of two-dimensional signals, it is obtained in such a way that the chromatographic dimension (time axis) where peak shifts are observed is removed. For instance, for a CE-MS signal, its Gram matrix has the dimensions (*number of m/z channels* × *number of m/z channels*). To compare a set of two-dimensional signals, the individual Gram matrices are then unfolded into a vector form and used to construct a similarity matrix. Based on the similarity matrix, the chromatographic data can be further explored and the samples compared using, for example, principal component analysis or hierarchical clustering methods. It was also shown that the proposed no-alignment strategies (including the one based on Gram matrices) also permit the construction of discriminant models. This feature opens up the possibility to use two-dimensional fingerprints (as they are), for instance, for medical diagnostics. The performance of the proposed method was illustrated on a capillary electrophoresis-mass spectrometry fingerprints of urine samples and compared with approaches that were based on a "blurred" or "fuzzy" correlation matrix. It was demonstrated that similar conclusions about the experiment could be drawn. It should be pointed out that no input parameters are required in this approach.

10.5   CONCLUDING REMARKS

Alignment of chromatographic signals when they are used as fingerprints for comparative analysis is a very important preprocessing step in the general experimental workflow. Since even small variations in the experimental conditions or fluctuations in the sample properties can affect the elution process, in many situations one cannot neglect the peak shift issue. Although using the fingerprinting approach and their further comparative analysis is very attractive from the practical point of view, signal alignment is not an easy task. Despite the relatively wide range of possibilities offered by the available alignment techniques, unfortunately they do not fully eliminate the risk of misalignment and any possible peak shape deformation. These are major concerns especially when the signals obtained using monochannel detectors are aligned. The relatively large alignment flexibility of algorithms and a certain level of ambiguity that are related to the cost function may negatively affect the final result of signal alignment. Therefore, the selection of the input parameters should be done with great care. Although in this respect, two-dimensional chromatographic fingerprints are far less prone to peak misalignments, they require more advanced alignment techniques and a larger computational effort.

REFERENCES

1. Daszykowski, M. and B. Walczak. 2006. Use and abuse of chemometrics in chromatography, *Trends in Analytical Chemistry*, 25(11): 1081–1096.
2. Stanimirova, I., C. Boucon, and B. Walczak. 2011. Relating gas chromatographic profiles to sensory measurements describing the end products of the Maillard reaction, *Talanta*, 83(4): 1239–1246.
3. Andersen, C.M. and R. Bro. 2010. Variable selection in regression—A tutorial, *Journal of Chemometrics*, 24(11–12): 728–737.
4. Wold, S., K. Esbensen, and P. Geladi. 1987. Principal component analysis, *Chemometrics and Intelligent Laboratory Systems*, 2(1–3): 37–52.
5. Massart, D.L. and L. Kaufman. 1989. *The Interpretation of Analytical Chemical Data by the Use of Cluster Analysis*, Krieger Publishing Co, New York.
6. Bratchell, N. 1987. Cluster analysis, *Chemometrics and Intelligent Laboratory Systems*, 6(2): 105–125.
7. Geladi, P. and B.R. Kowalski. 1986. Partial least-squares regression: A tutorial, *Analytica Chimica Acta*, 185: 1–17.
8. Barker, M. and W. Rayens. 2003. Partial least squares for discrimination, *Journal of Chemometrics*, 17(3): 166–173.
9. Brereton, R.G. and G.R. Lloyd. 2014. Partial least squares discriminant analysis: Taking the magic away, *Journal of Chemometrics*, 28(4): 213–225.
10. Brereton, R.G. 2011. One-class classifiers, *Journal of Chemometrics*, 25(5): 225–246.
11. Malmquist, G. and R. Danielsson. 1994. Alignment of chromatographic profiles for principal component analysis: A prerequisite for fingerprinting methods, *Journal of Chromatography A*, 687(1): 71–88.
12. Nielsen, N.-P.V., J.M. Carstensen, and J. Smedsgaard. 1998. Aligning of single and multiple wavelength chromatographic profiles for chemometric data analysis using correlation optimised warping, *Journal of Chromatography A*, 805(1–2): 17–35.
13. Tomasi, G., F. Savorani, and S.B. Engelsen. 2011. Icoshift: An effective tool for the alignment of chromatographic data, *Journal of Chromatography A*, 1218(43): 7832–7840.
14. Jellema, R.H. 2009. Variable shift and alignment. In *Comprehensive Chemometrics*, Elsevier, Amsterdam, the Netherlands.
15. Amigo, J.M., T. Skov, and R. Bro. 2010. ChroMATHography: Solving chromatographic issues with mathematical models and intuitive graphics, *Chemical Reviews*, 110(8): 4582–4605.

16. Daszykowski, M. and B. Walczak. 2007. Target selection for alignment of chromatographic signals obtained using monochannel detectors, *Journal of Chromatography A*, 1176(1–2): 1–11.

17. Montgomery, D.C. 2012. *Design and Analysis of Experiments*, John Wiley & Sons, Incorporated, Washington, DC.

18. Stanimirova, I., M. Kazura, D. de Beer, E. Joubert, A.E. Schulze, T. Beelders, A. de Villiers, and B. Walczak. 2013. High-dimensional nested analysis of variance to assess the effect of production season, quality grade and steam pasteurization on the phenolic composition of fermented rooibos herbal tea, *Talanta*, 115(October): 590–599.

19. Tomasi, G., F. Van Den Berg, and C. Andersson. 2004. Correlation optimized warping and dynamic time warping as preprocessing methods for chromatographic data, *Journal of Chemometrics*, 18(5): 231–241.

20. Pravdova, V., B. Walczak, and D.L. Massart. 2002. A comparison of two algorithms for warping of analytical signals, *Analytica Chimica Acta*, 456(1): 77–92.

21. Eilers, P.H.C. 2004. Parametric time warping, *Analytical Chemistry*, 76(2): 404–411.

22. Daszykowski, M., Y. Vander Heyden, C. Boucon, and B. Walczak. 2010. Automated alignment of one-dimensional chromatographic fingerprints, *Journal of Chromatography A*, 1217(40): 6127–6133.

23. Skov, T., F. Van Den Berg, G. Tomasi, and R. Bro. 2006. Automated alignment of chromatographic data, *Journal of Chemometrics*, 20(11–12): 484–497.

24. Eilers, P.H.C. 2004. Fast computation of trends in scatter plots, *Kwantitatieve Methoden*, 71: 38–45.

25. Hoffmann, N., K. Matthias, H. Neuweger, M. Wilhelm, P. Högy, K. Niehaus, and J. Stoye. 2012. Combining peak- and chromatogram-based retention time alignment algorithms for multiple chromatography-mass spectrometry datasets, *BMC Bioinformatics*, 13: 214.

26. Smith, C.A., E.J. Want, G. O'Maille, R. Abagyan, and G. Siuzdak. 2006. XCMS: Processing mass spectrometry data for metabolite profiling using nonlinear peak alignment, matching, and identification, *Analytical Chemistry*, 78(3): 779–787.

27. Krebs, M.D., R.D. Tingley, J.E. Zeskind, M.E. Holmboe, J.-M. Kang, and C.E. Davis. 2006. Alignment of gas chromatography–mass spectrometry data by landmark selection from complex chemical mixtures, *Chemometrics and Intelligent Laboratory Systems*, 81(1): 74–81.

28. Styczynski, M.P., J.F. Moxley, L.V. Tong, J.L. Walther, K.L. Jensen, and G.N. Stephanopoulos. 2007. Systematic identification of conserved metabolites in GC/MS data for metabolomics and biomarker discovery, *Analytical Chemistry*, 79(3): 966–973.

29. Robinson, M.D., D.P. De Souza, W.W. Keen, E.C. Saunders, M.J. McConville, T.P. Speed, and V.A. Likić. 2007. A dynamic programming approach for the alignment of signal peaks in multiple gas chromatography-mass spectrometry experiments, *BMC Bioinformatics*, 8: 419.

30. Prince, J.T. and E.M. Marcotte. 2006. Chromatographic alignment of ESI-LC-MS proteomics data sets by ordered bijective interpolated warping, *Analytical Chemistry*, 78(17): 6140–6152.

31. Prakash, A., P. Mallick, J. Whiteaker, H. Zhang, A. Paulovich, M. Flory, H. Lee, R. Aebersold, and B. Schwikowski. 2006. Signal maps for mass spectrometry-based comparative proteomics, *Molecular & Cellular Proteomics*, 5(3): 423–432.

32. Kaczmarek, K., B. Walczak, S. de Jong, and B.G.M. Vandeginste. 2002. Feature based fuzzy matching of 2D gel electrophoresis images, *Journal of Chemical Information and Computer Sciences*, 42: 1431–1442.

33. Daszykowski, M., I. Stanimirova, A. Bodzon-Kulakowska, J. Silberring, G. Lubec, and B. Walczak. 2007. Start-to-end processing of two-dimensional gel electrophoretic images, *Journal of Chromatography A*, 1158(1–2): 306–317.

34. Daszykowski, M., E. Mosleth Færgestad, H. Grove, H. Martens, and B. Walczak. 2009. Matching 2D gel electrophoresis images with Matlab "Image Processing Toolbox", *Chemometrics and Intelligent Laboratory Systems*, 96(2): 188–195.

35. Sinkhorn, R. 1964. A relationship between arbitrary positive matrices and doubly stochastic matrices, *The Annals of Mathematical Statistics*, 35(2): 876–879.

36. Danielsson, R., D. Bäckström, and S. Ullsten. 2006. Rapid multivariate analysis of LC/GC/CE data (single or multiple channel detection) without prior peak alignment, *Chemometrics and Intelligent Laboratory Systems*, 84(1–2): 33–39.

37. Ullsten, S., R. Danielsson, D. Bäckström, P. Sjöberg, and J. Bergquist. 2006. Urine profiling using capillary electrophoresis-mass spectrometry and multivariate data analysis, *Journal of Chromatography A*, 1117(1): 87–93.

38. Daszykowski, M., R. Danielsson, and B. Walczak. 2008. No-alignment-strategies for exploring a set of two-way data tables obtained from capillary electrophoresis-mass spectrometry, *Journal of Chromatography A*, 1192(1): 157–165.

11 Peak Purity and Resolution of Chromatographic Data

Sílvia Mas and Anna de Juan

CONTENTS

11.1 CHROMATOGRAPHY: CHEMICAL PROBLEMS, DATA STRUCTURE, AND UNDERLYING MODEL

Nowadays, many methods for the identification and determination of analytes in complex matrices are based on chromatographic techniques coupled with diverse detection systems. A chromatographic process can be defined as the separation of a mixture by sequential elution of their compounds. However, there are an increasing number of situations where it may not be feasible to obtain an adequate separation in a reasonable time, either because of the complexity of the mixture (sample) or because of the time-consuming procedures needed to provide optimal separation conditions. Moreover, problems such as impurities under main peaks, high noise level in some detection systems, and drifts in the baseline may decrease the quality of the final result of the analysis.

    Chemometrics can help to alleviate the problems mentioned above by means of many available tools. Preprocessing procedures are oriented to improve the signal quality and to remove uninformative variation, such as baselines. Exploratory approaches help to define the complexity of chromatographic data in terms of number and overlap of eluting compounds, to clarify elution patterns, and to detect impurities and minor compounds with methodologies much more performing than classical visual inspection. Finally, factor analysis–based resolution methods are the definitive tool to provide the mathematical separation of all compounds in the chromatographic run, providing elution and pure response profiles for each of them even when instrumental overlap among signals exists. Thus, resolution methods and chromatography aim at the same objective, providing information of the pure compounds of a mixture either by mathematical or by chemical means, respectively. Nowadays, the use of these chemometric methodologies is widely accepted in the chromatographic field, not to replace chemical ways to improve separation but as a means to support good analytical work when the complexity of the problem (linked to systems with a very high number of compounds or to massive data sets), quality of the instrumental signal, or low concentration of compounds requires more efficient ways to handle instrumental information.

The detector signals associated with chromatographic techniques can be univariate or multivariate. When a monochannel detector (e.g., UV, Flame Ionization Detector, conductivity) is used, only one number (e.g., absorbance at a single wavelength) is collected per each elution time. In this case, the measurement is univariate and the chromatogram is an array of numbers (vector). At present, multichannel detectors (e.g., UV-DAD, MS or fluorescence detectors) are mostly used, and they acquire a complete spectrum per elution time. Therefore, the measurement acquired from these detectors is multivariate and the chromatogram consists of a data table (matrix) with the rows containing the full spectra collected as a function of elution time, and the columns represent the chromatographic traces at each spectral channel.

When the chromatographic data tables are formed by a series of spectra collected as a function of time, the underlying model has the form of the Beer–Lambert law. The total raw mixed signal, represented by matrix $\mathbf{D}$, is the sum of the pure signal contributions coming from each eluted compound (see Figure 11.1a). Each one of these pure contributions ($\mathbf{D}_i$) can be expressed by the product of a dyad of profiles, $\mathbf{c}_i \mathbf{s}_i^T$, where $\mathbf{s}_i^T$ is the pure spectrum of the component weighted by the related elution profile, $\mathbf{c}_i$. This additive model of individual contributions (see Figure 11.1b) is equivalent to the compact representation of the bilinear chromatographic model, $\mathbf{D} = \mathbf{CS}^T$, where the columns in matrix $\mathbf{C}$ are the elution profiles and the rows in matrix $\mathbf{S}^T$ are their related pure spectra (see Figure 11.1c).

This general bilinear model can be easily extended to simultaneous analysis of a chromatographic run of a single sample with several detection systems in tandem or to analysis of multiple samples with some compounds in common. The data set structures including more than one data matrix are called augmented matrices or multisets. Multisets can be structured as (1) a row-wise augmented matrix, when data tables are appended beside each other, for example, analysis of a chromatographic run monitored simultaneously by different detection systems; (2) a column-wise augmented matrix, when data tables are one on top of each other, for example, analysis of multiple chromatographic runs of the same sample with a single detection system; and (3) row- and column-wise augmented matrices, when the multiset extends in the row and column directions. Figure 11.2 shows the different kinds of data set arrangements for simultaneous analysis of hyphenated data and their corresponding bilinear models.

The underlying model of the chromatographic measurement is identical to the model used in factor analysis (FA) methods, the goal of which is decomposing a data table ($\mathbf{D}$) into a bilinear model of factors, representing the different sources of variance of the data [1]. In the context of chromatographic data, the factors are the eluting compounds and the profiles defining each factor are the elution profile ($\mathbf{c}_i$) and the pure response of the compound ($\mathbf{s}_i^T$). This coincidence helps to understand the variety and extensive use of FA-derived methods for analysis of chromatographic data. These methods have been widely applied for the determination of the number of compounds, peak purity problems, resolution of overlapped compounds, or extension to simultaneous analysis of multiple runs to obtain qualitative and quantitative information.

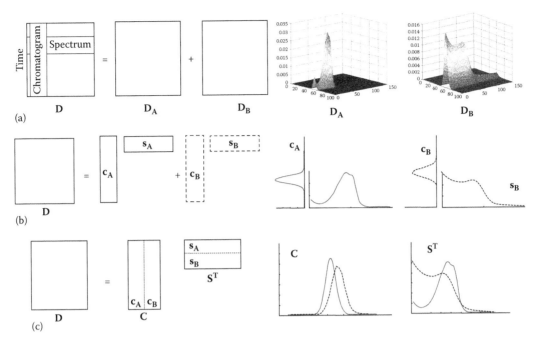

FIGURE 11.1 Measurement model of a two-component HPLC–DAD system expressed as (a) sum of pure signal contributions, (b) sum of dyads of pure concentration profile and spectrum, and (c) product of matrices of pure concentration profiles and spectra.

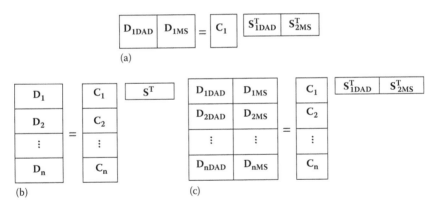

FIGURE 11.2 Augmented bilinear models. (a) Model for a row-wise augmented matrix, for example, a single chromatographic run with DAD and MS detection; (b) model for a column-wise augmented matrix, for example, several chromatographic runs with a common detection system; and (c) model for a row- and column-wise augmented matrix, for example, several chromatographic runs with DAD and MS detection in tandem.

FA-based methods that work with mathematical factors provide exploratory information (see Section 11.3), whereas methods oriented to the recovery of chemically meaningful factors attempt to achieve the real resolution of the chromatographic systems (see Section 11.4).

11.2   PREPROCESSING IN CHROMATOGRAPHY

In order to enhance the quality of chromatographic signals and, as a consequence, the performance of factor analysis methods, the use of preprocessing oriented to baseline correction, improvement of spectral signal, or data compression should be done before further analysis.

11.2.1 BASELINE CORRECTION

Chromatographic data analysis can be improved by decreasing noise in the detected signal and by suppressing background contributions; hence, baseline correction is a crucial step to reduce complexity and uninformative sources of variation. There are several algorithms that can be appropriate to overcome this problem. Many of the classical approaches were based on subtracting a fitted polynomial function following the baseline curvature. To do so, some reference baseline points were taken for fitting. Nowadays, there are approaches that may adapt better to the often irregular baselines found in chromatography. One of the most efficient and widely used is the asymmetric least squares (AsLS) method by Eilers [2,3].

AsLS is based on the recursive use of a Whittaker smoother to fit a baseline to the chromatogram. To do so, two parameters are used to control the baseline fitting (see Equation 11.1), one associated with the smoothness of the fit (λ) and the other with the penalty imposed to the measurement readings related to elution times providing positive residuals, that is, to measurements of elution times related to chromatographic peaks, which are above the fitted baseline (p). The error function, S, minimized is shown below:

$$S = \sum_i \omega_i \left(y_i - z_i \right)^2 + \lambda \sum_i \left(\Delta^2 z_i \right)^2 \tag{11.1}$$

where y is the signal to correct and z the fitted baseline, $\omega_i = p$ if $y_i > z_i$ or otherwise $\omega_i = 1 - p$, and $\Delta^2 z_i = z_i - 2z_{i-1} + z_{i-2}$.

AsLS can adapt to any kind of shape and intensity of baseline and does not work using a predefined mathematical function to define the baseline shape. The only requirement for AsLS to work is that the frequencies of the baseline and of the relevant signal features must be very different, which is the case in chromatography, between the broad baseline to be subtracted and the narrow chromatographic peaks. When having chromatographic data tables, AsLS can be applied sequentially to each of the chromatographic traces collected at the different spectral channels.

Figure 11.3 shows an example of HPLC–DAD chromatogram before and after baseline correction by AsLS.

11.2.2 SPECTRA PREPROCESSING AND DATA COMPRESSION

In a chromatographic signal, preprocessing in the spectral direction is useful to increase the signal-to-noise ratio for the peaks of interest and to remove unwanted artifacts or uninformative spectral regions. Therefore, the first step is selecting the spectral range of interest, which means often suppressing wavelength ranges in UV-DAD detectors where solvents and buffer solutions may absorb (typically 190–205 nm) or wavelength regions where the compounds of interest no longer absorb. This action removes unwanted sources of spectral variation and helps to reduce the data size.

Still, when chromatography is coupled to a UV-DAD detector, correction of mobile phase absorption may still be necessary in order to obtain more reliable analyte spectra. For this purpose, subtraction of the spectra of the eluent recorded at the beginning of the run or immediately before elution of the analytes of interest from the chromatogram spectra must be carried out.

In the case of fluorescence detection, it is also important to remove the possible scattering contributions, which can hamper the data analysis because of the unwanted nonlinear variation produced by this artifact in the spectral signal. Several methods are proposed in the literature based on removing specific bands of Raman and first and second Rayleigh scattering and replacing them with interpolated values [4,5].

The HPLC–MS chromatographic runs need to be compressed in the response direction by removing uninformative m/z channels related to noise. This compression is more necessary than ever before due to the high m/z resolution provided by the most advanced MS detection systems.

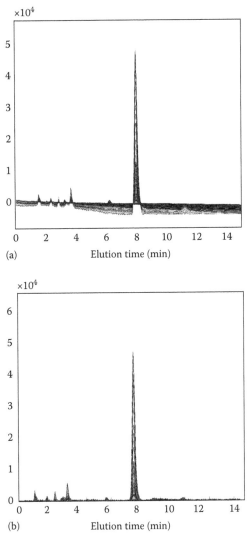

FIGURE 11.3 (a) HPLC-DAD raw data, (b) baseline corrected HPLC-DAD data by asymmetric least squares.

A classical way to compress MS data is performed by suppressing all m/z channels with mean intensity below a certain threshold. This threshold value could be defined as a percentage of the maximum intensity found in the chromatographic data analyzed (m/z channels with intensity lower than defined % of the maximum intensity channels are then removed).

Another approach used typically for MS compression consisted of binning a certain number of m/z channels to obtain a single numerical value [6]. Binning is a procedure that is less and less used because it implies a loss of spectral resolution and, hence, less capability to differentiate among very similar compounds and less accuracy in compound identification when the binned spectrum is compared with MS library spectra. To achieve a more MS-dedicated compression procedure that may provide a significantly larger decrease of m/z channels without losing spectral resolution, the so-called detection of regions of interest (ROI) technique has been proposed by Stolt et al. [7] and later on revisited by Bedia et al. [8]. This method is based on the idea that when an analyte elutes, there is a time elution range of data points with a high density of relevant measurements associated with very specific m/z channels. The regions where analytes are found and the related relevant m/z readings are called regions of interest (ROI). m/z channels related to ROIs are detected because a certain number of consecutive elution times (to be fixed by the user) present in these channels a signal with an intensity significantly higher than noise. The check for relevant m/z channels is done from the beginning to the end of the chromatogram scan by scan and only the relevant m/z channels are kept. When an m/z channel in a new scan is found to be significant, a last check is done to see whether it was already selected. To do so, a mass error tolerance is also set by the user to pool together all channels that are within the mass error allowed. Different from binning, data compression based on the search of ROI allows a much more efficient compression of the original LC–MS data with no loss of spectral resolution.

11.3 EXPLORATORY ANALYSIS OF CHROMATOGRAPHIC DATA: SAMPLE COMPLEXITY, ELUTION PATTERN, RESPONSE FEATURES, AND PEAK PURITY*

The exploratory analysis of chromatographic data provides relevant information about the system of interest that can also be used for further analysis (resolution and quantitative analysis). Exploratory analysis covers aspects such as (1) the estimation of the number of eluting compounds, including the detection of unexpected compounds, such as impurities and interferences; (2) knowledge of the elution pattern of compounds to detect selective elution regions and to define elution windows with different degree of compound overlap; and (3) the location of the most representative features or variables in both spectral and chromatographic direction.

Several exploratory methods for chromatographic data analysis have been suggested in the literature. The first option is to perform Principal Component Analysis (PCA) to estimate the overall complexity of the chromatographic run. Based on PCA and more dedicated to define the elution pattern and local elution complexity, local rank analysis methods, such as evolving factor analysis (EFA) [9–11] and fixed-size moving window evolving factor analysis (FSMW-EFA) [12,13], and evolutions of both methods were proposed. Finally, the family of purest variable selection methods, such as key set factor analysis (KSFA), simple-to-use interactive self-modeling mixture analysis (SIMPLISMA), orthogonal projection approach (OPA), needle search, and others [14–21], helps to provide the most relevant elution times and spectral variables of the data set, which provide in turn the most dissimilar spectra and chromatographic traces, respectively.

11.3.1 PRINCIPAL COMPONENT ANALYSIS

PCA is the most used chemometric method for exploratory analysis of multivariate data tables. PCA decomposes the original data set into a bilinear model of orthogonal (uncorrelated) mathematical factors (principal components, PCs) that describe the directions of maximum variance of the data [22,23]. In a chromatographic context, the original chromatographic data table (**D**) can be reproduced as the sum of the variation explained by every principal component (see Equation 11.2). Each PC is defined by a dyad of a score vector **t**, which can be interpreted as an abstract chromatogram and one loading vector p, which could be seen as the related abstract pure response:

$$\mathbf{D} = t_1 p_1' + t_2 p_2' + \cdots + t_R p_R' + \mathbf{E}$$ (11.2)

where **E** is the experimental error or the variance unexplained by the PCA model.

In a compact way, the PCA model is written as

$$\mathbf{D} = \mathbf{T}\mathbf{P}^T + \mathbf{E}$$ (11.3)

---

* See also Chapter 8.

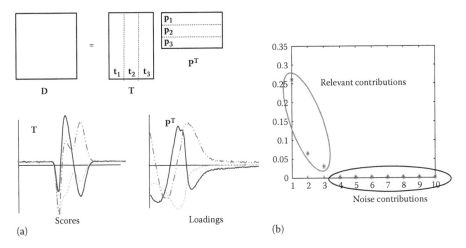

FIGURE 11.4 Application of PCA on an HPLC–DAD data matrix of a mixture of three compounds. (a) Scores and loadings profiles obtained. (b) Log (eigenvalue) plot.

The components in a PCA model are sorted according to decreasing order of variance and the amount of variance explained by every PC is described by the related eigenvalue. Therefore, the first component is associated with the largest eigenvalue and captures the major variance in the data. The second component has the second largest eigenvalue and explains the largest remaining variance not described by the first component and so on. A number of components equal to the minimum number of rows or columns can be calculated.

As mentioned before, the magnitude of eigenvalues is proportional to the variance explained by the related components. Therefore, eigenvalues related to chemical variation are large, whereas noise-related eigenvalues are small and similar among them. Thus, when PCA is applied to a chromatographic data set, the number of significant eigenvalues can be identified with the number of eluting compounds in the data set (both analytes and unexpected interferences).

Figure 11.4 shows the results obtained by the application of PCA on an HPLC–DAD data set of a mixture of three compounds. Upon plotting the eigenvalues, the number of components in the raw data can be estimated. Eigenvalues related to chemical contributions are large, whereas noise-related singular values are small and similar among them. Therefore, the break-off in the log (eigenvalue) plot in Figure 11.4a indicates that the number of significant contributions in the data set is three. The application of PCA also provides scores, looking like abstract elution profiles and loadings, representing their related abstract spectra (see Figure 11.4b). Because of the orthogonality condition imposed upon the calculation of principal components, we cannot expect that scores and loadings look like real elution profiles or pure spectra, but they certainly enhance the elution and spectral features most remarkably to describe the variation in the data set. It is worth noting at this point that the bilinear model provided by PCA is formally identical to the real model of the chromatographic measurement in Section 11.1. Since PCA aims at the optimal description of variance in the data, it makes sense that the number of relevant principal components is associated with the number of real compounds (relevant sources of variation) in the chromatographic run.

11.3.2 Local Rank Analysis Methods

PCA provides a first and global vision of the complexity of a chromatographic run, that is, determines the number of eluting compounds. However, it does not provide information on the elution pattern of the system and, in some instances, the presence of minor compounds with a low S/N ratio and located in a small time window of the chromatogram may be missed. To cope with these limitations and still use the power of the PCA method, local rank analysis tools were proposed.

Local rank analysis methods take advantage of the structured elution pattern of chromatographic processes, that is, of the sequential elution of compounds. This means that studying the chromatographic data set in a local and evolutionary way, the emergence and decay of the different eluting compounds, that is, their elution pattern, can be elucidated.

Evolving factor analysis (EFA) was the first local rank analysis method developed by Maeder and coworkers [10] to study sequential evolving processes, such as chromatographic data. Based on local and evolutionary analysis, EFA provides a useful tool to reveal the elution pattern of compounds in a chromatographic run.

The evolution of a chromatographic process is gradually measured by recording a new spectrum (detector response) at each elution time. Mimicking the experimental process, EFA performs PCA on submatrices of gradually increasing size in the time direction, enlarged by adding a new row (spectrum), one at a time. This procedure is performed from top to bottom of the data set (forward EFA) and from bottom to top (backward EFA) to investigate the emergence and the decay of the eluting compounds, respectively. Therefore, combining the information of forward and backward PCA, one can determine the complete elution pattern of each chemical compound.

To better understand the information provided by EFA, an example for a three-component chromatographic data table is shown in Figure 11.5a. All PCA analyses linked to EFA are displayed in a combined plot that represents adequately the results of forward and backward EFA as a function

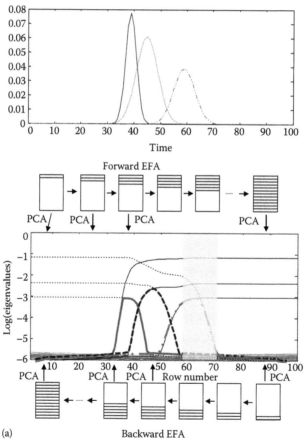

(a)

FIGURE 11.5 Top plot in both (a) and (b): concentration profiles of an HPLC–DAD data set. Bottom plot in (a) evolving factor analysis (EFA) analyses to the data set in top plot. Combined forward EFA (solid thin lines) and backward EFA (dashed thin lines) plot. Thick lines connecting forward and backward EFA lines describe elution profiles.   (*Continued*)

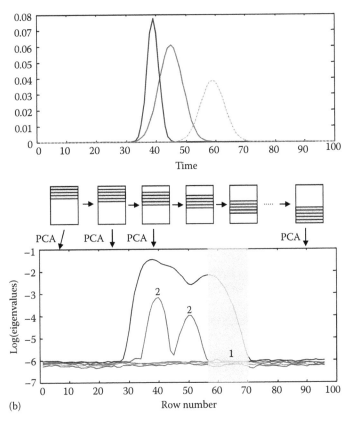

FIGURE 11.5 (*Continued*) Top plot in both (a) and (b): concentration profiles of an HPLC–DAD data set. The shaded zone marks the elution region for the third selective eluting compound. Bottom plot in (b): fixed-size moving window evolving factor analysis (FSMW-EFA) analysis. The shaded zone marks the selective window in the data set (rank 1).

of the elution time axis. The forward EFA curves (thin black solid lines) are built by connecting the log (eigenvalues) obtained from all PCA analyses of the submatrix enlarged in the forward direction. The different lines connect all first eigenvalues, second eigenvalues, and so forth. Likewise, the backward EFA curves (thin red dashed lines) are obtained by connecting the log (eigenvalues) obtained from the PCA analysis of the submatrix expanding in the backward direction. Eigenvalues linked to significant components become larger and clearly distinguishable from the eigenvalues associated with noise, which are smaller and at the bottom of EFA plots. In the forward direction, the emergence of a new eluting compound is detected by the appearance of a new eigenvalue above the noise level, whereas in the backward direction, the decay of an eluting compound is detected by the appearance of a new eigenvalue above the noise level. Combining the curve of the first eigenvalue from forward EFA and the curve from the last eigenvalue from backward EFA, the elution profile of the first eluting compound can be defined. Combining the curve from the second eigenvalue from forward EFA and the curve from the one before last eigenvalue from backward EFA, the profile of the second eluting compound is described, and so forth. Therefore, joining the suitable lines corresponding to forward and backward log (eigenvalues) allows the estimation of the elution window of each compound and a first estimation of the related elution profile.

The EFA plot does not only provide the elution pattern of all compounds but further relevant information. Knowing the elution windows of each compound, detection of selective elution regions (where a single compound elutes) can be defined, as well as a good local rank map of compound overlap, that is, definition of regions of two-compound overlap, three-compound overlap, and so on. This kind of information is extremely relevant for the mathematical resolution of chromatographic data sets, as will be explained in Section 11.4. The main assumption of EFA is the perfect sequence of elution of compounds, that is, the compound that appears first disappears first, the second appearing is the second disappearing, and so on. Indeed, this is most often the case, but there may be situations, for example, embedded peaks, that contravene this assumption. In these instances, the main question is, "which decaying eigenvalue belongs to which emerging eigenvalue?" To answer this question, an evolution of EFA, called exhaustive EFA (E-EFA), was proposed by Whitson and Maeder [24]. The exhaustive and systematic rank analysis of many combinations of submatrices of different rank result in an unambiguous identification of the correct connectivity of emergence and decay of concentration profiles. This modification of the parent method is oriented to clearly distinguish and correctly identify embedded from sequential elution patterns.

From the parent EFA method, other modifications widely used can be found in the literature. In contrast to analyzing a progressively increasing submatrix as in EFA, Keller and Massart [12] suggested a method called fixed-size moving window evolving factor analysis (FSMW-EFA), which uses a fixed-size window moving along the time direction or spectral direction to perform PCA. FSMW-EFA does not focus on a description of the evolution of the different components in a system as EFA does; rather, it focuses on determining the local rank of windows in either the elution or spectral direction.

FSMW-EFA is carried out by performing a series of PCA analyses on submatrices obtained by sliding a window of fixed size from the top of the matrix and downward one row at a time. The FSMW-EFA plot shows the eigenvalues obtained in all the PCA analyses as a function of elution time, as it was done in EFA. Visual examination of these plots gives a local rank map of the data set, that is, a representation of how many components overlap in the different elution regions of the data set (see Figure 11.5b). For each window analyzed, the number of eigenvalues exceeding the noise level threshold is used to determine the local rank. The local rank helps to identify selective zones in the data set and to know the degree of compound overlap along the elution direction. Thus, elution ranges where two eigenvalue lines arise from the noise level indicate the presence of two overlapped compounds. FSMW-EFA is particularly useful for the detection of selective elution regions for the different compounds, that is, zones where only one compound is present. When these are present, obtaining the pure spectra of the related components is straightforward.

FSMW-EFA was mainly oriented to detect impurities under main peaks and minor compounds. Looking at the way FSMW-EFA works, we can interpret the method as a way to zoom small elution windows of the chromatogram. In these small windows, the relative variance of a minor compound becomes proportionally bigger and easier to detect as compared with global PCA methods that take into account the whole chromatographic run.

One of the parameters to be chosen in FSMW-EFA is the window size, which must be always equal or slightly bigger than the total number of compounds in the system. When the window becomes smaller, there is a gain in accuracy to define the different local rank regions, whereas increasing slightly the window size may capture more variance related to minor compounds and increase the capacity of detection for these species. Some evolutions of FSMW-EFA, such as eigenstructure tracking analysis (ETA), tried to explore in a systematic way the effect of modification of the window size in the local rank study of the chromatographic runs [25].

It is important to consider that the estimation of the number of factors or significant eigenvalues in data sets and the identification with the number of eluting compounds may present difficulties in some situations, when there is a strong heteroscedastic noise, distortion of the linear detector response, or nonchemical contributions to the signal such as drift of baselines and other unidentified systematic variations [26]. In order to solve these problems, data preprocessing procedures or use of weighted PCA algorithms that take into account the noise level and structure are suggested.

11.3.3 PUREST VARIABLE SELECTION METHODS

This family of exploratory tools aims at selecting the purest variables from the complete chromatographic data set associated with the most dissimilar spectra (related to the purest rows) or elution profiles (related to the purest columns). These methods are denominated purest variable selection methods and can work irrespective of the presence or absence of an evolutionary structured concentration direction.

The mathematical background of these methods is very diverse, but the most widely used choose the most dissimilar rows or columns directly from the data set, such as simple-to-use interactive self-modeling mixture analysis (SIMPLISMA) [15,16] or the orthogonal projection approach (OPA) [17].

SIMPLISMA is based on the selection of most dissimilar rows or columns using a purity criterion. The purity of a variable (from a row or a column of a data set) is defined by its relative standard deviation, which can be calculated as shown in Equation 11.4:

$$p_i = \frac{\sigma_i}{\mu_i} \quad \text{per } i = 1,\ldots,n \text{ rows or columns} \tag{11.4}$$

where
σ_i is the standard deviation
μ_i is the mean value of a particular row or column

Therefore, a high value of p_i means high purity of the corresponding row or column. However, when the mean value tends to zero, that is, in variables mostly related to noise, the purity value may artificially show very high values. Therefore, to avoid this problem, the purity was redefined by the addition of a small offset to the denominator (δ, usually defined as a percent of the mean signal in the data set), as shown in Equation 11.5. In this way, high purity values for noise-related variables are no longer obtained:

$$p_i = \frac{\sigma_i}{\mu_i + \delta} \tag{11.5}$$

Thus, the column or row with the highest relative standard deviation as defined by Equation 11.5 is taken as the first purest variable. The selection of the next pure variable has to fulfill two criteria: having a high purity value and, at the same time, be as dissimilar as possible to the first selected variable. To achieve this goal, the purity of a variable is recalculated every time a new variable is selected by means of Equation 11.6:

$$p_i' = w_i \frac{\sigma_i}{\mu_i + \delta} \tag{11.6}$$

where w_i is a weight that accounts for the dissimilarity of a particular variable with the previously selected ones. This weight is calculated as the determinant of the dispersion matrix formed by the normalized column or row vectors of the already selected variables and that of the variable i to be tested. When the variable tested has correlated information with previously selected ones, the determinant and, hence, the weight w_i, approaches to zero, whereas the determinant gets higher when the variable tested is more dissimilar to the ones already chosen. The smaller the weight, the smaller the purity value according to Equation 11.6. The weight factor, w_i, for each variable is recalculated every time a new variable is selected.

The concept of looking at the dissimilarity among variables within a chromatogram is also the basis of the orthogonal projection approach (OPA) [17]. This method differs mainly from

SIMPLISMA in the fact that the first variable selected is the most dissimilar to the mean row or column vector and not the one with the highest purity, as described in Equation 11.5.

Both SIMPLISMA and OPA can work on either the elution or the spectral direction, but the most interesting information comes when they are applied in the direction of least compound overlap. For instance, in an HPLC–DAD data set, the component overlap in the elution direction is lower than that in the spectral direction. Therefore, selecting the purest elution times, a good approximation of the purest spectra related to the selected times can be obtained. However, in the case of an HPLC–MS set, these methods can be useful in both directions. Because of the sequential elution of compounds, the purest elution profiles may provide good approximations of purest MS spectra, whereas the high specificity of MS signals may also provide good approximations of elution profiles when the purest m/z values are selected.

These purest variable selection methods can also be used as an auxiliary tool to determine the number of compounds of the original chromatographic data. Thus, the selection of a new spectrum very similar to an already selected one may indicate that there is no new chemical information in the data set analyzed and, hence, provide another way to estimate indirectly the number of components of the system.

The main limitation of methods using the determinant of the dispersion matrix to set dissimilarity measures is that the values of the original data set must be positive. Whereas this tends to be the case in the usual detection systems in chromatography, one has to be aware that, in case of working with derivative data or other measurements including negative values, transforms of these data sets, such as the positive part of second inverted derivatives [27] should be used for the variable selection purpose.

Other purest variable selection methods work on the abstract space provided by principal components. The main representatives of this group are key set factor analysis (KSFA) [14,19,20] or the needle search [21,28].

The needle algorithm [21] is based on the use of target factor analysis and needle elution targets to track the most relevant and distinct chromatographic features in the data set. The algorithm starts performing PCA and selecting the rank of the original data matrix. The suitable scores ($\mathbf{T}$) are used as a representation of the chromatographic space. In order to find the most relevant chromatographic features, needle targets covering the whole elution range are used. These vectors represent ideal chromatographic profiles formed by a single length-one spike located at a particular elution time and null element in the rest of the elution time range. Therefore, for m elution times, the series of needle targets can be described as follows: $x_1 = (1, 0, 0, ..., 0, 0, 0)$; $x_2 = (0, 1, 0, ..., 0, 0, 0)$; ... $x_m = (0, 0, 0, ..., 0, 0, 1)$.

The goodness of each needle target to represent a relevant chromatographic feature is tested through their projection on the score space. The apparent error in the target test vector ($\mathbf{e_i}$) is then calculated as the difference between each needle vector and its projection onto the score space. The length of this apparent error ($\|\mathbf{e_i}\|$) is a measure of the closeness between the needle chromatograms and the real ones, defined by the score space. The information of all target testing carried out is displayed in a needle plot, which represents the sizes of the errors $\|\mathbf{e_i}\|$ as a function of the related elution times. The local minima in this plot will coincide with the locations of the real peak maxima, since the needle chromatograms with a spike located in these points approximate best the true elution profiles (see Figure 11.6).

The inverted peaks of the needle plot provide diverse information: their number equals the number of chromatographic peaks (compounds), the position of their minima agrees with the location of the chromatographic peaks, and the spectra in these locations are good approximations of the pure spectra of the eluting compounds.

As other purest variable selection methods, the needle search method indirectly provides information on the number of compounds of the data set [21]. In this case, the sequential drawing of the needle plots obtained using stepwise increasing rank values becomes a new graphical tool to estimate the rank of a data matrix. Thus, as soon as the real rank of the data is exceeded, no new clear inverted peak-like signal appears and the overall pattern of the plot becomes clearly noisy.

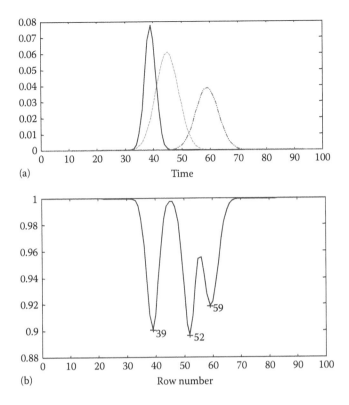

FIGURE 11.6 (a) Concentration profiles of an HPLC–DAD data set. (b) Related information derived by the needle algorithm. The inverted peaks of the needle plot provide double information: their number equals the number of chromatographic peaks and the position of their minima coincides with the location of the chromatographic peak.

All the valuable information provided by exploratory analysis of chromatographic data, that is, total number of compounds, local rank maps, purest spectra, and concentration profiles, can be used in further analysis with an attempt to achieve the real resolution of the chromatographic data and, hence, the recovery of the underlying chemically meaningful elution profiles and pure spectra. The methods used with this objective are denominated multivariate resolution methods.

11.4   RESOLUTION OF CHROMATOGRAPHIC DATA

11.4.1   Concept and Algorithms

Multivariate resolution methods are factor analysis–based methods, which aim at recovering the true underlying bilinear model ($\mathbf{D} = \mathbf{CS}^T$) of the chromatographic data set by using the sole information contained in the raw data set $\mathbf{D}$. In difference to exploratory approaches, resolution methods provide true elution profiles ($\mathbf{C}$) and pure spectra ($\mathbf{S}^T$) of compounds instead of abstract or approximate representations of them. Resolution methods may work using iterative and noniterative algorithms.

Noniterative resolution algorithms are one-step calculation algorithms that are based on mathematical operations performed using suitable combinations of small elution regions of the data set with particular properties to obtain either the elution profiles ($\mathbf{C}$) or the pure spectra ($\mathbf{S}$) of the compounds. The counterpart matrix of the bilinear model is afterward obtained by a single least-squares step. The properties of the elution windows combined are based on the presence or absence of particular components and are defined from the global and local rank information obtained by exploratory methods. As mentioned before, these methods can work well in the chromatographic context because the sequential structure of the elution pattern of compounds allows a straightforward correspondence among local rank regions and identification of the compounds present in them. Window factor analysis (WFA) [29,30] and its evolutional approach, orthogonal projection resolution (OPR) [31], subwindow factor analysis (SFA) [32,33], and the derived method parallel vector analysis (PVA) [34] or heuristic evolving latent projections (HELP) [35–37] are among the most significant approaches within this category. The correct definition of the elution windows is the critical point of these methods to avoid biased results, and their performance is more affected in high noise level systems or when minor compounds with a low S/N ratio have to be modeled. Although it is theoretically possible, these methods were not envisioned to deal with multiset structures, a common need in current data analysis.

Iterative resolution approaches are the most popular multivariate curve resolution methods due to their flexibility to cope with many kinds of data structures and chemical problems (e.g., data sets with partial or incomplete selectivity in the concentration or spectral domains) and their ability to accommodate external information in the resolution process. Therefore, prior knowledge about the data set (chemical or related to mathematical features) can be used in the optimization process, but it is not strictly necessary. Iterative methods require initial estimates of either $\mathbf{C}$ or $\mathbf{S^T}$ to be optimized. Depending on the kind of algorithm, both matrices are optimized during the resolution process or only one of them and, then, the counterpart is obtained by a single least-squares step. Likewise, the profiles of all components can be optimized simultaneously or be obtained one at a time.

In all iterative methods, the profiles in $\mathbf{C}$ or $\mathbf{S^T}$ are "tailored" according to the chemical and mathematical properties of each particular data set. This chemical and mathematical information is included in the iterative optimization process under the form of constraints. The iterative process stops when a convergence criterion (e.g., a preset number of iterative cycles is exceeded or the lack of fit goes below a certain value) is fulfilled. The most widely known iterative methods are iterative target transformation factor analysis (ITTFA), developed independently by Gemperline and by Vandeginste et al. [28,38], and multivariate curve resolution-alternating least squares (MCR-ALS), mainly promoted by Tauler and coworkers [39–44]. ITTFA works optimizing the elution profiles of compounds one at a time under constraints and recovering the spectra matrix in the bilinear model by a single least-squares step. MCR-ALS performs an alternating optimization of both concentration profiles and pure spectra in each iterative cycle. The rest of the manuscript will be oriented to describe in detail the application of the MCR-ALS algorithm to chromatographic data. This algorithm is one of the most used multivariate curve resolution methods because it is able to work with multiset structures formed by several chromatographic runs with single or multidetection systems and it can easily incorporate chemical and mathematical information under the form of constraints (local rank conditions or multiway models).

11.4.2 MCR-ALS: MODUS OPERANDI

11.4.2.1 Resolution of a Single Chromatographic Run

As mentioned before, MCR-ALS decomposes $\mathbf{D}$ into the bilinear model $\mathbf{CS^T}$ using an iterative algorithm based on two constrained linear least-squares steps. It requires initial estimates of the elution, $\mathbf{C}$, or of the spectral, $\mathbf{S^T}$, profiles to start the iterative optimization step that includes the calculation of $\mathbf{S^T}$ and $\mathbf{C}$ matrices in each iterative cycle as follows:

$$\mathbf{C^*} = \mathbf{DS}\left(\mathbf{S^T S}\right)^{-1} \tag{11.7}$$

$$\mathbf{S^T} = \left(\mathbf{C^{*T} C^*}\right)^{-1} \mathbf{C^{*T} D} \tag{11.8}$$

A reconstructed $\mathbf{D^*}$ matrix, obtained from the product of the calculated matrices $\mathbf{C^* S^{*T}}$ is then compared with the real $\mathbf{D}$ matrix and the iterative optimization continues until the convergence criterion,

based on a preset number of iterations or on the difference in model fit among consecutive iterations, is fulfilled. The quality of the model fit in the final MCR results is assessed by comparing the
reconstructed matrix $\mathbf{D}^*$ with the raw data matrix $\mathbf{D}$. Indicators for this purpose are the percentage
of lack of fit (lof %), as described in Equation 11.9:

$$\%\mathbf{lof} = 100 \times \sqrt{\frac{\Sigma e_{ij}^2}{\Sigma d_{ij}^2}} \tag{11.9}$$

where

e_{ij} is an element of the residual matrix, equal to $d_{ij} - d_{ij}^*$, that accounts for the variance unexplained by the MCR model

d_{ij} is an element of the raw $\mathbf{D}$ matrix

d_{ij}^* is the analogous element in the reconstructed $\mathbf{D}^*$ matrix and the percentage of variance
explained, r^2, given by Equation 11.10

$$\mathbf{r}^2 = \mathbf{100} \times \left(1 - \frac{\Sigma e_{ij}^2}{\Sigma d_{ij}^2} \right) \tag{11.10}$$

The first step in the optimization process is determining the number of eluting compounds in the
data set ($\mathbf{D}$). The number of elution compounds can be known beforehand or can be determined by
rank analysis methods (see Section 11.3). When in doubt, MCR models with different number of
components can be calculated. The analysis providing a satisfactory model fit according to the real
noise level of the data and the most interpretable chemically meaningful profiles will be adopted as
the final solution.

The next step involves generating initial estimates of elution profiles, $\mathbf{C}$, or pure spectra, $\mathbf{S}^T$,
with as many profiles as the number of components estimated for $\mathbf{D}$, to start the iterative resolution
process. The sequential structure of chromatographic data allows the use of initial estimates of elution profiles found with evolutionary factor analysis methods, such as EFA (see Section 11.3.2).
In chromatographic data sets, initial estimates of spectra can also be obtained by the selection of
the purest variables from the raw data set that represent the most dissimilar spectra (related to
the purest rows). In cases of specific detection responses, such as MS, the purest variable selection methods could be applied in the column direction and estimates of elution profiles would be
obtained (see Section 11.3.3).

The steps related to the application of the MCR-ALS algorithm are described above. To consider
that the resolution results of an analysis are good, two conditions have to be fulfilled: (1) a sufficiently high percentage of variance explained, and (2) obtaining concentration profiles and spectra
that are chemically meaningful and showing shapes in agreement with the variation seen in the
experimental raw data.

Although there is a real underlying bilinear model $\mathbf{D} = \mathbf{C}\mathbf{S}^T$ for the experimental data analyzed, there are other combinations of $\mathbf{C}$ and $\mathbf{S}^T$ matrices, which have different shapes (rotational
ambiguity) and/or intensities (intensity ambiguity) than the ones sought and that can reproduce
the $\mathbf{D}$ matrix with an identical fit [41,45–47]. This responds to the phenomenon of ambiguity that
is very specific to multivariate resolution methods. Equations 11.11 and 11.12 describe rotational
ambiguity:

$$\mathbf{D} = \mathbf{C} \left(\mathbf{T} \mathbf{T}^{-1} \right) \mathbf{S}^T \tag{11.11}$$

$$\mathbf{D} = \left(\mathbf{C} \mathbf{T} \right) \left(\mathbf{T}^{-1} \mathbf{S}^T \right) = \mathbf{C}' \mathbf{S}'^T \tag{11.12}$$

Here $\mathbf{T}$ is a transform matrix that may have very different forms only limited by the fulfillment of constraints. The profiles $\mathbf{C}' = \mathbf{CT}$ and $\mathbf{S}'^{T} = \mathbf{T}^{-1}\mathbf{S}^{T}$ are just another set of feasible solutions with shapes that are different from the true underlying elution and spectral profiles. Equation 11.13 represents the phenomenon of intensity (or scale) ambiguity:

$$\mathbf{D} = \sum_{i=1}^{n} \left(C_i \cdot k_i \right) \left(\frac{1}{k_i} S_i^{T} \right) \tag{11.13}$$

Here k_i is a scalar that can modify the relative intensity of the elution and spectrum profile of a dyad without changing the pure signal contribution of the related compound. Nowadays, there exist a variety of methods to estimate the ambiguity in the final MCR solutions, but a good application of MCR aims at reducing or suppressing this phenomenon whenever possible [45,46,48–50].

Multivariate resolution methods could potentially provide an infinite number of feasible and optimal solutions, hence the need to apply constraints in the optimization step in order to provide meaningful shapes for the profiles in $\mathbf{C}$ and $\mathbf{S}^{T}$ and to minimize as much as possible the ambiguity phenomenon. As mentioned before, constraints are chemical or mathematical properties that the profiles in $\mathbf{C}$ and/or $\mathbf{S}^{T}$ fulfill. They are applied during the iterative optimization process in such a way that the calculated profiles are modified to obey the constraint condition [40,42]. Constraints are applied optionally only when profiles naturally obey them and can be used allowing certain degrees of tolerance from the perfect fulfillment of the preset condition, for example, when profiles come from noisy data sets. When working with a chromatographic run, this means that elution profiles and spectra can be constrained in different ways, since their natural properties are different. The components within the $\mathbf{C}$ or $\mathbf{S}^{T}$ matrix can also be constrained in diverse ways when so needed, for example, constraining a background contribution vs. the elution peak of an analyte. Selecting the appropriate constraints and where and how to apply them is the most crucial point to ensure obtaining meaningful and reliable solutions. Taking into account the chromatographic nature of the data, particular properties of the elution direction allow for the introduction of constraints related to the peak shape or to the sequential elution pattern of compounds. Thus, the following constraints can be applied:

1. *Nonnegativity*: Forces the profiles to be formed by positive values and can be implemented replacing negative values by zeros or with softer algorithms, such as nonnegative least squares or fast nonnegative least squares [51]. It applies to both elution and spectral profiles, when coming from DAD, fluorescence, or MS detection.
2. *Unimodality*: It is a constraint that forces the presence of a single maximum per profile. It is the condition fulfilled by elution profiles in the analysis of chromatographic runs (peak-shaped signals). There are also different ways to implement this constraint according to the procedure used to suppress secondary maxima [43,52,53].
3. *Selectivity and local rank constraints*: They are mathematical properties of the data set and are associated with the concept of local rank, which describes how the number and distribution of components varies locally along the data set. Selectivity constraints can be applied in concentration and spectral windows when only one compound is present by setting the values of other compounds in the selective window to be zero [41,42]. This condition suppresses the ambiguity linked to the complementary profile of the dyad for the compound with selective information. Local rank constraints define situations where some of the compounds are absent in particular elution or spectral windows and they are set to null values in the related profiles. Selectivity is a particular case of local rank constraint. Special conditions of local rank, that is, compound overlap, in data sets allow for the complete suppression of ambiguity in the analysis of chromatographic data [54].

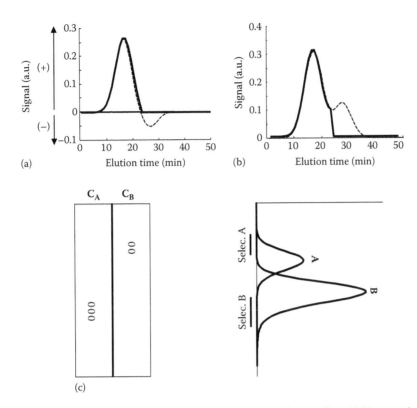

FIGURE 11.7 Common constraints applied to chromatographic elution profiles. (a) Nonnegativity, (b) unimodality, where thick profiles are constrained profiles and thin dashed lines represent elution profiles before being constrained, and (c) selectivity, included as null concentration values for the absent compounds.

Figure 11.7 shows the effect of these three constraints, which are the most applied in the analysis of single chromatographic runs.

The use of MCR methods does not have, in principle, limitation with respect to the total number of compounds present in the system. However, when chromatographic runs with a large number of components and an unspecific detection system, such as UV-DAD, have to be analyzed, the problem becomes complex and the ambiguity phenomenon more present. Taking advantage of the sequential elution of chromatographic compounds and in order to facilitate a better differentiation among the compounds involved in the chromatographic analysis, elution time windowing [55–57] could be carried out. This procedure consists of analyzing by MCR time windows of the total chromatogram separately. In this way, a complex resolution problem due to the presence of a high number of similar components becomes a set of simple subproblems with low component complexity. Such a procedure speeds up the analysis and decreases significantly the ambiguity in the recovered solutions.

It has to be mentioned that in chromatographic runs with very specific detection systems, such as MS, and after proper data compression by the ROI method (see Section 11.2.2), time windowing becomes unnecessary and there are works where the simultaneous resolution of more than 100 compounds has been reported [58].

11.4.2.2 Simultaneous Resolution of Several Chromatographic Runs and Other Complex Chromatographic Data

An outstanding advantage of MCR-ALS and other multivariate resolution methods is the possibility of working simultaneously with multiexperiment and/or multitechnique data structures, which gives a clear gain in the quality and quantity of information in the analysis of the chromatographic process.

As shown in the previous section, the main option for decreasing/suppressing ambiguities in MCR solutions is by means of the introduction of constraints. Some constraints, such as local rank, can suppress ambiguity, provided that some conditions of component overlap are fulfilled in the data set [54]. However, when those conditions of component overlap are not fulfilled, the quality of chromatographic profiles and pure spectra obtained from the MCR resolution is limited. A possible way to solve this drawback is by working with multiset structures (several chromatographic runs analyzed simultaneously or several detection systems combined). In this case, the diversity of information in the subsets appended, which may also be tailored to improve the compound overlap conditions, greatly decreases or suppresses ambiguity in the final results and provides, as a general trend, a more complete view of the chemical system under study.

There are different ways of organizing chromatographic runs for simultaneous analysis. Each data arrangement demands different requirements of the data sets appended and the mathematical model that describes the measurements is also different. In the introduction section, it was already shown how different individual data matrices can be arranged for simultaneous analysis (Figure 11.2). It is worth mentioning that the augmented matrices or multiset structures can be very flexible and can be formed by data matrices with different sizes and chemical information. Building a multiset is justified when data matrices analyzed together share some components. The only requirement to build a multiset is the presence of a single common mode among appended runs (either the concentration or the spectral mode). In the case of row-wise appended chromatograms, that is, a single sample injected and more than one detection system in tandem, only the elution mode needs to be common. This means that a single set of elution profiles should be valid to explain the full multiset structure. In these instances, if there was a time delay among the detectors connected, this should be synchronized before appending the chromatographic runs [57]. The rest of the information would be a row-wise augmented $\mathbf{S}^T$ matrix with the concatenated spectral signatures of the different detection techniques per each compound in the system.

The common case of column-wise appended chromatograms, that is, several runs analyzed using the same detection system, is worth being commented on. In this situation, the spectral mode must be common. This means that a single set of pure spectra will properly describe the identity of compounds in all chromatograms. This situation is easy, as long as detection is carried out in the same spectral range and runs are done in similar experimental conditions, for example, same mobile phase. The interest in this case lies in the freedom of the column-wise augmented elution mode. Since the bilinear model provides an augmented $\mathbf{C}$ matrix, with as many $\mathbf{C}_i$ submatrices as chromatograms in the data set, the elution pattern in the chromatographic runs can be completely different from one another. This implies that elution shifts, changes in peak shape, or elution reversals are not a problem when MCR is used to handle these multiset structures.

The operating procedure of MCR-ALS to analyze augmented data matrices (multisets) is the same as described in Section 11.4.2.1 for a single chromatographic run—the only difference being that $\mathbf{D}$ and $\mathbf{C}$ or $\mathbf{S}^T$ are now augmented matrices. Only some specificities regarding the treatment of multisets deserve further comment.

In the resolution of a column-wise augmented data matrix, the initial estimates can be either a single $\mathbf{S}^T$ matrix or a column-wise augmented $\mathbf{C}$ matrix. The column-wise concentration matrix is built by placing the initial $\mathbf{C}$-type estimates obtained for each chromatographic run one on top of the other. The appended initial estimates must be sorted into the same order as the initial data matrices in $\mathbf{D}$ and they must keep a correct correspondence of species, that is, each column in the augmented $\mathbf{C}$ matrix must be formed by appended concentration profiles related to the same chemical compounds. When no prior information about the identity of the compounds in the different data matrices is available, the correct correspondence of species can be estimated from the resolution results of each single matrix. For the sake of simplicity, the option most often taken is the use of single matrix of spectral estimates found by purest variable selection on the multiset.

The same constraints used in the resolution of a single data matrix can be applied to augmented data matrices [45,47]. In the case of unimodality, this condition would be applied to the elution profiles in each $\mathbf{C}_i$ submatrix separately. Again, the application of constraints is optional and, in a multiset context, this means that submatrices in the augmented direction (related to elution profiles of different runs or to instrumental responses of different detectors) can be constrained differently.

Moreover, in the simultaneous analysis of different chromatographic runs, some additional constraints can be used:

1. *Correspondence of species*: This constraint is only applied in column-wise augmented matrices (with an augmented concentration direction) and expresses the presence/absence of components in the analyzed experiments. This constraint has an important effect on suppression of ambiguities since it is a particular case of local rank constraint. The only difference is that the absence of a compound affects a full chromatographic run and not some restricted elution ranges [45,47,59].

2. *Trilinearity*: MCR relies on the recovery of the bilinear model linked to one or more chromatographic runs. However, other methods working with several sets of data structured as data cubes provide trilinear models, that is, a tryad of profiles defines a compound of the system. In the case of chromatography, the data cube is formed by three common directions (elution time, spectral channels, and samples). The consequence of using this model is that the elution profile of a particular compound should have the same location and shape in all chromatographic runs. In MCR, this kind of model structure may also be imposed as a constraint. In this situation, the shape of the elution profile for a particular component is forced to be identical in all runs analyzed. As a consequence, the concentration profile for a particular component in the different experiments will differ only in intensity but not in location or shape. In contrast to classical methods specially designed to work with complete trilinear systems, the implementation of this constraint in MCR-ALS is done compound-wise, that is, some or all of the compounds can be forced to have common profiles in the $\mathbf{C}$ matrix [45,60]. This allows having a high diversity of models in MCR, ranging from completely bilinear to partial or completely trilinear [61]. When MCR constrains a data set to obey a complete trilinear model, the elution profiles and pure spectra resolved are unique.

It is important to note that a trilinear model assumes that the elution profile of a particular compound always appears in the same location and has the same shape in all chromatographic runs. This is not often the case in real chromatographic separations. Therefore, the use of methods, such as PARAFAC [62,63], the generalized rank annihilation method (GRAM) [64], and direct trilinear decomposition (DTD) [65], should be limited to very clean and reproducible chromatographic elutions, such as a few cases in GC–MS systems [66]. When this is not the case, some authors opt for data preprocessing to correct for peak shifting of the same compound among the different runs [67–70]. Nevertheless, when complex systems with severe matrix effects (natural samples) or with minor compounds come into play, these methods cannot perform adequately because the differences in peak shape among runs cannot be properly corrected, that is, the trilinear model of the mathematical methods cannot describe adequately the true chromatographic behavior. To overcome this situation, variants of the methods that relax the trilinear condition keeping uniqueness in solutions were designed. Thus, PARAFAC2 is a variant of the PARAFAC technique, which has been applied to chromatographic data [71]. Strict trilinearity in PARAFAC2 model is not necessary, although deviations from it should not be significant and only related to regular shifts among runs [61,72].

The flexibility of MCR, which allows differences in the elution pattern among runs, opens a wide field of possibilities. For instance, mixtures of main compounds and impurities can be eluted under different conditions (e.g., pH or mobile phase composition) to change the elution pattern of the sample and help in a better detection and quantification of the unknown impurities [73]. Alternatively, total analysis time for routine determinations can be considerably shortened with fast chromatography methods if the decrease in chromatographic resolution is compensated by appending a "reference run" obtained with a longer analysis time and better chromatographic resolution [74,75].

In the same manner, analytes in complex samples, such as environmental samples, can be much better resolved from the interferences in the sample run if chromatograms with pure analyte standards are appended [76,77]. As will be seen in the application section, this freedom in the concentration direction allows for even more challenging multisets, combining chromatographic runs and other kinds of experiments, for example, kinetic spectral monitoring, to solve complex problems.

The direct consequence of analyzing several chromatographic runs together is the improvement of quality and accuracy of the resolved profiles (elution profiles and spectra). However, other benefits are obtained, such as the possibility to obtain quantitative information.

In contrast to multivariate calibration methods, the information related to a sample is not a vector (e.g., a spectrum) but a whole matrix. This fact and the separate resolved profiles provided by resolution methods allow having the so-called second-order advantage, which means that the concentration of an analyte in a sample can be determined in the presence of unknown interferences that do not need to be present in the calibration samples. This fact provides many benefits: on one hand, simplicity in the quantitative analysis since the identity of interfering compounds need not to be known. On the other hand, runs of standards can be designed as injections of pure compounds or simple mixtures of analytes, with the possibility to greatly increase the adequate local rank conditions for unique resolution of the analytes to be determined.

In multiset structures, the quantitative information is derived from the scale of the pure concentration profile of a compound in the different matrices. This scale can be represented by the area under the resolved elution profile or, in some instances, by their peak height. The traditional calibration approaches used in classical chromatography, that is, external calibration, standard addition, and internal standard, can also be performed using multivariate curve resolution [78]. Using these approaches and multivariate curve resolution, simultaneous determination of several analytes can be carried out with a single MCR analysis and, at the same time, there exists a clear analyte confirmation provided by the resolved spectra linked to each particular analyte. Figure 11.8 shows how to proceed to obtain the calibration lines and, hence, the quantitative information from the $\mathbf{C}$ matrix provided by multiset analysis. For a particular compound, the MCR concentration values per each calibration sample are generally obtained by integrating the suitable concentration profile in its related $\mathbf{C}_i$ submatrix. In the case of external calibration, these values are then regressed against the reference values of the calibration samples to build the calibration lines. To build calibration lines for the standard addition method, the peak areas of the MCR-ALS resolved elution profiles of the analytes in a sample are regressed against the added concentrations of these analytes to the analyzed sample. Finally, for internal standard calibration, peak areas of standards are divided by the peak area of an internal standard and regressed against the concentrations of each compound divided by the concentration of the internal standard assuming a linear relationship.

Recently, an innovation in terms of quantitative analysis for multiset structures has been introduced. It is the extension of the correlation constraint [79,80] to second-order data. This constraint works incorporating the step of regression between the peak areas found by ALS and the reference external information within the loop of the optimization of MCR. In this way, in every iteration, the peak areas are rescaled according to the regression model and a final result more reliable in terms of profile shape and quantitative information is obtained.

11.4.3 APPLICATIONS

The diversity of data arrangements that can be analyzed and the flexible application of constraints have promoted the use of MCR-ALS to solve many different kinds of problems in chromatographic data. The next section will describe a variety of applications of MCR-ALS in this field.

11.4.3.1 Qualitative and Quantitative Analysis of Compounds in Complex Samples

The combination of MCR methods with chromatography to perform qualitative and quantitative analysis of compounds in complex samples presents advantages in relation to traditional methods.

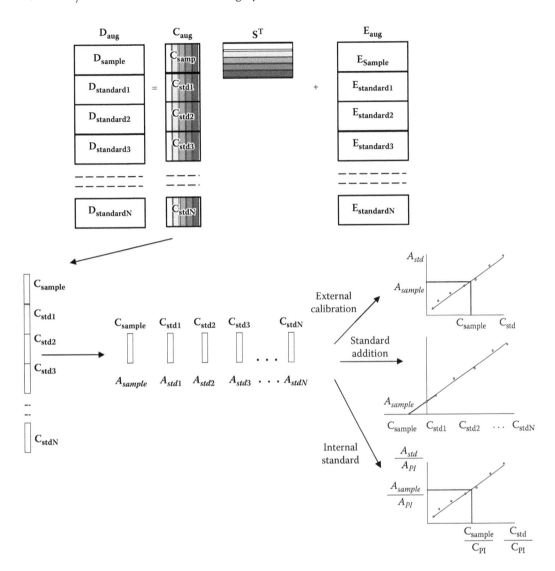

FIGURE 11.8 Multivariate curve resolution-alternating least squares (MCR-ALS) applied to the column-wise augmented matrix $\mathbf{D}_{aug}$, formed by a sample and a set of standards. $\mathbf{D}_{sample}$ is the unknown sample matrix. $\mathbf{D}_{standard}$ are the standard mixture samples data matrices. $\mathbf{C}_{aug}$ is the column-wise augmented elution profiles matrix. $\mathbf{C}_{sample}$ and $\mathbf{C}_{std}$ are the MCR-ALS resolved elution profiles matrices for the unknown and standard mixture samples, respectively. From the comparison of the areas of their resolved elution profiles in the different standard mixture samples, quantitative information from external calibration, standard addition, or internal standard strategies can be obtained. $\mathbf{S}^{\mathbf{T}}$ matrix is the pure spectra matrix of MCR-ALS resolved compounds, which allows their identification.

MCR methods allow resolving coeluted peaks and, as a consequence, this means a significant reduction of analysis time, with the consequent economic and environmental advantages of faster sample processing and saving of solvent consumption and other chemical reagents. In some instances, the use of multivariate resolution methods can avoid analytical complexity, such as the use of complex solvent gradients to achieve separations.

In terms of qualitative analysis and identification of species, MCR provides information on the identity of the compounds in a sample through the pure responses contained in the $\mathbf{S}^{\mathbf{T}}$ matrix that can be directly interpreted or conveniently matched with libraries of spectra. This is a common practice in the qualitative analysis of complex samples, the composition of which is relevant to understand their final properties. Besides, this identification is not subject to conditions of elution pattern, that is, comparison of retention times of analytes, since identification is carried out on the basis of the spectral signature recovered.

When quantitative analysis is the goal, the main benefit of using MCR methods is linked to the so-called second-order advantage, a property inherent to matrix instrumental data, which implies that analytes can be quantified in samples containing potential interferences even if those are absent in the calibration samples. MCR methods can handle with elution time shifts and band shape changes that usually occur from sample to sample. To perform quantitative analysis, MCR in multiset mode is used. In difference to multiway methods that work with several chromatographic runs organized in data cube structures and applying trilinear models, MCR can cope with shifts and peak shape changes among chromatographic runs, which provides a higher flexibility in the analysis. For all these reasons, MCR methods have been extensively applied to chromatography in qualitative and quantitative analysis. In the next paragraphs, some representative examples of these applications will be done.

Konof et al. [81] reported that analysis of Iranian olive fruit essential oil by conventional GC–MS yielded 90 detected compounds, while after MCR-ALS the df number of detected components increased to 141, resulting in a 56% increase. MCR-ALS was used to resolve and quantify 13 different biocides in sediment and wastewater samples by HPLC coupled to diode-array detection (HPLC–DAD) [72] and by HPLC coupled to mass spectrometry detection (HPLC–MS) [76]. Despite the very strong coelutions and matrix interferences, the analyzed biocides were properly resolved and their quantitation could be performed satisfactorily by external calibration. The difficulties encountered in practice for optimal reproducible quantitative determinations by HPLC–MS come from the combination of complex environmental matrices, embedded peaks, strong coelutions, and ion suppression effects. Interestingly, the fusion of DAD and MS detector signals improved the results and provided more reliable estimations compared to those obtained using only one of the two detection systems, as was demonstrated in [82]. Therefore, it is worth mentioning that multiresponse detection can help to distinguish components that can share an identical spectrum in a particular technique and be different according to another response detector, such as MS. In the combination of MS and DAD detection, MS can also benefit from the usual higher S/N ratio of the UV-DAD detection [57]. Other fusions of detectors also included DAD-fluorescence. In this case, the difference in compounds detected by the two techniques also helps in a better overall resolution of compounds.

In another study, PARAFAC, U-PLS/RTL and MCR-ALS coupled to HPLC-fast scanning fluorescence spectroscopy were employed for quantitation of fluoroquinolones in water samples [83]. The second-order advantage was exploited when analyzing samples containing uncalibrated interferences. PARAFAC and MCR-ALS were the algorithms that better exploited the second-order advantage when no peak time shifts occurred among samples. On the other hand, when the trilinearity was lost due to the occurrence of temporal shifts, MCR-ALS furnished the better results.

11.4.3.2 Process Monitoring

Chromatography and MCR have been extensively used in process monitoring. In these instances, aliquots of sample are collected along the process and analyzed chromatographically in order to obtain a better insight in the identification of products and in the description of process evolution. To do so, a column-wise augmented multiset including all chromatographic runs collected along the process is built and further analyzed by MCR [57,84].

Process monitoring by chromatography offers advantages over traditional spectroscopic monitoring, where a spectrum is collected in the solution sample (with all process components mixed together) as a function of the process variable (time, pH, etc.). Because of the action of chromatography, compounds are partially or completely separated in each stage of the process. Chromatography provides a data table (chromatogram) at each stage of the process and the process components show

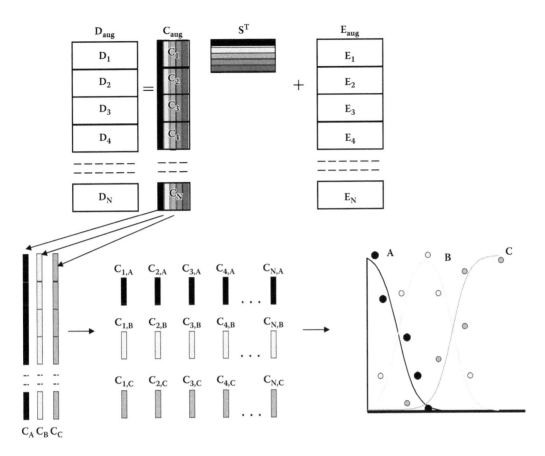

FIGURE 11.9 Data arrangement, resolution, and derivation of process profiles for a data set from a process monitored chromatographically.

differences both in spectral shape and in elution pattern. As a consequence, compounds with an identical kinetic evolution or with very similar spectra can be distinguished due to their different elution behavior. This procedure allows the identification of the products formed during the process (especially, if MS detection is used) and provides a description of the kinetic (or else) evolution of the compounds involved in the process from the evolution of their peak areas as a function of the process control variable (see Figure 11.9).

This strategy has been applied to study degradation processes of pesticides and has been proven to be particularly efficient in processes with complex pathways, where many components can evolve in an identical way [84,85]. From the process profiles obtained, kinetic or thermodynamic models can be fit that confirm the mechanism suggested by the chromatographic results and provide physicochemical parameters of interest. When the process analyzed is very complex and involves many compounds, the time-windowing strategy mentioned in Section 11.4.2.1 can also be used and several multisets related to different elution time windows can be analyzed separately. In the end, the peak areas of all compounds resolved in the different multiset analyses can be shown in a single plot describing the process evolution [57,86].

However, chromatographic process monitoring is generally limited by the poor process time axis used (i.e., the limited number of process chromatographic runs as compared with spectroscopic monitoring approaches). To overcome this limitation and to have a complete description of the kinetic mechanism of a process, Mas and coworkers have proposed a data fusion of UV–vis spectroscopic monitoring data with HPLC–DAD process runs and analysis by MCR algorithm with hard modeling constraint (HS-MCR) [87]. In this case, because multisets require a single mode in the structure in common, the concentration mode is completely free and allows joining chromatograms (which provide elution profiles as concentration profiles) with a kinetic spectroscopic monitoring experiment, which yields kinetic profiles as concentration profiles. The hard modeling constraint is applied only to the concentration profiles of the spectroscopic experiment and works fitting a postulated kinetic model to the selected concentration profiles in the **C** matrix and replacing the initial concentration profiles by those generated from the fitted kinetic model. This constraint provides the rate constants and, as any other constraint, it can be applied differently in each profile within **C** and in the different submatrices while using a multiset structure [88]. The latter situation adapts to the multiset in that study formed by spectroscopic experiments (model-based) and chromatographic runs (model-free). In this particular case, a comprehensive description of the kinetics linked to the photochemical degradation of the ketoprofen, with all successive steps, related rate constants, and photoproducts formed, has been provided. The combination of the two analytical methodologies (chromatographic/spectroscopic) resulted in the capacity of obtaining much better description of the process. The chromatographic information helped to differentiate among compounds, whereas the spectroscopic information provided a good time-dependence description of the process.

Within this kind of data fusion, a more advanced step has been given combining chromatographic runs collected along a process with DAD and MS detection and a kinetic spectroscopic experiment monitored by UV [89]. As inferred from the description, this gives rise to an incomplete multiset, where the block of the spectroscopic monitoring by MS is missing. Nevertheless, there exists a variant of MCR to handle incomplete multiset structures [90] that can handle the structure described. In this work, this implied an improvement in the identification of the identity of compounds from the MS spectral signatures and the expected good kinetic description of the system.

11.4.3.3   Multidimensional Chromatographic Techniques

Multidimensional chromatographic techniques like LC × LC and GC × GC have emerged as powerful techniques suitable for the separation of very complex mixtures because of their higher resolution and higher peak capacity. However, one of the main challenges in multidimensional chromatography is related to the difficulty of the analysis and interpretation of the enormous amount of data obtained in these cases. Additionally, complete separation of all detectable components can often not be achieved because of the extremely high complexity of natural samples and the limitations in experimental and instrumental conditions. Different multivariate resolution techniques can be used to address incomplete separation issues during multidimensional chromatographic analysis. However, the performance of most of these techniques depends on the reproducibility of retention times in both dimensions (trilinearity), which cannot be guaranteed during multidimensional chromatographic analysis.

Multidimensional chromatographic data can be structured in a data cube, where two modes refer to the two chromatographic dimensions and the third one is the spectral dimension. However, this structure does not strictly follow the trilinear model due to the elution time shifts observed especially in the second dimension. Thus, as mentioned before, MCR methods can be a very useful alternative for the analysis of this kind of chromatographic data without the assumption of the trilinearity condition. The optimal strategy for the resolution of the multidimensional chromatographic data by MCR methods is using a column-wise augmented multiset that keeps the common spectral mode and puts one under the other all second chromatograms related to each elution time in the first chromatographic direction. This kind of arrangement is valid to analyze a single multidimensional chromatographic run or several of them related to different samples (see Figure 11.10). In this strategy, retention time shifts, both within and between chromatographic runs, can be properly handled during the resolution process without any need to correct them before chemometric analysis. When several multidimensional chromatographic data matrices are

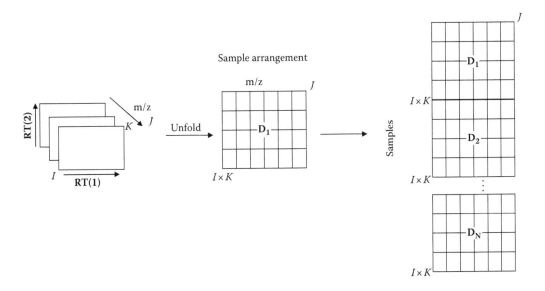

FIGURE 11.10 Scheme of construction of the data matrix **D** from multidimensional chromatographic technique with multichannel detector (e.g., GC × GC–MS).

analyzed by MCR, a 2D elution profile, the pure response and the quantitative information on the relative concentration in the different samples per compound are obtained.

Figure 11.10 shows an example of the construction of an augmented matrix **D** with data coming from a multidimensional chromatographic technique with multivariate detector (e.g., GC × GC–MS). The initial step is the unfolding of the GC × GC–MS chromatograms from a cube to a matrix that can be analyzed alone or simultaneously with matrices coming from other samples.

Several applications of MCR methods in multidimensional chromatographic data have been reported in the literature.

Bailey and Rutan [91] developed a method that combines iterative key set factor analysis (IKSFA) with MCR-ALS in order to extract the maximum information from multidimensional chromatographic data. In this work, urine samples were analyzed in a LC × LC-DAD system with replicates. The aim of the work was to resolve and quantify the nontargeted overlapped compounds. Firstly, the 2D chromatograms were divided in sections due to the complexity of the whole sample and also due to regions of detector saturation. In the section to be analyzed, the IKSFA starts by determining the number of compounds and spectral estimates are built. After that, MCR-ALS was applied in order to provide a resolution of the spectral different components, using nonnegativity and spectral selectivity constraints. Bailey and Rutan highlighted that this algorithm is insensitive to shifts of retention time and distortions of peak shape and, consequently, does not require alignment of the data before application. The use of nonnegativity and selectivity constraints is enough to obtain satisfactory results. An application of a similar methodology has been proposed by Parastar et al. [92], where MCR-ALS has been used to resolve and quantify a complex mixture of polycyclic aromatic hydrocarbons in Heavy Fuel Oil Sample by GC × GC–TOFMS. In this study, the results have been compared to those obtained by commercial software and to PARAFAC, showing an improvement of the results in terms of data fitting, elution process description, concentration relative errors, and relative standard deviations. Analogous studies about the potential of using the MCR-ALS method for the analysis of multidimensional chromatographic data have been performed [93–97]. Moreover, Navarro-Reig et al. [98] proposed the general use of the MCR-ALS method to take full advantage of LC × LC–MS data in the analysis of complex natural samples where strong coelutions and spectra overlap (isomeric species) are frequently encountered.

11.4.3.4 Fingerprint and "-Omics" Technologies

In recent years, fingerprint analysis has been increasingly applied in diverse fields, such as toxicology, pharmacology, disease diagnosis, food and nutrition science, environmental science, and all the "-omics" disciplines. Fingerprint analysis allows characterizing the composition of complex samples coming from the diverse fields mentioned above. The compositional profile (fingerprint) can be used to understand the effect of a biological, chemical, or physical perturbation on the samples or to describe the nature or behavior of a complex system. Sometimes fingerprinting is interesting for the sole purpose of identifying and defining the relative amount of the compounds (biomarkers) that are characteristics for a specific physiological status (e.g., disease or toxicity) or a specific designation of origin in foodstuffs. The general aims in all the studies using fingerprints include solving the problem under study (i.e., disease diagnosis, quality control, product origin, etc.) and identifying the compounds (markers) most relevant to reach that goal. An extensive set of compounds should be analyzed in order to establish a reliable fingerprint. The complexity of these analyses and the need to characterize the compounds in the fingerprint from the global measured signal justify the use of hyphenated chromatographic techniques combined with MCR methods.

Several examples of application of MCR in fingerprint analysis working with different kinds of chromatographic measurements can be found in the bibliography. These applications are used to tackle different problems, such as detection of adulteration of food products [99,100] or authentication of the origin of products [101–105], quality assessment in food or pharmaceutical field [106,107], and identification of biomarkers in lipid [108,109] or metabolites [110,111] that can indicate the effect of environmental or other stress agents in organisms.

As an example, MCR-ALS was used by Sánchez Pérez et al. [112] to detect changes in the concentration of tomato metabolites as a result of stress after treatment with carbofuran (insecticide). The authors reported that several metabolites present different evolutionary profiles over different sampling days depending on the presence of the pesticide. Also, a few components do not present any variation in their profiles. These findings suggest that the presence of pesticide causes changes over time in the behavior of certain endogenous tomato metabolites as the result of physiological stress. In another study, Wehrens et al. [113] presented an application of MCR in HPLC–DAD data in order to study the stability of isoprenoids in grape extracts. The authors showed that the addition of triethylamine prevents the degradation of grapes.

Due to the large number of compounds involved in such data, elution time-windowing strategy, as mentioned in Section 11.4.2.1, is usually used. Therefore, the fingerprint itself is obtained by analyzing sequentially several multisets formed by analogous small elution windows of the different chromatograms until all the elution regions have been covered [103,114]. Most recently, especially in -omic studies that use MS as detection system, data compression by ROI allows the analysis of chromatograms with a very high number of components without the need to use time-windowing strategies [115]. Figure 11.11 shows an example of the recently proposed method based on the concept of regions of interest (ROI) in liquid chromatography–mass spectrometry (LC–MS) together with multivariate curve resolution-alternating least squares (MCR-ALS). The effects of As(III) on the lipidomic profiles of two cyanobacteria species were assessed. Cyanobacteria were exposed to two concentrations of As(III) for a week, and lipid extracts were analyzed by ultrahigh-performance liquid chromatography/time-of-flight mass spectrometry in full scan mode. The data obtained were compressed by means of the ROI strategy and the resulting LC–MS data sets were analyzed by the MCR-ALS method. Since MS-ROI individual data matrices have different number of ROI m/z values, a preliminary step of ROI rearrangement to consider all those significant ROI m/z values (common and not common) is performed. Finally, comparison of profile peak areas resolved by MCR-ALS in control and exposed samples allowed the discrimination of lipids whose concentrations were changed due to As(III) treatment [109].

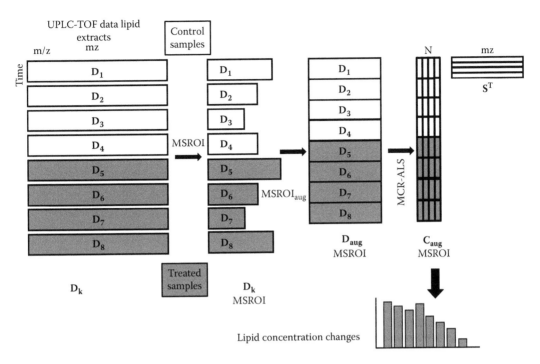

FIGURE 11.11 Example of MS-ROI data matrix augmentation and application of MCR-ALS to the simultaneous analysis proposed by Marques et al. [109].

In most of the works using MCR methods for fingerprint analysis, the compositional profile (fingerprint) obtained by MCR (based on peak areas) is frequently used as input information for other methods dedicated to data exploration. Thus, MCR quantitative fingerprint features can be submitted to PCA or to clustering methods to detect sample clusters that may suggest differences among different designations of origin [103]. Discrimination or classification methods, such as LDA or PLS-DA, also use MCR features to solve many diverse problems such as finding compositional patterns that were systematically covarying over multiple samples in relation to a specific physiological status.

Figure 11.12 shows a general scheme on how to use MCR-derived fingerprint information for a problem of the identity, authenticity, quality, and lot-to-lot consistency of natural products. Parastar et al. investigated fingerprints of secondary metabolites in citrus fruit peels [114]. The first step was segmenting the overlaid TIC of GC–MS into 18 samples in order to choose a number of chromatographic segments using local rank analysis and zero component regions. Afterward, the data were arranged in a bilinear way and then analyzed by MCR. As a result, the pure spectra of compounds in the citrus fruit peels for identification purposes and their relative concentration were obtained.

The fingerprint for each citrus fruit peel was formed by the peak areas of the secondary metabolites. The table of peak areas of samples was submitted to PCA and k-nearest neighbor (KNN) clustering methods to explore similarities and dissimilarities among different citrus samples according to their secondary metabolites. Four clusters were determined and the chemical markers (chemotypes) responsible for this differentiation were characterized by subsequent discriminant analysis using the counter-propagation artificial neural network (CPANN) method. As the authors concluded, the use of proposed strategy is more reliable and faster for the analysis of large data sets like chromatographic fingerprints of natural products compared to conventional methods.

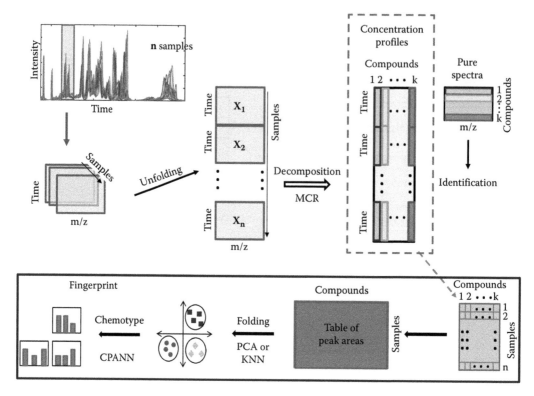

FIGURE 11.12 Overall data analysis procedure for the analysis of raw GC–MS fingerprints of secondary metabolites in citrus fruit peels proposed by Parastar et al. [114].

11.5   CONCLUSIONS

It has been shown how the use of chemometric tools, notably those based on factor analysis, can solve common and current challenging problems in chromatographic data sets. Preprocessing tools enhance the relevant information in the raw measurement by removing artifacts related to each specific detection technique, baseline contributions, and by compressing the data set when needed. Exploratory tools provide a first description of the chromatographic data set through the description of the total and local complexity of the sample. This means that detection of impurities can be done, a good elucidation of the elution pattern is obtained, and a first approximate description of the identity of compounds through the selection of the most dissimilar spectra in the run is achieved.

The information obtained from the exploratory analysis of chromatographic runs can be actively used in multivariate resolution methods. These are the most powerful tools to deal with chromatographic data because they provide the mathematical separation of compounds, that is, the definition of meaningful elution profiles and related pure responses. Such an output is highly relevant in complex situations encountered nowadays, such as the analysis of samples with a large number of overlapped compounds. The flexibility of multivariate resolution tools, such as multivariate curve resolution-alternating least squares (MCR-ALS), allows the analysis of data structures including several chromatographic runs and/or several response detection systems in tandem. Such scenario allows solving complex problems related to qualitative and quantitative analysis of complex samples, analysis of multidimensional chromatographic runs, chromatographic process monitoring, and fingerprint analysis linked to -omic problems or to food authentication and adulteration.

REFERENCES

1. Malinowski, E.R., 2002. *Factor Analysis in Chemistry*. Wiley, New York.
2. Eilers, P.H.C., 2004. Parametric time warping. *Anal. Chem.*, 76(2): 404–411.
3. Eilers, P.H.C. and Boelens, H.F., 2005. Baseline correction with asymmetric least squares smoothing. Leiden University Medical Centre Report 1, Leiden, the Netherlands.
4. Bahram, M., Bro, R., Stedmon, C., and Afkhami, A., 2006. Handling of Rayleigh and Raman scatter for PARAFAC modeling of fluorescence data using interpolation. *J. Chemometr.*, 20(3–4): 99–105.
5. Eilers, P.H. and Kroonenberg, P.M., 2014. Modeling and correction of Raman and Rayleigh scatter in fluorescence landscapes. *Chemom. Intel. Lab. Syst.*, 130: 1–5.
6. Tautenhahn, R., Böttcher, C., and Neumann, S., 2008. Highly sensitive feature detection for high resolution LC/MS. *BMC Bioinf.*, 9(1): 504.
7. Stolt, R., Torgrip, R.J., Lindberg, J., Csenki, L., Kolmert, J., Schuppe-Koistinen, I., and Jacobsson, S.P., 2006. Second-order peak detection for multicomponent high-resolution LC/MS data. *Anal. Chem.*, 78(4): 975–983.
8. Bedia, C., Tauler, R., and Jaumot, J., 2016. Compression strategies for the chemometric analysis of mass spectrometry imaging data. *J. Chemometr.*, 30(10): 575–588.
9. Maeder, M. and Zuberbuehler, A.D., 1986. The resolution of overlapping chromatographic peaks by evolving factor analysis. *Anal. Chim. Acta*, 181: 287–291.
10. Maeder, M., 1987. Evolving factor analysis for the resolution of overlapping chromatographic peaks. *Anal. Chem.*, 59(3): 527–530.
11. Maeder, M. and Zilian, A., 1988. Evolving factor analysis, a new multivariate technique in chromatography. *Chemom. Intel. Lab. Syst.*, 3(3): 205–213.
12. Keller, H.R. and Massart, D.L., 1991. Peak purity control in liquid chromatography with photodiode-array detection by a fixed size moving window evolving factor analysis. *Anal. Chim. Acta*, 246(2): 379–390.
13. Keller, H.R., Massart, D.L., and De Beer, J.O., 1993. Window evolving factor analysis for assessment of peak homogeneity in liquid chromatography. *Anal. Chem.*, 65(4): 471–475.
14. Malinowski, E.R., 1982. Obtaining the key set of typical vectors by factor analysis and subsequent isolation of component spectra. *Anal. Chim. Acta*, 134: 129–137.
15. Windig, W. and Guilment, J., 1991. Interactive self-modeling mixture analysis. *Anal. Chem.*, 63(14): 1425–1432.
16. Windig, W. and Markel, S., 1993. Simple-to-use interactive self-modeling mixture analysis of FTIR microscopy data. *J. Mol. Struct.*, 292: 161–170.
17. Sanchez, F.C., Toft, J., Van den Bogaert, B., and Massart, D.L., 1996. Orthogonal projection approach applied to peak purity assessment. *Anal. Chem.*, 68(1): 79–85.
18. Sánchez, F.C., Van den Bogaert, B., Rutan, S.C., and Massart, D.L., 1996. Multivariate peak purity approaches. *Chemom. Intel. Lab. Syst.*, 34(2): 139–171.
19. Schostack, K.J. and Malinowski, E.R., 1989. Preferred set selection by iterative key set factor analysis. *Chemom. Intel. Lab. Syst.*, 6(1): 21–29.
20. Van Zomeren, P.V., Darwinkel, H., Coenegracht, P.M.J., and De Jong, G.J., 2003. Comparison of several curve resolution methods for drug impurity profiling using high-performance liquid chromatography with diode array detection. *Anal. Chim. Acta*, 487(2): 155–170.
21. De Juan, A., Van den Bogaert, B., Sánchez, F.C., and Massart, D.L., 1996. Application of the needle algorithm for exploratory analysis and resolution of HPLC-DAD data. *Chemom. Intel. Lab. Syst.*, 33(2): 133–145.
22. Esbensen, K.H., 2000. *Multivariate Data Analysis in Practice*. CAMO, Oslo, Norway.
23. Massart, D.L., Vandeginste, B.G., Buydens, L.M.C., Lewi, P.J., and Smeyers-Verbeke, J., 1997. *Handbook of Chemometrics and Qualimetrics: Part A*. Elsevier Science Inc., Amsterdam, the Netherlands.
24. Whitson, A.C. and Maeder, M., 2001. Exhaustive evolving factor analysis (E-EFA). *J. Chemometr.*, 15(5): 475–484.
25. Toft, J. and Kvalheim, O.M., 1993. Eigenstructure tracking analysis for revealing noise pattern and local rank in instrumental profiles: Application to transmittance and absorbance IR spectroscopy. *Chemom. Intel. Lab. Syst.*, 19(1): 65–73.
26. Keller, H.R., Massart, D.L., Liang, Y.Z., and Kvalheim, O.M., 1992. Evolving factor analysis in the presence of heteroscedastic noise. *Anal. Chim. Acta*, 263(1–2): 29–36.
27. Windig, W. and Stephenson, D.A., 1992. Self-modeling mixture analysis of second-derivative near-infrared spectral data using the SIMPLISMA approach. *Anal. Chem.*, 64(22): 2735–2742.

28. Gemperline, P.J., 1984. A priori estimates of the elution profiles of the pure components in overlapped liquid chromatography peaks using target factor analysis. *J. Chem. Inf. Comput. Sci.*, 24(4): 206–212.

29. Malinowski, E.R., 1992. Window factor analysis: Theoretical derivation and application to flow injection analysis data. *J. Chemometr.*, 6(1): 29–40.

30. Schostack, K.J. and Malinowski, E.R., 1993. Investigation of window factor analysis and matrix regression analysis in chromatography. *Chemom. Intel. Lab. Syst.*, 20(2): 173–182.

31. Xu, C.J., Liang, Y.Z., and Jiang, J.H., 2000. Resolution of the embedded chromatographic peaks by modified orthogonal projection resolution and entropy maximization method. *Anal. Lett.* 33: 2105–2128.

32. Manne, R., Shen, H., and Liang, Y., 1999. Subwindow factor analysis. *Chemom. Intel. Lab. Syst.*, 45(1): 171–176.

33. Shen, H., Manne, R., Xu, Q., Chen, D., and Liang, Y., 1999. Local resolution of hyphenated chromatographic data. *Chemom. Intel. Lab. Syst.*, 45(1): 323–328.

34. Jiang, J.H., Šašić, S., Yu, R.Q., and Ozaki, Y., 2003. Resolution of two-way data from spectroscopic monitoring of reaction or process systems by parallel vector analysis (PVA) and window factor analysis (WFA): Inspection of the effect of mass balance, methods and simulations. *J. Chemometr.*, 17(3): 186–197.

35. Kvalheim, O.M. and Liang, Y.Z., 1992. Heuristic evolving latent projections: Resolving two-way multicomponent data. 1. Selectivity, latent-projective graph, datascope, local rank, and unique resolution. *Anal. Chem.*, 64(8): 936–946.

36. Liang, Y.Z., Kvalheim, O.M., Keller, H.R., Massart, D.L., Kiechle, P., and Erni, F., 1992. Heuristic evolving latent projections: Resolving two-way multicomponent data. 2. Detection and resolution of minor constituents. *Anal. Chem.*, 64(8): 946–953.

37. Jalali-Heravi, M. and Vosough, M., 2004. Characterization and determination of fatty acids in fish oil using gas chromatography–mass spectrometry coupled with chemometric resolution techniques. *J. Chromatogr. A*, 1024(1): 165–176.

38. Vandeginste, B.G., Derks, W., and Kateman, G., 1985. Multicomponent self-modelling curve resolution in high-performance liquid chromatography by iterative target transformation analysis. *Anal. Chim. Acta*, 173: 253–264.

39. Tauler, R., Casassas, E., and Izquierdo-Ridorsa, A., 1991. Self-modelling curve resolution in studies of spectrometric titrations of multi-equilibria systems by factor analysis. *Anal. Chim. Acta*, 248(2): 447–458.

40. Tauler, R., 1995. Multivariate curve resolution applied to second order data. *Chemom. Intel. Lab. Syst*, 30(1): 133–146.

41. Tauler, R., Smilde, A., and Kowalski, B., 1995. Selectivity, local rank, three-way data analysis and ambiguity in multivariate curve resolution. *J. Chemometr.*, 9(1): 31–58.

42. de Juan, A., Rutan, S.C., Maeder, M., and Tauler, R., 2009. MCR. In: Brown, S., Tauler, R., and Walczak, B. (Eds.). *Comprehensive Chemometrics*, Elsevier, Oxford, U.K., vol. 2, pp. 207–558.

43. Jaumot, J., de Juan, A., and Tauler, R., 2015. MCR-ALS GUI 2.0: New features and applications. *Chemom. Intel. Lab. Syst.*, 140: 1–12.

44. Gampp, H., Maeder, M., Meyer, C.J., and Zuberbühler, A.D., 1986. Calculation of equilibrium constants from multiwavelength spectroscopic data—IV: Model-free least-squares refinement by use of evolving factor analysis. *Talanta*, 33(12): 943–951.

45. Abdollahi, H. and Tauler, R., 2011. Uniqueness and rotation ambiguities in multivariate curve resolution methods. *Chemom. Intel. Lab. Syst.*, 108(2): 100–111.

46. Tauler, R., 2001. Calculation of maximum and minimum band boundaries of feasible solutions for species profiles obtained by multivariate curve resolution. *J. Chemometr.*, 15(8): 627–646.

47. Tauler, R. and Maeder, M., 2009. Two-way data analysis: Multivariate curve resolution–error in curve resolution. In: Brown, S., Tauler, R., and Walczak, B. (Eds.). *Comprehensive Chemometrics*, Elsevier, Oxford, U.K., vol. 2, pp. 2–20.

48. Gemperline, P.J., 1999. Computation of the range of feasible solutions in self-modeling curve resolution algorithms. *Anal. Chem.*, 71(23): 5398–5404.

49. Vosough, M., Mason, C., Tauler, R., Jalali-Heravi, M., and Maeder, M., 2006. On rotational ambiguity in model-free analyses of multivariate data. *J. Chemometr.*, 20(6–7): 302–310.

50. Sawall, M. and Neymeyr, K., 2014. A fast polygon inflation algorithm to compute the area of feasible solutions for three-component systems. II: Theoretical foundation, inverse polygon inflation, and FAC-PACK implementation. *J. Chemometr.*, 28(8): 633–644.

51. Bro, R. and De Jong, S., 1997. A fast non-negativity-constrained least squares algorithm. *J. Chemometr.*, 11(5): 393–401.

52. Bro, R. and Sidiropoulos, N.D., 1998. Least squares algorithms under unimodality and non-negativity constraints. *J. Chemometr.*, 12(4): 223–247.

53. De Juan, A., Vander Heyden, Y., Tauler, R., and Massart, D.L., 1997. Assessment of new constraints applied to the alternating least squares method. *Anal. Chim. Acta*, 346(3): 307–318.

54. Manne, R., 1995. On the resolution problem in hyphenated chromatography. *Chemom. Intel. Lab. Syst.*, 27(1): 89–94.

55. Jonsson, P., Johansson, A.I., Gullberg, J., Trygg, J., Grung, B., Marklund, S., Sjöström, M., Antti, H. and Moritz, T., 2005. High-throughput data analysis for detecting and identifying differences between samples in GC/MS-based metabolomic analyses. *Anal. Chem.*, 77(17): 5635–5642.

56. Domingo-Almenara, X., Perera, A., and Brezmes, J., 2016. Avoiding hard chromatographic segmentation: A moving window approach for the automated resolution of gas chromatography–mass spectrometry-based metabolomics signals by multivariate methods. *J. Chromatogr. A*, 1474: 145–151.

57. Mas, S., Carbó, A., Lacorte, S., De Juan, A., and Tauler, R., 2011. Comprehensive description of the photodegradation of bromophenols using chromatographic monitoring and chemometric tools. *Talanta*, 83(4): 1134–1146.

58. Gorrochategui, E., Jaumot, J., Lacorte, S., and Tauler, R., 2016. Data analysis strategies for targeted and untargeted LC-MS metabolomic studies: Overview and workflow. *Trends Anal. Chem.*, 82: 425–442.

59. Tauler, R. and Barceló, D., 1993. Multivariate curve resolution applied to liquid chromatography—Diode array detection. *Trends Anal. Chem.*, 12(8): 319–327.

60. Tauler, R., Marqués, I., and Casassas, E., 1998. Multivariate curve resolution applied to three-way trilinear data: Study of a spectrofluorimetric acid–base titration of salicylic acid at three excitation wavelengths. *J. Chemometr.*, 12(1): 55–75.

61. De Juan, A. and Tauler, R., 2001. Comparison of three-way resolution methods for non-trilinear chemical data sets. *J. Chemometr.*, 15(10): 749–771.

62. Harshman, R.A., 1970. Foundations of the PARAFAC procedure: Models and conditions for an "explanatory" multi-modal factor analysis. In *UCLA Working Papers in Phonetics*, Elsevier, Oxford, U.K., vol. 16, pp. 1–84.

63. Bro, R., 1997. PARAFAC. Tutorial and applications. *Chemom. Intel. Lab. Syst.*, 38(2): 149–171.

64. Sanchez, E. and Kowalski, B.R., 1986. Generalized rank annihilation factor analysis. *Anal. Chem.*, 58(2): 496–499.

65. Sanchez, E. and Kowalski, B.R., 1990. Tensorial resolution: A direct trilinear decomposition. *J. Chemometr.*, 4(1): 29–45.

66. Ortiz, M.C. and Sarabia, L., 2007. Quantitative determination in chromatographic analysis based on n-way calibration strategies. *J. Chromatogr. A*, 1158(1): 94–110.

67. Fraga, C.G., Prazen, B.J., and Synovec, R.E., 2001. Objective data alignment and chemometric analysis of comprehensive two-dimensional separations with run-to-run peak shifting on both dimensions. *Anal. Chem.*, 73(24): 5833–5840.

68. Comas, E., Gimeno, R.A., Ferré, J., Marcé, R.M., Borrull, F., and Rius, F.X., 2002. Time shift correction in second-order liquid chromatographic data with iterative target transformation factor analysis. *Anal. Chim. Acta*, 470(2): 163–173.

69. Tomasi, G., Van den Berg, F., and Andersson, C., 2004. Correlation optimized warping and dynamic time warping as preprocessing methods for chromatographic data. *J. Chemometr.*, 18(5): 231–241.

70. Skov, T., van den Berg, F., Tomasi, G., and Bro, R., 2006. Automated alignment of chromatographic data. *J. Chemometr.*, 20(11–12): 484–497.

71. Bro, R., Andersson, C.A., and Kiers, H.A., 1999. PARAFAC2-Part II. Modeling chromatographic data with retention time shifts. *J. Chemometr.*, 13(3–4): 295–309.

72. Bortolato, S.A. and Olivieri, A.C., 2014. Ultra performance liquid chromatography tandem mass spectrometry performance evaluation for analysis of antibiotics in natural waters. *Anal. Chim. Acta*, 842: 11–19.

73. De Braekeleer, K., De Juan, A., and Massart, D.L., 1999. Purity assessment and resolution of tetracycline hydrochloride samples analysed using high-performance liquid chromatography with diode array detection. *J. Chromatogr. A*, 832(1): 67–86.

74. Mas, S., Fonrodona, G., Tauler, R., and Barbosa, J., 2007. Determination of phenolic acids in strawberry samples by means of fast liquid chromatography and multivariate curve resolution methods. *Talanta*, 71(4): 1455–1463.

75. Peré-Trepat, E., Hildebrandt, A., Barceló, D., Lacorte, S., and Tauler, R., 2004. Fast chromatography of complex biocide mixtures using diode array detection and multivariate curve resolution. *Chemom. Intel. Lab. Syst.*, 74(2): 293–303.

76. Peré-Trepat, E., Lacorte, S., and Tauler, R., 2005. Solving liquid chromatography mass spectrometry coelution problems in the analysis of environmental samples by multivariate curve resolution. *J. Chromatogr. A*, 1096(1): 111–122.
77. Rodríguez-Cuesta, M.J., Boqué, R., Rius, F.X., Vidal, J.M., and Frenich, A.G., 2005. Development and validation of a method for determining pesticides in groundwater from complex overlapped HPLC signals and multivariate curve resolution. *Chemom. Intel. Lab. Syst.*, 77(1): 251–260.
78. Peré-Trepat, E., Lacorte, S., and Tauler, R., 2007. Alternative calibration approaches for LC–MS quantitative determination of coeluted compounds in complex environmental mixtures using multivariate curve resolution. *Anal. Chim. Acta*, 595(1): 228–237.
79. de Oliveira Neves, A.C., Tauler, R., and de Lima, K.M.G., 2016. Area correlation constraint for the MCR–ALS quantification of cholesterol using EEM fluorescence data: A new approach. *Anal. Chim. Acta*, 937: 21–28.
80. Tauler, R. and de Juan, A., 2015. Multivariate curve resolution for quantitative analysis. In: Olivieri, I.C., Escandar, G., Goicoechea, H.C., and de la Peña, A. M. (Eds.). *Fundamentals and Analytical Applications of Multi-Way Calibration*, vol. 29, p. 247. Elsevier, Oxford, U.K.
81. Konoz, E., Abbasi, A., Parastar, H., Moazeni, R.S., and Jalali-Heravi, M., 2015. Analysis of olive fruit essential oil: Application of gas chromatography-mass spectrometry combined with chemometrics. *Int. J. Food Prop.*, 18(2): 316–331.
82. Peré-Trepat, E. and Tauler, R., 2006. Analysis of environmental samples by application of multivariate curve resolution on fused high-performance liquid chromatography–diode array detection mass spectrometry data. *J. Chromatogr. A*, 1131(1): 85–96.
83. Alcaráz, M.R., Siano, G.G., Culzoni, M.J., de la Peña, A.M., and Goicoechea, H.C., 2014. Modeling four and three-way fast high-performance liquid chromatography with fluorescence detection data for quantitation of fluoroquinolones in water samples. *Anal. Chim. Acta*, 809: 37–46.
84. de Juan, A. and Tauler, R., 2007. Factor analysis of hyphenated chromatographic data: Exploration, resolution and quantification of multicomponent systems. *J. Chromatogr. A*, 1158(1): 184–195.
85. Gómez-Canela, C., Bolivar-Subirats, G., Tauler, R., and Lacorte, S., 2017. Powerful combination of analytical and chemometric methods for the photodegradation of 5-fluorouracil. *J. Pharm. Biomed. Anal.*, 137: 33–41.
86. Jayaraman, A., Mas, S., Tauler, R., and de Juan, A., 2012. Study of the photodegradation of 2-bromophenol under UV and sunlight by spectroscopic, chromatographic and chemometric techniques. *J. Chromatogr. B*, 910: 138–148.
87. Mas, S., Tauler, R., and De Juan, A., 2011. Chromatographic and spectroscopic data fusion analysis for interpretation of photodegradation processes. *J. Chromatogr. A*, 1218(51): 9260–9268.
88. de Juan, A., Maeder, M., Martınez, M., and Tauler, R., 2000. Combining hard-and soft-modelling to solve kinetic problems. *Chemom. Intel. Lab. Syst.*, 54(2): 123–141.
89. De Luca, M., Ragno, G., Ioele, G., and Tauler, R., 2014. Multivariate curve resolution of incomplete fused multiset data from chromatographic and spectrophotometric analyses for drug photostability studies. *Anal. Chim. Acta*, 837: 31–37.
90. Alier, M. and Tauler, R., 2013. Multivariate curve resolution of incomplete data multisets. *Chemom. Intel. Lab. Syst.*, 127: 17–28.
91. Bailey, H.P. and Rutan, S.C., 2011. Chemometric resolution and quantification of four-way data arising from comprehensive 2D-LC-DAD analysis of human urine. *Chemom. Intel. Lab. Syst.*, 106(1): 131–141.
92. Parastar, H., Radović, J.R., Jalali-Heravi, M., Diez, S., Bayona, J.M., and Tauler, R., 2011. Resolution and quantification of complex mixtures of polycyclic aromatic hydrocarbons in heavy fuel oil sample by means of GC × GC-TOFMS combined to multivariate curve resolution. *Anal. Chem.*, 83(24): 9289–9297.
93. Parastar, H., Jalali-Heravi, M., and Tauler, R., 2012. Comprehensive two-dimensional gas chromatography (GC × GC) retention time shift correction and modeling using bilinear peak alignment, correlation optimized shifting and multivariate curve resolution. *Chemom. Intel. Lab. Syst.*, 117: 80–91.
94. Parastar, H., Radović, J.R., Bayona, J.M., and Tauler, R., 2013. Solving chromatographic challenges in comprehensive two-dimensional gas chromatography–time-of-flight mass spectrometry using multivariate curve resolution–alternating least squares. *Anal. Bioanal. Chem.*, 405(19): 6235–6249.
95. Parastar, H. and Tauler, R., 2013. Multivariate curve resolution of hyphenated and multidimensional chromatographic measurements: A new insight to address current chromatographic challenges. *Anal. Chem.*, 86(1): 286–297.
96. Radović, J.R., Thomas, K.V., Parastar, H., Díez, S., Tauler, R., and Bayona, J.M., 2014. Chemometrics-assisted effect-directed analysis of crude and refined oil using comprehensive two-dimensional gas chromatography–time-of-flight mass spectrometry. *Environ. Sci. Technol.*, 48(5): 3074–3083.

97. Cook, D.W., Rutan, S.C., Stoll, D.R., and Carr, P.W., 2015. Two dimensional assisted liquid chromatography—A chemometric approach to improve accuracy and precision of quantitation in liquid chromatography using 2D separation, dual detectors, and multivariate curve resolution. *Anal. Chim. Acta*, 859: 87–95.

98. Navarro-Reig, M., Jaumot, J., van Beek, T.A., Vivó-Truyols, G., and Tauler, R., 2016. Chemometric analysis of comprehensive LC × LC-MS data: Resolution of triacylglycerol structural isomers in corn oil. *Talanta*, 160: 624–635.

99. De la Mata-Espinosa, P., Bosque-Sendra, J.M., Bro, R., and Cuadros-Rodriguez, L., 2011. Discriminating olive and non-olive oils using HPLC-CAD and chemometrics. *Anal. Bioanal. Chem.*, 399(6): 2083–2092.

100. Cuny, M., Vigneau, E., Le Gall, G., Colquhoun, I., Lees, M., and Rutledge, D.N., 2008. Fruit juice authentication by ^{1}H NMR spectroscopy in combination with different chemometrics tools. *Anal. Bioanal. Chem.*, 390(1): 419–427.

101. Jalali-Heravi, M., Parastar, H., and Ebrahimi-Najafabadi, H., 2010. Self-modeling curve resolution techniques applied to comparative analysis of volatile components of Iranian saffron from different regions. *Anal. Chim. Acta*, 662(2): 143–154.

102. Airado-Rodríguez, D., Durán-Merás, I., Galeano-Díaz, T., and Wold, J.P., 2011. Front-face fluorescence spectroscopy: A new tool for control in the wine industry. *J. Food Compos. Anal.*, 24(2): 257–264.

103. Salvatore, E., Cocchi, M., Marchetti, A., Marini, F., and de Juan, A., 2013. Determination of phenolic compounds and authentication of PDO Lambrusco wines by HPLC-DAD and chemometric techniques. *Anal. Chim. Acta*, 761: 34–45.

104. Ardila, J.A., Funari, C.S., Andrade, A.M., Cavalheiro, A.J., and Carneiro, R.L., 2015. Cluster analysis of commercial samples of *Bauhinia* spp. using HPLC-UV/PDA and MCR-ALS/PCA without peak alignment procedure. *Phytochem. Anal.*, 26(5): 367–373.

105. Pisano, P.L., Silva, M.F., and Olivieri, A.C., 2014. Exploration of liquid chromatographic-diode array data for Argentinean wines by extended multivariate curve resolution. *Chemom. Intel. Lab. Syst.*, 132: 1–7.

106. Aliakbarzadeh, G., Sereshti, H., and Parastar, H., 2016. Pattern recognition analysis of chromatographic fingerprints of *Crocus sativus* L. secondary metabolites towards source identification and quality control. *Anal. Bioanal. Chem.*, 408(12): 3295–3307.

107. Monago-Maraña, O., Durán-Merás, I., Galeano-Díaz, T., and de la Peña, A.M., 2016. Fluorescence properties of flavonoid compounds. Quantification in paprika samples using spectrofluorimetry coupled to second order chemometric tools. *Food Chem.*, 196: 1058–1065.

108. Gorrochategui, E., Casas, J., Porte, C., Lacorte, S., and Tauler, R., 2015. Chemometric strategy for untargeted lipidomics: Biomarker detection and identification in stressed human placental cells. *Anal. Chim. Acta*, 854: 20–33.

109. Marques, A.S., Bedia, C., Lima, K.M., and Tauler, R., 2016. Assessment of the effects of As (III) treatment on cyanobacteria lipidomic profiles by LC-MS and MCR-ALS. *Anal. Bioanal. Chem.*, 408(21): 5829–5841.

110. Farrés, M., Piña, B., and Tauler, R., 2015. Chemometric evaluation of *Saccharomyces cerevisiae* metabolic profiles using LC–MS. *Metabolomics*, 11(1): 210–224.

111. Navarro-Reig, M., Jaumot, J., García-Reiriz, A., and Tauler, R., 2015. Evaluation of changes induced in rice metabolome by Cd and Cu exposure using LC-MS with XCMS and MCR-ALS data analysis strategies. *Anal. Bioanal. Chem.*, 407(29): 8835–8847.

112. Pérez, I.S., Culzoni, M.J., Siano, G.G., García, M.D.G., Goicoechea, H.C., and Galera, M.M., 2009. Detection of unintended stress effects based on a metabonomic study in tomato fruits after treatment with carbofuran pesticide. Capabilities of MCR-ALS applied to LC-MS three-way data arrays. *Anal. Chem.*, 81(20): 8335–8346.

113. Wehrens, R., Carvalho, E., Masuero, D., de Juan, A., and Martens, S., 2013. High-throughput carotenoid profiling using multivariate curve resolution. *Anal. Bioanal. Chem.*, 405(15): 5075–5086.

114. Parastar, H., Jalali-Heravi, M., Sereshti, H., and Mani-Varnosfaderani, A., 2012. Chromatographic fingerprint analysis of secondary metabolites in citrus fruits peels using gas chromatography–mass spectrometry combined with advanced chemometric methods. *J. Chromatogr. A*, 125: 176–187.

115. Gorrochategui, E., Jaumot, J., and Tauler, R., 2015. A protocol for LC-MS metabolomic data processing using chemometric tools. *Protocol Exchange*. doi: 10.1038/protex.2015.1102. Accessed on November, 30, 2015.

12 Modeling of Peak Shape and Asymmetry

José Ramón Torres-Lapasió, Juan José Baeza-Baeza,
and María Celia García-Álvarez-Coque

CONTENTS

12.1 INTRODUCTION

Peak profiles in chromatography are the final result of different types of interactions within the column (e.g., partitioning, adsorption/desorption, ion-exchange, and size exclusion processes). To these, other intra-column effects (e.g., axial dispersion and diffusive migration through the pores in the packing materials) and extra-column dispersion within the connecting tubing and other components between injector and detector (column end fittings, frits, injection plug, and detector cell) should be added [1–3].

Martin and Synge, in a seminal article [4], derived a function based on theoretical considerations to describe chromatographic peak profiles in linear chromatography. A practical expression of the proposed function, in the absence of external dispersion sources, is

$$h(t) = h_0 \exp\left[(1-N)\left(\frac{t}{t_R} - \ln\left(\frac{t}{t_R}\right) - 1 \right) \right] \tag{12.1}$$

where

    $h(t)$ and h_0 are the heights at time t and at the peak maximum, respectively
    t_R is the retention time (i.e., the time at the peak maximum)
    N is the number of theoretical plates (plate count or efficiency)

Equation 12.1 tends to a Gaussian profile for sufficiently high N values:

$$h(t) = h_0 \exp\left[-\frac{1}{2}\left(\frac{t - t_R}{\sigma} \right)^2 \right] \tag{12.2}$$

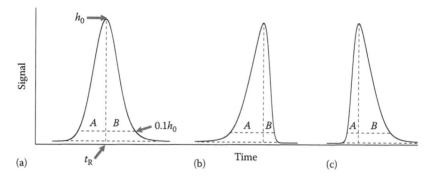

FIGURE 12.1 Chromatographic signals for (a) symmetrical, (b) fronting, and (c) tailing peaks. The height at the peak maximum (h_0), retention time (t_R), and left ($A_{0.1}$) and right ($B_{0.1}$) half-widths at 10% peak height are marked.

The efficiency N is inversely related to the peak variance (σ^2) (i.e., the squared standard deviation or half-width at 60.65% peak height):

$$N = \frac{t_R^2}{\sigma^2} \qquad (12.3)$$

The peak variance in Equation 12.3 only accounts for the intra-column dispersion (σ^2_{col}). In practice, chromatographic peaks are affected by both intra- and extra-column (σ^2_{ext}) dispersion:

$$\sigma^2 = \sigma^2_{col} + \sigma^2_{ext} \qquad (12.4)$$

Equation 12.2 describes well the elution profile of symmetrical and nonoverloaded chromatographic peaks (Figure 12.1a) [5,6]. Skewed nonideal peaks, either fronting (Figure 12.1b) or tailing (Figure 12.1c), are, however, very common in practice, due to the presence of complex solute-stationary phase interactions and diverse extra-column contributions, particularly relevant at short retention times. In addition, peak profiles (width and asymmetry) can change with the mobile phase composition, pH, ionic strength, and temperature, among other factors.

The accurate description of chromatographic peaks is needed to extract the relevant information of the signals for either isolated or overlapped peaks. This is required for the prediction, optimization, and deconvolution of signals and for the qualitative and quantitative analysis of samples. This chapter provides an overview of the mathematical models that have been reported for the description of chromatographic peak profiles. The application of models for the characterization, prediction, evaluation of the coelution level, and deconvolution of partially coeluting peaks, is also considered.

12.2 PEAK MODELS

According to Cavazzini et al. [7], "obtaining an analytical function that describes peak profiles at the exit of the chromatographic column is impossible, owing to the complex and numerous physicochemical processes involved." Indeed, asymmetrical peaks, which are often obtained in chromatography, cannot be described by simple functions. This is one of the reasons that explains the high number of authors involved in the development of mathematical functions for peak description (with different success), along decades. Some examples are given in References 8–40. Most models in the literature are empirical and lack of full theoretical justification, but can be useful for several purposes. The reader should address to published comparison studies to appraise the strong and weak points of the different peak models [6,30,41–45].

The compilation of about 90 peak models with almost 200 references, published by Di Marco and Bombi in 2001, is particularly interesting [44]. The study includes a detailed description of the models, the acceptable intervals for the fitting coefficients (i.e., model parameters) and independent variable, and the applications and properties of the models. Some of them have proved useful to describe signals from other instrumental methods (e.g., flow injection analysis, spectroscopic methods, voltammetry, mass spectrometry, and thermal analysis), and conversely, several functions proposed for other techniques have demonstrated suitable to fit chromatographic peaks. However, most reported peak models are not sufficiently accurate. In the two following subsections, a selection is made of the different types of peak models described in the literature (Tables 12.1 and 12.2), and the state of the art in peak modeling is shown.

In spite of the high number of peak models proposed in the literature, only a few have been used by other authors for developing applications. The most frequently used models are the Gaussian type. Some peak models have been more or less overlooked owing to their inaccuracy or complexity in the estimation of the peak parameters. However, as commented by Di Marco and Bombi [44], "some other functions possess good properties or have been demonstrated to be good ... their lack of popularity probably is merely due to the tendency of researchers to use traditional, well studied and frequently cited functions at the expense of the newer ones." These words are still true 15 years later.

Achievement of the peak model coefficients requires nonlinear fitting, which may give rise to convergence problems if the initial estimates of the parameters are far from the solution. As usual in optimization problems, the fitting quality is increased with the number of coefficients, but this makes convergence harder. This also increases the uncertainty in the estimated parameters and the flexibility of the model, giving rise occasionally to unexpected results. It is desirable that the model parameters be related to easily measurable peak properties, since this favors the chances of finding the right solution. When the retention time is an explicit parameter in the model, its accurate knowledge is particularly helpful to assure convergence. Very often, model parameters are just fitting coefficients without any physical meaning.

12.2.1 Modified Gaussian Models

The Gaussian function (Equation 12.2) seems to be a good starting point to obtain a model that describes the deviations from the ideal behavior of chromatographic peaks. Some of the earliest Gaussian modified models were developed by Haarhoff and van der Linde [8], Fraser and Suzuki [9,10], Grushka et al. [11,13], Grubner [12], Buys and de Clerk [14], and Chesler and Cram [15], in the 1960s and 1970s. A selection of some representative models is gathered in Table 12.1.

The bi-Gaussian model (Equation 12.5) describes asymmetrical peaks using two half-Gaussians with a common maximum, and different widths for the left and right peak regions. Long tailing or fronting peaks are, however, not well described since the front and rear parts of the peaks reach null values very rapidly. Various attempts have been made to provide asymmetry to the Gaussian by multiplying it with different polynomials. An example is the use of the Gram–Charlier series (Equation 12.6), extensively applied in the calculation of moments in chromatography (see Section 12.3). However, this function is highly complex and the number of n terms in the series is not univocally defined and depends on the peak skewness.

The exponentially modified Gaussian model (EMG) is probably the most popular to describe asymmetrical peaks in chromatography till date [32,46–51]. This model usually offers good fittings of moderately skewed peaks. The equation in Table 12.1 (Equation 12.7) includes explicitly the error function (*erf*), which is available in most graphical and mathematical software packages, or can be obtained through numerical approximations. Most equations reported in the literature for the EMG model include the error function implicitly, as a definite integral, instead [44].

A handier modified Gaussian model interprets the deviations from an ideal Gaussian peak as a change in the standard deviation with time, according to a polynomial function (Equation 12.8). Since the polynomial degree within the standard deviation term can be changed, the equation represents a family of models: the higher the degree, the more flexible the model, and the more the chances to fit the experimental data. The coefficients in the polynomial function are σ_0 (which describes a Gaussian that behaves as the asymmetrical peak in the maximum neighborhood) and c_1 and higher-order terms (which account for the peak skewness). For the simplest polynomial, these coefficients

TABLE 12.1
Some Representative Gaussian Modified Models

Name	Model and Equation[a]	Refs.
Bi-Gaussian	$h(t) = h_0 \exp\left[-\frac{1}{2}\left(\frac{t-t_R}{\sigma_1}\right)^2 \right]$ for $t < t_R$; $\quad h(t) = h_0 \exp\left[-\frac{1}{2}\left(\frac{t-t_R}{\sigma_2}\right)^2 \right]$ for $t \geq t_R$ $\quad$ (12.5)	[11]
Gram–Charlier series	$h(t) = h_0 \times \exp\left[-\frac{1}{2}\left(\frac{t-t_R}{w}\right)^2 \right]$ $\times \left[1 + \sum_{i=3}^{n} (-1)^i \frac{c_i}{i!} H_i(t) \right];$ $\quad H_i(t)$ is the i member of the Hermite polynomial $\quad$ (12.6)	[12,22]
Exponentially modified Gaussian	$h(t) = \frac{M_0}{2\tau} \exp\left[\frac{1}{2}\frac{\sigma_G^2}{\tau^2} - \frac{t-t_G}{\tau} \right] \times \left[1 - erf\left(\frac{\sigma_G}{\sqrt{2}\tau} - \frac{t-t_G}{\sqrt{2}\sigma_G} \right) \right]$ $\quad$ (12.7)	[32]
Polynomially modified Gaussian	$h(t) = h_0 \exp\left[-\frac{1}{2}\left(\frac{t-t_R}{\sigma_0 + c_1(t-t_R) + c_2(t-t_R)^2 + \cdots} \right)^2 \right]$ for PMG1 (linear function) $\quad$ (12.8) $\sigma_0 = 0.932\frac{A_{0.1}B_{0.1}}{A_{0.1}+B_{0.1}};$ $\quad c_1 = 0.466\frac{B_{0.1}-A_{0.1}}{A_{0.1}+B_{0.1}}$	[25]
Mixed exponential-PMG1	$h(t) = h_0 \exp\left[-\frac{1}{2}\left(\frac{t-t_R}{\sigma_0 + c_1(t-t_R)} \right)^2 \right]$ $\quad$ (12.9) Exponential decays: $\begin{cases} h = k_{1,\text{left}} \exp\{k_{2,\text{left}}(t-t_R)\} & \text{for } t < t_R - A_{0.1} \\ h = k_{1,\text{right}} \exp\{k_{2,\text{right}}(t-t_R)\} & \text{for } t > t_R + B_{0.1} \end{cases}$ $k_{1,\text{left}} = 0.1h_0 \exp(k_{2,\text{left}}A_{0.1});$ $\quad k_{2,\text{left}} = \frac{\sigma_0 A_{0.1}}{(\sigma_0 - c_1 A_{0.1})^3};$ $k_{1,\text{right}} = 0.1h_0 \exp(k_{2,\text{right}}B_{0.1});$ $\quad k_{2,\text{right}} = -\frac{\sigma_0 B_{0.1}}{(\sigma_0 + c_1 B_{0.1})^3}$	[36]
Parabolic variance Gaussian (PVG)	$h(t) = h_0 \exp\left[-\frac{1}{2}\frac{(t-t_R)^2}{\sigma_0^2 + c_1(t-t_R) + c_2(t-t_R)^2} \right]$ $\quad$ (12.10) $c_1 = \frac{B_{0.1}-A_{0.1}}{B_{0.1}A_{0.1}}s_0^2;$ $\quad c_2 = 0.217 - \frac{s_0^2}{B_{0.1}A_{0.1}}$	[33]
Parabolic-Lorentzian Gaussian (PLG)	$h(t) = h_0 \exp\left[-\frac{1}{2}\frac{1 + c_3(t-t_R) + c_4(t-t_R)^2}{\sigma_0^2 + c_1(t-t_R) + c_2(t-t_R)^2}(t-t_R)^2 \right]$ $\quad$ (12.11) $c_4 = \frac{c_3^2}{4} + c_5^2;$ $\quad \sigma_0^2 = \frac{4}{1+c_0^2}\frac{A_{0.6}^2 B_{0.6}^2}{(A_{0.6}+B_{0.6})^2}$ $c_1 = \frac{\sigma_0^2(B_{0.6}-A_{0.6})}{A_{0.6}B_{0.6}} + c_3 A_{0.6}B_{0.6} + c_4 A_{0.6}B_{0.6}(B_{0.6}-A_{0.6})$ $c_2 = 1 - \frac{\sigma_0^2}{A_{0.6}B_{0.6}} + c_3(B_{0.6}-A_{0.6}) + c_4\frac{A_{0.6}^3 + B_{0.6}^3}{A_{0.6}+B_{0.6}}$	[31,40]

[a] h_0, maximal height or related parameter; t_R, time at the peak maximum; σ, standard deviation or related parameter; M_0, peak area; σ_G and t_G, coefficients of the precursor Gaussian signal; τ, time constant; $c_0, c_1, c_2, \ldots$, fitting coefficients.

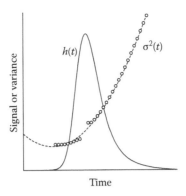

FIGURE 12.2 Chromatographic peak and experimental variance (arbitrary units). The dashed line depicts the parabolic fitting.

are expressed as a function of easily measurable parameters ($A_{0.1}$ and $B_{0.1}$, see Figure 12.1). In practice, parabolic (PMG2) or cubic (PMG3) functions can describe most chromatographic signals.

In a comparison study, Nikitas et al. pointed out that the PMG model was the only one able to describe almost any peak from tailing to fronting [45]. However, it has the drawback that the function does not work properly when the polynomial takes zero or negative values. Also, after reaching a minimum value, the predicted signal may grow outside the peak region. This is especially troublesome for the prediction of chromatograms, where the signals of individual peaks separated in time should be added to give composite signals. The artifact is more prominent for strongly asymmetrical signals ($B_{0.1}/A_{0.1} > 2.5$) and for long time windows. The proposal of a mixed exponential-PMG1 function (Equation 12.9), where the outer peak regions of the PMG1 model are replaced by exponential decays at both sides of the PMG1 peak, subjected to the condition of keeping the same slope at the connection points, solves perfectly the problem [36].

The experimental variance of a skewed chromatographic peak follows a parabolic trend (Figure 12.2). Based on this observation, two alternatives have been suggested to solve the drawbacks of the PMG model and enhance its fitting performance. The most simple includes merely a parabolic variance (Equation 12.10) [33]. The assumption of a parabolic trend does not allow full recovery of the baseline, leaving a small residual value that may be significant for highly asymmetrical peaks. This behavior was overcome by combining a parabola that describes the variance change in the peak region and a Lorentzian function that decreases the variance growth out of the peak region (Equation 12.11). The model shows an extraordinary fitting performance, even for highly asymmetrical peaks as is the case of some electrophoretic signals [52]. However, it has the drawback of its high flexibility, which makes the optimization process difficult when the initial estimates of the model parameters are far from the optimal ones [53]. To solve this problem, several semiempirical functions using half-width values at 60.65% and 10% peak height were proposed, which yielded prediction errors below 2% [40].

12.2.2 MODELS PROPOSED ON A NON-GAUSSIAN BASIS

Besides the wide variety of reported modified Gaussian models (Table 12.1), some authors have proposed several functions on a non-Gaussian basis, some of which are given in Table 12.2.

The extreme value (Equation 12.12), log-normal (Equation 12.13), generalized exponential (GEX, Equation 12.14), and Pap–Pápai (Equation 12.15) models are inspired in statistical distribution functions. The log-normal model assumes a normal distribution for the logarithm of the random variable. The extreme value model includes only three parameters, similar to the Gaussian function. The log-normal, GEX, and Pap–Pápai models all include four parameters.

TABLE 12.2

Models Proposed on Non-Gaussian Basis

Name	Model and Equation[a]		Refs.
Extreme value	$$h(t) = h_0 \exp\left[-\exp\left(-\frac{t - t_R}{w}\right) - \frac{t - t_R}{w} + 1\right]$$	(12.12)	[26]
Log-normal	$$h(t) = h_0 \exp\left[-\frac{\ln 2}{\alpha^2}\left[\ln\left(\frac{2\alpha(t - t_R)}{w} + 1\right)\right]^2\right]$$	(12.13)	[12,21]
Generalized exponential	$$h(t) = h_0\left(\frac{t - t_1}{t_R - t_1}\right)^{c_1} \exp\left\{\frac{c_1}{c_2}\left[1 - \left(\frac{t - t_1}{t_R - t_1}\right)^{c_2}\right]\right\}; \ c_1 > 0; \ c_2 > 0; \ t > t_1$$	(12.14)	[17]
Pap–Pápai	$$h(t) = h_0 \exp\left\{\left(\frac{4}{\alpha^2} - 1\right)\left[\ln\left(1 + \frac{2\alpha(t - t_R)}{\sigma(4 - \alpha^2)}\right) - \frac{2\alpha(t - t_R)}{\sigma(4 - \alpha^2)}\right]\right\}$$	(12.15)	[29]
Losev	$$h(t) = \frac{h_0}{\exp\left(-\dfrac{t - t_1}{a}\right) + \exp\left(\dfrac{t - t_1}{b}\right)}$$	(12.16)	[20]
Li	$$h(t) = \frac{h_0}{\left[1 + c_1 \exp\left[\dfrac{t_1 - t}{a}\right]\right]^{c_2} + \left[1 + c_3 \exp\left[\dfrac{t - t_2}{b}\right]\right]^{c_4} - 1}$$	(12.17)	[26,30]
Combined squared roots	$$h(t) = \frac{h_0}{2}\left(\sqrt{\left(\frac{t - t_R}{a} + 1 + c_1\right)^2 + c_2^2} + \sqrt{\left(\frac{t - t_R}{b} - 1 - c_3\right)^2 + c_4^2}\right.$$ $$\left. -\frac{a + b}{ab}\sqrt{(t - t_R)^2 + c_5^2} - c_1 - c_3\right)$$	(12.18)	[52]
Baker	$$h(t) = \frac{h_0}{1 + c_1(t - t_R)^2 + c_2(t - t_R)^3 + c_3(t - t_R)^4}$$	(12.19)	[5]
Giddings	$$h(t) = \frac{h_0}{2w}\sqrt{\frac{t_1}{t}} \times I_1\left(2\frac{\sqrt{tt_1}}{w}\right)\exp\left(-\frac{t + t_1}{w}\right); \ I_1(t) = \sum_{k=0}^{\infty}\frac{1}{k!(k+1)!}\left(\frac{t}{2}\right)^{2k+1}$$	(12.20)	[16]

[a] h_0, maximal height or related parameter; t_R, time at the peak maximum; w, parameter related to the peak width; α, parameter related with the asymmetry; t_1, initial time; t_1 and t_2, parameters related to the peak position; a, b, parameters related to the left and right half-widths; c_0, c_1, c_2, ..., fitting coefficients.

Other different approaches have been applied to model chromatographic peaks [44]. The models proposed by Losev (Equation 12.16) and Li (Equation 12.17) are examples of empirical functions, which break the peak into leading and tailing edges, each described with a sigmoidal function. A similar approach is followed by the CSR function (Equation 12.18), which is based on a combination of squared roots and has shown good performance to describe highly asymmetrical peaks. The main drawback of these three functions is that they need a large number of parameters to fit properly any peak.

Baker proposed instead a modified Lorentzian (Equation 12.19), which has a less pronounced decay than the Gaussian function. However, this model only allows the fitting of symmetrical peaks with high kurtosis and is not suitable for chromatography. Finally, the function proposed by Giddings (Equation 12.20) is included in several mathematical commercial software packages and is recommended for chromatography. Its high complexity makes it less attractive for routine applications.

With the exception of the Baker function, the functions in Table 12.2 are able to fit tailing peaks with good results, but only the Losev, Li, and CSR functions can fit fronting peaks. All functions allow an analytical integration and, therefore, the computation of moments, but the direct estimation of the half-widths is not possible. The log-normal function has been extensively used in chromatography, in some cases with better performance than the EMG model [21]. However, the Pap–Pápai equation can offer better fitting for both tailing and symmetrical peaks [31]. It should be mentioned that the practical use of the log-normal, GEX, and Pap–Pápai functions implies adopting restrictions to avoid logarithms of negative numbers or fractional exponential negative numbers.

Figure 12.3 compares the performance of several functions in the fitting of a tailing peak. As observed, the highest accuracy was achieved with the PVMG, PLMG, Pap–Pápai, and Li models.

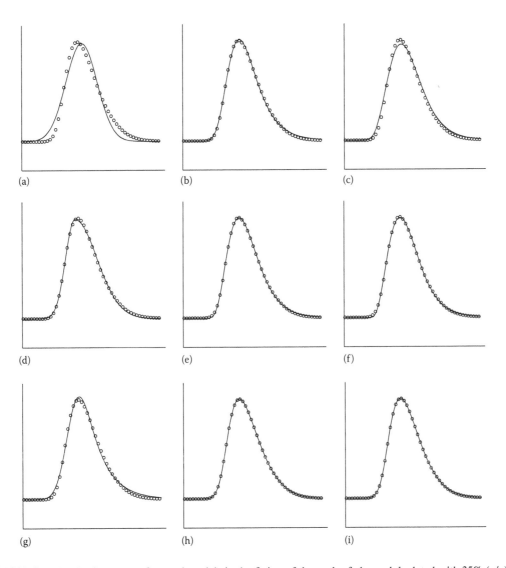

FIGURE 12.3   Performance of several models in the fitting of the peak of alprenolol, eluted with 35% (v/v) acetonitrile at pH 3 and 40°C from a Zorbax SB C18 column (asymmetry factor, $B_{0.1}/A_{0.1} = 2.1$). (a) Gaussian, (b) PMG1, (c) log-normal, (d) bi-Gaussian, (e) PVMG, (f) Pap–Pápai, (g) EMG, (h) PLMG, and (i) Li. The experimental points are depicted as (o) and the fitted function as a solid line.

The good performance of the PMG1 model should be also highlighted, considering the simplicity of this function. The nine functions applied in Figure 12.3 show good convergence and with current computers, the optimum is reached almost instantaneously.

12.3 AUTOMATIC PROCESSING OF CHROMATOGRAMS

In spite of the advances in the automatic processing of chromatographic peaks [53–67], the direct measurement of chromatographic properties, such as retention times, heights, standard deviations, widths, and half-widths, is not uncommon in routine analysis. The accurate estimation of peak properties is usually based on the calculation of statistical moments [62]: the zero moment (M_0) represents the peak area, the first moment (M_1) is the center of gravity of the distribution or mean time, the second central moment (M_2) represents the spread of the distribution or peak variance (σ^2), and the third central moment (M_3) represents the peak skewness:

$$M_0 = \int_0^\infty h(t)dt \tag{12.21}$$

$$M_1 = \frac{\int_0^\infty t\, h(t)dt}{M_0} \tag{12.22}$$

$$M_2 = \frac{\int_0^\infty (t-M_1)^2 h(t)dt}{M_0} \tag{12.23}$$

$$M_3 = \frac{\int_0^\infty (t-M_1)^3 h(t)dt}{M_0} \tag{12.24}$$

The moments M_1 to M_3 are independent of the amount of solute, since they are normalized to the peak area. For Gaussian peaks, M_3 has a zero value. Instrument manufacturers incorporate software to measure the moments, usually based on numerical methods. Also, the widths and half-widths (if provided) are measured at a fixed peak height (usually 50%). However, the uncertainties in the integration limits, together with imperfect baseline correction and noise, may lead to inaccurate moment estimation. Fitting of the chromatographic signal to appropriate peak models, through commercial or home-made software, attenuates the impact of these error sources, leading to accurate estimations in nonideal situations.

12.4 PEAK HALF-WIDTH AND WIDTH MODELING

The left and right half-widths of chromatographic peaks provide information about the peak width and skewness and, together with the values of retention times and peak heights or areas, allow the simulation of chromatograms based on peak models. In principle, half-widths (or widths) can be measured at any peak height ratio (e.g., 10%, 50%, or 60.65%). In the literature, measurement of half-widths at 10% peak height is preferred owing to the better estimation of peak asymmetry, since it is less sensitive to baseline noise and imperfect baseline subtraction.

The prediction of peak half-widths or other related properties (i.e., width, efficiency, and asymmetry) is less accurate than the prediction of retention times, for which there is an extensive literature [68]. Nevertheless, the development of models to predict peak shape parameters with optimization purposes (see Section 12.5.3), as a function of different experimental factors (e.g., mobile phase composition, pH, temperature, and flow rate), is preferable instead of the common practice of considering mean or arbitrary values for all peaks in a chromatogram. For this purpose, local or global models can be applied. In both cases, data from experimental designs are used to predict the variation in peak shape properties with the experimental factors.

12.4.1 LOCAL MODELS

The use of local models is a universal approach, where the information for the closest (usually two to four) available conditions to that predicted is used to obtain the peak profile parameters, through weighted means or interpolations. Good performance in peak profile prediction has been observed using local models based on efficiency and peak asymmetry values, measured for standards at different experimental conditions, and considering factors such as the concentration of organic solvent and additives such as amines, surfactants, and ionic liquids, pH, and temperature.

12.4.2 GLOBAL MODELS

A linear relationship can be expected between the variance and the squared retention time for compounds exhibiting Gaussian peaks and similar interaction kinetics with the stationary phase. From Equation 12.3, and considering the existence of extra-column dispersion (Equation 12.4):

$$\sigma^2 = \frac{1}{N}\left(t_R - t_{ext}\right)^2 + \sigma_{ext}^2 \tag{12.25}$$

where t_{ext} is the extra-column time (i.e., the time the solute needs to travel through the extra-column components). Equation 12.25 can be fitted using the (t_R, σ) data for a particular solute eluted at several experimental conditions, or for a set of solutes eluted at fixed or variable conditions.

When the extra-column contributions are nonsignificant, nearly linear relationships between the standard deviation and width ($w = A + B$) versus the retention time have been observed [69,70]:

$$\sigma = \sigma_0 + m_\sigma t_R \tag{12.26}$$

$$w = w_0 + m_w t_R \tag{12.27}$$

where
   σ_0 and w_0 are fitting parameters that include the extra-column peak broadening
   m_σ and m_w are the rates of increase of the standard deviation and width with the retention time

Similar models can be built for the half-widths to describe the peak asymmetry (Section 12.5.1). For wide retention time ranges, the trends are parabolic with a gentle curvature and can be assimilated to a linear trend [33,71–75]:

$$A = a_0 + a_1 t_R + a_{11} t_R^2 \approx A_0 + m_A t_R \tag{12.28}$$

$$B = b_0 + b_1 t_R + b_{11} t_R^2 \approx B_0 + m_B t_R \tag{12.29}$$

Since the retention time can be predicted with good accuracy using retention models as a function of diverse experimental factors, the half-widths and related properties can be predicted from these factors as well.

Figure 12.4 depicts the trends for the left and right half-widths for compounds yielding symmetrical (Figure 12.4a) [76] and asymmetrical (Figure 12.4b) [72] peaks, eluted at different mobile phase compositions in isocratic elution. The fitted trends (usually according to a linear regression) are useful for the characterization of column performance. The slope of the straight lines is larger for compounds experiencing slow adsorption–desorption processes, especially for the right half-width (Figure 12.4b). A scattered trend evidences the existence of different kinetic behaviors for the measured solutes [74].

Figure 12.5 depicts the change in peak half-widths during gradient elution versus the retention time. The trend for the left half-widths can be described by the following equation [73]:

$$\frac{A_0 - A}{A} = c_1 \left(\frac{t_R - t_g}{t_g} \right) + c_2 \left(\frac{t_R - t_g}{t_g} \right)^2 \tag{12.30}$$

where

A and t_g are the half-width and retention time in gradient elution
A_0 and t_R correspond to the isocratic elution using the organic solvent content at the beginning of each gradient

A similar equation is valid for the right half-width and width.

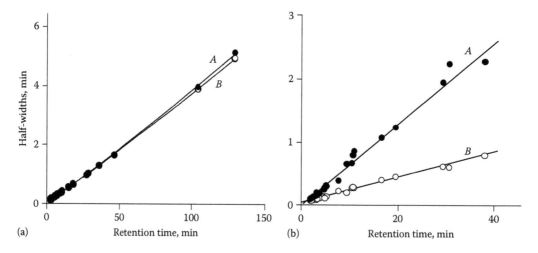

(a)   Retention time, min   (b)   Retention time, min

FIGURE 12.4   Half-width plots for (a) sulfonamides eluted from a cyano column with acetonitrile–water mixtures and (b) β-blockers eluted from a Kromasil C18 column with methanol–water mixtures. Half-widths: right (●) and left (○). (Reprinted from *J. Chromatogr. A*, 1217, Ruiz-Ángel, M.J., Carda-Broch, S., and García-Álvarez-Coque, M.C., Peak half-width plots to study the effect of organic solvents on the peak performance of basic drugs in micellar liquid chromatography, 1786–1798. Copyright 2010, with permission from Elsevier; Reprinted from *J. Chromatogr. A*, 1281, Ortiz-Bolsico, C., Torres-Lapasió, J.R., Ruiz-Ángel, M.J., García-Álvarez-Coque, M.C., Comparison of two serially coupled column systems and optimization software in isocratic liquid chromatography for resolving complex mixtures, 94–105. Copyright 2013, with permission from Elsevier [76].)

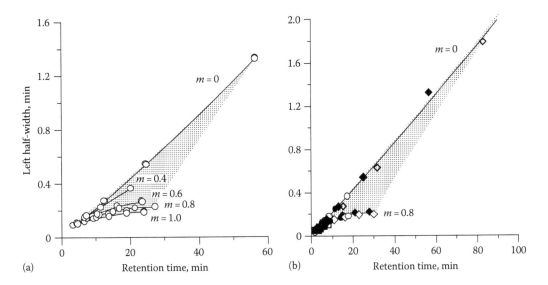

FIGURE 12.5 Left half-width plots at 10% peak height for peaks obtained in gradient elution with aceto-nitrile water for (a) xipamide and (b) a set of five diuretics (trichloromethiazide, □; althiazide, ●; furosemide, ○; xipamide, ◆; and ethacrynic acid, ◇). The lines correspond to different gradient slopes (m, % acetonitrile/min), and the points along the lines to different initial concentrations of acetonitrile. (Reprinted from *J. Chromatogr. A*, 1284, Baeza-Baeza, J.J., Ortiz-Bolsico, C., Torres-Lapasió, J.R., and García-Álvarez-Coque, M.C., Approaches to model the retention and peak profile in linear gradient reversed-phase liquid chromatography, 28–35, Copyright 2013, with permission from Elsevier.)

12.5 FIELDS OF APPLICATION

12.5.1 PEAK ASYMMETRY CHARACTERIZATION

Perfectly symmetrical chromatographic peaks are very rarely found in practice. Peak asymmetry is problematic because it contributes to a loss of efficiency and favors the overlap of neighboring peaks by placing more solute in the lower area of the peak rather than around the maximum. Also, the acquisition of information from the chromatographic signal is more complex. The asymmetry is the visible result of a number of chromatographic and instrumental issues, such as slow mass transfer, secondary chemical reactions, column overload, heterogeneity of the stationary phase surface and column packing, void volumes associated to the connections and fittings, among other factors [77]. All this explains the great effort made in method development to avoid situations that yield asymmetrical peaks, and the relevance of peak asymmetry to characterize system performance.

Statistically, peak asymmetry can be quantified by the Pearson's coefficient of skewness [62,78]:

$$\gamma = \frac{M_3}{\sigma^3} \tag{12.31}$$

where M_3 is the third central moment (Equation 12.24). Positive and negative γ values are obtained for fronting (Figure 12.1b) and tailing (Figure 12.1c) peaks, respectively. A zero value indicates that the tails on both sides of the mean are balanced out. This is the case for symmetrical distributions (i.e., Gaussian peaks), but it is also true for distributions where the asymmetries even out, such as one tail being long but thin, and the other being short but fat. Therefore, the coefficient γ is difficult to interpret. Moreover, in case of noisy baseline, the value of the skewness is highly dependent on the number of points used for the integration to get M_3.

Consequently, more intuitive parameters are preferred to evaluate the asymmetry. Some of them are the following:

1. The asymmetry factor, which measures the number of times the right half-width B is greater than the left half-width A [6,77]:

$$f_{asym} = \frac{B}{A}$$
(12.32)

This ratio is preferred instead of the inverse ratio, A/B, because peaks are more often tailing. In this way, $f_{asym} > 1$ for tailing peaks and $f_{asym} < 1$ for fronting peaks. As commented, the half-widths are more conveniently measured at 10% peak height. Peak asymmetry can be also calculated as follows (see Equations 12.28 and 12.29) [74]:

$$f_{asym} = \frac{m_B t_R + B_0}{m_A t_R + A_0}$$
(12.33)

which tends to a constant value for long enough retention times, where A_0 and B_0 become negligible. The ratio m_B/m_A represents the asymmetry factor of a highly retained compound, or the column component to peak asymmetry. The angle between the plots for A and B (Figure 12.4) is also a good measurement of peak asymmetry.

2. The tailing factor, defined as the ratio of the peak width to twofold the left half-width, measured at 5% peak height, is the descriptor recommended by the U.S. Pharmacopeia [79]:

$$T = \frac{A + B}{2A}$$
(12.34)

3. The asymmetry ratio, which shows the proportion of the peak width that can be ascribed to the asymmetry, has the advantage of providing the same value and opposite sign for fronting and tailing peaks:

$$r_{asym} = \frac{B - A}{A + B}$$
(12.35)

This parameter is found in the PMG1 model (Equation 12.8) [25].

Table 12.3 relates to each other the values of the peak asymmetry parameters described above, for bi-Gaussian peaks. For instance, for a peak with $\gamma = 0.50$, the right region is 2.33-fold larger, the

TABLE 12.3

Comparison of Parameters Used to Evaluate the Peak Asymmetry

γ	f_{asym}	T	r_{asym}
−0.26	0.67	0.84	−0.20
−0.13	0.82	0.91	−0.10
0	1.00	1.00	0.0
0.13	1.22	1.11	0.10
0.26	1.50	1.25	0.20
0.38	1.84	1.42	0.30
0.50	2.33	1.67	0.40

total width is 1.67-fold larger compared to a symmetrical peak taking the left half-width as reference, and 40% of the peak width corresponds to the asymmetry. Other less intuitive parameters can be found in the literature [6,77,80].

12.5.2 Efficiency Estimation

A common practice to determine the column performance is quantifying the height equivalent to a theoretical plate, H_{col}, which according to the Martin and Synge plate model is calculated as follows [4,81,82]:

$$H_{col} = \frac{L}{N_{col}} = \frac{\sigma_{col}^2}{t_{col}^2} L \qquad (12.36)$$

where
L is the column length
σ_{col} and N_{col} are the standard deviation and number of theoretical plates associated to the column
t_{col} is the time the solute needs to cross the column under isocratic conditions

The theoretical plate height and efficiency for a given column can be calculated by combining Equations 12.4 and 12.36, considering the extra-column contribution to the retention (t_{ext}):

$$H_{col} = \frac{\sigma_{tot}^2 - \sigma_{ext}^2}{\left(t_R - t_{ext}\right)^2} L \qquad (12.37)$$

$$N_{col} = \frac{\left(t_R - t_{ext}\right)^2}{\sigma_{tot}^2 - \sigma_{ext}^2} \qquad (12.38)$$

Both parameters are affected by the uncertainty in the measurement of the extra-column contribution to the total variance. Also, as commented, Gaussian peaks are assumed for sufficiently large efficiencies. A general solution to evaluate the efficiency, independent of the peak skewness, is offered by the moment method (see Equations 12.22 and 12.23):

$$N_{col} = \frac{M_1^2}{M_2} \qquad (12.39)$$

The computation of the efficiency through Equation 12.39 suffers from the same drawbacks as the moments M_1 and M_2: it is affected by the limits used in the integration, the baseline drift, and the noise. This, together with the need of digital curve fitting, has been the reason of the proposal of other approaches that allow a more simple estimation of the efficiency, based on the measurement of the half-widths above the baseline. The most generally accepted approach was developed based on the EMG model [83,84]:

$$N = \frac{41.7\left[t_R / \left(A_{0.1} + B_{0.1}\right)\right]^2}{\left(B_{0.1}/A_{0.1}\right) + 1.25} \qquad (12.40)$$

Equation 12.40 has been reported to yield errors <1.5% for asymmetry factors ($B_{0.1}/A_{0.1}$) in the range of 1.00–2.76 [83]. This equation, combined with Equations 12.28 and 12.29, is useful to predict the efficiency of solutes eluted at different retention times. The half-width equations also allow a proper peak capacity estimation, which requires knowledge on how peak half-widths change with the retention time [69,85].

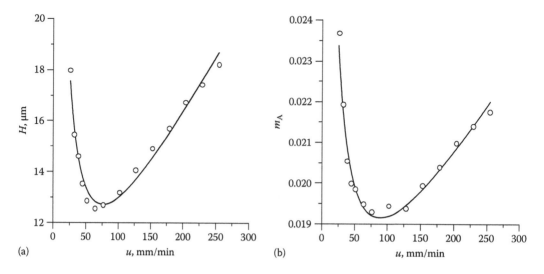

FIGURE 12.6 Chromatographic performance expressed as (a) plate height obtained from the peak of sulfamerazine and (b) slope of the left half-width plot (Equation 12.28) for the peaks of five sulfonamides. In both cases, a Spherisorb C18 column and 20% (v/v) acetonitrile were used. The lines in the plots correspond to the nonlinear fittings to (a) Equation 12.41 and (b) Equation 12.42. (Adapted from Baeza-Baeza, J.J. et al., *Chromatography*, 2, 625, 2015.)

Column characterization in liquid chromatography is also traditionally carried out by eluting probe compounds at different flow rates, and representing van Deemter plots [86]. These plots relate the plate height for a given solute, column and mobile phase composition to the linear mobile phase velocity, u (Figure 12.6a). The van Deemter equation can be reduced to

$$H = A + \frac{B}{u} + Cu \qquad (12.41)$$

Similar trends have been found for the slopes of the half-width plots (Figure 12.6b). For the left half-width (see Equation 12.28) [75]:

$$m_A = a + \frac{b}{u} + cu \qquad (12.42)$$

The same holds for the slope of the right half-width B (Equation 12.29). The sets of parameters (A, B, C) and (a, b, c) in Equations 12.41 and 12.42, although with different meaning, are useful to characterize chromatographic columns.

12.5.3 OPTIMIZATION OF PEAK RESOLUTION

The most efficient methodologies to find the optimal conditions in chromatography are based on the use of models that describe the solute retention and peak profile, upon changes in the experimental factors [68,74,87]. This is known as interpretive (interpretative or based on models) optimization. The models, fitted with a few experiments carefully designed, allow the prediction of the expected chromatograms for new conditions (Figure 12.7). The final aim is to maximize peak resolution in nonassayed experimental conditions. The optimization ranks the resolution for a grid of conditions in a selected range, from which the best are selected. Often, secondary aims are also included, such as short analysis times or lower cost.

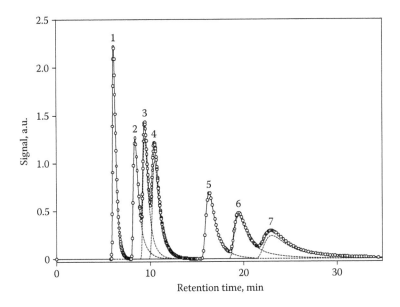

FIGURE 12.7 Predicted chromatogram for a mixture of β-blockers eluted with 18% (v/v) acetonitrile from a Nucleosil C18 column. The peaks were simulated based on the PMG1-exponential model using predicted retention times and half-widths. Compounds: (1) pindolol, (2) timolol, (3) acebutolol, (4) metoprolol, (5) esmolol, (6) celiprolol, and (7) oxprenolol. The dotted lines depict the individual signals for each compound, and the solid lines the global signal. Some predicted values for the global signals are also depicted (o). Signal is given in arbitrary units (a.u.) to get peaks with area = 1.

When the peak shape is considered in the optimization process, peaks are usually assumed to be Gaussian and the column efficiency is fixed to a previously selected value (usually a mean value). However, peaks are often skewed, and the efficiency depends on the solute nature and experimental conditions. Therefore, a reliable prediction of chromatographic resolution requires full knowledge of peak profiles [71,88–94].

Different resolution functions have been applied in the literature to measure the separation quality and guide the interpretive optimization of chromatographic performance [95,96]. Usually, all compounds in the sample are known and the search is based on the use of standards. Some approaches are also suitable for samples containing unknown compounds [97,98]. Almost all functions simplify reality by reducing the resolution to a combination of peak parameters. This explains that the optima found can differ significantly among criteria, and the analyst may find some considerably better than others [99].

The most commonly used function to measure the resolution is the so-called Snyder's resolution [79], which compares the distance between the peak maxima of two neighboring peaks with the average peak width:

$$R_S = \frac{t_{R,i+1} - t_{R,i}}{(w_i + w_{i+1})/2} \tag{12.43}$$

More complex functions have been proposed, based on the R_S function to account for the peak efficiency [95]. The following function considers differences in the efficiencies associated to each peak, N_i and N_{i+1}:

$$R_S = \sqrt{N_i N_{i+1}} \left(\frac{k_{i+1} - k_i}{k_{i+1} + k_i} \right) \frac{(k_i + k_{i+1})/2}{(1 + k_i)\sqrt{N_{i+1}} + (1 + k_{i+1})\sqrt{N_i}} \tag{12.44}$$

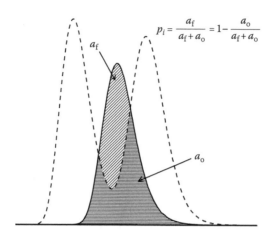

$$p_i = \frac{a_f}{a_f + a_o} = 1 - \frac{a_o}{a_f + a_o}$$

FIGURE 12.8 Measurement of the peak purity for a solute i (central peak), in the presence of two interferences whose chromatogram is delimited by a dashed line. Peak purity, p_i, can be computed either as a function of the free (a_f) or overlapped area (a_o).

We should note that the aim of a chromatographic optimization is finding conditions where the peaks are minimally overlapped. Therefore, a reliable estimation of the overlapping degree, considering the whole peak profile, should be ideally the best resolution measurement. As early as 1986, such an approach was suggested in a comprehensive book on chromatographic optimization [100]. However, the computation of the overlapping degree requires knowledge of the position, profile, and size for each peak in the chromatogram and is only feasible through digital processing. The availability of faster computers and the development of more practical peak models, 15 years later, revived the interest in this approach [99,101,102].

The complement of the overlapped fraction (the so-called peak purity, p_i) was preferred to estimate the resolution, as it correlates better with the lack of interference (Figure 12.8). The product of individual purities was also checked to offer the best evaluation of global resolution [101]. Defined in this way, individual and global peak purities range from zero for full overlapping to one for full resolution. However, it should be noted that the product is affected by the number of considered peaks and the distribution of the individual values. In addition, peak purities depend on the relative peak areas, although a general solution can be found by using normalized peak areas. More recently, two-order peak purity functions have been suggested for two-dimensional chromatography [103], and to account both time and spectral resolution [104].

In contrast to the R_S criterion (which is associated to a peak pair), the peak purity is an elementary measurement associated to each individual peak. This allows an easier weighting or exclusion of peaks, the optimization of the separation of particular compounds [102], and the development of approaches based on the number of visible peaks [105]. A derived concept is the "limiting peak purity" (maximal elementary value found for each solute), which indicates the maximal expectancy of resolution for each solute in the experimental domain: if the limiting value is small, the solute will remain unresolved under all conditions examined [106].

12.5.4 Deconvolution of Partially Overlapped Peaks

Chromatographic techniques often give rise to situations where reaching complete resolution is not possible. Fortunately, even in such cases, coelution is not always complete, and the individual signals can be retrieved by mathematical means (i.e., peak deconvolution). In some cases, this implies the fitting of the overall signal to a linear combination of models [44]. This is a key application, and hence, much research in peak modeling has been aimed to signal deconvolution [10,11,17,24,25,31,36,45,107–113].

Deconvolution accomplishes the mathematical resolution of the overlapped peaks by obtaining the parameters of the functions that describe each individual peak, which would yield the whole signal by linear combination. At moderate overlapping, deconvolution is feasible even with simple chemometrics. With higher overlapping levels, the results become ambiguous due to the existence of collinearity between parameters, which yields non-well-defined error surfaces. This is translated into multiple solutions, from which only one is correct. The quantification can be improved by enriching the information in the chromatogram through hyphenated techniques, in order to yield second-order data (e.g., a collection of spectra measured at different times) [104,114]. Through multivariate techniques, the analyst is able to retrieve the individual contributions for each compound. In the absence of selective regions, the solutions found will be biased.

When second-order instruments are not available, pseudo-second-order data can be built with a first-order instrument, by annexing measurements from several samples sharing the same solutes with different concentration patterns. The data are processed by application of multivariate analysis to the whole data set, instead of treating independently each experiment [110,111].

Comprehensive two-dimensional chromatography extends peak capacity through the coupling of two columns, ideally orthogonal, where peaks eluting from the first column are modulated into several fractions and reinjected into the second column. This increases the probability of resolution of highly complex samples. However, complete separation of all detectable components may be still incomplete. Therefore, it is interesting that the additional information found from both dimensions be simultaneously extracted and deconvolved and the pure profiles of the overlapping peak clusters reconstructed in both dimensions [115–117].

The complexity and diversity of the mathematical treatments needed to apply the approaches described in this chapter make their detailed description not possible. The citations given in the "Reference" section will be helpful to assist in their implementation.

REFERENCES

1. Neue, U. D. 1997. *HPLC Columns: Theory, Technology and Practice*. New York: Wiley.
2. Miyabe, K. 2009. Moment analysis of chromatographic behavior in reversed-phase liquid chromatography. *J. Sep. Sci.* 32: 757–770.
3. Gritti, F. and Guiochon, G. 2012. Mass transfer kinetics, band broadening and column efficiency. *J. Chromatogr. A* 1221: 2–40.
4. Martin, A. J. P. and Synge, R. L. M. 1941. A new form of chromatogram employing two liquid phases. *Biochem. J.* 35: 1358–1368.
5. Giddings, J. C. 1965. *Dynamics of Chromatography. Part I: Principles and Theory*. New York: Marcel Dekker.
6. Felinger, A. 1998. *Data Analysis and Signal Processing in Chromatography*. Amsterdam, the Netherlands: Elsevier.
7. Cavazzini, A., Dondi, F., Jaulmes, A., Vidal-Madjar, C., and Felinger, A. 2002. Monte Carlo model of nonlinear chromatography: Correspondence between the microscopic stochastic model and the macroscopic Thomas kinetic model. *Anal. Chem.* 74: 6269–6278.
8. Haarhoff, P. C. and van der Linde, H. J. 1966. Concentration dependence of elution curves in non-ideal gas chromatography. *Anal. Chem.* 38: 573–582.
9. Fraser, R. D. and Suzuki, E. 1966. Resolution of overlapping absorption bands by least squares procedures. *Anal. Chem.* 38: 1770–1773.
10. Fraser, R. D. and Suzuki, E. 1969. Resolution of overlapping bands: Functions for simulating band shapes. *Anal. Chem.* 41: 37–39.
11. Grushka, E., Meyers, M. N., and Giddings, J. C. 1970. Moment analysis for the discernment of overlapping chromatographic peaks. *Anal. Chem.* 42: 21–26.
12. Grubner, O. 1971. Interpretation of asymmetric curves in linear chromatography. *Anal. Chem.* 43: 1934–1937.
13. Grushka, E. 1972. Characterization of exponentially modified Gaussian peaks in chromatography. *Anal. Chem.* 44: 1733–1738.
14. Buys, T. S. and de Clerk, K. 1972. Bi-Gaussian fitting of skewed peaks. *Anal. Chem.* 44: 1273–1275.

15. Chesler, S. N. and Cram, S. P. 1973. Iterative curve fitting of chromatographic peaks. *Anal. Chem.* 45: 1354–1359.
16. Baker, C., Johnson, P. S., and Maddams, W. F. 1978. The characterization of infrared absorption band shapes. II. Experimental studies. *Spectrochim. Acta* 34A: 683–691.
17. Vaidya, R. A. and Hester, R. D. 1984. Deconvolution of overlapping chromatographic peaks using constrained non-linear optimization. *J. Chromatogr.* 287: 231–244.
18. Low, G. K. C., Haddad, P. R., and Duffield, A. M. 1984. Proposed model for peak splitting in reversed-phase ion-pair high-performance liquid chromatography with computer prediction of eluted peak profiles. *J. Chromatogr. B* 336: 15–24.
19. Plicka, J., Svoboda, V., Kleinmann, I., and Uhlirova, A. 1989. Mathematical modelling of the peak in liquid chromatography. *J. Chromatogr.* 469: 29–42.
20. Losev, A. 1989. A new lineshape for fitting x-ray photoelectron peaks. *Surf. Interface Anal.* 14: 845–849.
21. Olivé, J. and Grimalt, J. O. 1991. Log-normal derived equations for the determination of chromatographic peak parameters from graphical measurements. *Anal. Chim. Acta* 248: 59–70.
22. Olivé, J. and Grimalt, J. O. 1991. Gram-Charlier and Edgeworth-Cramer series in the characterization of chromatographic peaks. *Anal. Chim. Acta* 249: 337–348.
23. Le Vent, S. 1995. Simulation of chromatographic peaks by simple functions. *Anal. Chim. Acta* 312: 263–270.
24. Pietrogrande, M. C., Dondi, F., Felinger, A., and Davis, J. M. 1995. Statistical study of peak overlapping in multicomponent chromatograms: Importance of the retention pattern. *Chemom. Intell. Lab. Syst.* 28: 239–258.
25. Torres-Lapasió, J. R., Baeza-Baeza, J. J., and García-Álvarez-Coque, M. C. 1997. A model for the description, simulation and deconvolution of skewed chromatographic peaks. *Anal. Chem.* 69: 3822–3831.
26. Li, J. 1997. Development and evaluation of flexible empirical peak functions for processing chromatographic peaks. *Anal. Chem.* 69: 4452–4462.
27. Stromberg, A. G., Romanenko, S. V., and Romanenko, E. S. 2000. Systematic study of elementary models of analytical signals in the form of peaks and waves. *J. Anal. Chem.* 55: 615–625.
28. Lan, K. and Jorgenson, J. W. 2001. A hybrid of exponential and Gaussian functions as a simple model of asymmetric chromatographic peaks. *J. Chromatogr. A* 915: 1–13.
29. Pap, T. L. and Pápai, Z. 2001. Application of a new mathematical function for describing chromatographic peaks. *J. Chromatogr. A* 930: 53–60.
30. Li, J. 2002. Comparison of the capability of peak functions in describing real chromatographic peaks. *J. Chromatogr. A* 952: 63–70.
31. Caballero, R. D., García-Álvarez-Coque, M. C., and Baeza-Baeza, J. J. 2002. Parabolic-Lorentzian modified Gaussian model for describing and deconvolving chromatographic peaks. *J. Chromatogr. A* 954: 59–76.
32. Howerton, S. B., Lee, C., and McGuffin, V. L. 2003. Additivity of statistical moments in the exponentially modified Gaussian model of chromatography. *Anal. Chim. Acta* 478: 99–110.
33. Baeza-Baeza, J. J. and García-Álvarez-Coque, M. C. 2004. Prediction of peak shape as a function of retention in reversed-phase liquid chromatography. *J. Chromatogr. A* 1022: 17–24.
34. Moretti, P., Vezzani, S., Garrone, E., and Castello, G. 2004. Evaluation and prediction of the shape of gas chromatographic peaks. *J. Chromatogr. A* 1038: 171–181.
35. Steffen, B., Müller, K. P., Komenda, M., Koppmann, R., and Schaub, A. 2005. A new mathematical procedure to evaluate peaks in complex chromatograms. *J. Chromatogr. A* 1071: 239–246.
36. Vivó-Truyols, G., Torres-Lapasió, J. R., van Nederkaassel, A. M., Vander Heyden, Y., and Massart, D. L. 2005. Automatic program for peak detection and deconvolution of multi-overlapped chromatographic signals: Part II: Peak model and deconvolution algorithms. *J. Chromatogr. A* 1096: 146–155.
37. Romanenko, S. V., Stromberg, A. G., and Pushkareva, T. N. 2006. Modelling of analytical peaks: Peak properties and basic peak functions. *Anal. Chim. Acta* 580: 99–106.
38. Romanenko, S. V. and Stromberg, A. G. 2007. Modelling of analytical peaks: Peak modifications. *Anal. Chim. Acta* 581: 343–354.
39. Bolanĉa, T., Cerjan-Stefanović, S., Ukić, S., Rogošić, M., and Luŝa, M. 2009. Prediction of the chromatographic signal in gradient elution ion chromatography. *J. Sep. Sci.* 32: 2877–2884.
40. Baeza-Baeza, J. J., Ortiz-Bolsico, C., and García-Álvarez-Coque, M. C. 2013. New approaches based on modified Gaussian models for describing chromatographic peaks. *Anal. Chim. Acta* 758: 36–44.
41. Reh, E. 1995. Peak-shape analysis for unresolved peaks in chromatography: Comparison of algorithms. *Trends Anal. Chem.* 14: 1–5.

42. Tamisier-Karolak, S. L., Le Potier, I., Barlet, O., and Czok, M. 1999. Analysis of anions in aqueous samples by ion chromatography and capillary electrophoresis: A comparative study of peak modeling and validation criteria. *J. Chromatogr. A* 852: 487–498.

43. Stromberg, A. G., Romanenko, S. V., and Romanenko, E. S. 2000. Classification of mathematical models of peak-shaped analytical signals. *J. Anal. Chem.* 55: 1024–1028.

44. Di Marco, V. B. and Bombi, G. G. 2001. Mathematical functions for the representation of chromatographic peaks. *J. Chromatogr. A* 931: 1–30.

45. Nikitas, P., Pappa-Louisi, A., and Papageorgiou, A. 2001. On the equations describing chromatographic peaks and the problem of the deconvolution of overlapped peaks. *J. Chromatogr. A* 912: 13–29.

46. Foley, J. P. and Dorsey, J. G. 1984. A review of the exponentially modified Gaussian (EMG) function: Evaluation and subsequent calculation of universal data. *J. Chromatogr. Sci.* 22: 40–46.

47. Hanggi, D. and Carr, P. W. 1985. Errors in exponentially modified Gaussian equations in the literature. *Anal. Chem.* 57: 2394–2395.

48. Naish, P. J. and Hartwell, S. 1988. Exponentially modified Gaussian functions: A good model for chromatographic peaks in isocratic HPLC? *Chromatographia* 26: 285–296.

49. Jeansonne, M. S. and Foley, J. P. 1991. Review of the exponentially modified Gaussian (EMG) function since 1983. *J. Chromatogr. Sci.* 29: 258–266.

50. Berthod, A. 1991. Mathematical series for signal modeling using exponentially modified functions. *Anal. Chem.* 63: 1879–1884.

51. Kalambet, Y., Kozmin, Y., Mikhailova, K., Nagaev, I., and Tikhonov, P. 2011. Reconstruction of chromatographic peaks using the exponentially modified Gaussian function. *J. Chemom.* 25: 352–356.

52. García-Álvarez-Coque, M. C., Simó-Alfonso, E. F., Sanchis-Mallols, J. M., and Baeza-Baeza, J. J. 2005. A new mathematical function for describing electrophoretic peaks. *Electrophoresis* 26: 2076–2085.

53. Stevenson, P. G., Gritti, F., and Guiochon, G. 2011. Automated methods for the location of the boundaries of chromatographic peaks. *J. Chromatog. A* 1218: 8255–8263.

54. Yamaoka, K. and Nakagawa, T. 1974. Statistical moments in linear equilibrium chromatography. *J. Chromatogr. A* 93: 1–6.

55. Delley, R. 1984. The peak width of nearly Gaussian peaks. *Chromatographia* 18: 374–382.

56. Foley, J. P. 1987. Equations for chromatographic peak modeling and calculation of peak area. *Anal. Chem.* 59: 1984–1987.

57. Jeansonne, M. S. and Foley, J. P. 1989. Measurement of statistical moments of resolved and overlapping chromatographic peaks. *J. Chromatogr. A* 461: 149–163.

58. Garland, W. A., Crews, T., and Fukuda, E. K. 1991. Effect of noise on peak heights calculated using an exponentially modified Gaussian peak shape model. *J. Chromatogr. A* 539: 133–139.

59. Tamisier-Karolak, S. L., Tod, M., Bonnardel, P., Czok, M., and Cardot, P. 1995. Daily validation procedure of chromatographic assay using Gauss exponential modelling. *J. Pharm. Biomed. Anal.* 13: 959–970.

60. Dyson, N. 1999. Peak distortion, data sampling errors and the integrator in the measurement of very narrow chromatographic peaks. *J. Chromatogr. A* 842: 321–340.

61. Jin, D. and Pardue, H. L. 2000. Algorithms for time-dependent chromatographic peak areas. 1. Algorithms evaluated for fully resolved peaks. *Anal. Chim. Acta* 422: 1–10.

62. Poole, C. F. 2003. *The Essence of Chromatography*. Amsterdam, the Netherlands: Elsevier.

63. Christensen, J. H., Mortensen, J., Hansen, A. B., and Andersen, O. 2005. Chromatographic preprocessing of GC-MS data for analysis of complex chemical mixtures. *J. Chromatogr. A* 1062: 113–123.

64. Baeza-Baeza, J. J., Torres-Lapasió, J. R., and García-Álvarez-Coque, M. C. 2011. Approaches to estimate the time and height at the peak maximum in liquid chromatography based on a modified Gaussian model. *J. Chromatogr. A* 1218: 1385–1392.

65. Suvitaival, T., Rogers, S., and Kaski, S. 2014. Stronger findings from mass spectral data through multipeak modeling. *Bioinformatics* 15: 208.

66. Vanderheyden, Y., Broeckhoven, K., and Desmet, G. 2014. Comparison and optimization of different peak integration methods to determine the variance of unretained and extra-column peaks. *J. Chromatogr. A* 1364: 140–150.

67. Zabell, A. P. R., Foxworthy, T., Eaton, K. N., and Julian, R. K. 2014. Diagnostic application of the exponentially modified Gaussian model for peak quality and quantitation in high-throughput liquid chromatography-tandem mass spectrometry. *J. Chromatogr. A* 1369: 92–97.

68. Nikitas, P. and Pappa-Louisi, A. 2009. Retention models for isocratic and gradient elution in reversed-phase liquid chromatography. *J. Chromatogr. A* 1216: 1737–1755.

69. Shen, Y. and Lee, M. L. 1998. General equation for peak capacity in column chromatography. *Anal. Chem.* 70: 3853–3856.

70. Benická, E., Krupčík, J., Lehotay, J., Sandra, P., and Armstrong, D. W. 2005. Selectivity tuning in an HPLC multicomponent separation, *J. Liq. Chromatogr. Relat. Technol.* 28: 1453–1471.

71. Baeza-Baeza, J. J., Ruiz-Ángel, M. J., and García-Álvarez-Coque, M. C. 2007. Prediction of peak shape in hydro-organic and micellar-organic liquid chromatography as a function of mobile phase composition. *J. Chromatogr. A* 1163: 119–127.

72. Ruiz-Ángel, M. J., Carda-Broch, S., and García-Álvarez-Coque, M. C. 2010. Peak half-width plots to study the effect of organic solvents on the peak performance of basic drugs in micellar liquid chromatography. *J. Chromatogr. A* 1217: 1786–1798.

73. Baeza-Baeza, J. J., Ortiz-Bolsico, C., Torres-Lapasió, J. R., and García-Álvarez-Coque, M. C. 2013. Approaches to model the retention and peak profile in linear gradient reversed-phase liquid chromatography. *J. Chromatogr. A* 1284: 28–35.

74. Baeza-Baeza, J. J., Ruiz-Ángel, M. J., Carda-Broch, S., and García-Álvarez-Coque, M. C. 2013. Half-width plots, a simple tool to predict peak shape, reveal column kinetics and characterise chromatographic columns in liquid chromatography: State of the art and new results. *J. Chromatogr. A* 1314: 142–153.

75. Baeza-Baeza, J. J., Ortiz-Bolsico, C., and García-Álvarez-Coque, M. C. 2015. Prediction of peak shape and characterization of column performance in liquid chromatography as a function of flow rate. *Chromatography* 2: 625–641.

76. Ortiz-Bolsico, C., Torres-Lapasió, J. R., Ruiz-Ángel, M. J., and García-Álvarez-Coque, M. C. 2013. Comparison of two serially coupled column systems and optimization software in isocratic liquid chromatography for resolving complex mixtures. *J. Chromatogr. A* 1281: 94–105.

77. Pápai, Z. and Pap, T. L. 2002. Analysis of peak asymmetry in chromatography. *J. Chromatogr. A* 953: 31–38.

78. Choi, D. Y. and Row, K. H. 2004. Theoretical analysis of chromatographic peak asymmetry and sharpness by the moment method using two peptides. *Biotechnol. Bioprocess Eng.* 9: 495–499.

79. Snyder, L. R., Kirkland, J. J., and Glajch, J. L. 1997. *Practical HPLC Method Development*, 2nd edn. New York: John Wiley & Sons, p. 211.

80. Barber, W. E. and Carr, P. W. 1981. Graphical method for obtaining retention time and number of theoretical plates from tailed chromatographic peaks. *Anal. Chem.* 53: 1939–1942.

81. Colmsjö, A. L. and Ericsson, M. W. 1987. Assessment of the height equivalent to a theoretical plate in liquid chromatography. *J. Chromatogr. A* 398: 63–71.

82. Baeza-Baeza, J. J., Pous-Torres, S., Torres-Lapasió, J. R., and García-Álvarez-Coque, M. C. 2010. Approaches to characterise chromatographic column performance based on global parameters accounting for peak broadening and skewness. *J. Chromatogr. A* 1217: 2147–2157.

83. Foley, J. P. and Dorsey, J. G. 1983. Equations for calculation of chromatographic figures of merit for ideal and skewed peaks. *Anal. Chem.* 55: 730–737.

84. Jeansonne, M. S. and Foley, J. P. 1992. Improved equations for the calculation of chromatographic figures of merit for ideal and skewed chromatographic peaks. *J. Chromatogr. A* 594: 1–8.

85. Pous-Torres, S., Baeza-Baeza, J. J., Torres-Lapasió, J. R., and García-Álvarez-Coque, M. C. 2008. Peak capacity estimation in isocratic elution. *J. Chromatogr. A* 1205: 78–89.

86. van Deemter, J. J., Zuiderweg, F. J., and Klinkenberg, A. 1956. Longitudinal diffusion and resistance to mass transfer as causes of non-ideality in chromatography. *Chem. Eng. Sci.* 5: 271–289.

87. García-Álvarez-Coque, M. C., Torres-Lapasió, J. R., and Baeza-Baeza, J. J. 2006. Models and objective functions for the optimisation of selectivity in reversed-phase liquid chromatography. *Anal. Chim. Acta* 579: 125–145.

88. Yau, W. W. and Kirkland, J. J. 1991. Improved computer algorithm for characterizing skewed chromatographic band broadening. I. Method. *J. Chromatogr. A* 556: 111–118.

89. Youn, D. Y., Yun, S. J., and Jung, K. H. 1992. Improved algorithm for resolution of overlapped asymmetric chromatographic peaks. *J. Chromatogr. A* 591: 19–29.

90. Yau, W. W., Rementer, S. W., Boyajian, J. M., DeStefano, J. J., Graff, J. F., Lim, K. B., and Kirkland, J. J. 1993. Improved computer algorithm for characterizing skewed chromatographic band broadening. II. Results and comparisons. *J. Chromatogr. A* 630: 69–77.

91. Sekulic, S. and Haddad, P. R. 1988. Effects of peak tailing on computer optimisation procedures for high-performance liquid chromatography. I. Characteristics of tailed peaks under optimisation conditions. *J. Chromatogr. A* 459: 65–77.

92. Sekulic, S. and Haddad, P. R. 1989. Optimization strategies for solutes exhibiting peak tailing. Comparison of two approaches. *J. Chromatogr. A* 485: 501–515.

93. Song, D. and Wang, J. 2003. Modified resolution factor for asymmetrical peaks in chromatographic separation. *J. Pharm. Biomed. Anal.* 32: 1105–1112.

94. Bolanĉa, T., Cerjan-Stefanović, S., Ukić, S., Rogoŝić, M., and Luŝa, M. 2009. Application of a gradient retention model developed by using isocratic data for the prediction of retention, resolution, and peak asymmetry in ion chromatography. *J. Liq. Chromatogr. Relat. Technol.* 32: 1373–1391.

95. Foley, J. P. 1991. Resolution equations for column chromatography. *Analyst* 116: 1275–1279.

96. Tyteca, E. and Desmet, G. 2014. A universal comparison study of chromatographic response functions. *J. Chromatogr. A* 1361: 178–190.

97. Duarte, R. M. B. O. and Duarte, A. C. 2010. A new chromatographic response function for use in size-exclusion chromatography optimization strategies: Application to complex organic mixtures, *J. Chromatogr. A* 1217: 7556–7563.

98. Álvarez-Segura, T., Gómez-Díaz, A., Ortiz-Bolsico, C., Torres-Lapasió, J. R., and García-Álvarez-Coque, M. C. 2015. A chromatographic objective function to characterise chromatograms with unknown compounds or without standards available. *J. Chromatogr. A* 1409: 79–88.

99. López-Grío, S. J., Vivó-Truyols, G., Torres-Lapasió, J. R., and García-Álvarez-Coque, M. C. 2001. Resolution assessment and performance of several organic modifiers in hybrid micellar liquid chromatography. *Anal. Chim. Acta* 433: 187–198.

100. Schoenmakers, P. J. 1986. *Optimisation of Chromatographic Selectivity: A Guide to Method Development.* Amsterdam, the Netherlands: Elsevier.

101. Carda-Broch, S., Torres-Lapasió, J. R., and García-Álvarez-Coque, M. C. 1999. Evaluation of several global resolution functions for liquid chromatography. *Anal. Chim. Acta* 396: 61–74.

102. Torres-Lapasió, J. R. and García-Álvarez-Coque, M. C. 2006. Levels in the interpretive optimisation of selectivity in high-performance liquid chromatography: A magical mystery tour. *J. Chromatogr. A* 1120: 308–321.

103. Duarte, R. M. B. O., Matos, J. T. V., and Duarte, A. C. 2012. A new chromatographic response function for assessing the separation quality in comprehensive two-dimensional liquid chromatography. *J. Chromatogr. A* 1225: 121–131.

104. Torres-Lapasió, J. R., Pous-Torres, S., Ortiz-Bolsico, C., and García-Álvarez-Coque, M. C. 2015. Optimisation of chromatographic resolution using objective functions including both time and spectral information. *J. Chromatogr. A* 1377: 75–84.

105. Ortín, A., Torres-Lapasió, J. R., and García-Álvarez-Coque, M. C. 2011. Finding the best separation in situations of extremely low chromatographic resolution. *J. Chromatogr. A* 1218: 2240–2251.

106. Concha-Herrera, V., Vivó-Truyols, G., Torres-Lapasió, J. R., and García-Alvarez-Coque, M. C. 2005. Limits of multi-linear gradient optimisation in reversed-phase liquid chromatography. *J. Chromatogr. A* 1063: 79–88.

107. Wei, W., Wu, N. S., and Jiang, X. H. 1992. New approach for area determination of an overlapped pair of chromatographic peaks. *J. Chromatogr. A* 623: 366–370.

108. Papoff, P., Ceccarini, A., Lanza, F., and Fanelli, N. 1997. Enhancing the quality of information obtained by a comparison between experimental and deconvolved peak parameters in ion chromatography. *J. Chromatogr. A* 789: 51–65.

109. Jin, D. and Pardue, H. L. 2000. Algorithms for time-dependent chromatographic peak areas. II. Resolution of overlapped chromatograms for two- and three-component samples. *Anal. Chim. Acta* 422: 11–20.

110. Vivó-Truyols, G., Torres-Lapasió, J. R., Garrido-Frenich, A., and García-Álvarez-Coque, M. C. 2001. A hybrid genetic algorithm with local search. II. Continuous variables: Multibatch peak deconvolution. *Chemom. Intell. Lab. Syst.* 59: 107–120.

111. Vivó-Truyols, G., Torres-Lapasió, J. R., Caballero, R. D., and García-Álvarez-Coque, M. C. 2002. Peak deconvolution in one-dimensional chromatography using a two-way data approach. *J. Chromatogr. A* 958: 35–49.

112. Pai, S. C. 2004. Temporally convoluted Gaussian equations for chromatographic peaks. *J. Chromatogr. A* 1028: 89–103.

113. Yu, T. and Peng, H. 2010. Quantification and deconvolution of asymmetric LC-MS peaks using the bi-Gaussian mixture model and statistical model selection. *Bioinformatics* 11: 559.

114. Booksh, K. S. and Kowalski, B. R. 1994. Theory of analytical chemistry. *Anal. Chem.* 66: 782A–791A.

115. Kong, H., Ye, F., Lu, X., Guo, L., Tian, J., and Xu, G. 2005. Deconvolution of overlapped peaks based on the exponentially modified Gaussian model in comprehensive two-dimensional gas chromatography. *J. Chromatogr. A* 1086: 160–164.
116. Zeng, Z. D., Chin, S. T., Hugel, H. M., and Marriott, P. J. 2011. Simultaneous deconvolution and reconstruction of primary and secondary overlapping peak clusters in comprehensive two-dimensional gas chromatography. *J. Chromatogr. A* 1218: 2301–2310.
117. Hanke, A. T., Verhaert, P. D. E. M., van der Wielen, L. A. M., Eppink, M. H. M., van de Sandt, E. J. A. X., and Ottens, M. 2015. Fourier transform assisted deconvolution of skewed peaks in complex multi-dimensional chromatograms. *J. Chromatogr. A* 1394: 54–61.

13 Missing and Censored Data in Chromatography

Ivana Stanimirova

CONTENTS

Chromatography [1,2] is a versatile analytical technique, the results of which have influenced knowledge in many areas of the sciences. Due to its high accuracy, sensitivity, specificity, and the possibility for a rapid qualitative and quantitative analysis of complex natural mixtures, chromatography is usually the method of choice in a broad range of chemical, biological, medicinal, and pharmaceutical applications. The continuous demand for analysis of more complex natural samples, while reducing the cost and time required for a single sample analysis, stimulates further development of chromatographic techniques. The higher the complexity of the phenomenon being studied, the larger the number of samples collected and analyzed in the hope that the analysis will reveal the required information. The resulting data can be multivariate, multiblock, or multiway. Every sample in these chromatographic data may be represented as a unique sample fingerprint with its entire recorded instrumental signal or with a set of the identified and quantitatively determined compounds. Both representations are widely used in chromatographic practice. However, the first sample representation requires that some chemical components that differentiate defined groups of samples be found as well as their further qualitative and quantitative identification using other analytical techniques, while the second representation limits the interpretation to the number of compounds for which chromatographic standards are available. No matter which chromatographic sample representation is used, the collected data should be interpreted in an adequate way, taking into account all of the measured parameters. This requires the application of various chemometric methods for data exploration, classification, and modeling [3–5]. For example, a powerful tool for studying the (im)purity profiles (coelution of compounds) of different pharmaceutical and chemical products is now the combined use of a chromatographic hyphenated technique and a multivariate curve resolution (MCR) approach [6,7]. A methodology that is often applied to determine the authenticity or origin of unknown samples is a multivariate classification or discriminant model [8–10] using the entire chromatographic signals as unique sample fingerprints. The selection of an appropriate chemometric method depends on the type and complexity of the collected data that is in agreement with the aim of the study. The performance of any chemometric analysis, however, requires organizing of experimental measurements into a data matrix. To meet this requirement, the correspondence of peaks in the set of samples should be known or investigated using information from a multichannel detector. This information is also important when constructing a peak table, in which every compound that is present in the samples is represented by its characteristic peak area. A special case of such a peak table is when the amount of each compound is expressed as a part of the total sample content, for example, total phenolic, fatty, or amino acid content. Such data are

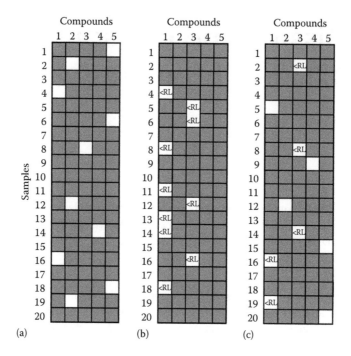

FIGURE 13.1 Patterns of 10% measurements that are (a) missing completely at random (MCAR), (b) not missing at random (NMAR), and (c) missing completely at random (5%) and missing not at random (5%).

known as compositional and their structure should be explored in agreement with the Aitchison geometry of their ratio sample space [11].

In analytical practice, it often happens that some measurements have not been recorded due to an insufficient sample amount or a high measurement cost or because the measured response is outside the instrument range. The collected quantitative data are then incomplete and must be treated with caution when they are interpreted and visualized using multivariate data analysis methods [12,13]. Some of the measurements may be missing completely at random (MCAR) [14], which means that their absence does not depend on the compounds being quantified. They usually do not form any particular pattern in the data, and they can have any value within the investigated measurement range (Figure 13.1a). Measurements that are below or above a given threshold value are not missing at random (NMAR) because their absence depends on the instrument range for these compounds. These measurements are also known as censored values and are labeled as left-censored measurements when their values are below the limit of detection or quantitation (reporting limit) as defined for a given compound (Figure 13.1b) or as right-censored measurements when their values are above the maximum value of the intensity that can be measured by the detector. Both types of missing measurements, MCAR and left-censored, can be present simultaneously in the chromatographic data (Figure 13.1c).

13.1 DEALING WITH DATA THAT CONTAIN MEASUREMENTS MISSING COMPLETELY AT RANDOM

Classic chemometric methods do not handle incomplete data, which makes any further data interpretation and visualization difficult. The usual practice for eliminating the incomplete data part by the deletion of the respective entire column or row from the data matrix leads to a large reduction of the good data when many measurements are missing, and therefore, many researchers still substitute the missing values with some constant values. The choice of this constant value depends on the mechanism of missingness. Data missing completely at random are usually substituted with the column or row means, while data that are below the reporting limit are often replaced by zeros or half of the respective reporting limit. These replacement methods are known to destroy the actual relationship among the measured variables in the data, which leads to biased solutions [12–14]. To solve this problem, methods that use the estimation of the mean and covariance or that are based on the construction of principal component analysis or regression models have been proposed to impute the missing information. A comprehensive overview of the main concepts for handling missing values will be presented here, and the advantages and disadvantages of the available algorithms will be discussed.

Over the past several years, the main efforts have been focused on developing algorithm(s) that can provide a relatively fast and stable solution with a low bias for a large percentage of missing data (usually over 60%). It is not surprising then that even though the number of algorithms that have been described in the literature is already exhaustive, new algorithms are still being proposed. Most of those algorithms solve the principal component analysis (PCA) problem, for example, finding the number of the score and loading vectors that sufficiently explain the latent structure of the incomplete experimental data. Algorithmically, this can be done in different possible ways [15–23], which implies some differences in the performance of the methods, which is especially noticeable when a large percentage of missing measurements must be handled. In general, the loss function for a measurement that is missing completely at random, MCAR, is zero. This means that the estimation of any model parameters for a given complexity should be done so that the predicted values for the MCAR measurements ideally fit the constructed model. Since no information about the values of these measurements is available, a reliable way of performing an analysis with them is to construct a model using only the observed measurements. The challenge then is to handle different complex patterns, in which every variable can contain missing measurements. The most commonly used methods are the so-called single imputation methods, for which only one value is filled in when the final model is built. Moreover, these models are said to ignore the uncertainty of the estimates and to give underestimated standard errors. These shortcomings are overcome with the multiple imputation algorithms, which, however, are inapplicable when a data matrix is described by a larger number of compounds in comparison with the number of samples. One of the most popular multiple imputation algorithms is the data augmentation (DA) method [24]. Among all of these, several important well-known concepts require attention and will be described here.

One of the most popular single imputation algorithms for handling MCAR measurements is the iterative algorithm (IA) [16,21]. It starts with the initialization of the missing elements with the corresponding means of the data row or column. After estimating of the model's (could be any multivariate model) parameters (score and loading vectors in PCA), the missing elements are refilled in with their predicted values, which were obtained in the interim model. This continues until the sum of squared residuals for the missing measurements in two consecutive iterations is below a given threshold value, for example, 10^{-8}. The IA algorithm is very flexible, and that is why it was used to obtain the model parameters of partial least squares regression (PLSR) [21,25] and soft independent modeling of class analogy (SIMCA) [26] methods for incomplete data. Another recently proposed approach to PCA or PLSR, the so-called trimmed score regression (TSR) [20], is related to the prediction of the missing part of the data only from the observed part using a regression model. After substituting the missing measurements with zeros and the initial estimation of a given number of PCA score and loading vectors using the centered substituted matrix, every incomplete row (sample) of the data matrix is reconsidered in order to obtain its trimmed score vector, that is, the score vector that is estimated only from the known variables (observed measurement data columns) and their respective loadings. Next, the estimation of the missing element from the data that is reconstructed with the trimmed score and loading vectors is replaced in the original centered matrix. Again, the convergence criterion is the difference in the sum of squared residuals for the missing measurements in two consecutive iterations. The equivalence of these two algorithms to other methods that have been proposed in the literature was reported [20]. When one principal component is obtained, the general iterative principal components imputation (GIP) performs the same as the iterative algorithm. Furthermore, the imputation by chained equation method (MICE) is equivalent to the known data regression (KDR), and finally, the same performance is also expected from the projection to model plane (PMP) method and a regularized version of the expectation-maximization approach using truncated total least squares regression (t-EM). An algorithm that is well known for its flexibility in handling any measurements that are missing in data completely at random is probabilistic principal component analysis (PPCA) [27,28]. This model is a formulation of conventional PCA from a Gaussian latent model. Additionally, an isotropic error or noise Gaussian model is specified. Within the steps of the iterative algorithm, the posterior distribution of the PCA latent parameters and the posterior distribution for missing measurements are updated interchangeably. An advantage of this algorithm is the automatic estimation of the PCA model complexity.

Figure 13.2 presents the performance of the TSR, IA, PPCA, and DA algorithms for an artificial data set of dimensions 50 × 20, which was simulated with a constant and proportional error structure. The proportional part was calculated as 20% of the respective elements of the error-free matrix, while the constant term was 5% of the maximum value of the error-free matrix. The simulated data represent a common chemical data structure with uncorrelated measurement errors. The simulation study was preferred here because of the possibility to estimate the error obtained from the prediction of the incomplete data part relative to the prediction of the original complete data set. Here, for each definite number of missing measurements, every algorithm was run 100 times for different patterns of MCAR measurements and the resulting mean square prediction errors (MSPEs) were averaged. The results showed negligible differences in the performance of the TSR, PPCA, DA, and IA algorithms for a data incompleteness of up to 25%–30%. Both TSR and PPCA have a considerably lower bias compared to IA as well as better convergence properties. The IA algorithm fails to achieve convergence in 7% of the cases for data with missing measurements of 45%. By contrast, PPCA is stable and even provides a solution for the highest rate (here 60%) of nondetects. The TSR algorithm is stable for data with up to 50% of MCAR measurements and has a failure rate of about 62% when 60% of the values are missing in the data. Compared to TSR, IA, and PPCA, the data augmentation algorithm showed a poorer performance with a failure rate of 15% for data with 15% of MCAR measurements. Of course, this tendency strongly depends on the correlation structure of

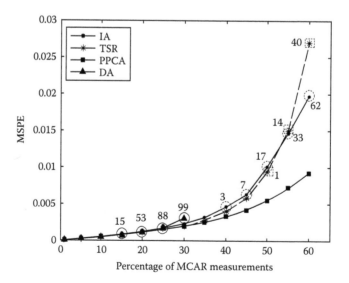

FIGURE 13.2 Curves of the mean square prediction error (MSPE) as a function of the percentage of the MCAR elements that are obtained from TSR, IA, PPCA, and DA. The failure rate in percentages is also presented.

the investigated data, but the general tendency, which is illustrated by this example, was observed in many simulated and real data examples.

As was mentioned earlier in the text, the multivariate curve resolution (MCR) method [29] is a basic chemometric tool that is applied when studying the (im)purity profiles of different pharmaceutical products. It offers a more general solution to the bilinear problem than PCA. The lack of the orthogonality of the factors and the possibility of the inclusion of various constraints such as nonnegativity, closure, unimodality, or others allows a unique or nearly unique solution to be obtained. This is achieved via the so-called alternating least squares algorithm. The alternating least squares (ALS) algorithm consists of estimating one matrix (e.g., contribution matrix) given the estimation of the other (e.g., composition matrix) in two cycles of iteration. The classic algorithm minimizes the sum of squared residuals and can easily be adapted to handle MCAR elements [30] in a similar way as in TSR. Simply, the projections in each cycle of the iteration are only done for the observed elements of each row or column of the original experimental data. Once the algorithm's convergence is achieved, a constraint MCR model of a definite complexity is obtained using only the observed data measurements.

Some other chemometric methods that are often used to explore and interpret chromatographic data have also been extended to process incomplete data. Among them are the Tucker [16] and parallel factor (PARAFAC) analysis [31] methods.

13.2 DEALING WITH MULTIVARIATE DATA THAT CONTAIN LEFT-CENSORED MEASUREMENTS

As was mentioned earlier, the aforementioned methods cannot process data that contain measurements that are below the reporting limit. The easiest way to handle the left-censored measurements, which is still recommended sometimes, is to replace them with half of the reporting limit. This has been shown in the other literature to be appropriate only for a small number of such elements per variable. Substitution of a large number of left-censored values in data, however, was reported to produce biased model solutions. Another inappropriate practice is the replacement of left-censored values with zeros under the assumption that their influence is insignificant for the final interpretation of the data structure. To avoid the distortion of the correlation structure of the data due to censoring, the predicted value for such an element should be below the reporting limit [12]. In contrast to the MCAR measurements, which have a zero loss function, the loss function for a left-censored element should allow a penalty for such an element. Specifically, the predicted value for a left-censored measurement above the reporting limit (upper limit) implies an error of a particular magnitude during the minimization of the sum of squared residuals, while a predicted value below the upper limit causes a zero error. Such a loss function is used for a left-censored element in the generalized nonlinear iterative partial least squares (GNIPALS) [32] approach to PCA. As the name of the algorithm suggests, in general, the MCAR elements and right-censored elements in data can also be processed by imposing the respective constraints while fitting the model. In GNIPALS, as in its classic NIPALS version [33], the score and loading vectors are obtained sequentially, but in contrast to NIPALS, they are obtained in the so-called convex optimization steps. In other words, after filling in the missing elements with their initial estimates and the random initialization of the first score and loading vectors, the first loading vector is reestimated iteratively minimizing the sum of squared residuals. Next, the local solution for the first score vector is found using the estimation for the loading vector minimizing the square loss function. The values of the predicted data with one factor are further examined and penalties are imposed on their data elements if necessary. The search for the subsequent score and loading vectors is performed in the residual space. Due to the sequential nature of the algorithm, the orthogonalization of the columns of the score matrix and a reoptimization of the final score and loading matrices that are used for the prediction are required.

Another way to handle missing measurements is by the inclusion of *a priori* knowledge about their measurement uncertainty [34]. The solutions that are obtained from the classic multivariate data methods are optimal under the assumption that the errors for all of the measurements are independent and identically distributed (i.i.d.) normal. It is known that the measurement uncertainty varies with the magnitude of the signal when analyzing natural complex environmental or biological samples. As a consequence of this, an analysis using classic chemometric methods tends to describe the compounds that are present in high concentrations/contents, while neglecting those in low concentrations/contents, although they may have the same signal-to-noise ratio. Therefore, in recent years, special attention has been paid to developing methods that incorporate the information about the magnitude of the measurement errors. There are several methods that have gained popularity in the literature. These are maximum likelihood PCA (MLPCA) [35], the weighted multivariate curve resolution method (MCR-WALS) [36], and positive matrix factorization (PMF) [37]. Even though the loss function in all of these methods is defined as the sum of squared weighted residuals, they differ algorithmically. Both MLPCA and MCR-WALS are performed using alternating least squares algorithms, while PMF uses an optimization procedure. A comparative study [38] of MCR-WALS and PMF has demonstrated that virtually the same results have been obtained from both methods for simulated and actual data sets. Each data measurement has a weight that is determined as the inverse of its measurement variance. In MLPCA, this weighting scheme implies that the direction of the projection onto the eigenvectors is weighted proportionally to the measurement variance so that the samples with large uncertainties are weighted more in order to have less influence. This maximum likelihood projection is reduced to its classic orthogonal PCA version for isotropic measurement errors. When very large uncertainties (e.g., error standard deviations of 10^5) are set for the MCAR measurements [22], the weighted loss function is zero. Therefore, it is not surprising that the IA, GNIPALS, and MLPCA algorithms perform the same for data with MCAR measurements [39]. Recently, an improvement of the MLPCA algorithm for dealing with data that contain MCAR elements was also presented [40]. In order to use MLPCA, MCR-WALS, and PMF for censored data, the uncertainty for the left- or right-censored measurements should be defined carefully. The uncertainty for a left-censored element should be set so that its predicted value is below the reporting limit in the final model [39]. In this context, the GNIPALS is easier to use, but it may be more time intensive at higher model complexities.

To illustrate the performance of generalized NIPALS, MLPCA, and different replacement methods, different censoring masks were created for fully observed environmental data [39]. The data set (28 × 17) is a collection of capillary gas chromatography measurements of the concentrations of 16 polycyclic aromatic hydrocarbons (PAHs) and the total amount of polychlorinated biphenyls (TPCB), which were gathered for samples of surface sediments that were collected at 28 sites in the harbor of Trieste. Different percentages of the measurements were then deleted below a certain threshold value (defined as the reporting limit) from different combinations of the variables, which account for 60% of the total number of data variables (e.g., 10 variables) and which have different pairwise correlations. Then, calculations for generalized NIPALS, MLPCA, and the classic principal components analysis, PCA, in which the censored elements were replaced with the respective reporting limits (RL) zeros or half of the reporting limits (RL/2) were performed. The results for the MSPE are presented in Figure 13.3. Percentages of left-censored measurements deleted from each variable (e.g., 5%, 10%, …, 60%) are also shown in percentages of the total number of measurements in the data (e.g., 2.10%, 6.30%, …, 36.71%), respectively.

The MSPE values are averaged over the three error values that were obtained for three different censoring masks. The best performance was observed for GNIPALS and MLPCA, especially for data that had a large percentage of censored measurements. Although substitution with half of the reporting limit can indeed be used when the percentage of left-censored elements per variable in the multivariate chemical data is relatively small, it gives biased results when a large number of variables are censored.

Up-to-date methods that incorporate measurement uncertainty information and GNIPALS can successfully process multivariate data that contain both MCAR and censored measurements simultaneously [39].

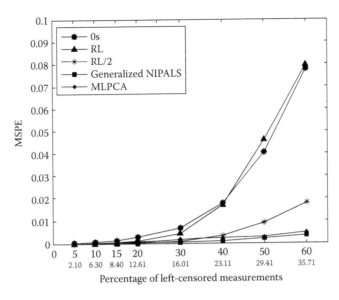

FIGURE 13.3 Curves of the mean square prediction error, MSPE, as a function of the percentage of the left-censored elements that are obtained from MLPCA, generalized NIPALS, and classic PCA, in which the censored elements were replaced with the reporting limits (RL) zeros or half of the reporting limits (RL/2).

Multivariate analysis and the interpretation of incomplete compositional data is still an ongoing challenge [11]. As was mentioned earlier in the text, the closure constraint (the sum of the contributions of the components that are preset in every sample is one) poses an additional problem in solving a bilinear problem. Because of the Aitchison geometry of the sample space, every sample of the original data is usually log-ratio transformed in order to obtain a solution using any classic multivariate method (e.g., PCA, LDA, or other) and the solution is then rotated and transformed back using an inverse transformation that is interpreted as compositions. Another method is to work directly with the compositions in order to obtain the method solution as compositions. The latter approach is well known in geology or social sciences as "end-member analysis" or "latent budget analysis," respectively. It is similar to the multivariate curve resolution-alternating least squares (MCR-ALS) method, which is widely used in chemistry. No matter what approach is used for the analysis, methods that can handle left-censored data are highly valued. In general, there are parametric and nonparametric methods [41,42] for single imputation, which should fulfill the closure constraint [11]. The log-transformations that are used in the first approach require the imputation of any nondetects beforehand. One of the most popular, fast, and easier nonparametric methods is multiplicative replacement. Multiplicative replacement is suitable for less than 10% of left-censored measurements (also known as "rounded zeros"). As its name suggests, the left-censored measurements are firstly substituted by the appropriate small values and the observed measurements are changed multiplicatively in order to fulfill the closure constraint. Substitution with a value equal to 65% of the reporting limit has been reported to not produce biased results. For large percentages of left-censored measurements, a modified version of the conventional expectation-maximization approach [43] applied to additive log-ratio (*alr*) transformed data has been shown to give imputation values that are in agreement with the data structure. However, this parametric approach is not suitable when every component comprises left-censored elements or when the assumption for multivariate *alr* normality is not fulfilled, for example, the data are strongly asymmetric, highly clustered, or discrete.

In order to use the second approach for chromatographic compositional data analysis, a modified weighted version of the "end member" model has been proposed in the literature. The modification [44] consists of incorporating information about the measurement uncertainty for all of the measurements in a similar way that is done in the weighted multivariate curve resolution-alternating least squares method (MCR-WALS) [36]. The possibility of setting the standard deviation for a measurement that is below the reporting limit so that its predicted value will be below the reporting limit allows unbiased contributions and composition profiles to be obtained, which fulfill the necessary constraints (usually nonnegativity and closure). This method [44] has been shown to be successful in interpreting the pure contributions due to the latent phospholipid fatty acids, PLFA, components that were obtained from compositional data that contain the entire PLFA profiles of water samples in terms of cultivated or noncultivated microbial communities.

13.3 DEALING WITH DATA THAT CONTAIN MISSING MEASUREMENTS AND OUTLYING SAMPLES

Even though a researcher can choose the appropriate method to analyze and interpret incomplete data, the solution may be unsatisfactory. This is especially noticeable when using supervised modeling methods for which the expected calibration trend or classification/discrimination rate is not observed. The problem might be that the investigated data contain samples for which some measurements have extreme values. Such extreme or outlying samples are known to strongly influence the solutions of all of the classic methods. A remedy for this problem is to identify and delete them followed by rerunning the chemometric analysis for clean data [45]. Another more straightforward way is to use a robust method for analysis, which allows the outliers to be downweighted in order to minimize their influence on the final model. The development of multivariate robust methods as counterparts of the classic approaches is an ongoing challenge. To date, several methods for robust PCA [46–48], robust estimators for PLS [49–51] and PCR regression, SIMCA [52], and PARAFAC [53] have been proposed in the literature. However, none of those methods can be used to identify outliers in incomplete data. The presence of outlying samples in incomplete data disturbs the correct fitting of the missing elements on one hand, while the presence of missing elements does not allow the adequate identification of outliers on the other. A solution to this problem is to extend the existing robust methods in order to process any incomplete data. The general framework for developing such methods, which has been described in the literature, consists of the implementation of a robust method within the steps of the iterative algorithm, IA. Several methods follow this framework, namely, expectation-maximization-spherical PCA (EM-SPCA) [54] and other variants of robust principal component analysis, expectation-maximization-spherical-SIMCA (EM-S-SIMCA) [26], expectation-maximization partial robust M-regression (EM-PRM) [25], a robust principal component regression [55], or robust parallel factor analysis [53]. These methods can process contaminated data that contain elements that are missing completely at random. The crucial step in all of these methods is the selection of a computationally efficient approach that has the best robust statistical properties.

The EM-SPCA approach [54] is based on spherical PCA (SPCA) [56]. It is the method of choice here due to its simplicity and high breakdown point. Firstly, the samples are projected onto a hypersphere of a unit radius from the robust center of the data (the L1-median center). Then, every sample is weighted according to its distance from the L1-median center so that the sample that is far from the center receives a small weight. Finally, applying classic PCA to the weighted data allows the robust score and loading vectors to be obtained. The outlier identification [54] is performed using the so-called distance–distance plot in which the Mahalanobis and orthogonal distance are calculated for each data sample.

Development of discriminant or classification rules based on which the assignment of a new sample to a given category is decided is particularly useful in the investigations of the authenticity of products or in metabolomic studies. For this purpose, a robust version of SIMCA using spherical PCA was proposed [52] and its convergence properties were investigated. The method was then further extended in order to handle missing elements [26]. A modified version of the leverage correction procedure was implemented in the algorithm in order to automatically choose a model's complexity. When a uniform design of the training and test sets using either the Kennard and Stone or duplex method is required, the selection of the training samples for the incomplete data should be done on the full rank score matrix that is obtained from the IA approach. The projection to model plane (PMP) method [17] was the method to predict the new samples with missing measurements.

The difference in the performance of the classic EM-SIMCA, which can handle missing measurements and its robust counterpart, is illustrated in Figure 13.4.

The low specificity of the model can be explained by the fact that the samples in the other groups (labeled by stars in Figure 13.4a) are assigned as belonging to the modeled group, which implies an incorrect estimation of the boundaries for the modeled group. By contrast, the EM-S-SIMCA model (in Figure 13.4b) that was constructed for the same group of samples presents a high sensitivity and specificity.

Expectation-maximization partial robust M-regression (EM-PRM) [25] was developed by embedding the PRM estimator within the steps of an iterative self-consistent procedure, IA, in order

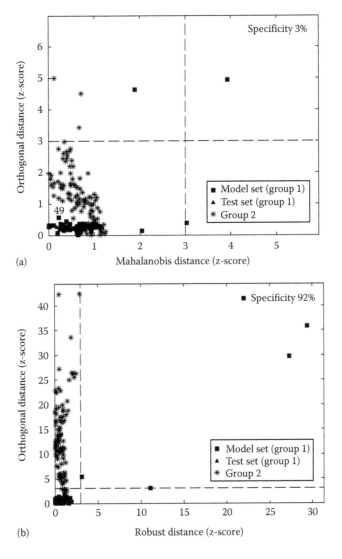

FIGURE 13.4 Distance–distance plots of the (a) EM-SIMCA and (b) EM-S-SIMCA models built for the contaminated training set containing 20% of MCAR measurements.

to minimize the loss function for missing elements. In recent years, the PRM-regression estimator has been shown to have the best bias properties [4,50] among the existing robust approaches to partial least squares, to be the most efficient at the normal model and to be less affected by wide-tailed distributions. In PRM, every sample is weighted according to its Mahalanobis distance in the space of the PRM scores and the residuals of the model, which implies that the influence of extreme samples is diminished. Figure 13.5a shows an EM-PLS model for a contaminated training set with 30% missing elements that is highly influenced by some extreme samples and in which the model trend is not associated with the trend that is associated with the majority of the samples in the training set.

The EM-PRM model [25] presented in Figure 13.5b presents better values of the root mean square (RMS) errors for model and test sets (Figure 13.5b) because the outlier influence is downweighted. This algorithm can only be applied to solve regression problems.

A similar idea has been used by other researchers in order to develop a robust principal component regression [55], other variants of robust principal component analysis [57], or robust parallel

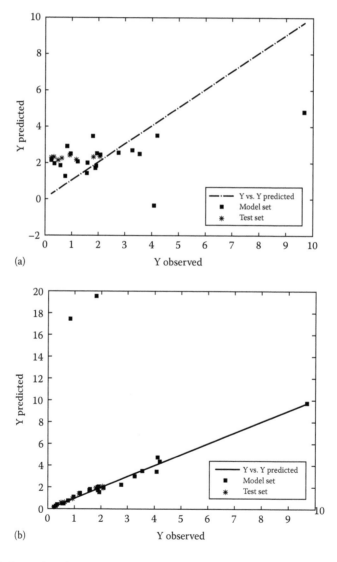

(a)

(b)

FIGURE 13.5 Y observed vs. Y predicted plots obtained from (a) EM-PLS and (b) EM-PRM for contaminated model set with 30% MCAR measurements.

factor analysis [53] that can process elements that are missing completely at random. Since several versions of principal component analysis are available, the advantages and disadvantages of their counterparts for incomplete data sets have also been discussed in detail.

In this chapter, the general concepts for dealing with incomplete data were presented and discussed. In general, the substitution of either measurements missing completely at random or not at random (censored measurements) is not recommended since this action produces biased solutions. Special attention is drawn to some new trends in developing and applying robust methods for incomplete data as well as to methods that incorporate the information about the magnitude of measurement uncertainties.

REFERENCES

1. Wixom, R.L., and C.W. Gehrke (Eds.). 2010. *Chromatography: A Science of Discovery*, John Wiley & Sons, Inc., New york.
2. Cserhati, T. 2008. *Multivariate Methods in Chromatography: A Practical Guide*, John Wiley & Sons, Inc., England, U.K.
3. Massart, D.L., B.G.M. Vandeginste, S.N. Deming, Y. Michotte, and L. Kaufman. 1988. *Chemometrics: A Textbook*, Elsevier Science Publishers B.V., Amsterdam, the Netherlands.
4. Lavine, B., and J. Workman. 2010. Chemometrics, *Analytical Chemistry*, 82 (12):4699–4711.
5. Lavine, B., and J. Workman Jr. 2013. Chemometrics, *Analytical Chemistry*, 85 (2):705–714.
6. Rutan, S.C., A. de Juan, and R. Tauler. 2009. Introduction to multivariate curve resolution. In *Comprehensive Chemometrics*, S.D. Brown, R. Tauler, B. Walczak (Eds.), Elsevier, Amsterdam, the Netherlands, Vol. 2, p. 249.
7. Daszykowski, M., and B. Walczak. 2006. Use and abuse of chemometrics in chromatography, *TrAC Trends in Analytical Chemistry*, 25 (11):1081–1096.
8. Krakowska, B., I. Stanimirova, J. Orzel, M. Daszykowski, I. Grabowski, G. Zaleszczyk, and M. Sznajder. 2015. Detection of discoloration in diesel fuel based on gas chromatographic fingerprints, *Analytical and Bioanalytical Chemistry*, 407:1159–1170.
9. Smilde, A.K., M.J. van der Werf, S. Bijlsma, B.J.C. van der Werff-van der Vat, and R.H. Jellema. 2005. Fusion of mass spectrometry-based metabolomics data, *Analytical Chemistry*, 77 (20):6729–6736.
10. Stanimirova, I., C. Boucon, and B. Walczak. 2011. Relating gas chromatographic profiles to sensory measurements describing the end products of the Maillard reaction, *Talanta*, 83 (4):1239–1246.
11. Pawlowsky-Glahn, V., and A. Buccianti (Eds.). 2011. *Compositional Data Analysis. Theory and Applications*, John Wiley & Sons Inc., U.K.
12. Helsel, D. 2005. *Nondetects and Data Analysis: Statistics for Censored Environmental Data*, John Wiley & Sons Inc., New York.
13. Little, R.J.A., and D.B. Rubin. 1987. *Statistical Analysis with Missing Data*, John Wiley & Sons Inc., New York.
14. Rubin, D.B. 1976. Inference and missing data, *Biometrika*, 63 (3):581–592.
15. Grung, B., and R. Manne. 1998. Missing values in principal component analysis, *Chemometrics and Intelligent Laboratory Systems*, 42 (1):125–139.
16. Walczak, B., and D.L. Massart. 2001. Dealing with missing data. Part I, *Chemometrics and Intelligent Laboratory Systems*, 58:15–27.
17. Arteaga, F., and A. Ferrer. 2002. Dealing with missing data in MSPC: Several methods, different interpretations, some examples, *Journal of Chemometrics*, 16:408–418.
18. Krzanowski, W.J. 1988. Missing value imputation in multivariate data using the singular value decomposition of a matrix, *Biometrical Letters*, 25:31–39.
19. Liu, Y., and S.D. Brown. 2013. Comparison of five iterative imputation methods for multivariate classification, *Chemometrics and Intelligent Laboratory Systems*, 120:106–115.
20. Folch-Fortuny, A., F. Arteaga, and A. Ferrer. 2015. PCA model building with missing data: New proposals and a comparative study, *Chemometrics and Intelligent Laboratory Systems*, 146:77–88.
21. Dempster, A.P., N.M. Laird, and D.B. Rubin. 1977. Maximum likelihood for incomplete data via the EM algorithm (with discussions), *Journal of the Royal Statistical Society Series B*, 39 (1):1–38.
22. Andrews, D.T., and P.D. Wentzell. 1997. Applications of maximum likelihood principal component analysis: Incomplete data sets and calibration transfer, *Analytica Chimica Acta*, 350:341–352.
23. Schneider, T. 2001. Analysis of incomplete climate data: Estimation of mean values and covariance matrices and imputation of missing values, *Journal of Climate*, 14 (5):853–871.

24. Tanner, M.A., and W.H. Wong. 1987. The calculation of posterior distribution by data augmentation (with discussion), *Journal of American Statistical Association*, 82:528–550.
25. Stanimirova, I., S. Serneels, P.J. Van Espen, and B. Walczak. 2007. How to construct a multiple regression model for data with missing elements and outlying objects, *Analytica Chimica Acta*, 581 (2):324–332.
26. Stanimirova, I., and B. Walczak. 2008. Classification of data with missing elements and outliers, *Talanta* 76 (3):602–609.
27. Tipping, M.E., and C.M. Bishop. 1999. Probabilistic principal component analysis, *Journal of the Royal Statistical Society Series B*, 61 (3):611–622.
28. Bishop, C.M. 2006. *Pattern Recognition and Machine Learning*, Springer Science, U.K.
29. de Juan, A., and R. Tauler. 2003. Chemometrics applied to unravel multicomponent processes and mixtures. Revisiting latest trends in multivariate resolution, *Analytica Chimica Acta*, 500 (1):195–210.
30. Beyad, Y., and M. Maeder. 2013. Multivariate linear regression with missing values, *Analytica Chimica Acta*, 796:38–41.
31. Tomasi, G., and R. Bro. 2005. PARAFAC and missing values, *Chemometrics and Intelligent Laboratory Systems*, 75 (2):163–180.
32. Ramon, J., and F. Costa. 2009. Handling missing values and censored data in PCA of pharmacological matrices, *StReBio'09*, Paris, France, ACM.
33. Wold, S., K. Esbensen, and P. Geladi. 1987. Principal component analysis, *Chemometrics and Intelligent Laboratory Systems*, 2 (1–3):37–52.
34. Wentzell, P.D. 2009. Other topics in soft-modeling: Maximum likelihood-based soft-modeling methods. In *Comprehensive Chemometrics*, S.D. Brown, R. Tauler, B. Walczak (Eds.), Elsevier, Amsterdam, the Netherlands, Vol. 2, p. 507.
35. Wentzell, P.D., D.T. Andrews, D.C. Hamilton, K. Faber, and B.R. Kowalski. 1997. Maximum likelihood principal component analysis, *Journal of Chemometrics*, 11:339–366.
36. Wentzell, P.D., T.K. Karakach, S. Roy, M.J. Martinez, C.P. Allen, and M. Werner-Washburne. 2006. Multivariate curve resolution of time course microarray data, *BMC Bioinformatics*, 7:343–361.
37. Paatero, P., and U. Tapper. 1994. Positive Matrix Factorization: A non-negative factor model with optimal utilization of error estimates of data values, *Environmetrics*, 5:111–126.
38. Stanimirova, I., R. Tauler, and B. Walczak. 2011. A comparison of positive matrix factorization and the weighted multivariate curve resolution method. Application to environmental data, *Environmental Science and Technology*, 45 (23):10102–10110.
39. Stanimirova, I. 2013. Practical approaches to principal component analysis for simultaneously dealing with missing and censored elements in chemical data, *Analytica Chimica Acta*, 796:27–37.
40. Folch-Fortuny, A., F. Arteaga, and A. Ferrer. 2016. Assessment of maximum likelihood PCA missing data imputation, *Journal of Chemometrics*, 30 (7):386–393.
41. Martín-Fernández, J.A., C. Barceló-Vidal, and V. Pawlowsky-Glahn. 2003. Dealing with zeros and missing values in compositional data sets using nonparametric imputation, *Mathematical Geology*, 35 (3):253–278.
42. Palarea-Albaladejo, J., J.A. Martín-Fernández, and J. Gómez-García. 2007. A parametric approach for dealing with compositional rounded zeros, *Mathematical Geology*, 39 (7):625–645.
43. Palarea-Albaladejo, J., and J.A. Martín-Fernández. 2008. A modified EM alr-algorithm for replacing rounded zeros in compositional data sets, *Computers and Geosciences*, 34:902–917.
44. Stanimirova, I., A. Woznica, T. Plociniczak, M. Kwasniewski, and J. Karczewski. 2016. A modified weighted mixture model for the interpretation of spatial and temporal changes in the microbial communities in drinking water reservoirs using compositional phospholipid fatty acid data, *Talanta*, 160:148–156.
45. Rousseeuw, P., and M. Leroy. 1987. *Robust Regression and Outlier Detection*, John Wiley & Sons Inc., New York.
46. Croux, C., and A. Ruiz-Gazen. 1996. A fast algorithm for robust principal components based on Projection Pursuit. In *COMPSTAT: Proceedings in Computational Statistics*, Physica-Verlag, Heidelberg, Germany, pp. 211–217.
47. Hubert, M., P. Rousseeuw, and S. Verboven. 2002. A fast method for robust principal components with application to chemometrics, *Chemometrics and Intelligent Laboratory Systems*, 60:101–111.
48. Maronna, R. 2005. Principal components and orthogonal regression based on robust scales, *Technometrics*, 47:264–273.
49. Serneels, S., C. Croux, P. Filzmoser, and P.J. Van Espen. 2005. Partial robust M regression, *Chemometrics and Intelligent Laboratory Systems*, 79:55–64.

50. Serneels, S., E. De Nolf, and P.J. Van Espen. 2006. Spatial sign preprocessing: A simple way to import moderate robustness to multivariate estimators, *Journal of Chemical Information and Modeling*, 46:1402–1409.

51. Hubert, M., and K. Vanden Branden. 2003. Robust methods for partial least squares regression, *Journal of Chemometrics*, 17:537–549.

52. Daszykowski, M., K. Kaczmarek, I. Stanimirova, Y. Vander Heyden, and B. Walczak. 2007. Robust SIMCA-bounding influence of outliers, *Chemometrics and Intelligent Laboratory Systems*, 87 (1):121–129.

53. Hubert, M., J. Van Kerckhoven, and T. Verdonck. 2012. Robust PARAFAC for incomplete data, *Journal of Chemometrics*, 26:290–298.

54. Stanimirova, I., M. Daszykowski, and B. Walczak. 2007. Dealing with missing values and outliers in principal component analysis, *Talanta*, 72:172–178.

55. Serneels, S., and T. Verdonck. 2009. Principal component regression for data containing outliers and missing elements, *Computational Statistics and Data Analysis*, 53 (11):3855–3863.

56. Locantore, N., J.S. Marron, D.G. Simpson, N. Tripoli, J.T. Zhang, and K.L. Cohen. 1999. Sociedad de Estadistica e Investigacion Operativa, *Test*, 8:1–74.

57. Serneels, S., and T. Verdonck. 2008. Principal component analysis for data containing outliers and missing elements, *Computational Statistics and Data Analysis*, 52:1712–1727.

Section IV

Classification, Discrimination, and Calibration

14 Linear Supervised Techniques

Łukasz Komsta and Yvan Vander Heyden

CONTENTS

14.1 INITIAL CONSIDERATIONS

The result of a chromatographic experiment can be often arranged as a *multivariate* dataset.* The linear supervised techniques are used when there is a need to obtain *a model* (an equation) allowing a *prediction* of a given property from such data.

A multivariate dataset arranged as an $n \times p$ matrix is often denoted as $\mathbf{X}$, with n rows (objects, samples) and p columns (variables, wavelengths, data acquisition points on chromatogram). In the considered case, a y value is given (known) for each object. This value is a property of the sample, and all y values are arranged into a vector $\mathbf{y}$ containing n elements. As this set is used to derive a model, it can be called a *calibration* dataset. The values of y are obtained with another analytical method, so they are *external* information and the whole problem is thus *supervised*. The objective is to find a function representing the dependence $\mathbf{y} = f(\mathbf{X})$.

When such a dependence is obtained, it can be used for two main tasks: to *interpret* this dependence in the context of its coefficients (making conclusions about its nature and theoretical phenomena lying in the background) and—most often—to be able to *predict* the y value for any new sample in the future. In the latter case, we assume that the obtained function has a *predictive ability*, and it does not only fit to the data but also *generalizes* the problem and allows to *interpolate* (and *extrapolate*) the obtained y value in the whole experimental domain.

Some examples of the above problem are as follows:

1. Prediction of some complex property from the vector of peak areas or from the whole chromatogram treated as a signal. This property is *complex* in the sense that many sample

---

* See Chapter 7 in this book.

ingredients have significant but different effect on this property and many interactions between these substances can occur. For example, the antioxidant capacity, toxicity, or other biological activities can be modeled and computed chemometrically from the sample chromatogram.

2. Prediction of the retention from the vector of numbers called *molecular descriptors*, characterizing the molecule of a solute. This is a QSRR (quantitative structure–retention relationship) approach—a special case of QSPR (quantitative structure–property relationship).*

3. Assigning a sample to a particular class (discrimination model): differentiating plant species from the chromatogram of an extract, distinguishing genuine drugs from counterfeits based on chromatographic data, recognition of the synthesis path of a drug from its chromatographic impurities profile, etc.[†]

14.2 LINEAR MODELS

The function fitted to the dataset can be linear and nonlinear. The multivariate linear models can be written as $y = \beta_0 + \beta_1 x_1 + \beta_2 x_2 + \beta_3 x_3 + \cdots$, that is, they are linear on first order in the regression parameters. If the dataset is centered (this is a standard preprocessing technique in chemometrics), the intercept term β_0 equals zero and can be omitted. Therefore, the calibration model can then be written in matrix notation as $\mathbf{y} = \mathbf{X}\boldsymbol{\beta}$, where $\mathbf{y}$ is a vector of calibration responses (obtained by a reference method or a priori known), $\mathbf{X}$ is a matrix of multivariate data, and $\boldsymbol{\beta}$ is a vector of equation coefficients. The goal is to make an estimation of these coefficients to find a model. The coefficients are often called shortly as an *estimate* (Figure 14.1).

Nonlinear models are not first order in the regression parameters. For instance, $y = \beta_0 + \log(\beta_1 x_1) + \log(\beta_2 x_2) \ldots$ are nonlinear. However, some can be transformed into linear models. A model as $y = \beta_0 + \beta_1 \log(x_1) + \beta_2 x_2^2 + \beta_3/x_3$ is also called linear models, because defining $x_1' = \log x_1$, $x_2' = x_2^2$, and $x_3' = 1/x_3$, a first-order model of x_1', x_2', x_3' is obtained.

Linear multivariate techniques can be used in most cases, as they can cope with a weak nonlinearity of linear variables. From the mathematical point of view, the single variables can have nonlinear dependences with y, but they often occupy a linear subspace[‡] in whole experimental multivariate space.

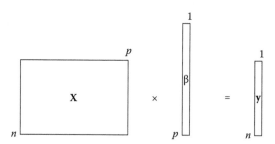

FIGURE 14.1 Graphical illustration of a multivariate calibration: the objective is to find a column vector $\boldsymbol{\beta}$ containing p coefficients of a linear equation. This equation, illustrated here as the matrix multiplication, computes $\mathbf{y}$ from $\mathbf{X}$ with the lowest possible error. The calibration dataset, consisting of n observations, is used to obtain the equation. It is desired that the equation has the predictive ability, allowing to predict y from any further $\mathbf{x}$ vector.

* See Chapters 17 and 18.
[†] See Chapter 19.
[‡] See Chapter 7.

14.2.1 ORDINARY LEAST SQUARES REGRESSION

In the simplest way, the modeling problem can be solved by the classical linear regression, called *simple linear regression* (SLR) or *ordinary least squares* (OLS) regression. The estimator $\boldsymbol{\beta}$ is obtained with the equation $\boldsymbol{\beta} = (\mathbf{X}^T\mathbf{X})^{-1}\mathbf{X}^T\mathbf{y} = \mathbf{X}^+\mathbf{y}$ (Figure 14.2). However, this method usually cannot be used in chemometrics, as it almost always yields unacceptable results.

The first reason is that in most cases the number of coefficients in vector $\boldsymbol{\beta}$ is larger than the number of calibration samples. As the problem can be treated analogously as solving a system of linear equations (where $\boldsymbol{\beta}$ are the unknowns), it leads to a system where there are less equations than unknowns. Such a system is unsolvable, that is, there is an *infinite* number of solution vectors, fully satisfying this equation (without any error), mostly without any sense.

Another problem is that if the $\mathbf{X}$ matrix is wide, it is not of full rank and therefore makes the crossproduct $\mathbf{X}^T\mathbf{X}$ singular. The solution cannot easily be computed as $(\mathbf{X}^T\mathbf{X})^{-1}\mathbf{X}^T\mathbf{y}$. If one computes it as $\mathbf{X}^+\mathbf{y}$ (pseudoinverse still can be computed), the result, called *minimum length* solution, is obtained. It is one of the infinity of solutions located closest to the origin in multivariate space (the vector has the shortest possible length). Again, in almost all cases, this solution has no sense at all.

OLS cannot be used even in the case of tall $\mathbf{X}$ matrix. Although there are more equations than unknowns and one can yield one compromise solution, this method suffers from common *multicollinearity* in chemometric datasets [1,2]. If strong multicollinearity occurs in matrix $\mathbf{X}$, the crossproduct can still become singular (or almost singular), as the rank is reduced. Even if the reverse is computable (or we try the pseudoinverse formula), the solution is unstable and also unusable. Thus, to use OLS regression, the modeling should be accompanied by a variable selection approach that reduces the number of remaining variables as much as possible (since one wants simple models) and eliminates the multicollinearity problem (see also further).

14.2.2 OVERFITTING AND MODEL COMPLEXITY

To understand why the above solution cannot be used as a good model, the reader must understand the phenomenon of *overfitting*. An overfitted model is a model that fits to the calibration data almost without an error, but the modeled dependence is without any sense (not connected with real phenomena) [3]. This occurs due to a too high *complexity* of the fitted model, which is often called as the *flexibility* of the model. The model fits not only to main trend in the data (what is to be modeled) but also to small details originating from noise or irrelevant information.

A simple two-dimensional example is as follows: a scatterplot is containing four points (Figure 14.3). One can fit a straight line by an ordinary regression and achieve quite good results and some

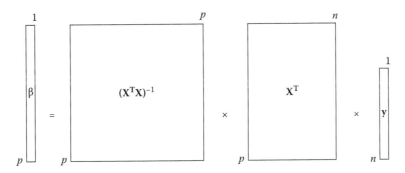

FIGURE 14.2 Graphical illustration of ordinary least squares (OLS) regression as a method to solve the problem given in Figure 14.1. The vector of coefficients, denoted as $\boldsymbol{\beta}$, is obtained by multiplying the inverse of the crossproduct of the $\mathbf{X}$ matrix (denoted as $(\mathbf{X}^T\mathbf{X})^{-1}$) by the transposed $\mathbf{X}$ matrix (denoted as $\mathbf{X}^T$), next multiplying the result with the $\mathbf{y}$ vector.

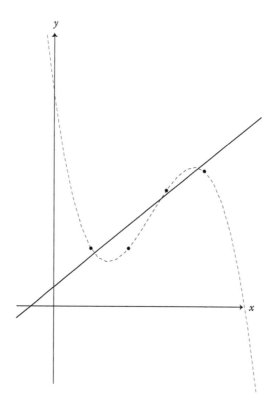

FIGURE 14.3 Very simple example of the overfitting phenomenon. The dependence (global or main tendency) between x and y variables is linear; however, the measurement error is quite large (or the second point is an outlier). The straight bold line does not fit the data very well, though it represents the global behavior with sufficient and optimal complexity. By fitting the cubic equation $y = \beta_0 + \beta_1 x + \beta_2 x^2 + \beta_3 x^3$ (dashed line), one obtains a model without any error (it fits to the points perfectly). However, this model has no predictive ability, as its complexity is too high and the experimental error affects the results.

fitting error. However, the same points can be used to fit a quadratic polynomial (not shown). It is a more flexible model, having higher complexity. Increasing still the degree of fitted polynomial (thus the complexity), one can build a cubic polynomial passing directly through the four points (the error is equal to zero). This curve is apparently ideal; however, it has no sense. The complexity is too high, the equation contains irrelevant information, and models experimental error. It has *no predictive ability*, as computing y from any x between these points (interpolation) or outside the calibration range (extrapolation) leads to meaningless and nonsensical results.

Occasionally, the quadratic dependence is really better in explaining this four-point graph than a line does. It is an example of finding the *optimal complexity*, when a linear model is not enough to fit to more complex dependence (it is *underfitted*), but increasing the degree too much still will result in an overfitted model.

14.2.3 THE PREDICTIVE ABILITY AND MODEL VALIDATION

A good model should have a good *predictive ability*, that is, the possibility to predict the y value for any x. The underfitting and overfitting phenomena reduce this ability to unacceptable levels. The predictive ability of a model can be presented as a Q^2 value. It is a value analogous to R^2 (coefficient of determination) of the fitted model. The coefficient of determination presents the fraction of the total variance of the *original* dataset explained by the model, whereas Q^2 presents analogous value but of the variance that *can be predicted*. It is determined from a validation set on cross-validation. Surprisingly, there is a possibility to achieve $Q^2 < 0$ in the case of extremely overfitted models. This means that the predictive ability of the model is even worse than using the mean of the y values as the prediction, regardless of the particular x variable. Q^2 value is computed from RMSEP (root mean squared error of prediction) or from PRESS (predicted residual error sum of squares), obtained during the *validation* of the model.

A careful validation of the model is absolutely necessary in chemometrics to ensure both lack of overfitting or a lack of fit and to estimate the real predictive ability. Validation is done using an *external test set*, that is, a set of samples, not used in any stage of model building, but selected only for validation. Having a dataset, a part of it (if possible about 30%) is excluded from model building and used to test the built model. The rest of the dataset is used as a calibration set. It is also very important that both the test and the calibration sets should be *representative*, that is, they must consist of maximally diverse samples. These samples can be chosen randomly if the dataset is large enough, but it does not always lead to a good split. Special algorithms (such as Kennard–Stone algorithm or Duplex algorithm) often are used to choose a set for validation or to split the dataset in a calibration and a validation set [4–6].

14.2.4 THE OPTIMAL MODEL COMPLEXITY AND CROSS-VALIDATION

The remaining samples of a calibration set are used to build the model. However, there is still a need to find the optimal complexity [7]. It can be achieved by cross-validation. In leave-one-out (LOO) cross-validation, each object is removed once from the dataset and the equation is built without it. Then, this object is predicted using the equation obtained. For each object, we obtain the error of prediction using the equation made without this object. We can then compute RMSECV (root mean squared error of cross-validation) and search for a model complexity with the lowest value of this error (Figure 14.4).

The LOO cross-validation is one of the more popular approaches. Others are k-fold cross-validation, which excludes not one object, but $1/k$ of the objects at once or Monte Carlo validation, when the training (calibration) set is divided many times randomly. The Monte Carlo approach is preferred for small datasets, whereas the LOO approach would lead to overoptimistic results [8]. In this situation, the training dataset can be split in a 50%/50% ratio (many times).

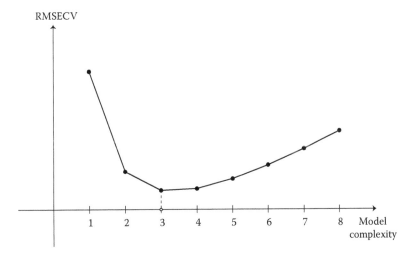

FIGURE 14.4 Typical example of the cross-validation plot. The optimal complexity has the lowest value of the RMSECV (root mean squared error of cross-validation, in this case $x = 3$). The increase of this error for lower complexity is caused by underfitting, whereas the increase for higher complexity is caused by overfitting.

Summarizing, the typical workflow of model validation is as follows:

1. Choose an external validation set and remove it from the entire dataset. The dataset without the validation samples is the training or calibration set.
2. Perform cross-validation on the training set for a range of possible model complexities. Evaluate the dependence between complexity and RMSECV error. The dependence should contain a minimum indicating optimal complexity. If the dependence falls and then from a given complexity does not increase anymore, choose the smallest complexity resulting in a small RMSECV.
3. Build the final equation on a full training set with the optimal complexity.
4. Test this equation on the external test set, obtaining RMSEP and Q^2 values. The Q^2 should be slightly worse than R^2 of the model. If it is considerably worse, the model did not pass the validation and is not predictive enough.

This workflow is absolutely necessary to ensure a good quality of the model. It should be emphasized that performing only the cross-validation step (without any external validation) and treating the RMSECV value as an estimate of the predictive power is incorrect. There are also proposals to make additional external loops of validation, as even the above workflow can show some "overoptimism" about the errors in certain cases [9].

14.3 REGRESSION TECHNIQUES

As mentioned above, the OLS approach often cannot lead to a good solution due to overfitting and too high complexity of the fitted model. Therefore, special techniques, which can control the model complexity, have to be used in model building. The reduction of the complexity can be achieved in various ways. In most cases, it is connected with searching for some directions in the multivariate space and then performing a regression in the subspace spanned by these directions. It should be underlined that there is no mathematical proof which regression technique works best for a given dataset. Thus, several techniques are often compared on the same dataset (for statistical discussion, see [10]). The similarity of the solution between two different techniques is an additional conclusion that a model works and the coefficients are not coincident.

14.3.1 Variable Selection

The simplest solution to reduce the dimensionality and decrease the flexibility seems to choose only a subset of the variables to use in modeling. Such technique is called variable selection or *feature selection*. In this case, it is assumed that only a small subset of variables is relevant and the eliminated variables do not contain any crucial information. Although this approach is quite often used in statistics, its use in chemometrics is quite rare.

As chemometric models are built on many variables and each of them can be included (or not) into the equation, there are 2^p possibilities of variable selection and this number is not *investigable* in any realistic time (e.g., when considering $p = 80$ variables, the number of possible combinations exceeds the number of microseconds which have passed since the beginning of the universe). The variable selection is a general optimization problem and there is no method for finding the exact optimal solution. Moreover, the variable selection often introduces new problems instead of solving the existing. Even if we restrict the number of variables, the OLS regression may still lead to overfitting because of various reasons (e.g., multicollinearity). The use of feature selection is justified only when after applying a regression on the full dataset does not yield any usable model. The examples of techniques used are LARS [11], LASSO [12] or UVEPLS [13], and the reader is referred to the literature for details (for instance, reviews on feature selection methods are [14,15]).

14.3.2 Ridge Regression

Historically, the first regression which copes with chemometric problems (still operating on the full dataset) is the *ridge regression* [16]. It is based on the formula $\boldsymbol{\beta} = (\mathbf{X}^T\mathbf{X} + \lambda\mathbf{I})^{-1}\mathbf{X}^T\mathbf{y}$, where λ is some arbitrary number added to the diagonal of the crossproduct, before computing the inverse (when $\lambda = 0$, the solution is equal to OLS). Adding the ridge parameter converts a singular matrix to a nonsingular and the inverse is then computable. Moreover, the ridge regression can be perceived as the *shrinkage* and *penalization* method, as it forces the solution to have a shorter length as a vector in multivariate space. Not entering fully into the mathematical background, shrinkage and penalization force the solution to be less complex, so increasing λ can be perceived in some simplified way as decreasing the model complexity.

The ridge regression is often used in model building as a comparison (reference) method. Its main advantage is the performance—it works well and gives very reliable models. The disadvantage is that a properly working model is obtained, but the regression coefficients cannot easily be interpreted.

Therefore, the most often applied techniques perform the reduction of the dataset during building of the model. The model building is done in an "indirect" way: some features are extracted from the dataset (most often projections onto some interesting directions), and then the model is made using these extracted features (e.g., PLS, PCR). It is opposite to ridge and ordinary regressions, which are "direct" techniques, that is, they operate on the original dataset.

14.3.3 Principal Component Regression

This technique is based on making a regression with the k first principal components of the original $\mathbf{X}$ matrix [17]. The principal component analysis is done on the matrix,* which decomposes the matrix $\mathbf{X}$ is decomposed into two matrices $\mathbf{X} = \mathbf{TP}^T$ (Figure 14.5a). The matrix $\mathbf{P}$ contains the orthonormal vectors of the rotated coordinate system, and as much as possible information is compressed into the first coordinates of the rotated system. In many cases, the first principal components contain almost all data information and all remaining components contain noise, artifacts, or some irrelevant trends. The idea is to make a regression on these first columns of matrix $\mathbf{T}$: $\mathbf{u} = (\mathbf{T}^T\mathbf{T})^{-1}\mathbf{T}^T\mathbf{y}$ (Figure 14.5b). We then obtain an equation based on scores; however, each score is a linear combination of the original variables. By simple multiplication by inverse (equal to transpose because of orthonormality) of $\mathbf{P}$, one can obtain the coefficients for equation of original variables: $\boldsymbol{\beta} = \mathbf{uP}^T$ (Figure 14.5c). The complexity of the model is varied by changing the number of principal components used for model building. Including all components in the model leads to the solution equal to OLS. This technique can be perceived as a regression on projected data[†]. For further discussion, the reader is referred to [18].

14.3.4 Partial Least Squares

The principal component regression is based on principal component analysis, which decomposes data in an *unsupervised* way. The information in $\mathbf{y}$ is not used to decompose the matrix and to obtain the rotated coordinate system. A step ahead is the PLS method (*Partial Least Squares* or *Projection to Latent Structures*) [19].

PLS works in a similar manner to PCR. However, the directions forming the rotated coordinate system are optimized not to explain the maximal variance inside $\mathbf{X}$ (what is done in PCA) but to explain the maximal *covariance* between $\mathbf{X}$ and $\mathbf{y}$ [20,21]. Regression is still done on the scores on the first directions found, thus basically the idea is very similar. However, this method *should* perform better when information about the predicted variable y is somehow hidden and the main

---

* See Chapter 8.
[†] See Chapter 7.

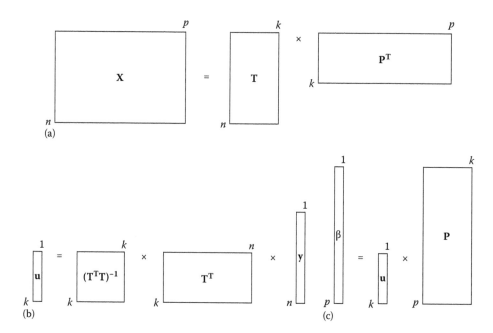

FIGURE 14.5 Graphical representation of principal component regression (PCR): (a) the matrix is first compressed to k principal components and (b) the classical regression is done on the scores, represented as **T** matrix. (c) After obtaining the k coefficients forming a vector **u**, one can compute the final estimate by multiplying **u** by matrix **P**, containing the loadings.

sources of variance are not related to this variable. The word "should" is again used in the context that there is no mathematical proof of that, so there is always a need to compare several techniques. PLS is the most often used approach in multivariate modeling and it can be considered as a *basic tool* and a technique of the *first choice*.

Performing PLS is not a simple task from the mathematical point of view, and the practitioner's way to its use is to understand the principle, but believe in already written routines inside various software. Moreover, there are several algorithms with various performances in particular conditions yielding the same result. SIMPLS [22] is one of the most frequently used algorithms for PLS regression. The interested reader is referred to the papers on computational details [23], comparison of algorithms [23,24], and wide discussion on the theoretical [25–27] or practical aspects in multivariate calibration [28–30]. The understanding of the PLS method can also be improved by comparing the classic PLS with a method extended to two **y** datasets, called LPLS [31]. A special variant of PLS, called O-PLS, also exists, which removes from the **X** matrix all variance which is not related (orthogonal) to **y** [32,33]. However, its performance is similar to a classic PLS with a proper data preprocessing (e.g., direct orthogonalization [34] or orthogonal scatter correction [35]).

14.3.5 Generalized Techniques and Modifications

The PLS regression can be understood as an approach situated between PCR (where **y** is not used at all during the direction finding) and OLS (where **y** is used in the strongest possible way). Theoretical work on this idea led to the introduction of several generalized regression methods. The PCR, PLS, and OLS techniques are the special cases and there is a possibility to obtain solutions in some "intermediate" way. In many practical cases, the model obtained by such an intermediate method shows a better predictive ability. The optimization of the parameters is done analogously to model complexity by searching for the minimal RMSECV value during cross-validation. The Cyclic Subspace

Regression [36–38] allows the user to set two independent complexity parameters, called j and l. The first parameter, j, must be equal or smaller than the second. The PLS solution is obtained when $l = r$, where r is the maximum possible complexity (matrix rank) and j allows to set the PLS model complexity. On the other hand, when both parameters are equal, the solution is equivalent to the PCR regression with given complexity. When both parameters are set to the maximum value (r), the solution is equivalent to OLS. The total number of possible models is equal to $(r^2 + r)/2$. The Continuum Power Regression [39] is another technique allowing to change the properties of the regression model continuously with one parameter α varying between OLS ($\alpha = 0$), PLS ($\alpha = 0.5$) and PCR ($\alpha = 1$). The reader interested in other approaches or generalizations is recommended to read on latent root regression [40,41], elastic component regression [42], and Tikhonov regularization [43,44].

14.3.6 Robust Linear Methods

The *robustness* of the modeling method is defined as its resistance to changes of the statistical distribution of the analyzed data. In chemometrics, this term is primarily related to the resistance of the method to *outliers*—observations that are distant from the main data cloud [45]. For instance, the univariate mean is sensitive to outliers, and inserting an outlying value changes the mean considerably. On the contrary, the median is an example of a statistic insensitive to the outliers' presence.

The classical regression methods (PLS, PCR, OLS) are not robust. The presence of outliers affects their performance strongly, as the outliers "pull on" the axes of the rotated coordinate system because the highest variance is seen along the direction between the outlier and the main data cloud. To retain the ability to model the datasets with outliers, but without the need to remove them, the regression methods were modified to be resistant to their presence. Such methods act like the median—they are not affected by the outliers and find the directions of the main data cloud, neglecting the outlying observations.

The most often used robust technique is robustified PLS [46,47]. It works just as other PLS methods, and there is no need to set any additional parameters. The reader interested in further study on the topic is referred to the following papers for more information about algorithm details [48], benchmark studies [49], or applications [50,51]. A robustified version of continuum regression [52] and M-regression [53] is also available.

14.4 SOFTWARE

The basic PLS algorithm is not very difficult to be programmed in a language (environment) that allows vector and matrix manipulations, such as MATLAB® and R. However, there are free and known packages for nonprogrammers, such as the "pls" package for R, available in the main package repository [54], as well as two toolboxes for MATLAB, called "Tomcat" [55] and "Libra" [56].

REFERENCES

1. De Levie, R. 2012. Collinearity in least-squares analysis, *Journal of Chemical Education*, 89 (1): 68–78.
2. Næs, T., and B.H. Mevik. 2001. Understanding the collinearity problem in regression and discriminant analysis, *Journal of Chemometrics*, 15 (4): 413–426.
3. Hawkins, D.M. 2004. The problem of overfitting, *Journal of Chemical Information and Computer Sciences*, 44 (1): 1–12.
4. Daszykowski, M., B. Walczak, and D.L. Massart. 2002. Representative subset selection, *Analytica Chimica Acta*, 468 (1): 91–103.
5. Rodionova, O.Y., and A.L. Pomerantsev. 2008. Subset selection strategy, *Journal of Chemometrics*, 22 (11–12): 674–685.

6. Tominaga, Y. 1998. Representative subset selection using genetic algorithms, *Chemometrics and Intelligent Laboratory Systems*, 43 (1–2): 157–163.

7. Hawkins, D.M., S.C. Basak, and D. Mills. 2003. Assessing model fit by cross-validation, *Journal of Chemical Information and Computer Sciences*, 43 (2): 579–586.

8. Xu, Q.-S., and Y.-Z. Liang. 2001. Monte Carlo cross validation, *Chemometrics and Intelligent Laboratory Systems*, 56 (1): 1–11.

9. Anderssen, E., K. Dyrstad, F. Westad, and H. Martens. 2006. Reducing over-optimism in variable selection by cross-model validation, *Chemometrics and Intelligent Laboratory Systems*, 84 (1–2 Special Issue): 69–74.

10. Frank, L.E., and J.H. Friedman. 1993. A statistical view of some chemometrics regression tools, *Technometrics*, 35 (2): 109–135.

11. Efron, B., T. Hastie, I. Johnstone, R. Tibshirani, H. Ishwaran, K. Knight, J.-M. Loubes et al. 2004. Least angle regression, *Annals of Statistics*, 32 (2): 407–499.

12. Fu, W.J. 1998. Penalized regressions: The bridge versus the lasso, *Journal of Computational and Graphical Statistics*, 7 (3): 397–416.

13. Centner, V., D.-L. Massart, O.E. De Noord, S. De Jong, B.G.M. Vandeginste, and C. Sterna. 1996. Elimination of uninformative variables for multivariate calibration, *Analytical Chemistry*, 68 (21): 3851–3858.

14. Goodarzi, M., Y.V. Heyden, and S. Funar-Timofei. 2013. Towards better understanding of feature-selection or reduction techniques for Quantitative Structure–Activity Relationship models, *TrAC—Trends in Analytical Chemistry*, 42: 49–63.

15. Goodarzi, M., B. Dejaegher, and Y. Vander Heyden. 2012. Feature selection methods in QSAR studies, *Journal of AOAC International*, 95 (3): 636–651.

16. Hoerl, A.E., and R.W. Kennard. 1970. Ridge regression: Biased estimation for nonorthogonal problems, *Technometrics*, 12 (1): 55–67.

17. Andrews, D.T., L. Chen, P.D. Wentzell, and D.C. Hamilton. 1996. Comments on the relationship between principal components analysis and weighted linear regression for bivariate data sets, *Chemometrics and Intelligent Laboratory Systems*, 34 (2): 231–244.

18. Daszykowski, M. 2007. From projection pursuit to other unsupervised chemometric techniques, *Journal of Chemometrics*, 21 (7–9): 270–279.

19. Wold, S., M. Sjöström, and L. Eriksson. 2001. PLS-regression: A basic tool of chemometrics, *Chemometrics and Intelligent Laboratory Systems*, 58 (2): 109–130.

20. Godoy, J.L., J.R. Vega, and J.L. Marchetti. 2014. Relationships between PCA and PLS-regression, *Chemometrics and Intelligent Laboratory Systems*, 130: 182–191.

21. Wentzell, P.D., and L.V. Montoto. 2003. Comparison of principal components regression and partial least squares regression through generic simulations of complex mixtures, *Chemometrics and Intelligent Laboratory Systems*, 65 (2): 257–279.

22. De Jong, S. 1993. SIMPLS: An alternative approach to partial least squares regression, *Chemometrics and Intelligent Laboratory Systems*, 18 (3): 251–263.

23. Dayal, B.S., and J.F. Macgregor. 1997. Improved PLS algorithms, *Journal of Chemometrics*, 11 (1): 73–85.

24. Manne, R. 1987. Analysis of two partial-least-squares algorithms for multivariate calibration, *Chemometrics and Intelligent Laboratory Systems*, 2 (1–3): 187–197.

25. Helland, I.S. 2001. Some theoretical aspects of partial least squares regression, *Chemometrics and Intelligent Laboratory Systems*, 58 (2): 97–107.

26. Næs, T. 1989. Leverage and influence measures for principal component regression, *Chemometrics and Intelligent Laboratory Systems*, 5 (2): 155–168.

27. Phatak, A., and S. De Jong. 1997. The geometry of partial least squares, *Journal of Chemometrics*, 11 (4): 311–338.

28. Gabrielsson, J., and J. Trygg. 2006. Recent developments in multivariate calibration, *Critical Reviews in Analytical Chemistry*, 36 (3–4): 243–255.

29. Geladi, P. 2002. Some recent trends in the calibration literature, *Chemometrics and Intelligent Laboratory Systems*, 60 (1–2): 211–224.

30. Wold, S., J. Trygg, A. Berglund, and H. Antti. 2001. Some recent developments in PLS modeling, *Chemometrics and Intelligent Laboratory Systems*, 58 (2): 131–150.

31. Sæbø, S., T. Almøy, A. Flatberg, A.H. Aastveit, and H. Martens. 2008. LPLS-regression: A method for prediction and classification under the influence of background information on predictor variables, *Chemometrics and Intelligent Laboratory Systems*, 91 (2): 121–132.

32. Trygg, J., and S. Wold. 2002. Orthogonal projections to latent structures (O-PLS), *Journal of Chemometrics*, 16 (3): 119–128.
33. Verron, T., R. Sabatier, and R. Joffre. 2004. Some theoretical properties of the O-PLS method, *Journal of Chemometrics*, 18 (2): 62–68.
34. Andersson, C.A. 1999. Direct orthogonalization, *Chemometrics and Intelligent Laboratory Systems*, 47 (1): 51–63.
35. Fearn, T. 2000. On orthogonal signal correction, *Chemometrics and Intelligent Laboratory Systems*, 50 (1): 47–52.
36. Kalivas, J.H. 1999. Cyclic subspace regression with analysis of the hat matrix, *Chemometrics and Intelligent Laboratory Systems*, 45 (1–2): 215–224.
37. Lang, P., A. Gironella, and R. Venema. 2013. Invariant subspaces and regression, *Communications in Statistics—Theory and Methods*, 42 (3): 491–504.
38. Lang, P., A. Gironella, and R. Venema. 2007. Properties of cyclic subspace regression, *Journal of Multivariate Analysis*, 98 (3): 625–637.
39. De Jong, S., B.M. Wise, and L.N. Ricker. 2001. Canonical partial least squares and continuum power regression, *Journal of Chemometrics*, 15 (2): 85–100.
40. Vigneau, E., D. Bertrand, and E.M. Qannari. 1996. Application of latent root regression for calibration in near-infrared spectroscopy. Comparison with principal component regression and partial least squares, *Chemometrics and Intelligent Laboratory Systems*, 35 (2): 231–238.
41. Vigneau, E., E.M. Qannari, and D. Bertrand. 2002. A new method of regression on latent variables. Application to spectral data, *Chemometrics and Intelligent Laboratory Systems*, 63 (1): 7–14.
42. Li, H.-D., Y.-Z. Liang, and Q.-S. Xu. 2010. Uncover the path from PCR to PLS via elastic component regression, *Chemometrics and Intelligent Laboratory Systems*, 104 (2): 341–346.
43. Kalivas, J.H. 2012. Overview of two-norm (L 2) and one-norm (L 1) Tikhonov regularization variants for full wavelength or sparse spectral multivariate calibration models or maintenance, *Journal of Chemometrics*, 26 (6): 218–230.
44. Stout, F., and J.H. Kalivas. 2006. Tikhonov regularization in standardized and general form for multivariate calibration with application towards removing unwanted spectral artifacts, *Journal of Chemometrics*, 20 (1–2): 22–33.
45. Rousseeuw, P.J., M. Debruyne, S. Engelen, and M. Hubert. 2006. Robustness and outlier detection in chemometrics, *Critical Reviews in Analytical Chemistry*, 36 (3–4): 221–242.
46. Gil, J.A., and R. Romera. 1998. On robust partial least squares (PLS) methods, *Journal of Chemometrics*, 12 (6): 365–378.
47. González, J., D. Peña, and R. Romera. 2009. A robust partial least squares regression method with applications, *Journal of Chemometrics*, 23 (2): 78–90.
48. Kruger, U., Y. Zhou, X. Wang, D. Rooney, and J. Thompson. 2008. Robust partial least squares regression: Part I, algorithmic developments, *Journal of Chemometrics*, 22 (1): 1–13.
49. Kruger, U., Y. Zhou, X. Wang, D. Rooney, and J. Thompson. 2008. Robust partial least squares regression: Part II, new algorithm and benchmark studies, *Journal of Chemometrics*, 22 (1): 14–22.
50. Daszykowski, M., Y. Vander Heyden, and B. Walczak. 2007. Robust partial least squares model for prediction of green tea antioxidant capacity from chromatograms, *Journal of Chromatography A*, 1176 (1–2): 12–18.
51. Kruger, U., Y. Zhou, X. Wang, D. Rooney, and J. Thompson. 2008. Robust partial least squares regression—Part III, outlier analysis and application studies, *Journal of Chemometrics*, 22 (5): 323–334.
52. Serneels, S., P. Filzmoser, C. Croux, and P.J. Van Espen. 2005. Robust continuum regression, *Chemometrics and Intelligent Laboratory Systems*, 76 (2): 197–204.
53. Serneels, S., C. Croux, P. Filzmoser, and P.J. Van Espen. 2005. Partial robust M-regression, *Chemometrics and Intelligent Laboratory Systems*, 79 (1–2): 55–64.
54. Mevik, B.H., and R. Wehrens. 2007. The pls package: Principal component and partial least squares regression in R, *Journal of Statistical Software*, 18 (2): 1–23.
55. Daszykowski, M., S. Serneels, K. Kaczmarek, P. Van Espen, C. Croux, and B. Walczak. 2007. TOMCAT: A MATLAB toolbox for multivariate calibration techniques, *Chemometrics and Intelligent Laboratory Systems*, 85 (2): 269–277.
56. Verboven, S., and M. Hubert. 2005. LIBRA: A MATLAB library for robust analysis, *Chemometrics and Intelligent Laboratory Systems*, 75 (2): 127–136.

15 Discriminant Analysis and Classification of Chromatographic Data

Alessandra Biancolillo and Federico Marini

CONTENTS

15.1 INTRODUCTION

Analyzing data, one could perform exploratory analysis to sum up the main characteristics of the system in order to enhance and ease its interpretation.* This allows, for example, pointing out *outliers*, or evaluating the relevance of the various variables in the data block [1]. Sometimes, a deeper level of interpretation of the system is needed. Consequently, after the exploratory analysis (which should always be performed in order to have an insight into the raw data) it is possible to look into *patterns* in data. One of the aims of *pattern recognition* is to assign objects to categories, namely, to *classify* them. For instance, in food analysis, one may be interested in predicting the geographical origin of some products [2–4], or checking for the presence of some specific compounds [5] in raw measurements (e.g., raw gas chromatograms). In other fields, like consumer science, one could be interested in predicting whether a product will be accepted by customers, knowing some sensory characteristics [6]. In medical science, it would be definitely relevant to build up a model that would allow predicting if a patient is healthy or ill based on a noninvasive test [7]. All these instances are examples of *classification* problems. Indeed, in these cases, the aim is to assign objects to a category (*geographical origin, lack/presence of a particular compound, healthy/ill*) on the basis of some specific features that are used to define the category itself.

Classification methods are *supervised*; consequently, in order to predict whether an object belongs to a specific *class* (i.e., category), it is necessary to create a model based on the measurements collected on a training set of objects whose class-belonging is known in advance.

---

* See also Chapter 8.

From a geometric point of view, classifying means finding the surfaces (in the multivariate space of the variables, where the objects lie) that separate the different category-regions.

15.1.1 Discriminant Classification versus Class-Modeling

Given the many possible ambits of application of pattern recognition strategies in the various areas of chemistry (and not only, of course), it is not surprising that, during the years, different classification methods have been proposed in the literature, each one with its own characteristics and peculiarities. In this respect, among the possible taxonomies that can be adopted to differentiate the various classification techniques, one which appears to be particularly relevant is the distinction between *discriminant* and *class-modeling* approaches [8].

Discriminant methods focus their attention on the dissimilarities among samples belonging to different classes, and operate by finding the decision boundaries that separate the regions of the multivariate space occupied by the categories under investigation. In particular, in the discriminant approach, the entire multivariate space of the objects is divided in a number of regions equal to the number of classes represented in the training set. For instance, considering a three-class problem, all the multivariate space will be divided into three regions through the definition of three boundaries (Figure 15.1a). Accordingly, the position occupied by a sample in the variable hyperspace will univocally determine the class it will be predicted to belong to. For example, an object which falls in the region of class1 will be predicted to belong to that category, another one falling in the region of class2 will be assigned to that other category and so on. Due to these characteristics, it is evident that, when a discriminant approach is adopted, a sample is always classified as belonging to one and only one of the classes represented in the training set.

On the other hand, class-modeling approaches focus on what makes the individuals of a particular class to be similar to one another and operate by modeling each class independently on the others in order to define a multivariate boundary which encloses the region of space where it is likely to find the samples of that category (consequently, the definition of the class-space for a specific category is not influenced by the possible presence of other classes). When considering the same problem involving three categories, introduced in the discussion of the discriminant approach, a different outcome may be expected from the application of a class-modeling technique: indeed, the three

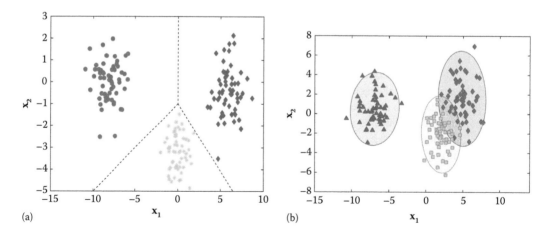

(a) (b)

FIGURE 15.1 Graphical representation of the main difference between discriminant (a) and class-modeling approaches (b). (a) Discriminant approaches divide the entire sample space in as many regions as the number of categories in the training set; each object is univocally assigned to a class. (b) Class-modeling approaches define the space of each category individually; consequently, class spaces may overlap (and objects may be accepted by more than one class) and a large portion of space is not occupied by any category.

regions associated to the different categories will occupy a limited portion of the multivariate space and, due to the fact that each boundary is calculated independently on the others, there can be the possibility that they overlap; moreover, a part of the space will not be assigned to any class (Figure 15.1b). This geometric representation translates to the fact that, when considering the classification of an unknown sample, different scenarios can occur. Indeed, an object could be assigned to a single class, if it falls inside the portion of space where a single category is mapped, or it could be accepted by more than one class ("confused" sample), if it falls in a region where two or more categories overlap (e.g., in the area in Figure 15.1b, where both class2 and class3 are mapped). Otherwise, it could be rejected by any class (i.e., fall in a portion of space where no category is mapped), and this, in turn, could mean that either it is an outlier or that it belongs to a different category that has not been modeled.

15.2 DISCRIMINANT CLASSIFICATION METHODS

As discussed in Section 15.1.1, discriminant classification methods divide the entire variable hyperspace in a number of regions equal to the number of classes represented in the training set used to build the model. Accordingly, independent of which particular method is chosen, the model-building step involves the calculation of a decision rule to be used to assign any future unknown sample to one of the available categories. In this context, it has been shown that, if one can assume that all the misclassifications are equally costly (i.e., if wrongly predicting that a sample from class1 belongs to class2 has the same severity as mispredicting a sample from class2 as coming from class1), the decision rule leading to the highest accuracy is the so-called Bayes' rule, which states that a sample should be assigned to the category it has the highest probability of belonging to [9]. Discriminant classification can then be seen as a two-step procedure

1. The *posterior* probability that an (unknown) object belongs to any of the categories represented in the training set is calculated.
2. Each object is assigned to the class it has the highest posterior probability of belonging to.

The posterior probability $p(g|\mathbf{x})$ that an object represented by the vector $\mathbf{x}$ belongs to the class g is calculated according to the Bayes' theorem, according to

$$p\left(g|\mathbf{x}\right) \propto f\left(\mathbf{x}|g\right)\pi_g \qquad (15.1)$$

where $f(\mathbf{x}|g)$ is the *likelihood* (i.e., the probability density of observing the vector of measurements $\mathbf{x}$ from a sample truly belonging to class g), while π_g is the *a priori* (i.e., estimated prior to carrying out any measurement) probability to observe an object coming from class g. The proportionality sign indicates that a normalization is required in order for the probability to meet the requirement of varying in [0,1].

The various discriminant classification methods differ among one another in the assumption they make in order to calculate the posterior probability according to Equation 15.1, or even in the fact that Bayes' rule is implicitly assumed, but never actively used in the definition of the classification rule, which is operationally based on other criteria (such as distances, as in *k*NN). In the following paragraphs, some of the most commonly used discriminant classification methods will be theoretically introduced and some hints about their possible application on chromatographic data will be given.

15.2.1 LINEAR DISCRIMINANT ANALYSIS

Linear discriminant analysis (LDA) was the first classification method introduced in the statistical literature, as it was originally proposed by R.A. Fisher in 1936 [10]. It is a parametric method as it assumes that, within each category, the multivariate observations are distributed according to a Gaussian probability density function, which means that, for a sample belonging to class g, the likelihood $f(\mathbf{x}|g)$, that is, the probability density function of observing a vector of measurement $\mathbf{x}$, takes the form:

$$f\left(\mathbf{x}|g\right) = \frac{1}{\left(2\pi\right)^{v/2}\left|\mathbf{S}_g\right|} e^{-0.5\left(\mathbf{x}-\bar{\mathbf{x}}_g\right)^T \mathbf{S}_g^{-1}\left(\mathbf{x}-\bar{\mathbf{x}}_g\right)} \tag{15.2}$$

where
  $\bar{\mathbf{x}}_g$ and $\mathbf{S}_g$ are the centroid and the variance/covariance matrix of the observations in the class, respectively
  v is the number of measured variables

In the case of linear discriminant analysis, a further assumption is made, that the variance/covariance matrices for the different categories under investigation are the same (i.e., that all the classes have the same dispersion):

$$\mathbf{S}_1 = \mathbf{S}_2 = \cdots = \mathbf{S}_g = \cdots = \mathbf{S}_G = \mathbf{S} \quad g = 1\cdots G \tag{15.3}$$

where $\mathbf{S}$ is estimated as the weighted average of the individual matrices $\mathbf{S}_g$:

$$\mathbf{S} = \frac{\sum_{g=1}^{G}\left(n_g - 1\right)\mathbf{S}_g}{N - G} \tag{15.4}$$

n_g and N being the number of samples in the gth category and in the whole data set.
  Since LDA is a parametric method, the posterior probability that a sample belongs to a particular class $p(g|\mathbf{x})$ can be calculated directly by combining the likelihood in Equation 15.2 with the prior probability for the category π_g (i.e., the probability that an unknown sample comes from the class, estimated prior to carrying out any measurement):

$$p\left(g|\mathbf{x}\right) \propto \pi_g f\left(\mathbf{x}|g\right). \tag{15.5}$$

In particular, by substituting Equation 15.2 in Equation 15.5, it is possible to explicitly write the posterior probability function of a sample for the gth class as

$$p\left(g|\mathbf{x}\right) = \frac{C_N \pi_g}{\left(2\pi\right)^{v/2}\left|\mathbf{S}\right|} e^{-0.5\left(\mathbf{x}-\bar{\mathbf{x}}_g\right)^T \mathbf{S}^{-1}\left(\mathbf{x}-\bar{\mathbf{x}}_g\right)} \tag{15.6}$$

where C_N is a normalization constant that is introduced in order to have probabilities varying between zero and one.
  As already anticipated in the introduction on discriminant methods, whenever it is possible to explicitly calculate the posterior probability that a sample belongs to the different categories, Bayes' rule suggests that these values should be used for the classification of unknown individuals. In particular, in order to minimize the classification error, the sample should be assigned to the class corresponding to the highest value of the posterior probability.
  In the case of LDA, the posterior probabilities are calculated according to Equation 15.6, so, in order to predict the class of an unknown sample $\mathbf{x}_u$, one should calculate the values

$$p\left(g|\mathbf{x}_u\right) \quad g = 1...G \tag{15.7}$$

and assign the individual to the category corresponding to the highest probability. However, the same considerations could be made if calculating any monotonic function of the probabilities in Equations 15.6 and 15.7. In particular, it is possible to define the so-called classification functions $\phi_g(\mathbf{x})$ as the natural logarithm of the posterior probabilities:

$$\phi_g(\mathbf{x}) = \ln p(g|\mathbf{x}) = \ln \frac{C_N}{(2\pi)^{v/2}|\mathbf{S}|} + \ln \pi_g - 0.5\mathbf{x}^T\mathbf{S}^{-1}\mathbf{x} + \bar{\mathbf{x}}_g^T\mathbf{S}^{-1}\mathbf{x} - 0.5\bar{\mathbf{x}}_g^T\mathbf{S}^{-1}\bar{\mathbf{x}}_g. \qquad (15.8)$$

Accordingly, Bayes' rule can be rephrased and, based on Equation 15.8, translated to a new decision that assigns any unknown sample to the category corresponding to the highest value of the classification function:

$$\max_g \phi_g(\mathbf{x}_u) \quad g = 1...G. \qquad (15.9)$$

In this context, it is worth stressing that, for a particular sample, the terms $\ln \dfrac{C_N}{(2\pi)^{v/2}|\mathbf{S}|}$ and $0.5\mathbf{x}^T\mathbf{S}^{-1}\mathbf{x}$ are the same for all the categories and cancel out in the comparison described by Equation 15.9. As a consequence, the classification functions for linear discriminant analysis may be defined in a simpler way as

$$\phi_g(\mathbf{x}) = \ln \pi_g + \bar{\mathbf{x}}_g^T\mathbf{S}^{-1}\mathbf{x} - 0.5\bar{\mathbf{x}}_g^T\mathbf{S}^{-1}\bar{\mathbf{x}}_g \qquad (15.10)$$

that is, as linear combinations of the measured variables.

Starting from Equation 15.10, it is then possible to define the boundaries that separate the region of the variable hyperspace associated to a category to the one associated to another class. Indeed, the decision surface separating the region of class $g1$ from the region of class $g2$ is characterized by the definition as being the portion of space where the probabilities of the two categories (or, better, the corresponding classification functions) are equal:

$$\phi_{g1}(\mathbf{x}) = \phi_{g2}(\mathbf{x}). \qquad (15.11)$$

Accordingly, by substituting Equation 15.10 into 15.11 one obtains

$$\ln \pi_{g1} + \bar{\mathbf{x}}_{g1}^T\mathbf{S}^{-1}\mathbf{x} - 0.5\bar{\mathbf{x}}_{g1}^T\mathbf{S}^{-1}\bar{\mathbf{x}}_{g1} = \ln \pi_{g2} + \bar{\mathbf{x}}_{g2}^T\mathbf{S}^{-1}\mathbf{x} - 0.5\bar{\mathbf{x}}_{g2}^T\mathbf{S}^{-1}\bar{\mathbf{x}}_{g2} \qquad (15.12)$$

which can be rearranged to

$$\left(\bar{\mathbf{x}}_{g1} - \bar{\mathbf{x}}_{g2}\right)^T \mathbf{S}^{-1}\mathbf{x} - 0.5\left(\bar{\mathbf{x}}_{g1}^T\mathbf{S}^{-1}\bar{\mathbf{x}}_{g1} - \bar{\mathbf{x}}_{g2}^T\mathbf{S}^{-1}\bar{\mathbf{x}}_{g2}\right) + \ln \frac{\pi_{g1}}{\pi_{g2}} = 0. \qquad (15.13)$$

Equation 15.13 indicates that in LDA the decision surfaces are linear in the measured variables (straight lines in 2D, planes in 3D, and hyperplanes in higher dimensions), justifying the name of the technique. This concept is graphically illustrated in Figure 15.2, where an example of the application of LDA to a three-class problem in two dimensions is shown.

The training data in the figure (filled symbols) are used to calculate the classification functions for the three categories as described in Equation 15.10; based on the classification functions, the three decision boundaries separating class 1 from 2, class 1 from 3, and class 2 from 3 are calculated according to Equation 15.13 and appear as linear surfaces (straight lines, since the example is two-dimensional).

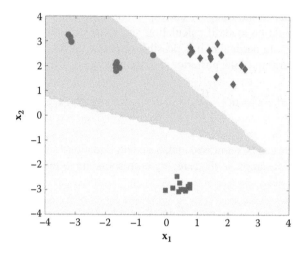

FIGURE 15.2 Linear discriminant analysis: illustration of the decision boundaries calculated by LDA for a problem involving three classes (red circles, blue squares, and green diamonds) in two dimensions. Light red, blue, and green areas indicate the regions assigned to the corresponding class.

15.2.2 QUADRATIC DISCRIMINANT ANALYSIS

When the within-class covariance matrices of the investigated categories $\mathbf{S}_g$ are too different from one another and can't be assumed to be the same, the use of linear discriminant analysis could lead to inaccurate predictions, as the decision boundaries wouldn't reflect the true probability distribution for the classes. In such cases, it is advisable to calculate the posterior probabilities in Equation 15.6 using the individual variance/covariance matrix for the category, $\mathbf{S}_g$. Accordingly, the classification functions in Equation 15.8 become

$$\phi_g(\mathbf{x}) = \ln p(g|\mathbf{x}) = \ln \frac{C_N}{(2\pi)^{\nu/2}|\mathbf{S}_g|} + \ln \pi_g - 0.5\mathbf{x}^T\mathbf{S}_g^{-1}\mathbf{x} + \bar{\mathbf{x}}_g^T\mathbf{S}_g^{-1}\mathbf{x} - 0.5\bar{\mathbf{x}}_g^T\mathbf{S}_g^{-1}\bar{\mathbf{x}}_g, \qquad (15.14)$$

which can be further simplified to

$$\phi_g(\mathbf{x}) = -\ln|\mathbf{S}_g| + \ln \pi_g - 0.5\mathbf{x}^T\mathbf{S}_g^{-1}\mathbf{x} + \bar{\mathbf{x}}_g^T\mathbf{S}_g^{-1}\mathbf{x} - 0.5\bar{\mathbf{x}}_g^T\mathbf{S}_g^{-1}\bar{\mathbf{x}}_g, \qquad (15.15)$$

considering that, also under these new assumptions, the term $\ln \dfrac{C_N}{(2\pi)^{\nu/2}}$ is the same for all the categories. Considering the individual variance/covariance matrices for the different categories in the discriminant model results in the classification functions to include a term $(\mathbf{x}^T\mathbf{S}_g^{-1}\mathbf{x})$, which is quadratic in the measured variables that, differently than in the case of LDA, can't be deleted.

As a consequence, the decision boundaries between pairs of categories, defined as in Equation 15.11, are also quadratic in the experimental variables:

$$-0.5\mathbf{x}^T\left(\mathbf{S}_{g1}^{-1} - \mathbf{S}_{g2}^{-1}\right)\mathbf{x} + \left(\bar{\mathbf{x}}_{g1}^T\mathbf{S}_{g1}^{-1} - \bar{\mathbf{x}}_{g2}^T\mathbf{S}_{g2}^{-1}\right)\mathbf{x} - 0.5\left(\bar{\mathbf{x}}_{g1}^T\mathbf{S}_{g1}^{-1}\bar{\mathbf{x}}_{g1} - \bar{\mathbf{x}}_{g2}^T\mathbf{S}_{g2}^{-1}\bar{\mathbf{x}}_{g2}\right) + \ln \frac{\pi_{g1}}{\pi_{g2}} - \ln \frac{|\mathbf{S}_{g1}|}{|\mathbf{S}_{g2}|} = 0. \ (15.16)$$

Due to these characteristics, the resulting classification method is called quadratic discriminant analysis (QDA), and the corresponding decision boundaries may be hyperspheres, hyperellipsoids, hyperparaboloids, or, in general, hyperquadrics [10–12].

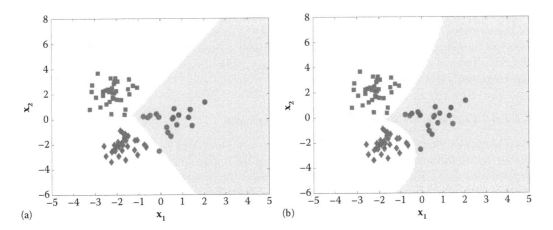

FIGURE 15.3 Comparison of the predictions of LDA (a) and QDA (b) for a problem involving three classes (red circles, blue squares, and green diamonds) in two dimensions. Light red, blue, and green areas indicate the regions assigned to the corresponding class.

An example of classification using QDA is reported in Figure 15.3 for a problem involving three categories in two dimensions.

In particular, the figure shows the different classifications that would be obtained by using LDA or QDA on the same data set: in the case of QDA, a larger portion of space is assigned to the red class, which is the more dispersed, allowing to classify correctly also the lowermost sample that, in the case of LDA, would have been predicted as green. Moreover, the different shape of the decision boundaries (linear and quadratic) is also evident from the figure.

15.2.3 FACTORS LIMITING THE PRACTICAL USE OF LDA AND QDA

The theoretical description reported in the previous paragraphs suggests that both linear and quadratic discriminant approaches could be valuable tools for the analysis of chemical (and in the context of this book in particular, chromatographic) data. Their parametric nature allows a straightforward model fitting as, differently, for example, from component-based models, there are no parameters or metaparameters to be adjusted so that, given the data, there is always a unique solution. In principle, their being parametric could also be a limitation, but it has been demonstrated that both methods—in particular LDA—are rather robust against moderate violations of their underlying probabilistic assumptions.

However, despite the advantages reported above, LDA and QDA suffer from severe drawbacks that limit their practical applicability to many real-world chemical problems. Indeed, as evidenced by Equations 15.10 and 15.15, model building in LDA and QDA requires the inversion of a pooled ($\mathbf{S}$) or individual ($\mathbf{S}_g$) within-class covariance matrix, respectively. In order for these matrices to be invertible, they shouldn't be "ill-conditioned": from an experimental standpoint, this statement translates to the requirement that the number of samples should be at least equal than the number of measured variables and that the variables themselves are as uncorrelated as possible, both conditions which are rarely met in chemistry in general and in modern chromatographic experiments in particular. Moreover, while in the case of LDA, where a pooled covariance matrix has to be calculated, these requirements apply to the data set as a whole; for QDA, they appear to be more stringent, as they have to be satisfied individually by each of the category subsets.

Accordingly, it is apparent that, in the context of the analysis of chromatographic data, LDA and QDA can't be applied directly on the whole experimental fingerprints, which contain many highly correlated variables, while—always taking into account the limitations described above—they can, in principle, be used for the analysis of not too large peak tables.

On the other hand, chromatographic profiles can still be analyzed by LDA or QDA after some sort of feature compression. In this respect, the possibility of using a reduced set of factors (e.g., either principal components or partial least squares [PLS] latent variables, see Section 15.2.4), which share the characteristics of being orthogonal and spanning a low-dimensional subspace, represents the most reasonable and most frequently used strategy.

15.2.4 PARTIAL LEAST SQUARES DISCRIMINANT ANALYSIS

As discussed in the previous paragraph, one of the main limitations to the use of linear (and quadratic) discriminant analysis in chemical problems is the impossibility to apply such a method on problems involving ill-conditioned data matrices. One of the possible ways to overcome this limitation is the use of a discriminant classification method that relies on the use of a bilinear approach to extract a few orthogonal components from the data and uses these components to build the decision rule. In this respect, the approach that appears to be more widely used nowadays in the chemometric literature is partial least squares discriminant analysis (PLS-DA).

PLS-DA was introduced in the chemometric literature to extend the advantages of the PLS algorithm* [13] already exploited in the case of regression (where it allowed to overcome similar limitations related to multicollinearity and samples to variable ratio) to the domain of classification problems [14–16].

In order to be able to use a regression method to deal with classification problems, it is necessary to find a suitable coding for building a response matrix $\mathbf{Y}$ accounting for class belonging. This is generally accomplished by introducing a dummy binary matrix having as many rows as the number of samples and as many columns as the number of categories to be modeled. For each sample, the corresponding row of the dummy $\mathbf{Y}$ matrix contains all zeros except for the column which corresponds to the class it belongs to. For instance, in a problem involving five categories, for all the samples belonging to the first class, the corresponding rows of the dummy $\mathbf{Y}$ matrix will be

$$\mathbf{y}_{class1} = \begin{bmatrix} 10000 \end{bmatrix}, \tag{15.17}$$

whereas all the samples from the second category will be coded as

$$\mathbf{y}_{class2} = \begin{bmatrix} 01000 \end{bmatrix}, \tag{15.18}$$

and so on:

$$\mathbf{y}_{class3} = \begin{bmatrix} 00100 \end{bmatrix}$$
$$\mathbf{y}_{class4} = \begin{bmatrix} 00010 \end{bmatrix} \tag{15.19}$$
$$\mathbf{y}_{class5} = \begin{bmatrix} 00001 \end{bmatrix}.$$

Once the dummy matrix is defined in this way, then classification is accomplished by building a regression model relating the $\mathbf{Y}$ to the descriptor matrix $\mathbf{X}$

$$\mathbf{Y} = \mathcal{F}(X) + \mathbf{E}_Y = \hat{\mathbf{Y}} + \mathbf{E}_Y \tag{15.20}$$

(where $\mathcal{F}(\mathbf{X})$ indicates a generic functional relationship and $\mathbf{E}_Y$ are the Y-residuals) and looking at the predicted values of the responses $\hat{\mathbf{Y}}$ [16–19].

---

* See also Chapter 14.

Here it should be stressed that, so far, all these considerations apply to all the regression-based classification methods and not only to PLS-DA. In this context, the peculiarity of PLS-DA is that the regression model in Equation 15.20 is calculated using partial least squares regression (PLS-R).

PLS-R belongs to the family of linear regression models, so that, in Equation 15.20, the predicted responses are assumed to be linear combinations of the measured variables:

$$\widehat{\mathbf{Y}} = \mathcal{F}(\mathbf{X}) = \mathbf{XB} \tag{15.21}$$

$\mathbf{B}$ being the matrix collecting the weighting coefficients (*regression coefficients*). Within the family of linear regression models described by Equation 15.21, what characterizes PLS-R is that the regression coefficients are estimated in a biased way, that is, not using all the information available in $\mathbf{X}$. Indeed, the PLS algorithm involves projecting the data in $\mathbf{X}$ onto a relevant subspace of orthogonal latent variables corresponding to the directions where there is maximum covariance with the responses:

$$\mathbf{T} = \mathbf{XR} \tag{15.22}$$

where $\mathbf{R}$ is the weight matrix containing the coefficients of the projection, while the coordinates of the samples onto the new axes (*scores*) are collected in the matrix $\mathbf{T}$. These scores are then used as independent variables in the regression equation to predict the $\mathbf{Y}$:

$$\widehat{\mathbf{Y}} = \mathbf{TQ}^T \tag{15.23}$$

$\mathbf{Q}$ being the Y-loadings, that is, the coefficients relating the X-scores to the response(s). By combining Equations 15.21 through 15.23, it is evident that

$$\mathbf{B} = \mathbf{RQ}^T. \tag{15.24}$$

As anticipated, PLS-DA operates by calculating a regression model between the measured variables and the dummy response matrix using the PLS algorithm, as described in Equations 15.21 through 15.24. This means that the vector of descriptors collected on a particular sample $\mathbf{x}_i$ is used to predict the values of the corresponding responses $\widehat{\mathbf{y}}_i$, which constitutes the basis for its classification. However, while the target ("true") values of the dummy $\mathbf{Y}$ are binary coded, the corresponding predictions (which result from the application of a quantitative regression tool) are real-valued. For instance, in the case of the five classes problem exemplified in Equations 15.17 through 15.19, the predicted responses for a particular individual could be

$$\widehat{\mathbf{y}}_i = \begin{bmatrix} 0.02 & -0.13 & 0.87 & 0.06 & -0.03 \end{bmatrix}. \tag{15.25}$$

It is then necessary to define a rule allowing to predict the class of the samples under investigation, based on the predicted values of the response vector. In this respect, the simplest rule, which anyway is quite often used and has proved to be successful on many occasions, is to assign the sample to the category corresponding to the highest value of $\widehat{\mathbf{y}}_i$ [14,17]: based on this criterion, in the case of Equation 15.25, the sample would have been assigned to the third class, as the maximum value of the predicted response $\widehat{\mathbf{y}}_i$ (0.87) corresponds to the third entry of the vector.

When there are only two categories involved, due to the symmetry of the classification problem, it is customary to use only a scalar value ($y = 1$ for class1 and $y = 0$ for class2) instead of a vector ($\mathbf{y} = [1\ 0]$ and $\mathbf{y} = [0\ 1]$ for class1 and class2, respectively) for coding. In such a case, the simple

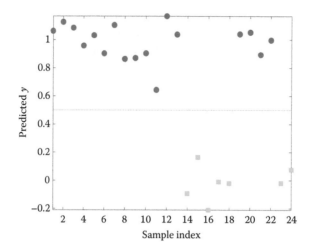

FIGURE 15.4 Partial least squares discriminant analysis (PLS-DA): illustration of the simplest classification rule for the two-class problem. If the predicted Y is higher than the threshold, then the sample is assigned to the red class ($y = 1$) otherwise, if it falls below the line, the individual is assigned to the green class ($y = 0$).

classification rule described above becomes: assign a sample to class1 if $\hat{y} > 0.5$, otherwise it is predicted to belong to class2. This situation is graphically illustrated in Figure 15.4.

Figure 15.4 illustrates how PLS-DA classification can be carried out in the case of a problem involving two categories based on the values of the predicted response. Since class1 (red circles) is coded as $y = 1$ and class2 is coded as $y = 0$, the classification threshold is set to $\hat{y} = 0.5$ (blue dashed line): if a sample falls above the horizontal line ($\hat{y} > 0.5$), it is predicted as belonging to class1, otherwise ($\hat{y} < 0.5$) it is assigned to class2. Here it should be stressed that in cases when there is a severe imbalance between the number of training samples of the two categories, or if there is an asymmetric classification risk (i.e., classification errors in one direction are costlier than in the other as it happens, for instance, in medical diagnosis, where false negatives are to be avoided more than false positives), then the theoretical threshold of 0.5 can result inadequate, and a more appropriate empirical threshold should be computed.

More in general, together with the simple approach illustrated in the previous lines, other different possibilities have been proposed in the literature in order to build the classification rule based on the predicted response vector. In particular, a possible approach involves using the PLS regression step only for feature reduction, in order to be able to apply LDA or other classifiers on the vector of predicted responses or on the corresponding X-scores [18,19].

15.2.5   OTHER DISCRIMINANT CLASSIFICATION METHODS

Together with the methods described in the previous paragraphs, several other chemometric tools with different characteristics can be used for discriminant classification of chromatographic data: kernel-based approaches (support vector machines [20] and kernel-PLS-DA [21]), classification trees [22], and their evolution (random forests [23]).*

Among the remaining ones, it is worth mentioning the k nearest neighbors (kNN) algorithm [24,25], as it has a rather simple decision rule and, at the same time, it has proved to give acceptably good results on many data sets. kNN is a nonparametric method, so that it doesn't make explicit use of the calculation of a posterior probability to define the classification rule and the decision

---

* See Chapter 16.

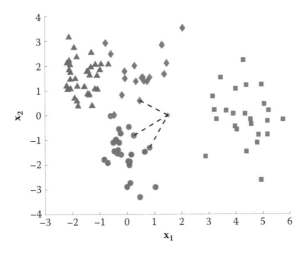

FIGURE 15.5 Graphical illustration of the kNN classification criterion with $k = 3$. Dashed lines connect the unknown sample (gray star) to its three nearest neighbors. Since two of them belong to the red class and one to the green, the unknown is assigned to the red class.

boundaries. On the other hand, the classification of an unknown individual is based on the definition of its k closest training samples (nearest neighbors) according to a predefined metrics (which in most of the cases is the Euclidean distance). In particular, the unknown sample is assigned to the category the majority of its k nearest neighbors belongs to. The classification rule is graphically illustrated in Figure 15.5 for a problem involving four classes in two dimensions; in the example a value of $k = 3$ is chosen.

In order to decide which class the unknown sample (indicated as a gray star in the figure) should be assigned to, it is necessary to identify its three nearest neighbors (which are pointed at with dashed lines). Since two of the three nearest neighbors belong to the red class and the remaining one to the green class, the unknown sample is assigned to the red category.

Due to the nature of the classification rule, kNN results in nonlinear decision boundaries, whose shape (and nonlinearity) strongly depends on the choice of the number of nearest neighbors to be considered. This concept is graphically illustrated in Figure 15.6, where the different classifications resulting from the data set shown in Figure 15.5 when choosing one, three, or five nearest neighbors are reported.

It is evident from the figure that, as may be expected, when using a single nearest neighbor the resulting decision surfaces are more irregular and with a higher extent of nonlinearity, whereas, by increasing the value of k, the surfaces become smoother. Accordingly, the choice of the value of k is a fundamental part of model selection, and it is normally accomplished based on cross-validation.

Since the kNN classification rule is based on the calculation of distances, the method could be in principle applied to all kinds of chromatographic data (both peak tables and whole profiles). However, it has been demonstrated in the literature that in the case of high-dimensional data (i.e., when the number of variables is relatively high) the distance measures tend to shrink and, in particular, the ratio between the distances to the nearest and the farthest samples tends to one [26]. Therefore, in such situations, it may be advisable to apply kNN classification on the PCA or the PLS-DA scores in order to obtain more reliable results.

Another method that is worth mentioning when considering discriminant classification approaches that could be useful for the analysis of chromatographic data is multilinear partial least squares discriminant analysis (NPLS-DA) [1]. Indeed, all the techniques described so far are applicable to data matrices, that is, two way arrays of data, where customarily the rows correspond to the number of samples and the columns correspond to the number of variables. In the context of chromatographic

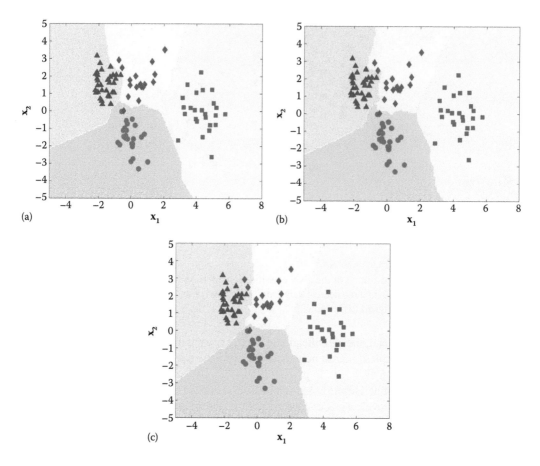

FIGURE 15.6 Comparison of the classifications of the *k*NN algorithm with (a) $k = 1$, (b) $k = 3$, and (c) $k = 5$.

applications, similar data matrices can be the result of extracting a peak table or be obtained by considering, for each individual, the whole chromatographic profile at a particular wavelength or *m*/*z* value, or the total ion current. However, when dealing with multidimensional or hyphenated chromatography, each sample is characterized by a landscape, so that when data are collected on more than one individual, a so-called multiway array is obtained.* An example of multiway (in particular, three-way as there are three sources of variability) data structure resulting from a chromatographic experiment is reported in Figure 15.7.

When the data have a multiway structure, one possibility is to rearrange the elements of the array in order to reorganize them in the form of a matrix (unfolding), as illustrated in Figure 15.7. However, the use of the conventional two-way methods on the unfolded three-way arrays can lead to a poorer interpretability and, from a prediction standpoint, to less accurate results and/or overfitting. Accordingly, when dealing with such kind of data, the use of suitable modeling approaches, which explicitly take into account the data structure (multiway methods), is recommended. In this context, NPLS-DA represents the multiway generalization of the PLS-DA algorithm described in Section 15.2.4. In particular, it works by decomposing the multiway array of measurement to extract components having maximum covariance with the dummy **Y**, and then using the resulting scores as predictors (i.e., it makes use of multilinear PLS [27] instead of conventional PLS to predict the responses).

---

* See also Chapter 23.

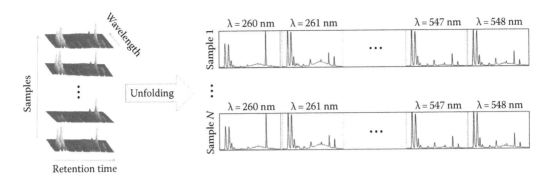

FIGURE 15.7   Illustration of how hyphenated chromatography can generate a multiway data set. When spectrochromatographic landscapes are collected on N samples (e.g., with a diode array), then a three-way data set (data cube) is obtained (having dimensions N samples $\times$ J retention times $\times$ K wavelengths). The figure also shows the unfolding procedure needed to process the three-way data with standard two-way methods: to "matricize" the array, the K slices corresponding to all the chromatograms collected on the samples at a particular wavelength are juxtaposed one after another.

15.3   CLASS-MODELING METHODS

As stated in Section 15.1.1, class-modeling techniques define the region of space associated to each category individually and independently of any other possible class. These characteristics make the modeling approach particularly suitable to deal with *asymmetric classification* problems, that is, those situations where one is interested only in a single group and, therefore, the aim is to recognize whether a sample comes from that specific category or not [28,29].

Contrarily to what happens with discriminant approaches, the number of class-modeling techniques available in the literature is relatively limited and only two or three of them are routinely used. Accordingly, in the following paragraph, only one of these class modeling tools, soft independent modeling of class analogies (SIMCA), which is by far the most commonly used, will be discussed in detail.

15.3.1   SOFT INDEPENDENT MODELING OF CLASS ANALOGIES

SIMCA, originally introduced by Svante Wold in 1976 [30,31], is the oldest and the most commonly used class-modeling technique in the chemometric literature. Actually, one could affirm that the concept of class-modeling itself emerged as a result of the introduction of the SIMCA approach in the literature. As its name suggests, SIMCA is based on building independent models for each category, which summarizes the analogies (i.e., the systematic variation) among the members of the class by means of a soft (i.e., bilinear) model. In detail, the SIMCA assumes that the systematic variation which characterizes a particular category can be captured by a principal component model of appropriate dimensionality:

$$\mathbf{X}_g = \mathbf{T}_g \mathbf{P}_g^T + \mathbf{E}_g, \tag{15.26}$$

where
   $\mathbf{X}_g$ is the matrix containing only the training data from the gth class
   $\mathbf{T}_g$ and $\mathbf{P}_g$ are the scores and loadings matrices and the residuals are collected in the matrix $\mathbf{E}_g$

The bilinear model in Equation 15.26 then represents the starting point for the classification of the samples. As already pointed out, class modeling techniques operate in a rather different way with respect to the discriminant approaches, as they are based on verifying whether a sample is fitted well ("accepted") by the model of the category or not ("rejected"). In that, they share many similarities with outlier detection approaches from which, indeed, they borrow the machinery.

Accordingly, in the case of SIMCA, once the PCA decomposition in Equation 15.26 has been carried out, the residuals (and, in some versions, the scores) are used to build outlier detection tests which, in turn, are exploited to decide whether a sample is accepted by the class or not.

In the original formulation of SIMCA, only the samples' residuals were taken into account to decide whether the tested individual should be accepted or not by the model of the category. Operationally, the sample to be classified (characterized by the row vector of measurements $\mathbf{x}_u$) is projected onto the PC space of the category defined as in Equation 15.26:

$$\mathbf{t}_{u(g)} = \mathbf{x}_u \mathbf{P}_g^T, \tag{15.27}$$

where the loadings $\mathbf{P}_g$ calculated from the training samples are used to extract the scores $\mathbf{t}_{u(g)}$. The residuals for the new sample $\mathbf{e}_{u(g)}$ are then estimated as the difference between the measured vector $\mathbf{x}_u$ and its principal component estimate $\hat{\mathbf{x}}_u$:

$$\mathbf{e}_{u(g)} = \mathbf{x}_u - \hat{\mathbf{x}}_u = \mathbf{x}_u - \mathbf{t}_{u(g)} \mathbf{P}_g^T. \tag{15.28}$$

In order to assess whether the tested sample should be accepted or not, its residuals $\mathbf{e}_{u(g)}$ are transformed in a variance

$$s_{u(g)}^2 = \frac{\sum_{j=1}^{v} e_{uj(g)}^2}{v - A} \tag{15.29}$$

where A is the number of components in the PC model of the class, while $e_{uj(g)}$ is the jth component of the residual vector $\mathbf{e}_{u(g)}$.

$s_{u(g)}^2$ is then compared to the average residual variance of the training samples of the category $s_{0(g)}^2$:

$$s_{0(g)}^2 = \frac{\sum_{i=1}^{n_g} \sum_{j=1}^{v} e_{ij(g)}^2}{(v - A)(n_g - A - 1)} \tag{15.30}$$

$e_{ij(g)}$ being the (i,j)th element of the residual matrix for the training samples of the class and n_g being their number. The comparison is carried out by means of an F test, so that the acceptance criterion is formulated as

$$s_{u(g)} \leq s_{0(g)} \sqrt{F_{0.95}(v - A, n_g - A - 1)} \tag{15.31}$$

where $F_{0.95}(v-A, n_g-A-1)$ is the 95th percentile of the F distribution with $v-A$ and n_g-A-1 degrees of freedom, respectively. Accordingly, the tested sample is accepted by the class if its residual variance is not statistically different from the average residual variance of the training individuals, otherwise it is rejected.

This original formulation of SIMCA was almost immediately modified to account for the possibility of a new sample to be outlying not only in terms of residuals, but also by being far from the training samples of the category in the scores space [31]. A so-called augmented distance to the model $d_{u(g)}^{aug}$ was then defined as

$$d_{u(g)}^{aug} = \sqrt{s_{u(g)}^2 + \sum_{a=1}^{A} \left(t_{ua(g)} - \vartheta_{a(g)}^{\lim}\right)^2 \frac{s_{0(g)}^2}{s_{a(g)}^2}} \tag{15.32}$$

where

$t_{ua(g)}$ is the score of the tested sample onto the ath component

$\vartheta_{a(g)}^{\lim}$ can be either the lower or upper score boundary for that component (depending on the value of $t_{ua(g)}$)

$s_{a(g)}^2$ is the variance of the scores of the training individual along the same factor

Equation 15.32 corresponds to identifying a box in the space of the relevant principal components, where it is likely to find the samples of the category; in general, for each component, the limits of this box $\vartheta_{a(g)}^{\lim}$ are assumed to be the minimum and maximum value of the scores of the training sample, even if the possibility of restricting or enlarging the interval has also been accounted for. If along a component the score of the tested sample is within the limit, the quantity $\left(t_{ua(g)} - \vartheta_{a(g)}^{\lim}\right)$ is equal to zero, otherwise it corresponds to the distance of the score to its nearest limit. The quantity $\dfrac{s_{0(g)}^2}{s_{a(g)}^2}$ is introduced to put the residual standard deviation and the scores distance on a comparable basis. In this modified version of SIMCA, the classification of a sample is still based on an F test, so that the individual is accepted if

$$d_{u(g)}^{aug} \leq s_{0(g)}\sqrt{F_{0.95}\left(v - A, n_g - A - 1\right)} \tag{15.33}$$

where all the terms have the same meaning as in Equation 15.31.

More recently, an alternative version of the SIMCA algorithm, inspired by the outlier detection strategies employed in multivariate statistical process control, has been proposed and it is currently the most commonly used implementation of this technique. In this alternative SIMCA algorithm, the distance of a new sample to the model is also defined as a combination of an orthogonal (based on the residuals after PCA projection) and a scores distance, but these two distances are expressed by means of the so-called Q and T^2 statistics.

The statistical variable Q is defined as the sum of squared residuals:

$$Q_{u(g)} = \sum_{j=1}^{v} e_{uj(g)}^2 \tag{15.34}$$

while T^2 is the Mahalanobis distance of the sample to the center of the PC space:

$$T_{u(g)}^2 = \sum_{a=1}^{A} \frac{t_{ua(g)}^2}{s_{a(g)}^2}. \tag{15.35}$$

Accordingly, following this alternative approach, the distance of a new sample to the SIMCA model of the category $d_{r,u(g)}$ is calculated as

$$d_{r,u(g)} = \sqrt{\left(\frac{T_{u(g)}^2}{T_{0.95(g)}^2}\right)^2 + \left(\frac{Q_{u(g)}}{Q_{0.95(g)}}\right)^2} = \sqrt{\left(T_{r,u(g)}^2\right)^2 + \left(Q_{r,u(g)}\right)^2} \tag{15.36}$$

where the comparability between the two statistics is ensured by considering their "reduced" version, that is, by normalizing $T_{u(g)}^2$ and $Q_{u(g)}$ by the 95th percentiles of the corresponding distributions under the null hypothesis ($T_{0.95(g)}^2$ and $Q_{0.95(g)}$, respectively) [32]. Then, in order for a sample to be accepted by the model of a particular category, it should satisfy

$$d_{r,u(g)} \leq \sqrt{2}, \tag{15.37}$$

otherwise it is rejected.

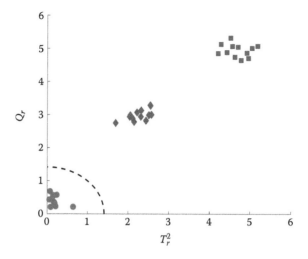

FIGURE 15.8 Graphical representation of the SIMCA space for a particular category (red circles) through the $T^2_{r(g)}$ vs. $Q_{r(g)}$ plot. The dashed black line indicates the acceptance threshold: samples falling below the line are accepted, while those mapped above are rejected. Blue squares and green diamonds indicate two other categories, different than the one modeled.

The definition of the SIMCA model according to Equations 15.36 and 15.37 allows a straightforward visualization of the classification outcomes, by plotting each sample in the space spanned by $T^2_{r(g)}$ and $Q_{r(g)}$ (see Figure 15.8).

Since the two axes of the plot are $T^2_{r(g)}$ and $Q_{r(g)}$, the dashed black line corresponding to $d_{r(g)} \leq \sqrt{2}$ represents the boundary of the class space: if an individual falls below the line, it is accepted by the category, otherwise it is rejected. In the example reported in Figure 15.8, all the samples of the modeled category (red class) are correctly accepted, whereas all the individuals from the other categories (green and blue) are correctly rejected.

It is worth mentioning that recently a modified version of the SIMCA algorithm has been introduced in the literature, in order to perform modeling classification directly on multiway arrays. In this so-called NSIMCA [33], a multiway component algorithm (which can be either PARAFAC or Tucker3 [1])* is used to obtain the sample scores and residuals, which are then, in turn, combined to define a distance to the model analogous to what is described in Equation 15.36.

15.4 CONCLUDING REMARKS

When chromatographic data are collected on a set of samples with the aim of obtaining a qualitative information, that is, with the scope of achieving a classification of the individuals, there are many tools available that differ among one another in complexity, applicability, and also in the kind of outcomes they produce (depending on whether a modeling or a discriminant approach is adopted). In this respect, the nature of the data to be analyzed, that is, whether one wishes to work with peak tables, retention profiles, or spectrochromatographic landscapes, plays a fundamental role in guiding the choice of the most appropriate technique.

---

* See Chapter 23.

REFERENCES

1. A. Smilde, R. Bro, P. Geladi, *Multi-Way Analysis*, John Wiley & Sons, Chichester, U.K., 2004.

2. N. E. Tzouros, I. S. Arvanitoyannis, Agricultural produces: Synopsis of employed quality control methods for the authentication of foods and application of chemometrics for the classification of foods according to their variety or geographical origin, *Crit. Rev. Food Sci. Nutr.*, 41 (2001) 287–319.

3. E. Borras, J. Ferrer, R. Boque, M. Mestres, L. Aceña, O. Busto, Data fusion methodologies for food and beverage authentication and quality assessment—A review, *Anal. Chim. Acta*, 891 (2015) 1–14.

4. M. Silvestri, L. Bertacchini, C. Durante, A. Marchetti, E. Salvatore, M. Cocchi, Application of data fusion techniques to direct geographical traceability indicators, *Anal. Chim. Acta*, 769 (2013) 1–9.

5. M. Lipp, Determination of the adulteration of butter fat by its triglyceride composition obtained by GC. A comparison of the suitability of PLS and neural networks, *Food Chem.*, 55 (1996) 389–395.

6. T. Naes, P. Brockoff, O. Tomic, *Statistics for Sensory and Consumer Science*, John Wiley & Sons, Chichester, U.K., 2010.

7. D. Huang, Y. Quan, M. He, B. Zhou, Comparison of linear discriminant analysis methods for the classification of cancer based on gene expression data, *J. Exp. Clin. Cancer Res.*, 28 (2009) 149–156.

8. C. Albano, W. Dunn III, U. Edlund, E. Johansson, B. Nordén, M. Sjöström, S. Wold, Four levels of pattern recognition, *Anal. Chim. Acta*, 103 (1978) 429–443.

9. R. O. Duda, P. E. Hart, D. G. Stork, *Pattern Classification*, 2nd edn., John Wiley & Sons, New York, 2001.

10. R. A. Fisher, The use of multiple measurements in taxonomic problems, *Ann. Eugen.*, 7 (1936) 179–188.

11. G. J. McLachlan, *Discriminant Analysis and Statistical Pattern Recognition*, John Wiley & Sons, New York, 1992.

12. G. W. Brown, Discriminant functions, *Ann. Math. Statist.*, 18 (1947) 514–528.

13. S. Wold, H. Martens, H. Wold, The multivariate calibration problem in chemistry solved by the PLS method. In: A. Ruhe, B. Kågström (Eds.), *Proceedings of Conference on Matrix Pencils*, Lecture Notes in Mathematics, Springer Verlag, Heidelberg, Germany, 1983, pp. 286–293.

14. S. Wold, C. Albano, W. Dunn III, U. Edlund, K. Esbensen, P. Geladi, S. Hellberg, E. Johansson, W. Lindberg, M. Sjöström, Multivariate data analysis in chemistry. In: B. R. Kowalski (Ed.), *Chemometrics: Mathematics and Statistics in Chemistry*, NATO ASI Series C, vol. 138, Reidel Publishing Company, Dordrecht, the Netherlands, 1984, pp. 17–95.

15. M. Sjöström, S. Wold, B. Söderström, PLS discriminant plots. In: E. S. Gelsema, L. N. Kanal (Eds.), *Pattern Recognition in Practice II*, Elsevier, Amsterdam, the Netherlands, 1986, pp. 461–470.

16. M. Barker, W. Rayens, Partial least squares for discrimination, *J. Chemometr.*, 17 (2003) 166–173.

17. L. Ståhle, S. Wold, Partial least squares analysis with cross-validation for the two-class problem: A Monte Carlo study, *J. Chemometr.*, 1 (1987) 185–196.

18. U. G. Indahl, H. Martens, T. Naes, From dummy regression to prior probabilities in PLS-DA, *J. Chemometr.*, 21 (2007) 529–536.

19. H. Nocairi, E. M. Qannari, E. Vigneau, D. Bertrand, Discrimination on latent components with respect to patterns. Application to multicollinear data, *Comput. Stat. Data Anal.*, 48 (2004) 139–147.

20. V. N. Vapnik, *The Nature of Statistical Learning Theory*, 2nd edn., Springer, Berlin, Germany, 1995.

21. B. Walczak, D. L. Massart, Application of radial basis functions—Partial least squares to non-linear pattern recognition problems: Diagnosis of process faults, *Anal. Chim. Acta*, 331 (1996) 187–193.

22. L. Breiman, J. H. Friedman, R. A. Olhsen, C. J. Stone, *Classification and Regression Trees*, Wadsworth International Group, Belmont, CA, 1984.

23. L. Breiman, Random forests, *Machine Learning*, 45 (2001) 5–32.

24. T. M. Cover, P. E. Hart, Nearest neighbor pattern classification, *IEEE Trans. Inf. Theory*, 13 (1967) 21–27.

25. D. Coomans, D. L. Massart, Alternative k-nearest neighbour rules in supervised pattern recognition. Part 1. k-Nearest neighbour classification by using alternative voting rules, *Anal. Chim. Acta*, 136 (1982) 15–27.

26. K. Beyer, J. Goldstein, R. Ramakrishnan, U. Shaft, When is "nearest neighbor" meaningful? In: C. Beeri, P. Buneman (Eds.), *Database Theory—ICDT'99*, Lecture Notes in Computer Science 1540, Springer Verlag, Heidelberg, Germany, 1998, pp. 217–235.

27. R. Bro, Multi-way calibration. Multilinear PLS, *J. Chemometr.*, 10 (1997) 47–61.

28. M. Forina, P. Oliveri, S. Lanteri, M. Casale, Class-modeling techniques, classic and new, for old and new problems, *Chemometr. Intell. Lab. Syst.*, 93 (2008) 132–148.

29. P. Oliveri, G. Downey, Multivariate class modeling for the verification of food-authenticity claims, *Trends Anal. Chem.*, 35 (2012) 74–86.

30. S. Wold, Pattern recognition by means of disjoint principal components models, *Pattern Recogn.*, 8 (1976) 127–139.
31. S. Wold, M. Sjöström, SIMCA: A method for analyzing chemical data in terms of similarity and analogy. In: B. Kowalski (Ed.), *Chemometrics: Theory and Application*, ACS Symposium Series, vol. 52, American Chemical Society, Washington, DC, 1977, pp. 243–282.
32. H. H. Yue, S. J. Qin, Reconstruction-based fault identification, *Ind. Eng. Chem. Res.*, 40 (2001) 4403–4414.
33. C. Durante, R. Bro, M. Cocchi, A classification tool for N-way array based on SIMCA methodology, *Chemometr. Intell. Lab. Syst.*, 106 (2011) 73–85.

16 Nonlinear Supervised Techniques

Geert Postma, Lionel Blanchet, Frederik-Jan van Schooten, and Lutgarde Buydens

CONTENTS

16.1 INTRODUCTION

Classifiers such as linear discriminant analysis (LDA)* assume a linear separation boundary between the classes. This does not necessarily need to be the case. The boundary between the classes can be curved or can have an even more complex shape. One of the reasons for this could be different data distributions underlying each class with different variances of and covariances between the variables. Ideally, a transformation should be able to turn the problem into a linear one, for example, a log transform. Alternatively, some classifiers can cope with this, for example, quadratic discriminant analysis (QDA), given that the number of samples is larger than the number of variables. This type of data or data with more complex separation boundaries can also be targeted by k-nearest neighbor (KNN), which models the separation using the nearest neighbors of a reference data set with known class information.

There are other classifiers that are specifically suited to nonlinear separation boundaries between classes (although most of them also perform very well in case of linear separation boundaries). These are, among others, neural networks (NN), kernel-based classifiers, such as support vector machines (SVM) and kernel partial least squares (KPLS), and random forests (RF). Especially kernel-based classifiers can handle large numbers of variables, a property that is frequently associated with chromatographic data.

---

* See Chapter 15.

16.2 KERNELS AND KERNEL-BASED CLASSIFIERS

Kernel-based classifiers are based on the principle that by applying some kind of transformation of the data, a new data space is created in which the nonlinear separation boundary is transformed into a linear one. In this new data space, a linear classifier is applied (see Figure 16.1).

A possible transformation function could be a polynomial transformation. For instance, if we apply a polynomial transformation of degree 2 on a sample $\mathbf{x}_i$ with two variables: $x_{i,1}$ and $x_{i,2}$, this will result in five new variables: $x_{i,1}, x_{i,2}, x_{i,1} * x_{i,2}, x_{i,1}^2$, and $x_{i,2}^2$; the data are mapped from a two-dimensional feature space to a five-dimensional feature space. However, for high-dimensional data, this will result in an explosion in a number of new variables. Overfitting becomes then more likely: far more variables are to be fitted than there are data points, possibly resulting in an unreliable model [1]. Although methods such as PLS* are able to handle such situations, there are also alternative approaches to solve this issue.

One of the options is to take the inner-product of the transformed data [2]. This will result in a so-called kernel matrix of size $n \times n$ (n is the number of samples in the data set). In formula:

$$K(\mathbf{x}_i, \mathbf{x}_j) = \theta(\mathbf{x}_i) \cdot \theta(\mathbf{x}_j) \tag{16.1}$$

in which $\theta(\mathbf{x}_i)$ represents the transformation function on $\mathbf{x}_i$. The most simple kernel is the linear kernel, being the in-product of $\mathbf{x}_i$ and $\mathbf{x}_j$ ($\mathbf{x}_i \cdot \mathbf{x}_j$, i.e., no transformation on $\mathbf{x}_i$ and $\mathbf{x}_j$). Other well-known kernel functions are, for instance, the polynomial kernel of the degree q:

$$K(\mathbf{x}_i, \mathbf{x}_j) = (p + \mathbf{x}_i \cdot \mathbf{x}_j)^q \tag{16.2}$$

the Gaussian kernel (or radial basis kernel, RBF):

$$K(\mathbf{x}_i, \mathbf{x}_j) = \exp\left(-\|\mathbf{x}_i - \mathbf{x}_j\|^2 / 2\sigma^2\right) = \exp\left(-\gamma \|\mathbf{x}_i - \mathbf{x}_j\|^2\right) \tag{16.3}$$

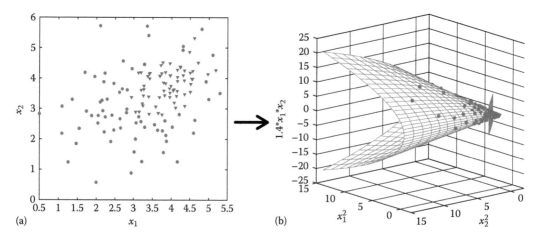

(a)   (b)

FIGURE 16.1 Example of a kernel-like data transformation. The blue triangles represent data from healthy individuals; the red circles are data from a diseased patient group. (a) Original data and (b) situation after a kernel-like data transformation. Due to the data transformation, a linear decision boundary between both classes can be constructed, as indicated.

---

* See Chapter 14.

Also, the Euclidean distance calculation, resulting in the distance matrix, can be seen as a kernel transformation. In fact, a kernel function $K(\mathbf{x}_i, \mathbf{x}_j)$ can exist without knowing the exact transformation function $\Theta(\mathbf{x}_i)$ on the data.

A drawback of the kernel transformation is that the original variables are gone, and this seriously hampers the interpretation of the classification model that is constructed using these kernels: it is not possible anymore to see which variables are important in the model. There is, however, a solution for this using so-called pseudo-samples. The transformation by a kernel function has been introduced in SVMs [2,3]. Later on they are introduced in other algorithms, such as PLS [4–6] and kernel Fisher discrimination [7].

16.2.1 Support Vector Machines

Suppose we have a two-class classification problem. For this, a set of n training samples $(\mathbf{x}_i,$ $i: 1 \dots n)$ is required of which we know the class, coded with -1 and 1, respectively (stored in the $\mathbf{y}$ vector). Using these samples, we need to construct a model, a classification function, with which future samples with unknown class can be classified.* In the simplest case, the samples of both classes are nicely separated; see Figure 16.2 for two variables. Then a decision line (in case of two variables; a plane or hyperplane in case of three or more variables) can be drawn between both classes, and this line can be used as classification function: samples on the left side of the line are classified as belonging to the -1 class, samples on the other side as belonging to the class coded with ones.

The question is: what is the most optimal line? SVM uses the following principle: look for a line or hyperplane, so that the distance to the nearest samples of either class is the largest. In other words, look for a line with a margin as large as possible around it. This margin can be defined by parallel lines or hyperplanes on either side, touching the nearest samples. These samples are called support vectors. In formulas, the decision hyperplane can be defined as

$$\mathbf{wx} + \mathbf{b} = 0 \tag{16.4}$$

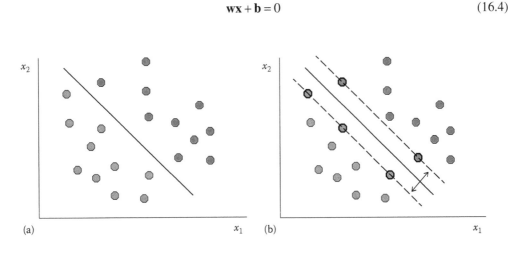

(a)   (b)

FIGURE 16.2   An example of a linearly separable classification problem of samples belonging to two classes (indicated with blue and red dots). (a) With a possible decision border, (b) the most ideal solution based on the SVM margin, indicated by the dashed lines. The black encircled dots indicate the support vectors. .

---

* See also Chapter 15.

in which $\mathbf{w}$ is a weight vector of the variables in $\mathbf{x}$ and $\mathbf{b}$ is an offset vector. The classification rules then are: if $\mathbf{wx}_i + \mathbf{b} > 0$, then sample $\mathbf{x}_i$ belongs to class 1 and if $\mathbf{wx}_i + \mathbf{b} < 0$, then sample $\mathbf{x}_i$ belongs to class -1. The supporting (hyper)planes on either side are defined by

$$\mathbf{wx} + \mathbf{b} = -1 \tag{16.5}$$

$$\mathbf{wx} + \mathbf{b} = +1 \tag{16.6}$$

It can be derived that the distance or margin between both supporting planes is

$$d = 2/\|\mathbf{w}\|_2 \tag{16.7}$$

in which $\|\mathbf{w}\|_2$ is the length of the weight vector $\mathbf{w}$. The goal is to maximize d or minimize $\|\mathbf{w}\|_2$ or its squared version $\|\mathbf{w}\|_2^2$ (multiplied with ½). This minimization has certain constraints: we want that the samples are correctly classified. In formula

$$y_i\left(\mathbf{wx}_i + \mathbf{b}\right) \geq 1 \quad \text{for } i = 1,\ldots,n \tag{16.8}$$

This minimization can be solved using Lagrange multipliers α_i ($\alpha_i \geq 0$):

$$L\left(\mathbf{w},b,\alpha\right) = \frac{1}{2}\|\mathbf{w}\|^2 - \sum_1^n \alpha_i\left(y_i\left(\mathbf{wx}_i + \mathbf{b}\right) - 1\right) \tag{16.9}$$

By taking the proper derivatives and resubstituting them in Formula 16.9, this results in the minimization with respect to α of

$$\frac{1}{2}\sum_{i=1}^n\sum_{j=1}^n y_i y_j \alpha_i \alpha_j \mathbf{x}_i \cdot \mathbf{x}_j - \sum_{i=1}^n \alpha_i \tag{16.10}$$

under the conditions $\alpha_i \geq 0$ and $\sum \alpha_i y_i = 0$. The solution vector α ($=(\alpha_1, \ldots, \alpha_n)$) can be obtained using quadratic programming. The derivative of (16.9) with respect to $\mathbf{w}$ leads to $\mathbf{w} = \sum y_i \alpha_i \mathbf{x}_i$, which, inserted in the classification rules, provides the classification function:

$$f\left(x\right) = \text{sign}\left(\sum_1^n \alpha_i y_i\left(\mathbf{x}_i \cdot \mathbf{x}\right) + \mathbf{b}\right) \tag{16.11}$$

in which *sign*() means the conversion of the sign of the outcome of the calculation between brackets to -1 or $+1$, indicating the class of $\mathbf{x}$. The nice properties of this solution are that it is unique and that the solution vector α has only nonzero values for those training samples in the data that act as support vectors: the solution is called sparse. And this solution is not in terms of the variables (as is $\mathbf{w}$ or a regression vector resulting from, e.g., PLS-DA*) but in terms of the samples, which can be an advantage from a computational point of view, as frequently (with chromatographic data) the number of samples is lower than the number of variables. However, as stated earlier, using α the interpretation is seriously hampered.

The above-derived classification rule holds for the ideal case: both classes are nicely separated. In practice, this is frequently not the case (see Figure 16.3). This leads to a slightly modified goal

---

* See Chapter 15.

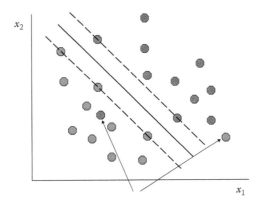

FIGURE 16.3   The nonideal case. The arrows indicate the samples at the wrong side of the decision line.

for the position of decision line: a margin as large as possible with an error as low as possible. The influence of the samples on the wrong side of the separation (hyper)plane is introduced in the formulas with a so-called slack variable ξ, multiplied by a constant C with which we can tune how much the erroneous samples are taken into account. Instead of minimization of $1/2\|\mathbf{w}\|_2^2$ the next formula is minimized:

$$\frac{1}{2}\|w\|^2 + C\sum_1^n \xi_i \quad \text{under the constraint } y_i\left(\mathbf{w}\cdot\mathbf{x}_i + \mathbf{b}\right) \geq 1 - \xi_i \quad \text{for } i = 1,\ldots,n \qquad (16.12)$$

The resulting classification function has the same form as Formula 16.11. However, it is influenced by the value of C, which has to be tuned. A too high value of C could force the algorithm to take all erroneous samples into account, resulting in a model that overfits the data: a model perfectly suited to the data with which the model is constructed but performing worse for new data.

As stated earlier, the classification rule, and so the model, is based on the in-product of the samples in $\mathbf{X}$. It is also possible to first apply a transformation function $\Theta(\mathbf{x})$ on $\mathbf{x}$, for example, in order to take some nonlinearity in the data into account. And this can further be generalized by applying a kernel transformation (see Section 16.1). The drawback of this procedure is that several choices have to be made: the C value, the type of kernel that is to be applied, and frequently an associated kernel parameter (the degree of the polynomial, the σ or γ of the RBF kernel, etc.). This requires an additional optimization step. A simple solution is to apply a grid search, by varying the values of the parameters with fixed steps (e.g., 5 or 10) and evaluating the performance of the resulting classifier at each of the value combinations. This makes the development of an SVM model time-consuming. However, once the model is developed, the class prediction for a new sample is straightforward and fast. As explained earlier, another drawback of the kernel transformation is the hampered interpretation of the model. There are procedures to solve this. Since the original development of SVM (also called C-SVM), several alternative implementations of SVM have been developed, such as nu-SVM [8] and least-squares SVM (LS-SVM) [9]. Their models can be calculated faster, by preventing the time-consuming quadratic programming. In general, the results are comparable, but, for example, LS-SVM holds that the solution is less sparse: the α vector hardly contains zeros [10].

The output of an SVM classifier is binary: yes or no. It is possible to obtain associated uncertainties (see, e.g., Luts et al. [10]). If the class sizes differ (strongly), then this should be accounted for. There are methods that use a bias term to correct for this, but according to Luts et al. this comes down to training the classifier using resamplings of the data with equal class sizes.

16.2.2 Kernel PLS-DA

As stated before, in (nonlinear) kernel PLS [5] and direct kernel PLS (DK-PLS) [6], a kernel transformation on $\mathbf{X}$ is applied by applying the kernel function on every sample combination of $\mathbf{X}$. The next step then is to use PLS on this kernel. First of all, the kernel needs to be mean-centered:

$$\mathbf{K}^c = \mathbf{K} - \bar{\mathbf{K}}_{.j} - \bar{\mathbf{K}}_{i.} + \bar{\mathbf{K}}_{ij} \quad i, j = 1\ldots n \tag{16.13}$$

in which
  $\mathbf{K}^c$ is the (double) mean-centered version of $\mathbf{K}$
  $\bar{\mathbf{K}}_{.j}$ is a matrix with column means
  $\bar{\mathbf{K}}_{i.}$ is a matrix with row means
  $\bar{\mathbf{K}}_{ij}$ is a matrix with the overall mean of $\mathbf{K}$ at each position

PLS is applied on $\mathbf{K}^c$ and the class vector $\mathbf{y}$ according to:

$$\mathbf{y}_{(n\times1)} = \mathbf{K}^c_{(n\times n)}\mathbf{b}^K_{(n\times1)} + \mathbf{f}_{(n\times1)} \tag{16.14}$$

in which
  $\mathbf{b}^K$ is the regression vector of size $n \times 1$ on the kernel $\mathbf{K}^c$ for the optimized number of latent variables
  $\mathbf{f}$ are the residuals

The class of new samples can be predicted by kernel transformation of these samples using the original data matrix $\mathbf{X}$, mean-centering the resulting kernel matrix (or vector) and multiplying it with the regression vector $\mathbf{b}^K$:

$$\mathbf{C}^c = \mathbf{C} - \bar{\mathbf{K}}_{.j} - \bar{\mathbf{C}}_{i.} + \bar{\mathbf{K}}_{ij} \quad i, j = 1\ldots n \tag{16.15}$$

$$\hat{\mathbf{y}}_{(p\times1)} = \mathbf{C}^c_{(p\times n)}\mathbf{b}^K_{(n\times1)} \quad \text{KPLS} \tag{16.16}$$

in which $\mathbf{C}$ is the kernel matrix calculated from the new samples using the original data matrix $\mathbf{X}$ and a certain kernel function, $\mathbf{C}^c$ is its mean-centered version, and $\hat{\mathbf{y}}$ contains the predicted y values for the p new samples. The predicted class for instance can be obtained by applying the sign function on $\hat{\mathbf{y}}$.

16.2.3 Retrieving the Variables in Kernel-Based Classifiers

Kernel-based classifiers such as SVM and KPLS-DA have nice properties such as high classification accuracies and the ability to handle nonlinear data, but one of the drawbacks is the interpretation. As stated earlier, due to the kernel transformation on the data the original variables are gone. If, for instance, a Euclidian distance calculation is used as kernel function, the new variables are the distances of each sample to all the other samples (including itself). And the regression vectors ($\mathbf{b}^K$ in case of PLS and $\boldsymbol{\alpha}$ in case of SVM) are in terms of these new variables. In case of normal PLS-DA, one would investigate the PLS regression vector $\mathbf{b}$ to see which variables are important for the model. This is not possible anymore with the kernel function. There is, however, a solution [11]. This solution is based on the approach described by Gower and Harding [12], using so-called pseudo-samples. As presented in [11], pseudo-samples carry all their weight in one variable. This means that these pseudo-samples have a value of 0 for all variables, except for one variable. For example, $[1, 0, 0, \ldots, 0]$ is a $(1 \times m)$ pseudo-sample with a value 1 for variable 1 and a value 0

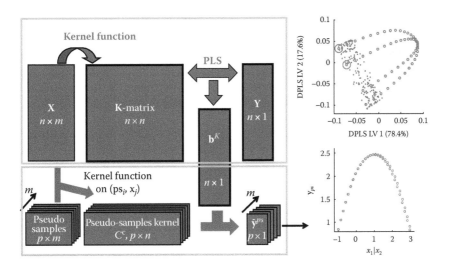

FIGURE 16.4 Procedure for the application pseudo-samples in case of KPLS(-DA). A similar procedure is used for SVM. In that case, instead of the regression vector $\mathbf{b}^K$ of PLS the $\boldsymbol{\alpha}$ vector of SVM is used. The example figures on the right concern a two-class classification problem in which the data of each class are coded with ones respectively twos. The upper-right figure is obtained by projecting the rows of the pseudo-sample kernels in the PLS latent variable space (by multiplying with the PLS first and second loading vectors). The data of each class are identified by red and light blue dots.

for all other variables. It is also possible to replace the value of 1 in the pseudo-sample by p different equidistant values z in the range of the variable. Then a total of p pseudo-samples are obtained:

$$\begin{bmatrix} z_1,0,0,\ldots,0 \\ z_2,0,0,\ldots,0 \\ z_3,0,0,\ldots,0 \\ \vdots \\ z_p,0,0,\ldots,0 \end{bmatrix}_{(p\times m)} \tag{16.17}$$

Using these pseudo-sample matrices for each variable as unknown samples, predicting their y values using the KPLS or SVM models, and plotting these y values (ones series for each variable), one can visualize the influence of each variable on the model. The procedure and two examples are presented in Figure 16.4. The examples show that the trajectories of pseudo-samples move from the first class to the second and back again, visualizing that the corresponding variables contribute to the discrimination of both classes. Another application of this procedure is presented in Smolinska et al. [13]. The GC-MS data set used in this chapter will be used throughout the chapter as an example. Briefly, it consists of the relative concentrations of 29 metabolites in 38 cerebrospinal fluid samples from rats. The cohort is here divided into two groups: clinically isolated syndrome of demyelination and multiple sclerosis (MScl).

This methodology also works for NNs [14] and RFs [15].

16.2.4 PARAMETER OPTIMIZATION AND VALIDATION

As stated earlier, kernel-based techniques require some optimization of the parameters. One could start with the frequently used RBF kernel, but there are more kernel types. This means that the kernel type needs to be selected and also the associated parameter (e.g., σ of the RBF kernel or the degree of the polynomial kernel). Also the parameter of the classification method itself needs to be optimized (the value of C of SVM or the number of latent variables of PLS). This optimization (using, for instance, a grid search) needs to be performed in a proper validation setting (see also Chapter 14 on classification and validation). This means that during the optimization, the performance of the classifier for each combination of parameters and kernel type needs to be calculated using, for example, a cross-validation. Once the choice of the best performing combination of parameters and kernel type is made, this choice again needs to be validated using an independent (not used) test set of data or using an outer cross-validation.

Wang et al. [16] claim that in case of SVM, the optimal value for σ of the RBF kernel can be found using Fisher LDA, at least for data sets with little noise. In general, the value for σ should be in the range of the distances between the data points. Liu et al. [17] elaborated this approach in a slightly different way and tested it on 17 different data sets. Their approach performed almost equally well as grid search, requiring a fraction of the calculation time of the latter. Luts et al. [10] provide a set of guidelines on the training and use of SVM.

16.2.5   EXAMPLES, KPLS, SVM

There are a lot of applications of kernel-based techniques for the prediction of retention indices of series of compounds on certain LC or GC columns (quantitative structure retention studies; see for instance [18]). Also RFs are used for this purpose [19]. These applications are based on regression. Remember that PLS, and so KPLS, originally is a regression technique; there is also a regression version of SVM called support vector regression. Tang et al. [20] applied a version of KPLS for the prediction of five radiation dose classes (from no radiation to very severe radiation) using LC-MS measurements of amino acids in plasma. They obtained an average accuracy of 94% at 72 h after radiation exposure. In fact, they also performed a regression, predicting the actual radiation dose after which they grouped the predicted dose values into the five radiation classes. Martelo-Vidal and Vázquez [21] classified red wines from two Spanish Designation of Origin using their phenolic

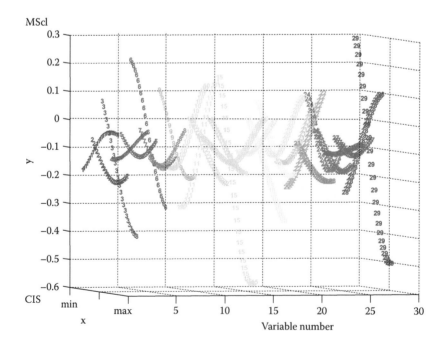

FIGURE 16.5   Pseudo-sample trajectories of 29 selected GC-MS variables based on a KPLS model (RBF kernel) for the discrimination of clinically isolated syndrome of demyelination (CIS) and MScl in rats. (See also Smolinska et al. [13].)

compound profiles, which were measured with reversed-phase HPLC using diode array detection. Using SVM with an RBF kernel they obtained 100% classification accuracy, whereas a linear kernel resulted in 73% accuracy. Other linear techniques resulted in 76% (Fisher LDA) and with 84% accuracy (QDA).

Smolinska et al. [13] applied SVM and KPLS on GC-MS data in combination with proton nuclear magnetic resonance (^{1}H-NMR) data for a discrimination (classification) of patients diagnosed with MScl from patients who are in an early stage of MScl. They introduced an advanced procedure in which they used an SVM-based procedure for the selection of the most relevant metabolites from the original data sets, constructed kernels using these reduced variable sets, and then intelligently combined these kernels, after which PLS-DA was applied. Using the selected variables of only the GC-MS data, already an accuracy of 85% was obtained (using a linear classifier classification was not possible). After combining the kernels, this increased to 89%. Using the variable visualization procedure, described earlier, the most important variables in the final model could be visualized, as well as how they nonlinearly contributed to the model (e.g., either increased or decreased or even showed more complex behavior). But even using the GC-MS data alone this is possible. This is visualized in Figure 16.5. This figure is based on 29 selected GC-MS variables described in the paper. Clearly, it can be seen that some variables increase for MScl, whereas others strongly decrease (e.g., variables 3, 6, 15, and 29) or hardly show an effect (e.g., variables 4 and 26).

16.3   RANDOM FOREST

16.3.1   WHAT IS A TREE FOR A DATA ANALYST?

The concept of classification is in essence very simple and can be generally formulated in equally simple questions. Are you a man or a woman? Is this fruit an apple or a pear? Is my salary high enough? Such questions can be formulated on binary variables (e.g., gender) or based on a continuous scale (e.g., your salary). In the latter case, a threshold needs to be defined to split the different options into two classes. A classification tree [22,23] follows the same procedure and repeats it in a hierarchical fashion using multiple variables. Each question is denoted as a node. Each node can be answered in two ways leading to the next two questions: the child nodes. Translated into the world of chromatography, a classification tree will use the different peak areas (or intensities) to decide if a given sample belongs either to one class or another. Figure 16.6 presents a toy example.

Starting from the bottom of the tree, a sample is examined according to the first question: Is the peak of ethanol larger than 40 a.u. of area? If yes, does it contain resveratrol? If no, does it contain some fructose? Following a complete path until the end of the tree leads the sample into a "leave," which corresponds to a specific class, that is, allows predicting the class of the sample. Note that multiple leaves can correspond to the same classes.

The concept of a classification tree is rather intuitive; the challenge is to construct such tree on a real problem. Classification trees are, as other supervised methods, constructed through a training phase. The main objective is to define a set of thresholds on specific variables that allow separating the global population in subgroups corresponding to the classes of interest. The most common approach is to use the Gini impurity index $G(S)$ and the information gain (IG). In classification, with an outcome class $k = 1, 2, ..., C$, the Gini impurity index [24] is defined as follows:

$$G(S) = 1 - \sum \pi_k^2 \qquad (16.18)$$

where π_i is the relative frequency of class k in S, and where S is the node being evaluated. Pure nodes will have a very low impurity; nodes containing multiple classes will lead to high impurity.

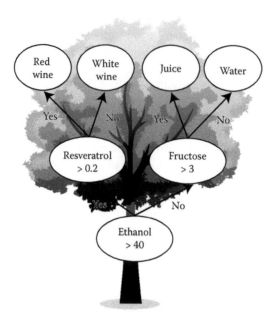

FIGURE 16.6   Schematic example of a classification tree.

Differences in the Gini information indices between the parent node and the child nodes (weighted by their probabilities) are termed *information gain*:

$$IG = G_P - \frac{\text{True}}{\text{Total}} G_T - \frac{\text{False}}{\text{Total}} G_F \qquad (16.19)$$

where
   G_P, G_T, and G_F are the Gini impurity indices of the parent and children nodes
   True is the number of samples passing the current question
   False is the number of samples failing the test
   Total is the total number of samples

The largest the IG will indicate what is the most relevant question to ask, that is, which variable and which threshold should be used to split the population.

   Figure 16.7 illustrates how the Gini impurity is used to evaluate all possible splits. This simple data set consists of three classes of 10 samples described by 2 variables. The impurity of the parent node, the complete data, is equal to 0.67. In Figure 16.7a, three possible splits on variable 1 (ethanol in our previous example) are evaluated. Clearly, the split at 40 divides the population in a more meaningful way; the corresponding *IG* is 0.33. However, it is also possible to use variable 2 to split the population. Figure 16.7b represents three possible splits. The best split, in that case, is around 30 with an *IG* of 0.22, which is still lower than the one obtained in variable 1. Hence, the best question to ask here is if a sample has a value superior to 40 on variable 1. Note that this split already generates a terminal node (a leaf) where all samples belong to the same class (circular points). On the next node, variable 2 will be the most informative and will permit separating the squares from the triangles. In practice, the number of split points to evaluate is much larger—each midpoint between two values observed in the data is generally considered as a potential split point; hence, data consisting of 100 samples lead to 98 possible split points for each variable.

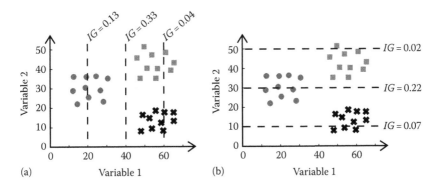

(a)   Variable 1   (b)   Variable 1

FIGURE 16.7   Illustration of the use of the Gini impurity index and of the *IG* to select the most optimal split in the construct of a classification tree. (a) Three examples of possible splits with associated IG on variable 1 and (b) on variable 2.

16.3.2   FROM A TREE TO A FOREST

Classification trees are relatively simple to build, straightforward to apply, and easy to interpret. However, they suffer from a poor predictive power. The main reason is that classification trees tend to overfit the training data; they only describe the specificities of these data and cannot be generalized to other samples. Breiman introduced an adaptation of the classification tree working around this defect [25]. The general idea is to construct hundreds of classification trees on different random subsets of the training data. Taking these trees altogether, the chance for overfitting is drastically reduced. This method is logically called random forest.

One central element in this procedure is the generation of data subsets. A bootstrapping procedure [26], with replacement, selects *n* samples to train a tree. The training set for any given tree is usually set to represent 67% of the complete data, whereas 33% is left out in the so-called out-of-bag (OOB) set to test the performance of the classification tree, as represented in Figure 16.8. As indicated earlier, this procedure is repeated hundreds of times; the error of prediction measured on the OOB samples is aggregated to estimate the overall error of prediction of the forest. The quality of the RF model can be assessed looking at the evolution of the OOB error in function of the number of trees. If the data contain relevant information, the OOB error should progressively reduce as the number of trees increases, since the chance of overfitting is getting smaller, until it reaches a plateau. Note that because of the bootstrapping procedure and the multitude of trees used in RF the risk of overfitting is very limited, hence pruning is not applied to the decision trees.

16.3.3   HOW TO PREDICT?

The ultimate goal of a random forest is to predict to class labeling for new samples. To do so, a new sample has to go through each tree forming the forest. Obviously, the decisions taken by different trees might differ. The final classification is a consensus of all trees, but one can also look at the distributions of the decisions as a form of probability of membership to each class.

16.3.4   HOW TO VISUALIZE AND INTERPRET A RANDOM FOREST MODEL

As mentioned earlier, prediction involves using every tree of the forest; additional information can be extracted during that procedure facilitating a visualization of the model. A proximity matrix can be calculated by evaluating the number of trees in which two samples are put in the same terminal node. Two samples in the same terminal node are seen as similar by the tree. If all trees lead to the same conclusion, the similarity between the two samples must be very high. The proximity matrix summarizes this

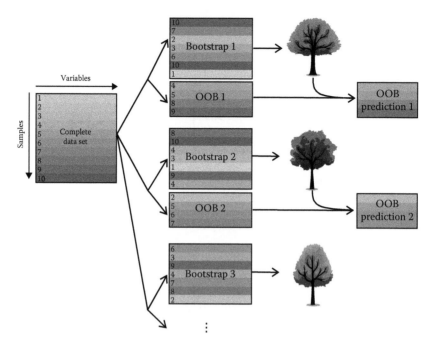

FIGURE 16.8   Overview of the bootstrapping procedure underlying random forest.

information in a matrix of dimension n by n (where n is the number of samples). The proximity values correspond to the co-occurrences in a terminal node divided by the number of trees in the forest, and are therefore bounded between zero and one. The proximity matrix can, therefore, be seen as a similarity matrix. A dissimilarity matrix can straightforwardly be obtained (by calculating $1 - $ proximity) and used as input in a multidimensional scaling analysis (MDS) [27] or a classical PCA.

For comparison purpose, the GC-MS data mentioned in the K-PLS section is re-analyzed using RF. Figure 16.9 illustrates the visualizations associated with an RF model. In Figure 16.9a, the score plot obtained via MDS on the proximity matrix clearly shows two separate groups. The projection of the test sets is consistent with the training set and represents adequately the performance obtained on the test set: 90% correct prediction rate.

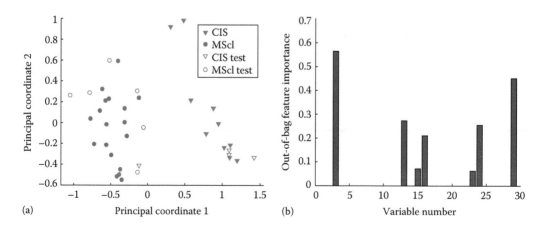

FIGURE 16.9   (a) Score plot obtained by MDS on the proximity matrix of RF of the GCMS data (empty symbols represent the test samples). (b) Variable importance of the RF model (after variable selection).

The interpretation of an RF model requires evaluating the importance of the variables. The most usual procedure is to evaluate the change in mean of the error of a tree (or misclassification rate) when the values for these variables are randomly permuted in the OOB samples [28]. If the variable is important, a permuted version of the values seen in the OOB samples should increase the error. The resulting variable importance values can be used to rank the variables. Note that negative values are possible here—they indicate that the corresponding variable is irrelevant to the model.

Figure 16.9b demonstrates that RF can also be used to interpret the results from a chemical point of view. Indeed, the variable importance can be estimated showing here that a handful of metabolites are sufficient to produce a good classifier.

16.3.5 APPLICATIONS OF RF IN CHROMATOGRAPHY

Especially in the last 10 years, there are an increasing number of publications on the application of random forests on chromatographic data. For instance, Lin et al. demonstrate the usefulness of RF on metabolomics data—three compounds were associated with a metabolic syndrome. Interestingly, all three compounds are part of the same pathway [29]. Li et al. [30] applied RF for the discrimination of dogs suffering from canine degenerative mitral valve disease (the most common form of heart disease in dogs) and healthy age-matched and gender-matched dogs using metabolomic GC-MS and LC-MS data obtained from serum. The "out of bag" error rate was 20.69%. The RF analysis identified 30 metabolites that contributed significantly to the model. Mahdavi et al. [31] applied RF to the classification of rice treated with an insecticide versus untreated rice using GC-MS data in order to identify metabolic changes in the rice due to the insecticide. The classification accuracy was 93% on an independent test set. Ai et al. demonstrate the applicability of RF clustering (i.e., unsupervised) to the analysis of the fatty acid composition of edible oils obtained by GC-MS [32]. Chen et al. [33] compared RF with SVM with linear kernel, PLS, and LDA on a GC-MS urine data set having 187 peaks (representing 35 identified and 152 unidentified metabolites). RF always performed best in repeated cross-validations with a far lower standard deviation. The latter is logical, as RF itself is already based on a large set of classifiers. However, the comparison seems not completely fair as apart from RF the other classifiers are all linear classifiers.

16.4 FINAL REMARKS

The question could be: Should one use this kind of nonlinear techniques? For example, Argyris et al. [34] state: "It is a general admission that if the number of features in an examined data set is large one may not need to map data to a higher dimensional space. Respectively, the non-linear mapping … does not significantly—or at all—improve the final performance. Thus, using the linear kernel may prove to be good enough, by adjusting C." However, in their paper, there is no reason to suggest a nonlinear relation. An example of this situation could be present in Guan et al. [35]. The results on the negative-mode LC-MS data using SVM with a polynomial kernel were slightly worse than those using a linear kernel. Even better or similar results were obtained by classifying the published data using just PLS-DA. Luts et al. [10] state in their guidelines "Consider the use of a linear kernel if the number of variables is far larger than the number of samples." However, there are also a lot of publications claiming the advantage of KPLS and SVM using nonlinear kernels. Another advantage of the use of kernels is that they are capable of modeling data sets with interacting variables and these interactions can be investigated (see Engel et al. [36]).

The basic SVM, using a linear kernel, is widely known to be an excellent linear classification technique, which frequently outperforms PLS-DA (e.g., Mahadevan et al. [37]). Another advantage of SVM is that it is not sensitive to outliers (a property that can easily be explained from the fact that it only uses a subset of the samples to define the separation hyperplane).

Menze et al. [38] compared RF with PLS-DA using NMR spectral data. They concluded that RF was better for variable selection; PLS-DA had a better classification performance. Gromski et al.

[39] lists the pros and cons of PLS-DA, SVM, and RF. All can handle noisy data, and are robust to overfitting; RF is good in handling thousands of variables and missing values, is robust to overtraining, and no preprocessing of the data is required. RF and SVM are robust to outliers. Additionally, Amaratunga et al. [40] states that RFs automatically compensate for different class sizes (which could indeed negatively influence the classification outcome of SVMs).

The ability of RF to handle nonnumerical data offers the possibility to include additional (nonchromatographic) information in the analysis, for example, a code corresponding to the sampling sites.

ACKNOWLEDGMENT

Parts of the figures are based on designs from Freepik.com.

REFERENCES

1. J.A. Westerhuis, H.C.J. Hoefsloot, S. Smit, D.J. Vis, A.K. Smilde, E.J.J.v. Velzen, J.P.M.v. Duijnhoven, F.A.v. Dorsten, Assessment of PLSDA cross validation, *Metabolomics*, 4 (2008) 81–89.
2. B.E. Boser, I.M. Guyon, V.N. Vapnik, A training algorithm for optimal margin classifiers, *Proceedings of the Fifth Annual Workshop on Computational Learning Theory, COLT'92*, Pittsburgh, Pennsylvania, ACM, New York, 1992, pp. 144–152.
3. V. Vapnik, *The Nature of Statistical Learning Theory*, Springer, New York, 1995.
4. B. Walczack, D.L. Massart, The radial basis functions—Partial least squares approach as a flexible non-linear regression technique, *Analytica Chimica Acta*, 331 (1996) 177–185.
5. R. Rosipal, L.J. Trejo, Kernel partial least squares regression in reproducing kernel Hilbert space, *Journal of Machine Learning Research*, 2 (2001) 97–123.
6. K.P. Bennett, M.J. Embrechts, An optimization perspective on kernel partial least squares regression, in: J.A.K. Suykens, G. Horvath, S. Basu, C. Micchelli, J. Vandewalle (Eds.) *Advances in Learning Theory: Methods, Models and Applications*, IOS Press, Amsterdam, the Netherlands, 2003, p. 227.
7. S. Mika, G. Rätsch, J. Wetson, B. Schölkopf, K.R. Müller, Fisher discriminant analysis with kernels, *NNSP'99*, Madison, WI, 1999, pp. 41–48.
8. B. Schölkopf, A.J. Smola, R.C. Williamson, P.L. Bartlett, New support vector algorithms, *Neural Computation*, 12 (2000) 1207–1245.
9. J.A.K. Suykens, J. Vandewalle, Least squares support vector machine classifiers, *Neural Processing Letters*, 9 (1999) 293–300.
10. J. Luts, F. Ojeda, R.V.d. Plas, B.D. Moor, S.V. Huffel, J.A.K. Suykens, A tutorial on support vector machine-based methods for classification problems in chemometrics, *Analytica Chimica Acta*, 665 (2010) 129–145.
11. G.J. Postma, P.W.T. Krooshof, L.M.C. Buydens, Opening the kernel of kernel partial least squares and support vector machines, *Analytica Chimica Acta*, 705 (2011) 123–134.
12. J.C. Gower, S.A. Harding, Nonlinear biplots, *Biometrika*, 75 (1988) 445–455.
13. A. Smolinska, L. Blanchet, L. Coulier, K.A.M. Ampt, T. Luider, R.Q. Hintzen, S.S. Wijmenga, L.M.C. Buydens, Interpretation and visualization of non-linear data fusion in kernel space: Study on metabolomic characterization of progression of multiple sclerosis, *PLoS ONE*, 7 (2012) e38163.
14. R. Kewley, M. Embrechts, C. Breneman, Data strip mining for the virtual design of pharmaceuticals with neural networks, *IEEE Transactions on Neural Networks*, 11 (2000) 668–679.
15. P. Cortez, M.J. Embrechts, Using sensitivity analysis and visualization techniques to open black box data mining models, *Information Sciences*, 225 (2013) 1–17.
16. W. Wang, Z. Xu, W. Lu, X. Zhang, Determination of the spread parameter in the Gaussian kernel for classification and regression, *Neurocomputing*, 55 (2003) 643–663.
17. Z. Liu, M.J. Zuo, X. Zhao, H. Xu, An analytical approach to fast parameter selection of Gaussian RBF kernel for support vector machine, *Journal of Information Science and Engineering*, 31 (2015) 691–710.
18. T. Cserháti, M. Szőgyi, Application of multivariate regression models in chromatography. New advances, *European Chemical Bulletin*, 1 (2012) 274–279.
19. M. Cao, K. Fraser, J. Huege, T. Featonby, S. Rasmussen, C. Jones, Predicting retention time in hydrophilic interaction liquid chromatography mass spectrometry and its use for peak annotation in metabolomics, *Metabolomics*, 11 (2015) 696–706.

20. X. Tang, M. Zheng, Y. Zhang, S. Fan, C. Wang, Estimation value of plasma amino acid target analysis to the acute radiation injury early triage in the rat model, *Metabolomics*, 9 (2013) 853–863.

21. M.J. Martelo-Vidal, M. Vázquez, Polyphenolic profile of red wines for the discrimination of controlled designation of origin, *Food Analytical Methods*, 9 (2016) 332–341.

22. G. De'ath, K.E. Fabricius, Classification and regression trees: A powerful yet simple technique for ecological data analysis, *Ecology*, 81 (2000) 3178–3192.

23. L. Breiman, *Classification and Regression Trees*, Wadsworth International Group, Belmont, CA, 1984.

24. G.K. Gupta, *Introduction to Data Mining with Case Studies*, PHI Learning Private Limited, Delhi, India, 2014, p. 734.

25. L. Breiman, Random forests, *Machine Learning*, 45 (2001) 5–32.

26. R. Wehrens, H. Putter, L.M.C. Buydens, The bootstrap: A tutorial, *Chemometrics and Intelligent Laboratory Systems*, 54 (2000) 35–52.

27. H. Abdi, Metric multidimensional scaling, in: N.J. Salkind (ed.), *Encyclopedia of Measurement and Statistics*, Sage Publications, Inc., Thousand Oaks, CA, 2007, 598–605.

28. R. Genuer, J.M. Poggi, C. Tuleau-Malot, Variable selection using random forests, *Pattern Recognition Letters*, 31 (2010) 2225–2236.

29. Z. Lin, C.M. Vicente Gonçalves, L. Dai, H.-m. Lu, J.-h. Huang, H. Ji, D.-s. Wang, L.-z. Yi, Y.-z. Liang, Exploring metabolic syndrome serum profiling based on gas chromatography mass spectrometry and random forest models, *Analytica Chimica Acta*, 827 (2014) 22–27.

30. Q. Li, L.M. Freeman, J.E. Rush, G.S. Huggins, A.D. Kennedy, J.A. Labuda, D.P. Laflamme, S.S. Hannah, Veterinary medicine and multi-omics research for future nutrition targets: Metabolomics and transcriptomics of the common degenerative mitral valve disease in dogs, *OMICS: A Journal of Integrative Biology*, 19 (2015) 461–470.

31. V. Mahdavi, M.M. Farimani, F. Fathi, A. Ghassempour, A targeted metabolomics approach toward understanding metabolic variations in rice under pesticide stress, *Analytical Biochemistry*, 478 (2015) 65–72.

32. F.-f. Ai, J. Bin, Z.-m. Zhang, J.-h. Huang, J.-b. Wang, Y.-z. Liang, L. Yu, Z.-y. Yang, Application of random forests to select premium quality vegetable oils by their fatty acid composition, *Food Chemistry*, 143 (2014) 472–478.

33. T. Chen, Y. Cao, Y. Zhang, J. Liu, Y. Bao, C. Wang, W. Jia, A. Zhao, Random forest in clinical metabolomics for phenotypic discrimination and biomarker selection, *Evidence-Based Complementary and Alternative Medicine (eCAM)*, 2013 (2013) 298183.

34. A. Argyris, J.-J. Filippi, D. Syvridis, Support vector machine classification of volatile organic compounds based on narrowband spectroscopic data, *Journal of Chemometrics*, 29 (2015) 38–48.

35. W. Guan, M. Zhou, C.Y. Hampton, B.B. Benigno, L.D. Walker, A. Gray, J.F. McDonald, F.M. Fernández, Ovarian cancer detection from metabolomic liquid chromatography/mass spectrometry data by support vector machines, *BMC Bioinformatics*, 10 (2009) 259.

36. J. Engel, G.J. Postma, I.v. Peufflik, L. Blanchet, L.M.C. Buydens, Pseudo-sample trajectories for variable interaction detection in Dissimilarity Partial Least Squares, *Chemometrics and Intelligent Laboratory Systems*, 146 (2015) 89–101.

37. S. Mahadevan, S.L. Shah, T.J. Marrie, C.M. Slupsky, Analysis of metabolomics data using support vector machines, *Analytical Chemistry*, 80 (2008) 7562–7570.

38. B.H. Menze, B.M. Kelm, R. Masuch, U. Himmelreich, P. Bachert, W. Petrich, F.A. Hamprecht, A comparison of random forest and its Gini importance with standard chemometric methods for the feature selection and classification of spectral data, *BMC Bioinformatics*, 10 (2009) 213.

39. P.S. Gromski, H. Muhamadali, D.I. Ellis, Y. Xu, E. Correa, M.L. Turner, R. Goodacre, A tutorial review: Metabolomics and partial least squares-discriminant analysis—A marriage of convenience or a shotgun wedding, *Analytica Chimica Acta*, 879 (2015) 10–23.

40. D. Amaratunga, J. Cabrera, Y.-S. Lee, Enriched random forests, *Bioinformatics*, 24 (2008) 2010–2014.

Section V

Retention Modeling

17 Introduction to Quantitative Structure–Retention Relationships

Krzesimir Ciura, Piotr Kawczak,
Joanna Nowakowska, and Tomasz Bączek

CONTENTS

17.1 QUANTITATIVE STRUCTURE–RETENTION RELATIONSHIPS

Quantitative structure–chromatographic retention relationships (QSRRs) depend on the chromatographic retention of the analyte including the chemical structure and physicochemical properties of stationary and mobile phases in liquid chromatography [1–4].

QSRRs represent the statistically determined correlation between chromatographic parameters and values (descriptors) that characterize the structure of the examined analytes. QSRRs are used in the description of molecular mechanisms responsible for the chromatographic separation system; they also evaluate the complex physical and chemical properties of the analytes and the chromatographic stationary phase in the prediction of chromatographic retention of new structures (Figure 17.1).

In order to analyze the QSRR, two sets of data are collected: a set of parameters describing the retention of the analyte series and a set of structural parameters of separated molecules (descriptors).

For the data retention parameter, the data most commonly used are the logarithm of the retention factor ($\log k$). Alternatively, $\log k$ can be extrapolated to pure water as a hypothetical mobile phase ($\log k_w$), in order to clearly distinguish between the hydrophobic properties of the analytes. Less commonly used is retention time (t_R). The descriptors reflecting the physical and chemical properties of analytes can be obtained experimentally or by calculation methods.

The simplest approach is the regression retention QSRR (R_t) to the calculated value of the partition coefficient n-octanol/water ($C \log P$) [5–8]. This approach uses the following equation:

$$R_t = k_1 + k_2 C \log P \tag{17.1}$$

where k_1 and k_2 are the coefficients of the regression equation.

In the other QSRR model, retention depends on the specific descriptors obtained using molecular modeling. The equation used differentiates stationary phases in liquid chromatography by the

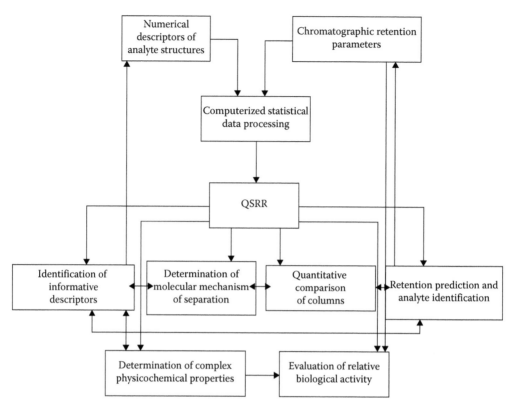

FIGURE 17.1 Quantitative structure–activity relationship methodology and objectives. (Based on Kaliszan, R., *Chem. Rev.*, 107, 3212, 2007.)

chemical properties of the carrier and the ligand. It describes the mechanism of interaction of the analyte with a bed of studied column.

On the basis of the work by Bączek and colleagues [9,10], it was demonstrated that the use of following descriptors provides reliable results:

μ—Total dipole moment, which describes the impact of dipole–dipole interactions between the analyte and the induced chromatographic mobile phase and the stationary phase

δ_{Min}—The biggest negative excess charge in the molecule, which specifies the local polarity of the analyte and the ability to participate in polar interactions

A_{WAS}—The space available in contact with the solvent dispersion, which characterizes the strength of interaction (such as London) between analyte particles and the chromatographic liquid phase

For such a model, an equation of the following form is used [11]:

$$\log k_w = k_1' + k_2'\mu + k_3'\delta_{Min} + k_4'A_{WAS}$$ (17.2)

where $k_1'k'$, $k_2'k'$, $k_3'k'$, and $k_4'k'$ are regression coefficients.

Another QSRR model is based on linear free energy relationships. The concept is extended from the atomic properties to the sphere of intermolecular interactions and is described as linear dependence of solvation energy (LSER), acting with a thermodynamic kind of dependence, which combines physical and chemical process models with the concepts of thermodynamics [1].

Suitable assumptions allow for the omission of the stringent laws of thermodynamics, assuming that there is an existing relationship between the models of chemical processes and the principles of thermodynamics without the derivation of mathematical dependencies [11].

This QSRR model assumes the existence of a linear relationship between the standard retention parameters and the change of the free energy associated with the separation in the chromatographic process [12]. According to this model, a chromatographic column can be considered as a free energy converter that converts the difference in chemical potential resulting from the structure of analytes into quantitative differences in the properties of the retention [1].

This approach has been introduced and developed by Abraham et al. [13], where the LSER general equation becomes as follows:

$$\log k = \log k_0 + rR_2 + vV_x + s\pi_2^H + a\sum\alpha_2^H + b\sum\beta_2^H \tag{17.3}$$

where
R_2 is an excess of molar refraction of the analyte
V_x is the volume of the McGowan's molecular algorithm
π_2^H is a descriptor bipolarity/polarizability
$\sum\alpha_2^H$ is a measure of transfer of a proton to the analyte
$\sum\beta_2^H$ is a similar measure to receive a proton
$\log k_0$ is a constant value
r, v, s, and b are regression coefficients constituting a set of complementary properties of the chromatographic system consisting of chromatographic stationary and mobile phases

Currently, QSRR is widely used for the comparison of chromatographic columns (stationary phase) [14].

17.2 QSRR ON TLC

QSRR has its origins in thin-layer chromatography (TLC) [15]. It is a powerful theoretical tool for the description and prediction of molecular systems in chromatographic research. The QSRR approach is still developing over the past few decades.

Numerous descriptors can be applied in QSRR studies, such as empirical, semiempirical, and nonempirical parameters, which quantitatively differentiate molecules. Among chromatographic parameters, the extrapolated value of R_M, called R_M^0 or in case of an aqueous mobile phase it could also be R_M^W, is commonly used [16]. For this calculation, another chromatographic parameter such as the slope of Soczewinski equation can be exploited. For generalization of the R_M value, the principal components analysis (PCA) can be applied [17]. Moreover, single values of R_M and R_f can also be used for QSRR studies [18].

In the QSRR approach, the relationship between chromatographic parameters and some molecular descriptors can be examined by the linear modeling methods,* such as linear regression (LR), multiple linear regression (MLR), principal component regression, and partial least squares (PLS) regression. MLR and PLS are the most frequently used methods. The advantages of MLR are simplicity and easy interpretation of established models. Additionally, the obtained models are directly related to the original data. MLR has also some disadvantages, the most important being poor prediction and inability to treat interrelated variables and missing data. On the contrary, PLS gives a partial solution, which is more robust than MLR. It can analyze highly collinear data; additionally, the number of descriptors can be greater than the number of analytes, whereas in MLR this is unacceptable [19]. The multivariate analyses, such as PCA and

* See Chapter 14 of this book.

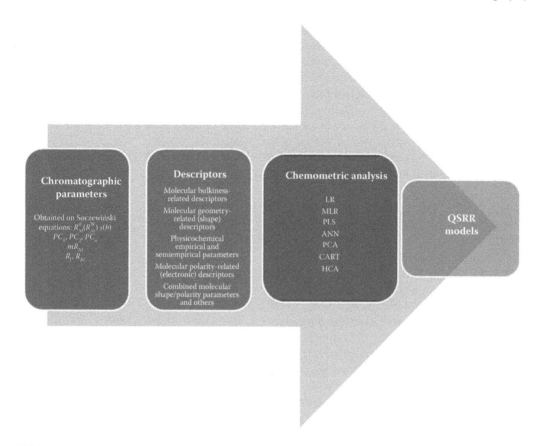

FIGURE 17.2 Chromatographic, molecular, and chemometric methods used for QSRR analysis on TLC.

hierarchical cluster analysis (HCA), are also often used to provide data overview in QSRR studies.* The other chemometric techniques, such as artificial neural networks (ANNs) [20] or regression trees (CART), [21] can be applied in QSRR studies but they are not very popular. Figure 17.2 shows parameters for both chromatographic and molecular techniques, as well as chemometric methods used for QSRR analysis on TLC.

Most papers regarding QSRR on TLC describe the investigation of reversed phases (RP-TLC) where the mechanism of retention is based on partitioning. On the other hand, papers regarding normal phases (NP-TLC) or modified adsorbents are relatively rare [22]. The QSRR methodology was also applied in studies where modifications of mobile phases were used: micellar liquid chromatography (MLC) [23] and salting-out thin-layer chromatography (SO-TLC) [24].

17.3 REVERSED-PHASE THIN-LAYER CHROMATOGRAPHY

Among chromatographic techniques, the RP-TLC is one the most popular methods for lipophilicity estimation, because it is low-cost, rapid, and requires small amounts of substances. Moreover, the time of analysis is relatively short and consumption of chemicals is comparatively low. Furthermore, RP-TLC analysis needs no complicated instrumentation and is easy to conduct [16]. Several commercial RP-TLC plates are available, such as C_{18}, C_8, bonded silica gel, and cyano (CN) bonded plates, but the C_{18} bonded silica gel is predominantly used. Moreover, silica gel impregnated with oil substance can be used as RP-TLC stationary phase [25].

---

* See Chapter 8 of this book.

The simple regression models between chromatographic parameters and lipophilicity can be found in scientific articles that have been listed in Table 17.1. The calculated log P values are frequently used, but there are also papers that mention the use of experimental log P values determined for the binary n-octanol–water system [16].

Extensive work based on models with different substances and several RP-TLC approaches for lipophilicity estimation have been published by Komsta and coworkers. A single TLC run, extrapolation of retention to 0% organic modifier in the eluent, principal component analysis of retention matrices, parallel factor analysis (PARAFAC) on three-way array, and PLS regression using nine concentrations of six modifiers (acetonitrile, acetone, 1,4 dioxane, propan-2-ol, methanol, and tetrahydrofuran) were compared. Unpredictably, good correlations were obtained for the single TLC runs. However, this method is undervalued in the literature. On the other hand, the recently proposed advanced chemometric processing, such as PCA, PARAFAC, and PLS, did not show any visible advantages compared to classical methods. Among the tested organic modifiers of mobile phases, methanol and dioxane were the best, while acetonitrile gave the worst and unacceptable correlation of retention with lipophilicity.

Jevrić et al. presented a study where retention parameters obtained on C_{18} plates for s-triazine derivatives were correlated with the same structural descriptors. The s-triazine derivatives are an attractive group of chemical compounds from the medicinal chemistry point of view because they have important biological activities, such as cytotoxic, anticancer, and antibacterial properties. Mixtures of methanol–water and acetonitrile–water were used as mobile phases. The QSRR equations were established by using MLR.

$$R_M^0 \text{MeOH} = -1.943 + 0.851 mi \log P + 0.09 E_t; \quad n = 14, \ r = 0.955 \tag{17.4}$$

$$R_M^0 \text{ACN} = -0.826 + 0.536 A \log P + 1.367 pK_a; \quad n = 14, \ r = 0.982 \tag{17.5}$$

where
   mi log P, A log P are the partition coefficients calculated by applying different theoretical
      procedures
   pK_a is the dissociation constant
   E_t is the total energy

These results confirmed that hydrophobic interactions highly affect the retention parameters of the investigated s-triazine compounds [71].

Strahinja and coworkers examined the chromatographic behavior of estrane derivatives on C_{18} bonded silica gel plates and methanol–water mobile phase. Three chemometric methods (PCA, MLR, and PLS) were applied in this study. The best obtained MLR model is presented here:

$$R_M^0 = 0.002038 \cdot \text{GE} + 0.01496148 \cdot \text{MR} - 0.017372 \cdot \text{MP} + 1.571067; \quad n = 18, \ r = 0.996 \tag{17.6}$$

where
   GE is the Gibbs free energy
   MR is the molar refractivity
   MP is the melting point

However, the MLR model showed the best characteristics according to internal validation (the highest r), and the external validation of the proposed models showed that the PLS model has the best predictive power. The descriptors included in the obtained MLR and PLS models were of a similar nature [72].

The QSRR approach was used in order to investigate the relationship between the structural descriptors of 23 newly synthesized N-substituted 2-alkylidene-4-oxothiazolidines and their

TABLE 17.1

Application of RP-TLC in Lipophilicity Measurements

Analytes	Stationary Phase	Mobile Phase	References
Bis-salicylic acid derivatives	C_{18}	Methanol–water	[26]
		1,4-Dioxane–water	
Cephalosporins	C_{18}	2-Propanol–water	[27]
		1,4-Dioxane–water	
		Acetone–water	
		Acetonitrile–water	
		Methanol–water	
N-3-Methyl succinimide derivatives	C_{18}	1,4-Dioxane–water	[28]
		Acetonitrile–water	
Bile acids and their derivatives	C_{18}	Methanol–water	[29]
Bile acids	CN	Methanol–water	[30,31]
	C_2	Methanol–acetonitrile–water	
		Acetone–water	
		1,4-Dioxane–water	
Nonsteroidal anti-inflammatory drugs	C_{18}	2-Propanol–water	[32]
		1,4-Dioxane–water	
		Acetone–water	
		Acetonitrile–water	
		Methanol–water	
Estradiol derivates	C_{18}	Acetonitrile–water	[33]
		Methanol–water	
γ-Hydroxybutyric acid	C_{18}	Methanol–TRIS buffer–acetic acid	[34]
Natural and synthetic coumarins	C_{18}	Tetrahydrofuran–water	[35]
		Acetonitrile–water	
		Methanol–water	
Aldopentose derivatives	Silica gel impregnated with paraffin oil	Acetone–water	[36]
		1,4-Dioxane–water	
6-Mercatopurine derivatives	C_{18}	Acetonitrile–water	[37]
Angiotensin-converting enzyme inhibitors and their metabolites	C_{18}	Methanol–water	[38]
		Ethanol–water	
		Acetone–water	
Pyrrolidin-2-one and pyrrolidine derivatives	C_{18}	Acetonitrile–0.1 M Tris buffer pH 7.0	[39]
Thiobarbiturates	C_{18}	Methanol–water	[40]
Synthetic dyes	C_{18}	Methanol–water	[41]
	C_{18w}		
	CN		
Precursors of peraza crown ethers	C_{18}	Methanol–water	[41]
Acetylcholinesterase inhibitors	C_{18}	1,4 Dioxane–citric buffer pH 3.0	[42]
	C_8		
	C_2		
	CN		
	NH_2		
3-Hydroxy-1,2 benzoisoxazoles	C_{18}	Methanol–water	[43]
Estrane derivatives	C_{18}	Methanol–water	[44]
Dehydroepiandrosterone derivates	C_{18}	Acetone–water	[45]
		1,4-Dioxane–water	

(Continued)

TABLE 17.1 (*Continued*)
Application of RP-TLC in Lipophilicity Measurements

Analytes	Stationary Phase	Mobile Phase	References
Artificial and natural sweeteners	C_{18}	2-Propanol–water	[46]
	C_8	Ethanol–water	
	C_2	Acetone–water	
	CN	Acetonitrile–water	
	NH_2	Methanol–water	
Compounds with a phenanthrene skeleton	C_{18}	Acetone–water	[16]
	CN	Acetonitrile–water	
		Acetone–petroleum ether	
α-Adrenergic and imidazoline receptor ligands	C_{18}	Methanol–dioxane	[47]
Amine-acrylate derivatives	C_{18}	Methanol–water	[48]
Selected pairs of isomeric organic compounds	C_{18}	Methanol–water, 9:1	[49]
		Ethanol–water, 9:1 (v/v)	
Pesticides	C_{18}	Methanol–water	[50]
Nicotinic acid derivatives	C_{18}	Methanol–water	[51]
Urea pesticides	C_8	Methanol–water	[52]
Salicylic acid and its derivatives	C_{18}	Methanol–water	[53]
	C_8		
	CN		
Steroid compounds	C_{18}	Methanol–water	[54]
		Acetonitrile–water	
Phenolic drugs	C_{18}	Methanol–water	[55]
	C_8		
Guanidine and imidazoline derivatives	C_8	Methanol–water	[56]
α-Adrenergic and imidazoline receptor ligands	C_{18}	Methanol–water	[57]
	CN	Tetrahydrofuran–ammonia–water	
1,3-Oxazolidine derivatives	C_{18}	Methanol–water	[58]
Thiosemicarbazides and their cyclic analogues	C_{18}	Methanol–water	[59]
Cyanoacetamide derivatives	C_{18}	2-Propanol–water	[60]
		Ethanol–water	
		Acetone–water	
		Tetrahydrofuran–water	
Ginger compounds	C_{18}	Acetonitrile–water	[61]
γ-Butyrolactone derivative	C_{18}	Acetonitrile–TRIS buffer pH 7.4	[62]
Thiosemicarbazide derivatives	C_{18}	1,4-Dioxane–water	[63]
	CN	Acetone–water	
		Acetonitrile–water	
		Methanol–water	
		Tetrahydrofuran–water	
Benzoxazine derivatives containing a thioxo group	C_{18}	Acetone–phosphate buffer	[64]
Fat-soluble vitamins	C_{18}	Methanol–water	[65]
	C_8		
2-Amino-1-cyclohexanol derivatives	C_{18}	Methanol–water	[66]
			(*Continued*)

TABLE 17.1 (Continued)
Application of RP-TLC in Lipophilicity Measurements

Analytes	Stationary Phase	Mobile Phase	References
Psychotropic drugs	C_{18}	Tetrahydrofuran–water	[67]
		1,4-Dioxane–water	
		Acetone–water	
		Acetonitrile–water	
		Methanol–water	
N,N-disubstituted-2-phenylacetamide derivatives	C_{18}	Tetrahydrofuran–water	[68]
		1,4-Dioxane–water	
		Acetone–water	
		Acetonitrile–water	
		Methanol–water	
		Ethanol–water	
		1-Propanol–water	
		2-Propanol–water	
Diquinothiazines	C_{18}	Acetone–TRIS buffer	[69]
Polydentate Schiff bases	C_{18}	Acetonitrile–water	[70]
		Methanol–water	
		Tetrahydrofuran–water	

chromatographic parameters establish by RP-TLC. Two stationary phases, C_{18} and CN, as well as three organic modifiers of eluent systems—methanol, acetonitrile, and tetrahydrofuran—were examined. Simple models based on calculated lipophilicity were obtained [73]. Another work published by the same authors concerns application of molecular descriptors using a chemometric approach. This study is based on MLR and PLS. The obtained QSRR equations show differences between retention on C_{18} and CN stationary phases.

$$R_M^0\left(\text{CN}\right) = 2.869 + 0.591A\log P + 0.113\text{HBD} - 0.000315E_{HOMO}; \quad n = 23,\ r = 0.964 \quad (17.7)$$

$$R_M^0\left(C_{18}\right) = 2.212 + 0.472C\log P - 0.09568\text{SP} - 0.07562\text{HBA}; \quad n = 23,\ r = 0.919 \quad (17.8)$$

where
   models obtained on chromatographic systems included water–tetrahydrofuran mobile phase
   HBD is the H bond donor
   HBA is the H bond acceptor
   SP is the solubility parameter

Finally, it appears that descriptors incorporated in the presented MLR and PLS models belong to the same class. The fact is that the calculated lipophilicity parameter was selected out of a large set of designated molecular descriptors in all models, which suggested the importance of lipophilicity and its impact on the retention behavior of tested substances [19].

Chromatographic properties of 23 disazo and trisazo 4,4′-diaminobenzanilide-based direct dye molecules were determined with RP-TLC. Silica gel plates were impregnated by mixtures of hexane and paraffin oil (95:5 $v{:}v$) before usage. Mobile phases included isopropanol and ammonia in different proportions. Numerous molecular descriptors that take into account size, shape, symmetry, electronic structure, atom or group distribution, and hydrophobicity of the dyes were derived from the optimized three-dimensional geometries. Results of MLR and ANN analysis show that the R_M^0

values can be effectively expressed by a combination of hydrophobic and polarity dye structural parameters.

$$R_M^0 = -4.22(\pm0.39) + 0.46(\pm0.10)\log P - 1.05(\pm0.23)n\text{COOHPh}; \quad n = 21, \ r = 0.907 \quad (17.9)$$

During MLR calculations, two outliners were excluded. The ANN analysis was performed in order to check the possibility of nonlinear influences. The architecture of the alternative ANN model was (2 inputs + 1 bias):(3 hidden-layer nodes + 1 bias):(1 output). This nonlinear model slightly improved the value of r^2 compared to the aforementioned MLR model [20].

17.4   NORMAL-PHASE THIN-LAYER CHROMATOGRAPHY

The number of scientific papers on QSRR analysis of NP-TLC is significantly lower than that on RP-TLC. The main reason is the fact that generally parameters different than $\log P$ influence the retention of NP-TLC. However, this chromatographic method can be useful when it comes to lipophilicity estimation and can be also applied in QSAR study.

Table 17.2 provides a list of scientific articles where correlations between chromatographic data and $\log P$ value are presented.

Both NP-TLC and RP-TLC were used in order to predict lipophilicity of seco-androstene derivatives. Mobile phases included water–dioxane and water–acetonitrile (RP-TLC), as well as toluene–dioxane and toluene–acetonitrile (NP-TLC). Significant linear correlation was found between RP-TLC retention constants (R_M^0) and calculated $\log P$ values and also between NP-TLC retention constants (C_0) and $C \log P$. Furthermore, the obtained chromatographic lipophilicity parameters were applied during QSAR study. Finally, the relationships between chromatographic data and computational absorption constants, affinity for plasma proteins, volume of distribution, and logarithm of blood–brain permeation were presented [78].

Kovačević and coworkers presented QSRR analysis focused on identification of the most important descriptors affecting normal-phase chromatographic behavior of 1,2-O-cyclohexylidene xylofuranose derivatives. Mobile phases that contained cyclohexane and four diluents—ethyl acetate, acetone, dioxane, and tetrahydrofuran—were examined. Silica gel was used as the stationary phase. The proposed models suggested that the retention of this class of compounds is strongly correlated

TABLE 17.2
Application of NP-TLC in Lipophilicity Measurements

Analytes	Stationary Phase	Mobile Phase	Reference
Bile acids	Silica gel	Toluene–butanol	[74]
		Toluene–ethanol	
Angiotensin-converting enzyme inhibitors and their metabolites	Silica gel	Ethyl methyl ketone–ethanol	[75]
		Carbon tetrachloride–ethanol	
		Toluene–ethanol	
Compounds with a phenanthrene skeleton	Silica gel	Acetone–water	[16]
	CN	Acetonitrile–water	
		Acetone–petroleum ether	
Carbohydrate derivatives	Silica gel	Benzene–ethyl acetate	[76]
		Benzene–1,4 dioxane	
		Benzene–acetone	
		Benzene–tetrahydrofuran	
Biphenylamine derivatives	Silica gel	Petroleum ether–ethyl acetate	[77]

with thermodynamic descriptors such as critical pressure, Gibbs energy distribution, or ideal gas thermal capacity, and molecular bulkiness descriptors such as polarity parameter and total energy [79].

A QSRR study on bile acids based on topological indices and electrotopological states as molecular descriptors was presented by Pyka and Dołowny. Two stationary phases, silica gel and diol, were tested. Mobile phases included three components: n-hexane, ethyl acetate, and acetic acid with different composition. Finally, the topological index C and Gutman's index M^v were found to highly affect the retention of the tested normal-phase chromatographic systems [18].

The biologically active steroids were used as model compounds, in order to find and describe the physicochemical parameters of solutes that govern the retention on aluminum oxide–covered plates. The eluents contained dichloroethane–dioxane mixtures in various proportions. After completing a stepwise regression analysis, one statistical model was found to be significantly sound.

$$R_M^0 = -0.46 + (0.86 \pm 0.16)\text{H–Do}; \quad r = 0.806, \; n = 17 \tag{17.10}$$

This equation suggests that the hydrogen-donor capacity of steroids had the highest influence on their retention. Furthermore, the PCA confirmed this finding [80].

Nowakowska and coworkers presented a QSRR study of natural steroid hormones on aluminum plates. Four mobile phases of acetonitrile–that, acetonitrile–DMSO, acetone–petroleum ether, and acetone–water were investigated. Significant QSRR equations were found for three tested eluent systems, except for the mixture of acetonitrile–water.

$$R_M^0 (\text{acetone–water}) = 0.006(\pm0.001)\text{Heat}_{Form} - 0.695(\pm0.0176); \quad r = 0.894, \; n = 8 \tag{17.11}$$

$$R_M^0 (\text{acetone–water}) = -0.022(\pm0.0045)\text{H-Ac} - 0.751(\pm0.0138); \quad r = 0.894, \; n = 8 \tag{17.12}$$

$$R_M^0 (\text{acetonitrile–DMSO}) = 0.321(\pm0.0289)\text{H-Ac} - 2.735(\pm0.0875); \quad r = 0.977, \; n = 8 \tag{17.13}$$

$$R_M^0 (\text{acetonitrile–DMSO}) = -0.019(\pm0.003)\text{Tot}_{Energy} - 4.499(\pm0.5268); \quad r = 0.901, \; n = 8 \tag{17.14}$$

$$R_M^0 (\text{acetone–petroleum ether}) = 0.815(\pm0.0251)\text{H-Ac} - 2.076(\pm0.0076); \quad r = 0.972, \; n = 8 \tag{17.15}$$

$$R_M^0 (\text{acetone–petroleum ether}) = -0.02(\pm0.0002)\text{Heat}_{Form} - 2.687(\pm0.4057); \quad r = 0.950, \; n = 8 \tag{17.16}$$

These results show that in the set of steroid hormones, there is a high influence of bulkiness-related descriptors such as heat of formation and total energy. Moreover, the number of hydrogen-acceptors also significantly influences the retention parameter R_M^0. These findings were supported by PCA too. This analysis suggested that when acetonitrile is used as an organic modifier of mobile phase, the maximum charge of atoms and the number of hydrogen-donors can explain the slight differences between the chromatographic parameters of tested steroid hormones [81].

17.5 NONCOMMERCIAL SELF-PREPARED STATIONARY PHASES

The QSRR approach can also be applied to present the retention models on noncommercial self-prepared stationary phases.

A glass-backed plate of rice starch was used in the QSRR study of 2,4-dioksotetrahydro-1,3–thiazole derivatives. Three binary mobile phases containing aqueous ammonia and an organic modifier (methanol, acetone, or dioxane) were examined. Finally, the PLS models for two tested eluents (acetone and dioxane) were established. In general, descriptors that highly influence the retention are lipophilicity and electronic characteristics of the molecule. These results are not surprising because the thiazoles are specific solutes that are capable of polar interactions [82].

Another example of the use of QSRR methodology to examine the retention mechanism of a self-prepared stationary phase is work presented by Csiktusnádi-Kiss et al. The TLC plates were prepared as follows: 8 g of 20–80 mm diameter alumina particles coated with 2.5% polyethylene was mixed with water to obtain a slurry. The slurry was spread on glass plates of 20 × 20 cm. Next the plates were dried at room temperature during the night. Tetrazolium salts were used as model substances, and retention was observed on the TLC plates and compared with a similar HPLC system. In order to find a relationship between the physicochemical parameters of solutes and their retention data, PCA and cluster analysis were applied. The result of PCA shows that the TLC system is strictly different from the HPLC system for tetrazolium salts. The TLC retention parameters belong to the same cluster with polarizability, refractivity, heat formation, and van der Waals surface values of the samples [83].

17.6 SALTING-OUT THIN-LAYER CHROMATOGRAPHY

SO-TLC is an example of a planar chromatographic technique based on the use of concentrated aqueous solutions of inorganic salts as mobile phase and sorbents with high polarity such as silica gel, cellulose, or aluminum oxide. Under these conditions, nonspecific hydrophobic interactions govern the chromatographic mechanism.

The QSRR study of some oral hypoglycemic drugs on SO-TLC was published. The tested group included seven compounds. Silica gel was used as the stationary phase, whereas the mobile phases contained aqueous solutions of ammonium sulfate and acetonitrile in different proportions. One- and two-parameter QSRR equations were established. One-parameter QSRR equation showed that the R_M^0 values significantly affect aqueous solubility. The two-parameter models that were obtained suggest that the most important factor that influences retention is lipophilicity [84]:

$$R_M^0 = 0.445 + 0.022\text{MR} + 0.115\log P; \quad r = 0.880, \ n = 7 \quad (17.17)$$

$$R_M^0 = 0.262 + 0.164\log P - 0.002E\_vdw; \quad r = 0.868, \ n = 7 \quad (17.18)$$

Tosti and coworkers presented linearity between retention parameters of five macrolide antibiotics that were determined on the SO-TLC and $C\log P$ values calculated by the computer software KNOWWIN. The experiments were carried out on cellulose plates with an aqueous solution of ammonium sulfate as eluent [85].

The investigation of lipophilicity and hydrophobicity parameters of selected macrolide antibiotics obtained in several SO-TLC systems was introduced. Four inorganic salts—sodium chloride, calcium chloride, ammonium chloride, and ammonium sulfate—were examined. Furthermore, chromatographic analysis was conducted on three types of stationary phases: silica gel, cellulose plates, and basic aluminum oxide. Correlations between computational $\log P$ and $\log D$ values were found and illustrated as simple regression models. The obtained QSRR equations showed that basic aluminum oxide plates can be successfully used to predict $C\log D$ values in pH 7.4. For ionizing compounds, the distribution coefficient D is used to take into consideration the pH dependence of partitioning. The obtained results suggest that the basic aluminum oxide as the stationary phase stabilizes the compounds in non-ionized form, which consequently improves the QSRR equations, in comparison with other tested stationary phases. Additionally, the obtained chromatographic parameters were applied during QSAR study. The presented QSAR models can be used for prediction of antimicrobial activities against *Streptococcus pyogenes*, *Streptococcus pneumoniae*, and *Listeria monocytogenes*. These results were also confirmed by PCA [24].

Flieger et al. presented chemometric analysis of the retention data from SO-TLC in relation to the structural parameters and biological activity of the chosen sulphonamides. Several aqueous solutions of inorganic salts were tested on silica gel plates. Each of the obtained equations contains a descriptor related to the size and lipophilicity of the solutes. This is consistent with the theory that nonspecific hydrophobic interactions are decisive in this chromatographic technique. In addition to observing the relation between the structural parameters, chromatographic behavior, and pharmacological activity of the studied compounds, multidimensional cluster analysis has also been conducted. It was based on the 3D scatterplot of the chromatographic parameter R_M^0 (NaCl) and the molecular parameters of molar refractivity and calculated log P (mi log P). The first cluster included sulphonamides, which are typical bacteriostatic drugs. The second cluster contained sulphonamides used as locally administered drugs because they are characterized by poor solubility or ionization at physiologic pH [86].

17.7   MICELLAR LIQUID CHROMATOGRAPHY

A competitive study of HPLC, over-pressured-layer chromatography (OPLC), and TLC techniques with micellar mobile phases was proposed to evaluate the lipophilicity of 21 newly synthesized 1,2,4-triazoles. These compounds can be useful in medicine or agriculture because of their antifungal activity. Mobile phases were used containing buffered safety data sheet–acetonitrile in different proportions in the case of MLC and buffer–acetonitrile and buffer–tetrahydrofuran in the case of RP-TLC. As stationary phases in micellar TLC and OPLC, CN bonded silica gel was applied, while C_8 bonded silica gel was used in RP-TLC. The obtained linearity between computational log P values, especially with x log $P2$ and x log $P3$, shows that MLC can be a promising tool for lipophilicity estimation of triazoles. Moreover, OPLC seems to be a very attractive technique because of the significant reduction in reagent consumption and analysis time [23]. The same authors presented work where micellar TLC and OPLC were used in order to evaluate the lipophilicity of N-phenyltrichloroacetamide derivatives [87].

REFERENCES

1. Kaliszan, R. 1987. *Quantitative Structure—Chromatographic Retention Relationships.* New York: John Wiley & Sons.
2. Kaliszan, R. 1997. Quantitative structure–retention relationships. *Anal. Chem.* 64:619–631.
3. Kaliszan, R. 1997. *Structure and Retention in Chromatography: A Chemometric Approach.* Amsterdam, the Netherlands: Harwood Academic Publishers.
4. Kaliszan, R. 2007. QSRR: Quantitative structure-(chromatographic) retention relationships. *Chem. Rev.* 107:3212–3246.
5. Al-Haj, M.A., Kaliszan, R., and Buszewski, B. 2001. Quantitative structure–retention relationships with model analytes as a means of an objective evaluation of chromatographic columns. *J. Chromatogr. Sci.* 39:29–38.
6. Bączek, T. and Kaliszan, R. 2002. Combination of linear solvent strength model and quantitative structure–retention relationships as a comprehensive procedure of approximate prediction of retention in gradient liquid chromatography. *J. Chromatogr. A* 962:41–55.
7. Jiskra, J., Claessens, H.A., Cramers, C.A., and Kaliszan, R. 2002. Quantitative structure–retention relationships in comparative studies of behavior of stationary phases under high-performance liquid chromatography and capillary electrochromatography conditions. *J. Chromatogr. A* 977:193–206.
8. Kaliszan, R., Bączek, T., Buciński, A., Buszewski, B., and Sztupecka, M. 2003. Prediction of gradient retention from the linear solvent strength (LSS) model, quantitative structure-retention relationships (QSRR), and artificial neural networks (ANN). *J. Sep. Sci.* 26:271–282.
9. Bączek, T. 2008. Computer-assisted optimization of liquid chromatography separations of drugs and related substances. *Curr. Pharm. Anal.* 4:151–161.
10. Bączek, T., Kaliszan, R., Novotna, K., and Jandera, P. 2005. Comparative characteristics of HPLC columns based on quantitative structure–retention relationships (QSRR) and hydrophobic-subtraction model. *J. Chromatogr. A* 1075:109–115.

11. Kaliszan, R., van Straten, M.A., Markuszewski, M., Cramers, C.A., and Claessens, H.A. 1999. Molecular mechanism of retention in reversed-phase high-performance liquid chromatography and classification of modern stationary phases by using quantitative structure-retention relationships. *J. Chromatogr. A* 855:455–486.

12. Kaliszan, R. 1999. Chromatography and capillary electrophoresis in modelling the basic processes of drug action. *Trends Anal. Chem.* 18:400–410.

13. Sadek, P.C., Carr, P.W., Doherty, R.M., Kamlet, M.J., Taft, R.W., and Abraham, M.H. 1985. Study of retention processes in reversed-phase high-performance liquid chromatography by the use of the solvatochromic comparison method. *Anal. Chem.* 57:2971–2978.

14. Héberger, K. 2007. Quantitative structure–(chromatographic) retention relationships. *J. Chromatogr. A* 1158:273–305.

15. Kaliszan, R. and Foks, H. 1977. The relationship between the RM values and the connectivity indices for pyrazine carbothioamide derivatives. *Chromatographia* 10:346–349.

16. Ciura, K., Nowakowska, J., Piku, P., Struck-Lewicka, W., and Markuszewski, M.J. 2015. A comparative quantitative structure-retention relationships study for lipophilicity determination of compounds with a phenanthrene skeleton on cyano-, reversed phase-, and normal phase-thin layer chromatography stationary phases. *J. AOAC Int.* 98:345–353.

17. Komsta, Ł., Skibiński, R., Berecka, A., Gumienieczek, A., Radkiewicz, B., and Radoń, M. 2010. Revisiting thin-layer chromatography as a lipophilicity determination tool—A comparative study on several techniques with a model solute set. *J. Pharm. Biomed. Anal.* 53:911–918.

18. Pyka, A. and Dołowy, M. 2004. Application of structural descriptors for the evaluation of physicochemical properties of bile acids. *Acta Pol. Pharm. Drug Res.* 6:407–413.

19. Dabić, D., Natić, M., Džambaski, Z., Marković, R., Milojković-Opsenica, D., and Tešić, Ž. 2011. Quantitative structure-retention relationship of new N-substituted 2-alkylidene-4-oxothiazolidines. *J. Sep. Sci.* 18:2397–2404.

20. Funar-Timofei, S., Fabian, W.M.F., Simu, G.M., and Suzuki, T. 2006. Quantitative structure-retention relationships (QSRR) for chromatographic separation of disazo and trisazo 4,4′-diaminobenzanilide-based dyes. *Croat. Chem. Acta* 2:227–236.

21. Komsta, Ł. 2008. Quick prediction of the retention of solutes in 13 thin layer chromatographic screening systems on silica gel by classification and regression trees. *J. Sep. Sci.* 15:2899–2909.

22. Komsta, Ł. 2008. A functional-based approach to the retention in thin layer chromatographic screening systems. *Anal. Chim. Acta* 1–2:66–72.

23. Janicka, M., Stępnik, K., and Pachuta-Stec, A. 2012. Quantification of lipophilicity of 1,2,4-triazoles using micellar chromatography. *Chromatographia* 9–10:449–456.

24. Ciura, K., Nowakowska, J., Rudnicka-Litka, K., Kawczak, P., Bączek, T., and Markuszewski, M.J. 2016. The study of salting-out thin-layer chromatography and their application on QSRR/QSAR of some macrolide antibiotics. *Monatshefte für Chemie Chem. Mon.* 2:301–310.

25. Kovačević, S.Z., Podunavac Kuzmanović, S.O., Jevrić, L.R., and Lončar, E.S. 2014. Assessment of chromatographic lipophilicity of some anhydro-D-aldose derivatives on different stationary phases by QSRR approach. *J. Liq. Chromatogr. Relat. Technol.* 4:492–500.

26. Djaković-Sekulić, T., Perisić-Janjić, N., and Djurendi, E. 2009. Retention data from reverse-phase high-performance thin-layer chromatography in characterization of some bis-salicylic acid derivatives. *Biomed. Chromatogr.* 8:881–887.

27. Dąbrowska, M., Komsta, Ł., Krzek, J., and Kokoszka, K. 2015. Lipophilicity study of eight cephalosporins by reversed-phase thin-layer chromatographic method. *Biomed. Chromatogr.* 11:1759–1768.

28. Perisic-Janjic, N., Kaliszan, R., Milosevic, N., Uscumlic, G., and Banjac, N. 2013. Chromatographic retention parameters in correlation analysis with in silico biological descriptors of a novel series of N-phenyl-3-methyl succinimide derivatives. *J. Pharm. Biomed. Anal.* 72:65–73.

29. Sârbu, C., Kuhajda, K., and Kevresan, S. 2001. Evaluation of the lipophilicity of bile acids and their derivatives by thin-layer chromatography and principal component analysis. *J. Chromatogr. A* 1–2:361–366.

30. Pyka, A., Dołowy, M., and Gurak, D. 2005. Lipophilicity of selected bile acids, as determined by TLC. IV. Investigations on CNF 254 stationary phase. *J. Liq. Chromatogr. Relat. Technol.* 17:2705–2717.

31. Pyka, A. and Dołowy, M. 2005. Lipophilicity of selected bile acids as determined by TLC. III. Investigations on RP2 stationary phase. *J. Liq. Chromatogr. Relat. Technol.* 11:1765–1775.

32. Starek, M., Komsta, Ł., and Krzek, J. 2013. Reversed-phase thin-layer chromatography technique for the comparison of the lipophilicity of selected non-steroidal anti-inflammatory drugs. *J. Pharm. Biomed. Anal.* 85:132–137.

33. Djaković-Sekulić, T., Ačanski, M., and Perisić-Janjić, N. 2002. Evaluation of the predictive power of calculation procedure for molecular hydrophobicity of some estradiol derivates. *J. Chromatogr. B Anal. Technol. Biomed. Life Sci.* 1:67–75.

34. Malawska, B. and Tabor, A. 1998. Determination of the lipophilicity of active anticonvulsant N-substituted amides of alpha-arylalkylamine-gamma-hydroxybutyric acid. *Acta Pol. Pharm. Drug Res.* 6:461–465.

35. Rabtti, E.H.M.A., Natic, M.M., Milojkovic-Opsenica, D.M. et al. 2012. RP TLC-based lipophilicity assessment of some natural and synthetic coumarins. *J. Braz. Chem. Soc.* 3:522–530.

36. Livaja-Popovic, D., Loncar, E., Jevric, L., and Malbasa, R. 2012. Reversed-phase thin-layer chromatography behavior of aldopentose derivatives. *Hem. Ind.* 3:365–372.

37. Czyrski, A. and Kupczyk, B. 2013. The determination of partition coefficient of 6-mercaptopurine derivatives by thin layer chromatography. *J. Chem.* 1–4.

38. Odovic, J., Stojimirovic, B., Aleksic, M., Milojkovic-Opsenica, D., and Tesic, Z. 2006. Reversed-phase thin-layer chromatography of some angiotensin converting enzyme (ACE) inhibitors and their active metabolites. *J. Serb. Chem. Soc.* 6:621–628.

39. Kulig, K. and Malawska, B. 2003. Estimation of the lipophilicity of antiarrhythmic and antihypertensive active 1-substituted pyrrolidin-2-one and pyrrolidine derivatives. *Biomed. Chromatogr.* 5:318–324.

40. Kepczyńska, E., Obłoza, E., Stasiewicz-Urban, A., Bojarski, J., and Pyka, A. 2007. Lipophilicity of thiobarbiturates determined by TLC. *Acta Pol. Pharm. Drug Res.* 4:295–302.

41. Onişor, C., Blăniţă, G., Coroş, M., Bucşa, M., Vlassa, M., and Sârbu, C. 2010. A comparative study concerning chromatographic retention and computed partition coefficients of some precursors of peraza crown ethers. *Cent. Eur. J. Chem.* 6:1203–1209.

42. Szymański, P., Skibiński, R., Liszka, M., Jargieło, I., Mikiciuk-Olasik, E., and Komsta, Ł. 2013. A TLC study of the lipophilicity of thirty-two acetylcholinesterase inhibitors—1,2,3,4-Tetrahydroacridine and 2,3-dihydro-1H-cyclopenta[b]quinoline derivatives. *Open Chem.* 6:927–934.

43. Sławik, T., Skibiński, R., Paw, B., and Działo, G. 2009. Reversed-phase TLC study of the lipophilicity of some 3-hydroxy-1,2-benzisoxazoles substituted in the benzene ring. *Acta Chromatogr.* 2:251–258.

44. Petrovic, S.M., Loncar, E., Kolarov, L., and Pejanovic, V. 2002. Correlation between retention and 1-octanol-water partition coefficients of some estrane derivatives in reversed-phase thin-layer chromatography. *J. Chromatogr. Sci.* 40:569–574.

45. Perišić-Janjić, N., Djaković-Sekulić, T., Stojanović, S., and Penov-Gaši, K. 2004. Evaluation of the lipophilicity of some dehydroepiandrosterone derivates using RP-18 HPTLC chromatography. *Chromatographia* S1:201–205.

46. Briciu, R.D., Kot-Wasik, A., Wasik, A., Namieśnik, J., and Sârbu, C. 2010. The lipophilicity of artificial and natural sweeteners estimated by reversed-phase thin-layer chromatography and computed by various methods. *J. Chromatogr. A* 23:3702–3706.

47. Erić, S., Pavlović, M., Popović, G., and Agbaba, D. 2007. Study of retention parameters obtained in RP-TLC system and their application on QSAR/QSPR of some alpha adrenergic and imidazoline receptor ligands. *J. Chromatogr. Sci.* 3:140–145.

48. Wang, Q.S., Zhang, L., Yang, H.Z., and Liu, H.Y. 1999. Lipophilicity determination of some potential photosystem II inhibitors on reverse-phase high-performance thin-layer chromatography. *J. Chromatogr. Sci.* 37:41–44.

49. Stefaniak, M., Niestrój, A., Klupsch, J., Śliwiok, J., and Pyka, A. 2005. Use of RP-TLC to determine the log P values of isomers of organic compounds. *Chromatographia* 1–2:87–89.

50. Pyka, A. and Miszczyk, M. 2004. Chromatographic evaluation of the lipophilic properties of selected pesticides. *Chromatographia* 1–2:37–42.

51. Pyka, A. and Klimczok, W. 2005. Study of lipophilicity and application of selected structural descriptors in QSAR analysis of nicotinic acid derivatives. Investigations on RP18WF254 plates. Part II. *J. Planar Chromatogr. Mod. TLC* 104:300–304.

52. Dołowy, M., Miszczyk, M., and Pyka, A. 2014. Application of various methods to determine the lipophilicity parameters of the selected urea pesticides as predictors of their bioaccumulation. *J. Environ. Sci. Health B* 10:730–737.

53. Pyka, A., Rusek, D., Bocheńska, P., and Gurak, D. 2009. Use of RP-TLC and theoreticalcomputational methods to compare the lipophilicity of salicylic acid and its derivatives. *J. Liq. Chromatogr. Relat. Technol.* 2:179–190.

54. Pyka, A. 2009. Use of selected topological indexes for evaluation of lipophilicity of steroid compounds investigated by RP-HPTLC. *J. Liq. Chromatogr. Relat. Technol.* 20:3056–3065.

55. Pyka, A. and Gurak, D. 2007. Use of RP-TLC and theoretical computational methods to compare the lipophilicity of phenolic drugs. *J. Planar Chromatogr. Mod. TLC* 5:373–380.

56. Filipic, S., Elek, M., Nikolic, K., and Agbaba, D. 2015. Quantitative structure-retention relationship modeling of the retention behavior of guanidine and imidazoline derivatives in reversed-phase thin-layer chromatography. *J. Planar Chromatogr. Mod. TLC* 2:119–125.
57. Mohamed Shenger, M.S., Filipic, S., Nikolic, K., and Agbaba, D. 2014. Estimation of lipophilicity and retention behavior of some alpha adrenergic and imidazoline receptor ligands using RP-TLC. *J. Liq. Chromatogr. Relat. Technol.* 20:2829–2845.
58. Sârbu, C., Casoni, D., Darabantu, M., and Maiereanu, C. 2004. Quantitative structure-retention and retention-activity relationships of some 1,3-oxazolidine systems by RP-HPTLC and PCA. *J. Pharm. Biomed. Anal.* 1:213–219.
59. Hawryl, A.M., Popiołek, Ł.P., Hawryl, M.A., Świeboda, R.S., and Niejedli, M.A. 2015. Chromatographic and calculation methods for analysis of the lipophilicity of newly synthesized thiosemicarbazides and their cyclic analogues 1,2,4-triazol-3-thiones. *J. Braz. Chem. Soc.* 8:1617–1624.
60. Vastag, G., Apostolov, S., Nakomčić, J., and Matijević, B. 2014. Application of chemometric methods in examining of retention behavior and lipophilicity of newly synthesized cyanoacetamide derivatives. *J. Liq. Chromatogr. Relat. Technol.* 17:2529–2545.
61. Bakht, M.A., Alajmi, M.F., Alam, P., Alam, A., Alam, P., and Aljarba, T.M. 2014. Theoretical and experimental study on lipophilicity and wound healing activity of ginger compounds. *Asian Pac. J. Trop. Biomed.* 4:329–333.
62. Bajda, M., Guła, A., Więckowski, K., and Malawska, B. 2013. Determination of lipophilicity of γ-butyrolactone derivatives with anticonvulsant and analgesic activity using micellar electrokinetic chromatography. *Electrophoresis* 20–21:3079–3085.
63. Hawryl, A., Kuśmierz, E., Pisarczyk, P., Wujec, M., and Waksmundzka-Hajnos, M. 2012. Determination of the lipophilicity of some new thiosemicarbaside derivatives by reversed-phase thin-layer chromatography. *Acta Chromatogr.* 2:271–290.
64. Petrlíková, E. and Waisser, K. 2011. A TLC study of the lipophilicity of new antimycobacterial active benzoxazine derivatives containing a thioxo group. *J. Planar Chromatogr. Mod. TLC* 3:196–200.
65. Pyka, A. 2009. Evaluation of the lipophilicity of fat-soluble vitamins. *J. Planar Chromatogr. Mod. TLC* 3:211–215.
66. Pekala, E. and Marona, H. 2009. Estimating the lipophilicity of a number of 2-amino-1-cyclohexanol derivatives exhibiting anticonvulsant activity. *Biomed. Chromatogr.* 5:543–550.
67. Hawryl, A., Cichocki, D., and Waksmundzka-Hajnos, M. 2008. Determination of the lipophilicity of some psychotropic drugs by RP-TLC. *J. Planar Chromatogr. Mod. TLC* 5:343–348.
68. Perišić-Janjić, N., Vastag, G., Tomić, J., and Petrović, S. 2007. Effect of the physicochemical properties of N,N-disubstituted-2-phenylacetamide derivatives on their retention behavior in RP-TLC. *J. Planar Chromatogr. Mod. TLC* 5:353–359.
69. Nowak, M. and Pluta, K. 2006. Study of the lipophilicity of novel diquinothiazines. *J. Planar Chromatogr. Mod. TLC* 108:157–160.
70. Perušković, D.S., Darić, B., Blagus, A. et al. 2015. Influence of organic modifiers on RP-TLC determination of lipophilicity of some polydentate Schiff bases. *Monatshefte für Chemie Chem. Mon.* 1:1–6.
71. Jevric, L., Koprivica, G., Misljenovic, N., Tepic, A., Kuljanin, T., and Jovanovic, B. 2011. Estimation of the correlation between the retention of s-triazine derivatives and some molecular descriptors. *Acta Period. Technol.* 42:231–239.
72. Kovačević, S.Z., Jevrić, L.R., Podunavac Kuzmanović, S.O., and Lončar, E.S. 2013. Chemometric estimation of the RP TLC retention behaviour of some estrane derivatives by using multivariate regression methods. *Cent. Eur. J. Chem.* 12:2031–2039.
73. Dabić, D., Natić, M., Džambaski, Z., Stojanović, M., Marković, R., Milojković-Opsenica, D., and Tešić, Ž. 2011. Estimation of lipophilicity of n-substitued 2-alkyldiene-4-oxothiazolidines by means of reversed-phase thin-layer chromatography. *J. Liq. Chromatogr. Relat. Technol.* 10–11:791–804.
74. Posa, M., Raseta, M., and Kuhajda, K. 2011. A contribution to the study of hydrophobicity (lipophilicity) of bile acids with an emphasis on oxo derivatives of 5β-cholanoic acid. *Hem. Ind.* 2:115–121.
75. Odovic, J., Stojimirovic, B., Aleksic, M., Milojkovic-Opsenica, D., and Tesic, Z. 2009. Normal-phase thin-layer chromatography of some angiotensin converting enzyme (ACE) inhibitors and their metabolites. *J. Serb. Chem. Soc.* 6:677–688.
76. Karadžić, M.Ž., Jevrić, L.R., Podunavac Kuzmanović, S.O., Lončar, E.S., and Kovačević, S.Z. 2015. Lipophilicity estimation of some carbohydrate derivatives in TLC with benzene as a diluent. *J. Liq. Chromatogr. Relat. Technol.* 17:1593–1600.
77. Margabandu, R. and Subramani, K. 2010. Experimental and theoretical study on lipophilicity and antibacterial activity of biphenylamine derivatives. *Int. J. ChemTech Res.* 3:1501–1506.

78. Milošević, N.P., Stojanović, S.Z., Penov-Gaši, K., Perišić-Janjić, N., and Kaliszan, R. 2014. Reversed- and normal-phase liquid chromatography in quantitative structure retention–property relationships of newly synthesized seco-androstene derivatives. *J. Pharm. Biomed. Anal.* 88:636–642.

79. Kovačević, S.Z., Jevrić, L.R., Kuzmanović, S.O.P., Kalajdžija, N.D., and Lončar, E.S. 2013. Quantitative structure-retention relationship analysis of some xylofuranose derivatives by linear multivariate method. *Acta Chim. Slov.* 2:420–428.

80. Cserháti, T. and Forgács, E. 1998. Effect of molecular parameters on the retention of steroid drugs on alumina support. *J. Pharm. Biomed. Anal.* 5:497–503.

81. Nowakowska, J., Ciura, K., Struck-Lewicka, W., Pikul, P., Markuszewski, M.J., and Kawczak, P. 2016. Study of the chromatographic behavior of selected steroid hormones on aluminum oxide plates based on quantitative structure-retention relationships. *JPC J. Planar Chromatogr. Mod. TLC* 2:113–120.

82. Arbu, C.S., Janji, C., and Si, N.P. 2009. Quantitative structure–retention study of some 2,4-dioksotetra-hydro-1,3–thiazole derivatives using the partial least squares method. *Turk. J. Chem.* 33:149–157.

83. Csiktusnádi-Kiss, G.A., Forgács, E., Markuszewski, M., and Balogh, S. 1998. Application of multivariate mathematical-statistical methods to compare reversed-phase thin-layer and liquid chromatographic behaviour of tetrazolium salts in Quantitative Structure-Retention Relationships (QSRR) studies. *Analusis* 10:400–406.

84. Mohamed, A.M.I., Mohamed, F.A.-F., Ahmed, S.A.-R., Aboraia, A.S., and Mohamed, Y.A.S. 2014. Salting-out thin-layer chromatography and computational analysis of some oral hypoglycemic drugs. *Biomed. Chromatogr.* 8:1156–1162.

85. Tosti, T., Drljević, K., Milojković-Opsenica, D., and Tešić, Ž. 2005. Salting-out thin-layer chromatography of some macrolide antibiotics. *J. Planar Chromatogr. Mod. TLC* 106:415–418.

86. Flieger, J., Świeboda, R., and Tatarczak, M. 2007. Chemometric analysis of retention data from salting-out thin-layer chromatography in relation to structural parameters and biological activity of chosen sulphonamides. *J. Chromatogr. B Anal. Technol. Biomed. Life Sci.* 1–2:334–340.

87. Janicka, M. and Pietras-Ożga, D. 2010. Chromatographic evaluation of the lipophilicity of N-phenyltrichloroacetamide derivatives using micellar TLC and OPLC. *J. Planar Chromatogr. Mod. TLC* 6:396–399.

18 Topological Indices in Modeling of Chromatographic Retention

Małgorzata Dołowy, Katarzyna Bober, and Alina Pyka-Pająk

CONTENTS

18.1 DEVELOPMENT AND SURVEY OF DIFFERENT CALCULATION METHODS OF TOPOLOGICAL INDICES

As has been well described in the guide to the International Union of Pure and Applied Chemistry (IUPAC) nomenclature, a topological index is a numerical value combined with chemical constituents for the correlation of a chemical structure with different physicochemical properties or biological activities of various classes of organic compounds [1]. According to the chemical graph theory, whose focus of interest is developing topological indices of chemical constituents, a topological representation of a molecule is a molecular graph. Each graph represents the carbon-atom skeleton of an organic molecule (an object). The vertices of this graph represent the carbon atoms, and its edges show the carbon–carbon bond (covalent bonds) [2,3]. Two vertices connected by an edge are said to be *adjacent*. There are many different types of topological indices. Most of the proposed topological indices are either based on vertex adjacency relationship (atom–atom connectivity), for example, Gutman's index (M), Platt's index (F), or Gordon–Scantleburry's index (N_2), or related to topological distances, that is, Wiener's index (W), Balaban's index, the Pyka indices (A, ^0B, ^1B, ^2B, ^2B_q, ^3B, ^3B_q, and C), optical index (I_{opt}), Rouvraya–Crafforda's index (R), polarity number (p), stereoisomeric topological index (I_{STI}), Hosoya's index (also called Z-index), molecular topological index Schultza (MTI), and others [4,5]. The topological distance used in the calculation of these indices is defined as the smallest number of edges (bonds), and thus the shortest path between the vertices (atoms), in the graph. Topological indices based on atom–atom connectivity are usually described as the total sum of some combination of degrees of the adjacent vertices [6]. Other types of topological indices include the centric topological indices developed by Balaban [7] and by Balaban and Motoc [8] for acyclic graphs, which have been calculated on the basis of their center, and also the topological indices based on information theory. Several topological indices have been used in research on the prediction of various physicochemical and biological properties of many compounds. The pioneering topological indices that have been successfully applied in the determination of the selected physicochemical properties of mixtures consisting of organic compounds belonging to various groups, such as boiling point, molar volume, molar refraction, and also heat of formation and vaporization, are Wiener's index, polarity number, and Platt's index. In recent years, the manual calculation methods are substituted by software programs such as Dragon [9]. Dragon is a very popular software used by many researchers to calculate a wide range of molecular descriptors: topological descriptors, connectivity indices, topological charge indices, geometrical descriptors, and many more. Scientists can choose the most suitable descriptors for their research. The Dragon software is still updated and now calculates a few thousand descriptors on the basis of the most common molecule file formats. The main advantage of Dragon is fast computation of descriptors avoiding time-consuming quantum-chemical calculations.

Recently, there has been significant interest in determining topological indices in different research areas, especially in pharmaceutical analysis in drug discovery, for the determination of pharmacological activity of new candidates for drugs. A literature survey confirms that of all topological descriptors, topological indices are very important and they play a crucial role in current chemistry [10]. Topological indices are widely used in quantitative structure–retention relationships (QSRRs), quantitative structure–property relationships (QSPRs), and quantitative structure–activity relationships (QSARs) of molecules, as well as in drug design. The increasing application of topological indices in various research areas is causing a need to find more informative topological indices among available ones or to develop new ones. For this purpose, some novel calculating methods for the hyper-Wiener index and edge-Wiener index have been elaborated [11]. Another topological index used in QSRR is the semi-empirical topological index introduced by Brazilian researchers [12]. It was applied for the analysis of some different groups of chemical compounds. The index marked as I_{ET} is calculated with the use of the following equation:

$$I_{ET} = \sum_i \left(C_i + \delta_i\right) \tag{18.1}$$

where
C_i is the value attributed to each carbon atom i in the molecule
δ_i is the sum of the logarithm of the value of each adjacent carbon atom

The molecules are represented by molecular graphs based on the chemical graph theory, where the carbon atoms (C_i) are vertexes of this graph.

As can be observed, one of the most important properties that have been extensively studied is chromatographic retention, hence the chromatographic behavior of molecules during their separation by means of various chromatographic techniques. Therefore, general tendencies, practices, and conclusions of the retention models in thin-layer chromatography (TLC), high-performance liquid chromatography (HPLC), gas chromatography (GC), capillary zone electrophoresis (CZE), and micellar electrokinetic capillary chromatography (MEKC) obtained using topological indices for organic compounds from different classes are reported in this study. In addition to this, QSAR analysis using topological indices in chemical and pharmaceutical analysis is described.

18.2 TOPOLOGICAL INDICES USED IN MODELING RETENTION DATA IN TLC

Among various chromatographic methods, TLC is still widely used as a rapid, low-cost, and simple method in the separation of many biologically important compounds. The use of topological indices for the prediction of retention (retardation factor, R_f) or related value (R_M) is helpful in the confirmation or identification of different organic compounds, especially in the case of absence of their retention data.

One of the first papers that have focused on the use of topological indices in TLC identification of selected tocopherol was prepared by Śliwiok and coworkers [13]. He observed that the values of topological indices Randić: $^0\chi^\nu$, $^1\chi^\nu$, $^2\chi^\nu$ and Pyka: ^0B, C and D obtained for four investigated compounds (α-, β-, γ-, δ-) of tocopherols increased as the R_f decreased. For the separation of the D and L enantiomers of the studied tocopherols on glass-backed chiral plates, a linear correlation between R_M values and calculated topological index I_t was found. It was stated that biparametric equations, consisting of two topological indices, are more suitable for the prediction of R_M values than the monoparametric variety. In continuation of this study, Pyka and Niestrój [14] and also Pyka and Śliwiok [15] confirmed the significant role of topological index ^0B in the prediction of R_M values of (α-, β-, γ-, δ-) tocopherols separated by reversed-phase high-performance thin-layer chromatography (RP-HPTLC). However, it was stated that the most accurate prediction of R_M values of examined tocopherols was achieved by the use of two parametric equations employing the dipole moment of the applied mobile phase in RP-HPTLC analysis of tocopherols and one of the investigated topological indices: $^2\chi^\nu$, C, ^0B, or the sum of the net electron charge (ΣNEC). Further works by Pyka and Dołowy [16] and Dołowy [17] show the utility of topological indices and also some electrotopological states to describe retention parameters (R_f and R_M, respectively) and lipophilic property of selected free and conjugated bile acids, such as cholic, deoxycholic, lithocholic, chenodeoxycholic, glycocholic, glycodeoxycholic, and glycolithocholic acids obtained under various chromatographic conditions using TLC, HPTLC, and RP-TLC plates. Of all applied indices—Gutman's index (M^ν), Randić indices ($^0\chi^\nu$, $^1\chi^\nu$, $^2\chi^\nu$), Pyka (χ_{012}, A, ^0B, ^1B, C), and Wiener (W)—the index C gives good correlation with both chromatographic parameters R_f and R_M on non-impregnated and impregnated TLC and RP-TLC plates. Thus, it can be used in the prediction of chromatographic separation of these compounds. The correlation between index C and R_M is linear. In the case of R_f and C values, the quadratic relationship is observed. For silica gel 60 plates with a concentrating zone, the most accurate prediction of the R_M values of examined bile acids was achieved by a monoparametric linear equation between R_M and the topological index M^ν or W, A, M^ν, $^0\chi^\nu$, χ_{012}^ν, respectively [18]. The following work of Pyka [19] on chromatographic separation of selected steroids (i.e., corticosterone, corticosterone acetate, 11-dehydrocorticosterone acetate, hydrocortisone, cortisone, 11-dehydrocorticosterone) by RP-TLC method indicates that from amongst topological indices such as Randić indices ($^0\chi$, $^0\chi^\nu$, $^1\chi^\nu$, $^1\chi$), Pyka (A, ^0B), Wiener (W), and Balaban (I_B), the most significant linear correlation ($r > 0.96$) with chromatographic R_M values was obtained for topological index ^0B [19]. A further paper prepared by Pyka et al. [20] concerning the use of traditional structural descriptors including topological indices like Randić indices ($^0\chi$, $^1\chi$, $^2\chi$ $^0\chi^\nu$, $^1\chi^\nu$, $^2\chi^\nu$), Pyka (A, ^0B, ^1B, C, D), Gutman's indices (M^ν, M), Wiener (W), Balaban (I_B), and Rouvray (R) in QSRR and QSAR analyses of thirteen 5,5-disubstituted barbiturates confirms that the best prediction of R_M values of studied barbiturates on RP-TLC plates was obtained by the use of a two-parametric equation employing the dipole moments of mobile phase and one topological index from among the topological indices $^0\chi$, $^1\chi$, $^0\chi^\nu$, $^1\chi^\nu$, R, W, A, ^1B, and three-parametric equations employing the dipole moments of mobile phase and two topological indices I_B and $^1\chi$ as well as I_B and $^1\chi^\nu$. Other calculated topological indices such as Gutman's indices (M^ν, M) and Randić indices of zero order ($^0\chi$, $^0\chi^\nu$) are the most universal for QSAR analysis of barbiturates studied.

Pyka and coworkers confirmed that of the selected topological indices, such as Randić indices ($^0\chi$, $^1\chi$, $^2\chi$ $^0\chi^\nu$, $^1\chi^\nu$, $^2\chi^\nu$, $^4\chi_c$, $^4\chi_c^\nu$), Pyka (A, ^0B, ^1B), Gutman's indices (M^ν, M), Wiener (W), Rouvray (R), and also $S_{i(O)}$ index, only topological index ($S_{i(O)}$) and Randić indices $^4\chi_c$, $^4\chi_c^\nu$ allow for estimation of the chromatograms obtained for separated essential oil components: (+) menthol, borneol, geraniol, linalool, carvone, camphor, and (1R)-(−) fenchone) in the TLC system [21]. The results of the following work by Pyka and Śliwiok [22] indicate that some topological indices are useful for QSRR study of nicotinic acid esters. The most accurate prediction of the R_M values of esters examined was achieved by using a two-parametric equation relating the dipole moment (μ) and numerical value of topological indices ^0B, ^1B with the electrotopological state dssC.

In further investigations, Pyka [23] studied different disubstituted isomeric derivatives of benzene, for example, o-, m-, and p-isomers of chlorotoluene, nitrotoluene, cresol, toluidine, ethylphenol, chloroaniline, chlorphenol, nitrophenol, nitroaniline, and aminophenol, by the use of TLC. Based on the obtained results it was stated that the process of separation, and hence the occurrence of hydrogen bonds in benzene derivatives on both adsorbents, can be accurately described using two graphical relationships between R_M values and topological index ^1B, which was calculated for these compounds.

In the case of polynuclear aromatic hydrocarbons (PAHs) and their quinones studied by Pyka in another paper [24] in continuation of the tests which use topological indices, the correlation between the R_M values and the dipole moment (μ_{mph}) of the mobile phase used with the number of carbons in the molecule and total dipole moments of molecules with numerical values of topological indices—Gutman's index (M), Randić indices ($^0\chi$, $^1\chi$, $^2\chi$, $^0\chi^v$, $^1\chi^v$, $^2\chi^v$), Wiener (W), Pyka (A, ^0B, ^1B, χ_{012}), and Balaban index (I_B)—has been described. It was stated that the best prediction of the R_M values of the compounds investigated was achieved by the use of five-parameter equations relating the dipole moment of mobile phases (μ_{mph}), the topological indices (^0B, ^1B), and two other parameters [24].

The investigations of a homologous series of higher fatty acids by Pyka [25] with RP-TLC methods indicate that of two kinds of calculated topological indices based on the distance matrix (Wiener index [W], Pyka [A, ^0B, ^1B]) and on the adjacency matrix (Gutman's index [M], Randić indices [$^0\chi$, $^1\chi$, $^0\chi^v$, $^1\chi^v$]), the most accurate prediction of the R_M values of the examined compounds can be obtained by the use of monoparametric equations containing one topological index based on the adjacency matrix [25]. In continuation of these investigations by Pyka and coworkers [26], Niestrój et al. [27] indicate that the relationships (four-parameter equations) between R_M values of higher fatty acids, hydroxyl fatty acids, and their esters and selected topological indexes mentioned earlier, as well as the dipole moment (μ_{mph}) of the mobile phase or hydrogen-bond donors (#HD), and molecular weight (M_w), the electrotopological states of the atoms attached to the carboxyl group of these compounds enable accurate prediction of fatty acids and their esters in RP-TLC.

Another work by Pyka [28] confirms that in the case of benzyl alcohols, separated by thin-layer adsorption chromatography, R_M and topological indexes of these alcohols, such as Gutman's (M), Randic ($^0\chi^v$, $^1\chi^v$, $^2\chi^v$, $^3\chi^v$), Wiener (W), Pyka (A, ^0B, ^1B, ^2B, ^2B_q, ^3B, ^3B_q, C, D, χ_{012}), and Balaban index (I_B), can be performed by new polyparametric equations. It was stated that the interdependence between R_M and selected topological indexes described by Pyka [28] can be used in the formulation of a rule for the identification of the studied isomeric alcohols.

The next two papers by Pyka [29,30] describe how the selected derivatives of phenol have been separated by partition TLC. The investigations of isomeric phenol derivatives have shown that there is interdependence between R_M and topological indexes (I_t) which exist in the shape of polyparametric equations [29]. In the case of m- and p-alkoxyphenols, the best correlation equations for prediction of R_M values separately for meta- and para-alkoxyphenols were obtained by the use of the topological indexes $^2\chi^v$, C, and χ_{012} [30].

In order to predict the elution sequence of natural phenolic compounds having various pharmacological activities and examined by TLC, the multivariate regression model containing connectivity indices proposed by Kier and Hall ($^m\chi_t$), such as $^3\chi_c^v$, $^4\chi_{pc}^v$, and $^4\chi_{pc}^v$, is useful [31].

Another study performed by Pyka in 1994 [32] shows that it is possible to predict R_M values of selected benzoic acid derivatives separated by adsorption TLC using multiparametric equations and numerical values of topological indices (I_t)—Gutman (M), Randić ($^0\chi^v$, $^1\chi^v$, $^2\chi^v$, $^3\chi^v$), Wiener (W), Pyka (A, ^0B, ^1B, ^2B, ^2B_q, ^3B, ^3B_q, C, D, χ_{012})—and other parameters like Hammett constant (σ), hydrophobic constant (π), and partition coefficient (log P).

The next two research studies reported by Gutman and Pyka [33] and Pyka [34], which have focused on the use of topological indices for predicting the separation of various isomeric compounds, indicate that in the case of enantiomers and stereoisomers optical index (I_{opt}) and stereoisomeric topological index (I_{STI}) are suitable for distinguishing both. The two new topological indexes I_{opt} and I_{STI} developed by Pyka enable the distinction between optical isomers of D and L (index I_{opt})

of hydroxyl acids (i.e., lactic acid) and amino acids [35]. Numerical values of the stereoisomeric index (I_{STI}) were helpful in describing the stereoisomeric menthols and thujols with hydroxyl groups in axial and equatorial positions [36].

Another topological index proposed by Pyka, the optical topological index I_{opt} [37], has been correlated with R_M values of hydroxyl acids and amino acids separated on a chiral stationary phase. The second paper prepared by Pyka confirms that I_{STI} and its modification, index ${}^0B_{STI}$, indicate good linear correlation with R_F and R_M, respectively, of menthol and thujol stereoisomers in adsorption TLC [36].

In order to characterize the TLC behavior of selected isomer fatty acids (9-octadecanoic, 11-octadecanoic) of *cis* and *trans* (Z-E) configuration, Pyka [37], in 1997, used the new chromatographic index I_{CHR}. The results obtained are presented as the relationships between R_{MW} values and topological index ${}^0B$ or I_{CHR}, respectively [37]. The dependencies suggested by Pyka make it possible to characterize the chromatographic behavior of the isomeric acids on the basis of their chemical constitution.

Similar RP-TLC studies of these higher fatty acid isomers have been performed by Pyka and Bober in 2001. In this work, the authors confirmed the usefulness of three-parameter equations containing the dipole moments or permittivity of the mobile phase and the impregnation percentage of stationary phases, as well as one topological index based on the distance matrix of all examined: Gutman (M), Randic (${}^0\chi^\nu$, ${}^1\chi^\nu$), Wiener (W), Pyka (A, ${}^0B$, ${}^1B$, ${}^2B$), and Balaban index (I_B) [38]. Significant correlation of R_M and R_F of selected positional isomers α- and β-naphtol, naphthylamine, and also propyl, isopropyl, elaidate, and oleate is indicated in the paper reported by Pyka in 1991. Correlation between R_M values and ${}^1\chi^\nu$ and also with ${}^2B_q$ is the best. However, R_F values correlate well with ${}^2\chi^\nu$, C, D, and χ_{012} [39].

Numerous examples of the application of topological indexes to predict the retention parameters and also for the determination of lipophilicity and other physicochemical constants of various organic compounds separated by planar chromatography have been reviewed by Pyka in 2001 [40] and also in 2010 [41].

Table 18.1 summarizes the solutes, chromatographic conditions, detection, and retention models using topological indices applied in QSRR study by TLC.

18.3 PREDICTION OF CHROMATOGRAPHIC RETENTION OF VARIOUS CLASSES OF COMPOUNDS IN HPLC BY MEANS OF TOPOLOGICAL INDICES

HPLC is a very popular and useful method of analysis of many classes of compounds, including those with biological activity. There are also many classes of compounds investigated by taking into consideration the relationships between different HPLC retention factors and topological indexes. Some examples are presented in Table 18.2. The most often used and also the most popular topological indexes for this purpose are Randić indices, which represent connectivity indices of different orders [22,42–50]. Other topological indices used are Wiener index [22,43,45,46,49,51], lesser known Harary index calculated for anthocyanins [43], Pyka indices [22,45], Gutman index calculated for pesticides [44] and anilides [45], and many others taking into consideration the topological properties of chemical compounds. There are also several computer programs that are very useful for the calculation of topological indices. Many researchers use these programs in their works. The most popular is Dragon software used for the calculation of indices, for example, of pesticides [52–54], arylpropionic acid derivatives [55], polyaromatic hydrocarbons [56], anthraquinoids [57], and natural compounds [58]. The other, less popular software programs are used for the calculation of indices of polyaromatic compounds and taxanes (HyperChem) [56,59], polycyclic aromatic compounds (Cerius²QSAR+ software) [60], or steroids (MOLCONN-X) [49]. Computer programs can calculate several indices at the same time, even a few thousands for a group of compounds.

TABLE 18.1

Topological Indices Used in the Prediction of Chromatographic Retention in TLC

Class of Compounds/Solutes	Topological Index	Retention Model	Method, Stationary Phase, Mobile Phase, Detection	Reference
α-, β-, γ-, δ-Tocopherols	Randić: $^0\chi^v$, $^1\chi^v$, $^2\chi^v$ Pyka: ^{0}B, C, and D	$R_f = f(\text{topological index})$, linear	TLC, kieselguhr G plates, methanol–water, visualization by ferric chloride in methanol (0.2%), and dipyridyl in methanol (0.5%)	[13]
D,L-Tocopherols	I_t	$R_M = f(I_t)$, linear	Chiralplates, 2-propanol–water–methanol, detection at 254 nm	[15]
α-, β-, γ-, δ-Tocopherols	Topological indices based on connectivity (M, $^1\chi^v$), on distance matrix: W, ^{0}B, MTI, and topological indices based on information theory: I_{AC}, $\bar{I}_{AC}$	$R_f = f(^0B)$, significant correlation	RP-HPTLC, RP-18F$_{254}$ plates, ethanol–water, ferric chloride in methanol (0.2%), and dipyridyl in methanol (0.5%)	[14]
Tocopherols	Randić: $^2\chi^v$ Pyka: C, ^{0}B	R_M, linear	RP-HPTLC, RP-18F$_{254}$ plates, ethanol, methanol, n-propanol, ethanol–water, n-propanol–water, visualization by solutions of ferric chloride in methanol (0.2%), and solution of dipyridyl in methanol (0.5%)	
Bile acids (free and conjugated with glycine)	Gutman's index (M^v), Randić indices ($^0\chi^v$, $^1\chi^v$, $^2\chi^v$), Pyka (χ_{012}-A, ^{0}B, ^{1}B, C), Wiener (W)	$R_f = f(C)$ quadratic $R_M = f(C)$ linear	TLC, HPTLC, silica gel and diol modified silica gel plates, n-hexane–ethyl acetate–acetic acid, visualization by aqueous solution of 10% H$_2$SO$_4$, and 10% ethanolic solution of phosphomolybdic acid	[16]
Bile acids (free and conjugated with glycine)	Gutman's index (M^v), Randić indices ($^0\chi^v$, $^1\chi^v$, $^2\chi^v$), Pyka (χ_{012}-A, ^{0}B, ^{1}B, C), Wiener (W)	$R_f = f(C)$, quadratic $R_M = f(C)$, linear	TLC plates (silica gel and silica gel/kieselguhr) and modified (impregnated with inorganic salts, i.e., CuSO$_4$, MnSO$_4$, FeSO$_4$, NiSO$_4$), n-hexane–ethyl acetate–acetic acid, visualization by aqueous solution of 10% H$_2$SO$_4$	[17]
Bile acids (free and conjugated with glycine)	Gutman's index (M^v), Randić indices ($^0\chi^v$, $^1\chi^v$, $^2\chi^v$), Pyka (χ_{012}-A, ^{0}B, ^{1}B, C), Wiener (W)	$R_M = f(W, A, M^v, \chi^v, \chi^v_{012})$, linear	TLC, silica gel plates with concentrating zone, n-hexane–ethyl acetate–methanol–acetic acid, visualization by methanol solution of 10% H$_2$SO$_4$	[18]

(Continued)

TABLE 18.1 (Continued)
Topological Indices Used in the Prediction of Chromatographic Retention in TLC

Class of Compounds/Solutes	Topological Index	Retention Model	Method, Stationary Phase, Mobile Phase, Detection	Reference
Selected steroids (corticosterone, corticosterone acetate, 11-dehydrocorticosterone acetate, hydrocortisone, cortisone, 11-dehydrocorticosterone)	Randić indices ($^0\chi$, $^0\chi^v$, $^1\chi^v$, $^1\chi$), Pyka (A, ^0B, χ_{012}, χ_{012}^v), Wiener (W), Balaban (I_B)	$R_M = f(^0B)$, linear	RP-TLC, silica gel/kieselguhr plates for RP-TLC impregnated with paraffin oil in hexane, methanol–water, visualization in UV light (254 nm)	[19]
Barbiturates (5,5-disubstituted barbituric acid derivatives)	Randić indices ($^0\chi$, $^1\chi$, $^2\chi$, $^0\chi^v$, $^1\chi^v$, $^2\chi^v$), Pyka (A, ^0B, ^1B, C, D), Gutman's indices (M^v, M), Wiener (W), Balaban indices (I_B), Rouvray (R)	Two- or three-parametric equations $R_M = f(I_B, ^1\chi, \mu)$ and $R_M = f(I_B, ^1\chi, \mu, ^1\chi^v)$, linear	RP-TLC, silica gel plates for RP-TLC, methanol–water, visualization in UV light (254 nm)	[20]
Selected essential oil components: (+) menthol, borneol, geraniol, linalool, carvone, camphor, (1R)–(–) fenchone	Randić indices ($^0\chi$, $^1\chi$, $^2\chi$, $^0\chi^v$, $^1\chi^v$, $^2\chi^v$, $^4\chi$, $^4\chi^v$), Pyka (A, ^0B, ^1B), Gutman's indices (M^v, M), Wiener (W), Rouvray (R), and also $S_{i(o)}$ index	$R_M = f(S_{i(o)}, ^4\chi_C, ^1\chi_C^v)$, linear	TLC, silica gel plates, benzene, detection using 5% solution of K_2CrO_4 in 40% H_2SO_4	[21]
Nicotinic acid esters	Pyka (A, ^0B, ^1B, ^2B), Wiener (W), Randić indices ($^0\chi^v$, $^1\chi^v$, $^2\chi^v$), indices based on information theory (I_{SA}, $\bar{I}_{SA}$)	Two-parametric equations $R_M = f(^0B, \mu)$ and $R_M = f(^1B, \mu)$ with electrotopological state dssC, linear	TLC, silica gel 60F$_{254}$ and silica gel 60F$_{254}$/kieselguhr TLC plates, n-hexane–acetone, visualization in UV light (254 nm)	[22]
Disubstituted isomeric derivatives of benzene (including-o, -p, and m-isomers)	Pyka (^1B)	Graphical relationship between R_M data and index ^1B, significant correlation	TLC, silica gel 60F$_{254}$ and aluminum oxide TLC plates, n-heptane–benzene–diethyl ether, detection with K_2CrO_4 in 40% H_2SO_4, and rhodamine B (at $\lambda = 365$ nm)	[23]
PAHs and their quinones	Gutman's index (M), Randić indices ($^0\chi$, $^1\chi$, $^2\chi$, $^0\chi^v$, $^1\chi^v$, $^2\chi^v$), Wiener (W), Pyka (A, ^0B, ^1B, χ_{012}), and Balaban index (I_B)	Multiple correlation equations between R_M and ^0B, ^1B, μ_{mph} with two other parameters	TLC, silica gel 60F$_{254}$ TLC plates, benzene–acetone, detection under UV light ($\lambda = 254$ nm)	[24]
Homologous series of higher fatty acids, higher alcohols, and methyl esters of higher fatty acids	Gutman's index (M), Randić indices ($^0\chi$, $^1\chi$, $^0\chi^v$, $^1\chi^v$), Wiener (W), and Pyka (A, ^0B, ^1B)	Monoparametric equations between R_M and one of topological indices: $^0\chi$, $^1\chi$, $^0\chi^v$, $^1\chi^v$, or M are linear	RP-TLC, plates for RP-TLC precoated with kieselguhr F$_{254}$ impregnated with paraffin oil, methanol–water and ethanol–water, visualization using bromothymol blue and fuchsin	[25]

(Continued)

TABLE 18.1 (*Continued*)
Topological Indices Used in the Prediction of Chromatographic Retention in TLC

Class of Compounds/Solutes	Topological Index	Retention Model	Method, Stationary Phase, Mobile Phase, Detection	Reference
Selected isomeric aromatic alcohols	Topological indexes (I_t): Gutman's (M), Randić ($^0\chi^v$, $^1\chi^v$, $^2\chi^v$, $^3\chi^v$), Wiener (W), Pyka (A, ^{0}B, ^{1}B, ^{2}B, ^{2}B$_q$, ^{3}B, ^{3}B$_q$, C, D, χ_{012}), Balaban index (I_B)	Polyparametric equations type: $R_M = a \cdot I_{ti} + b \cdot I_{tj} + c$, $R_M = a \cdot I_{ti} + b \cdot I_{tj} + c \cdot I_{tk} + d$, significant	NP-TLC, silica gel 60 plates, carbon tetrachloride–diethyl ether–ethyl acetate, detection of spots by basic solution of thymol blue	[28]
Phenol derivatives Isomeric phenol derivatives	Gutman's (M), Randić ($^0\chi^v$, $^1\chi^v$, $^2\chi^v$, $^3\chi^v$), Wiener (W), and Pyka (A, ^{0}B, ^{1}B, ^{2}B, ^{2}B$_q$, ^{3}B, ^{3}B$_q$, C, D, χ_{012})	Simple linear and multiple equations type: $R_M = a \cdot I_t + b$, $R_M = a \cdot I_t^2 + b \cdot I_t + c$ are significant	RP-TLC, impregnating silica gel G with silicone DC 200, veronal acetate buffer (pH 7.4)–acetone, tetradiazotized benzidine	[29]
m- and p-Alkoxyphenols	Randić ($^2\chi^v$, $^3\chi^v$), Wiener (W), Pyka (A, ^{0}B, ^{1}B, ^{2}B, ^{2}B$_q$, ^{3}B, ^{3}B$_q$, C, D, χ_{012}), Balaban (I_B)	Linear and second- or third-degree polynomial equations between R_M and one topological index, the best correlations are with $^2\chi^v$, $C\chi_{012}$	RP-TLC, chromatographic plates precoated with cellulose and impregnated with ethyl oleate in diethyl ether, ethanol–water mixtures, alkaline solution of permanganate was used as visualizing reagent	[30]
Natural phenolic derivatives	Kier and Hall connectivity indices $^3\chi_c^v$, $^4\chi_{pc}^v$, and $^4\chi_{pc}$	Between R_M and connectivity indices, linear relationship	NP-TLC, silica gel plates, benzene–dioxane–acetic acid, benzene–methanol–acetic acid, diazotized benzidine, and sodium nitrate solution as visualizing reagents	[31]
Benzoic acid derivatives	Gutman's (M), Randić ($^0\chi^v$, $^1\chi^v$, $^2\chi^v$, $^3\chi^v$), Wiener (W), and Pyka (A, ^{0}B, ^{1}B, ^{2}B, ^{2}B$_q$, ^{3}B, ^{3}B$_q$, C, D, χ_{012})	Multiparametric equations between R_M and calculated indexes were performed	RP-TLC, plates precoated with cellulose, isobutanol–ammonia mixture as mobile phase, basic solution of bromothymol or bromocresol as visualizing reagent	[32]
D and L isomers of hydroxyl acids and amino acids	Optical topological index (I_{opt}) and modified indexes A', ^{1}B', ^{0}B'	Simple or multiple regression equations between R_M and calculated indexes were found	TLC on chiral stationary-phase ChirHPTLC plates, acetonitrile–water was used as mobile phase for D and L amino acids, basic solution of thymolphthalein was a visualizing reagent Amino acids were separated using methanol–water–acetonitrile and visualized by ninhydrin (0.3%)	[35]

(Continued)

TABLE 18.1 (*Continued*)
Topological Indices Used in the Prediction of Chromatographic Retention in TLC

Class of Compounds/Solutes	Topological Index	Retention Model	Method, Stationary Phase, Mobile Phase, Detection	Reference
Stereoisomeric menthols and thujols	Stereoisomeric topological index (I_{STI}) and modified index $^0B_{STI}$	Simple regression equations between R_F or R_M and calculated indexes were found $R_F = a + b \cdot I_t$, $R_M = a + b \cdot I_t$	TLC, silica gel plates, n-hexane–ethanol for isomeric menthols, benzene was used as a mobile phase for isomeric thujols, detection by basic solutions of thymol blue and bromothymol	[36]
Isomers of fatty acids with *cis* and *trans* (Z-E) configuration	Chromatographic index I_{CHR}, ^0B	Relationship between the R_M values and chromatographic index I_{CHR} and ^0B was significant	RP-TLC, silica gel RP-18F$_{254}$, silica gel F$_{254}$ impregnated with paraffin oil, glacial acid was used as a mobile phase, the isomers were detected with iodine vapor	[37]
Positional isomers α- and β-naphtol and naphthylamine, propyl, isopropyl, elaidate, and oleate	Gutman's index (M), Randić ($^1\chi^v$, $^2\chi^v$), Wiener (W), Pyka (A, ^0B, ^1B, ^2B, ^2B_q, C, D, χ_{012}), and Balaban index (I_B)	Correlation between R_M values and $^1\chi^v$ or ^2B_q are the best, R_F values correlate well with $^2\chi^v$, C, D, and χ_{012}	TLC, plates precoated with kieselguhr G impregnated with paraffin oil, methanol–water, or 1-propanol–water were used as mobile phases, visualization with a solution of fuchsin (0.005%)	[39]

TABLE 18.2
Topological Indices Used in the Prediction of Chromatographic Retention in HPLC

Class of Compounds/Solutes	Topological Index	Retention Model	Stationary Phase, Mobile Phase	Reference
Saturated aliphatic alcohols	MPEI: molecular polarizability index; OEI: odd–even index; SX$_{ICH}$: eigenvalues of bond-connecting matrix	Between chromatographic (Kovat's) retention index and topological indices studied, multiple linear regression (MLR)	Six different stationary phases, data taken from literature	[61]
Saturated aliphatic alcohols	Molecular connectivity indices ($^1\chi$): path and cluster	Chromatographic (Kovat's) retention index and topological indices, MLR, artificial neural network (ANN)	Six different stationary phases, data taken from literature	[64]
Single- and multi-ring aromatic hydrocarbons	Valence connectivity indexes ($^1\chi^v$, $^2\chi^v$)	Retention time and connectivity indexes, MLR	[3-(2,4-Dinitroanilino)]-propyl-silica) column, n-pentane, and methylene chloride	[42]
Different classes of drug-like compounds	Electrotopological state (E-state), molecular connectivity indices	Chromatographic (Kovat's) retention index and topological descriptors, MLR, ANN	C-8, phosphoric acid–triethylamine in water or phosphoric acid in trimethylamine and acetonitrile in water	[62]
Anthocyanins	Wiener index, connectivity indices ($^1\chi$, $^2\chi$, $^3\chi$), Harary indices (H, H', H'')	Between retention time and topological indexes, linear relation	N/A	[43]
Pesticides	Gutman's (M, M^v) indices, connectivity indices ($^0\chi$, $^1\chi$, $^2\chi$, $^0\chi^v$, $^1\chi^v$, $^2\chi^v$)	Logarithm of retention factor (log k) and topological indices, linear relation	C18, methanol–water	[44]
Para substituted anilides	Rouvray (R), Wiener (W), and Pyka (A, ^{0}B, ^{1}B, ^{2}B) indices, Gutman's (M, M^v) indices, connectivity indices ($^0\chi$, $^1\chi$, $^2\chi$, $^0\chi^v$, $^1\chi^v$, $^2\chi^v$)	Logarithm of retention factor (log k) and topological indices, MLR	N/A	[45]
Chalcones	Wiener (W), valence and connectivity indices, kappa-shaped indexes	Capacity factor (k) and logarithm of capacity factor (log k) with topological indices, partial least squares (PLS)	–NH$_2$, DIOL, –CN, ODS, C8, methanol–water	[51]
Amino acids	Connectivity indices ($^0\chi$, $^1\chi$, $^2\chi$, $^3\chi$, $^4\chi$), Wiener index (W)	Retention factor (k) and connectivity indices, linear and MLR	C18, trifluoroacetic acid–water and trifluoroacetic acid–acetonitrile	[46]
Pesticides	Descriptors obtained by use of Dragon software	Between retention time and indices, MLR, ANN	N/A	[52]

(Continued)

TABLE 18.2 (Continued)
Topological Indices Used in the Prediction of Chromatographic Retention in HPLC

Class of Compounds/Solutes	Topological Index	Retention Model	Stationary Phase, Mobile Phase	Reference
Nucleic acids	Connectivity indices ($^0\chi$, $^1\chi$, $^2\chi$, $^3\chi$, $^4\chi$, $^5\chi$), Balaban index	Retention factor (k), MLR with topological indices	C18, sodium phosphate buffer–methanol	[47]
Diverse drugs	Connectivity indices ($^5\chi^v$, $^6\chi^v$), electrotopological state (E-state)	Binding affinity constant (k), all-possible subsets regression with topological indices	N/A	[48]
Steroids	Connectivity indices ($^{1-10}\chi$, $^{1-10}\chi^v$, $^{3,4}\chi_{c/cp}$, $^{3,4}\chi_{c/cp}^v$), topological equivalence indexes, electrotopological indices, Bonchev–Trinajstić indices, Platt's and Wiener numbers	Retention time and topological indices, MLR	C18, methanol–tetrahydrofuran–NaH$_2$PO$_4$ monohydrate or methanol–tetrahydrofuran–water	[49]
Amino acids	Connectivity indices ($^0\chi$, $^1\chi$, $^2\chi^v$)	Retention time and topological indices, ANN	C18, acetonitrile–acetate buffer	[50]
Pesticides	Descriptors obtained by use of Dragon software	Logarithm of retention factor (log k) and descriptors, ANN	N/A, acetonitrile–water	[53]
Arylpropionic acid derivatives	Descriptors obtained by use of Dragon software	Logarithm of retention factor (log k) and descriptors, ANN	N/A Acetonitrile–water	[55]
Polyaromatic hydrocarbons	Descriptors obtained by use of HyperChem and Dragon software	Retention time and descriptors, MLR	C18, water–trifluoroacid, and acetonitrile–trifluoroacid	[56]
Anthraquinoids	Descriptors obtained by use of Dragon software	Relative retention factor and descriptors, MLR, and PLS	N/A	[57]
PAHs	Descriptors obtained by use of Cerius²QSAR + software	Between logarithm of retention index (log I) and descriptors, PLS	C18, acetonitrile–water	[60]
Nicotinic acid esters	Connectivity indices ($^0\chi^v$, $^1\chi^v$, $^2\chi^v$), Wiener (W), Pyka (A, ^{0}B, ^{1}B, ^{2}B) indices, and indices based on information theory (I_{SA}, $\overline{I_{SA}}$)	Retention time and topological indices, MLR	LiChrospher Si 60, benzene–methanol	[22]

(Continued)

TABLE 18.2 (Continued)
Topological Indices Used in the Prediction of Chromatographic Retention in HPLC

Class of Compounds/Solutes	Topological Index	Retention Model	Stationary Phase, Mobile Phase	Reference
Atrazine (pesticide)	Descriptors obtained by use of Dragon software	Retention factor and descriptors, MLR, and PLS	C8, C12, and C18, different mobile phases consist of acetonitrile, methanol, propanol, tetrahydrofuran, acetone, dioxane mixed with water	[54]
Taxanes	Descriptors obtained by use of HyperChem software	Retention time and descriptors, MLR, ANN	XR-ODS, water–acetonitrile	[59]
Suspected sports doping–related compounds	Descriptors obtained by use of Dragon software	Retention time and descriptors, PLS	C18, water–acetonitrile	[63]
Natural products with general formula $C_xH_yO_z$	Topological polar surface area	Retention time and topological polar surface area, MLR, ANN	C18, acetonitrile–formic acid	[58]

It is very useful but also requires more sophisticated method of description of the relationships between indices and other parameters. The easiest description of such relationships is linear and multiple linear regression (MLR) used for predicting the retention parameter of amino acids [46], pesticides [44,52,54], nucleic acids [47], steroids [49], aliphatic alcohols [61], aromatic and polyaromatic hydrocarbons [42,56], and many classes of drug-like compounds [62], anthocyanins [43], anilides [45], anthraquinoids [57], nicotinic acid esters [22], taxanes [63], and natural product [58]. The partial least squares (PLS) model is also often used as a predictive model [51,54,57,60,63]. The advanced computer models are increasingly being used for the description of relationships and prediction of retention parameters. They allow for more detailed descriptions of relationships investigated and also give the possibility of prediction of retention parameters with better results, on the basis of known values of topological parameters obtained from calculation; hence, they are only theoretical. The artificial neural network (ANN) is very often used for such advanced descriptions of relationships obtained [47,50,53,55,58,64]. Another, rarely used model of prediction is the all-possible subsets regression model applied in the case of investigation of drugs [48]. This model identifies all the possible regression models derived from all possible combinations of descriptors, and then determines the best predictor model and evaluates it statistically. Another problem while considering the relationships between retention parameter and topological indexes is the kind of parameter used. In some cases, the retention time (RT) is used. The relationships between RT and topological indexes are considered for anthocyanins [43], pesticides [52], steroids [49], amino acids [50], nicotinic acid esters [22], aromatic and polyaromatic hydrocarbons [42,56], taxanes [59], sports doping–related compounds [63], and natural products [58]. For some group of compounds, the retention factor k or log k is used [44–47,51,53–55,60]. Some of the researchers investigating diverse drugs call it the "binding affinity constant k" instead of using the term "retention factor" [48]. The value of k is calculated according to the equation $k = (t_R - t_m)/t_m$, where t_R is the RT of the solute and t_m is the RT of unretained compound [46]. The relative retention factor is used for the description of relationships between retention parameter and topological indexes in the case of anthraquinoids [57]. It is calculated according to the equation: $Rt_{X/Anq} = (t_{G(X)} - t_0)/(t_{G(Anq)} - t_0)$, where $t_{G(X)}$ and $t_{G(Anq)}$ are the RT of a compound and that of anthraquinone, respectively. In the case of saturated aliphatic alcohols [61,64] and many different classes of drug-like compounds [62], the chromatographic (Kovat's) retention index (RI) is considered. This index allows converting RTs into system-independent constants. Most of the mentioned investigations have the possibility to predict the RT or RI of the compound investigated on the basis of known values of topological indexes and other topological parameters obtained by calculation. Some advanced computer programs have turned out to be very useful for this purpose. The usefulness of topological indices in the prediction of retention in HPLC analysis was already presented considering particular classes of compounds [65–67]. The example correlation equations were presented on the basis of available literature. Polyakova and Row [68] also consider the possibility of utilizing topological and molecular descriptors as predictive models in QSRR. They divided descriptors into classes, depending on their origin. Among many parameters they considered topological indexes such as Wiener, Randić, Balaban, as well as Kier and Hall indexes, and also topological electronic indices, structural information content index, and others. The possibility of applying different computer programs for the calculation of topological parameters was also described. The authors showed the usefulness and importance of conducting QSRR analysis with the use of topological and other molecular parameters.

18.4 USE OF TOPOLOGICAL INDICES IN STRUCTURE–RETENTION RELATIONSHIP STUDY OF DIFFERENT COMPOUNDS IN GC

The use of topological indices in structure–retention relationship studies is widely investigated in GC. Among the correlations between chromatographic retention and molecular descriptors, for example, topological indices provide significant information on the effect of molecular structure, on RT, and on the possible mechanism of elution. Moreover, the development of these relationships is

important since standards are not available. Many topological indices have been proposed in establishing the structure–retention relationship for classes of organic compounds investigated by GC
with different structural features. QSRR approaches in GC are usually based on the use of topological indices in the prediction of RIs (i.e., Kovat's index). It is a popular dependent variable in QSRR
study because of its reproducibility and accuracy.

One of the most useful topological indices that have been applied in QSRR analysis of organic
compounds was that based on Wiener's index, named Lu index. This index was proposed by Lu
and coworkers [69] for modeling properties including RIs of GC of hydrocarbons and halogenated
hydrocarbons. The linear relationships constructed by Lu et al. between RI and Lu index with high
correlation coefficients (>0.99) indicate that Lu index can be used not only in hydrocarbons but also
in heteroatom-containing and multiple bond–containing organic compounds [69].

Another work by Lu and coworkers [70] confirmed that the chromatographic RI strongly depends
on interactions between eluents and stationary phases. For this reason, a single index cannot give
simple and accurate models for RI as the polarity of columns increases. It was stated that the final
model for describing the RI of 90 saturated esters on seven stationary phases containing non-, low-,
medium-, and high-polarity polysiloxanes is the one generated by using the Lu index and several
distance-based atom-type topological indices (DAIs) in the form of MLR equations.

Another work shows the use of molecular connectivity indices to predict the gas chromatography
of Kovat's relative retention indices of 132 volatile organic compounds on 12 (4 apolar and 8 polar)
stationary phases [71].

The Balaban, Wiener, and electrotopological state and molecular shape indices were determined
and correlated with RIs of linear alkylbenzene isomers with C_{10}–C_{14} linear alkyl chains [72]. Of all
the indices, the Wiener index (W) is the only one that showed good single linear correlation with the
RIs and a correct elution sequence.

Another study by Heinzen and coworkers [12] indicates the usefulness of a semi-empirical topological method for the prediction of the chromatographic retention in GC of *cis*- and *trans*-alkene
isomers and alkanes.

Two papers prepared by Santiuste confirm the utility of topological indices in studies of the
relationships between retention data in GC and solute structures within chemical groups for a wide
variety of compounds belonging to hydrocarbons, derivatives of hydrocarbons, and chemical derivatives of benzene [73,74]. Three topological indices (first-order connectivity index, Wiener's index,
and Balaban index) were studied in these papers.

In 2001, Feng and Du determined the linear regression equation between RI and topological
index ^mQ. This model can elucidate the change rule of RI for amines [75].

The novel semi-empirical topological index (I_{ET}) developed by Junkes and coworkers [76], which
has been used earlier for modeling of chromatographic retention in GC of *cis*- and *trans*-linear
alkene isomers and alkane, has been also used to predict the RIs of a series of saturated alcohols
examined on different stationary phases of low to medium polarity. The statistical analysis of the
obtained correlation confirms that the application of the QSRR model to stationary phases of low
polarity (OV-1) has predictive ability for external data.

Krawczuk and coworkers reported that the electrotopological index (TI^E) and Balaban index
(I_B) can be successfully used in the prediction of polychlorinated biphenyls' (PCBs) retention data
despite the temperature program applied in the GC analysis [77].

A paper by Zenkevich and Marinichev [78] noted that a linear model between topological indices
Wiener (W) and Hosoya (Z) respectively and GC RI is helpful to predict the order of GC elution
of isomer compounds belonging to heptanes, hexanes, hexynes, cycloalkanes, pentylbenzenes, and
mono- and disubstituted chloro- and bromobutanes.

Many researchers have analyzed the medical importance of constituents of essential oils by GC.
The QSRR models were built for a series of 96 essential oils containing volatile constituents and
were used to predict their GC RTs (t_R) [79]. Different chemometric tools have been applied to

build various QSRR models using 13 nonzero E-state indices developed by Kier and coworkers in 1990 and selected molecular connectivity indices (Chi indices), Randić topological indices, Balaban index, and Wiener topological index [79].

A good linear relationship between GC RI and electronic or topological descriptors including topological indices was found by linear regression analysis for PAHs. The significant correlation with RI showed molecular polarizability (α), second-order molecular connectivity, and Kier and Hall index ($^2\chi$) [80].

The QSRR study of 37 phenolic derivatives reported by Garkani-Nejad [81] indicates that the Kier symmetry index (S0k) is the most important factor affecting the retention behavior of these compounds. Another work indicates that GC RI of methylalkanes produced by insects can be modeled by molecular structural descriptors like the Balaban topological index [82].

In a further paper, Lu index and DAI combined with electronegativity χep (χDAI) were used to predict GC relative RTs of 81 organic pesticides with diverse heteroatoms. The MLR technique was applied to predict the relationships between relative RTs of these compounds on the four kinds of chromatographic columns [83].

A paper by Liu et al. [84] illustrates that two novel topological indices, the modified polarizability effect index and the modified inner polarizability effect index, produce good correlations with Kovat's RI of complex compounds with polar multifunctional groups [84]. A second paper of these authors confirms the applicability of polarizability effect index, odd–even index, and steric effect index to predict the GC RI of saturated esters on seven different stationary phases [85]. In 2003, Ren developed MLR models using Xu and AI topological indices to correlate Kovat's RI of a mixed set of aliphatic aldehydes and ketones on different polar stationary phases [86].

Another work illustrates the use of MLR equations based on Gutman index, Balaban-type index, and valence connectivity index chi-2, chi-0, and chi-1 in the prediction of GC RI of O-, N-, and S-heterocyclic compounds [87].

Can et al. [88] have applied ANN for predicting 80 polycyclic aromatic sulfur compounds (PASHs) examined by GC. It was stated that among different descriptors generated for these compounds using Dragon software, only 3D-Balaban index and the sum of Kier–Hall electrotopological state may be satisfactorily used in the prediction of RI of PASHs.

Another work illustrates that MLR can be used to map the topological indices of simple molecular connectivity of Chi indices ($^4\chi^d$, $^1\chi^{a,b}$, $^0\chi^{a,b}$) to the GC RIs of 656 flavor compounds examined by GC [89].

The utility of the semi-empirical topological index (I_{ET}) developed by Heinzen in the prediction of chromatographic retention of branched alkanes is demonstrated in the paper prepared by Junkes and coworkers [90]. A linear relation was confirmed between the topological index (I_{ET}) and chromatographic RI of branched alkenes.

Satisfactory linear correlation between GC RIs and connectivity index $^1\chi^f$ of various alkanes and alcohols was also found by Randic and coworkers [91].

A paper concerning GC analysis of selected alcohols and esters illustrates that the topological indices R_{AF} and Δ_A have the best predictive power of GC RTs of alcohols and esters, respectively [92].

The paper of Rojas et al. [93] indicates that in the modeling of RI of 1208 flavors and fragrances on the OV-101 glass capillary column, the solvation connectivity index of first order is strongly correlated with the RI.

A multiple linear model using various molecular descriptors enabling the prediction of the behavior of PCB in capillary GC within a wide range of separation conditions was built by D'Archivio et al. [94].

Different structural parameters, such as topological indices and electrotopological states and others, have been used to find the relation between GC RTs of 20 substituted coumarins examined on two different stationary phases with low polarity [95].

Brazilian researchers investigated many classes of compounds taking into consideration the semi-empirical electrotopological index I_{SET} [76,96–99]. This index is calculated according to the following equation:

$$I_{SET} = \sum_{i,j} \left(SET_i + \log SET_j \right) \qquad (18.2)$$

where
    i is the overall sum of the atoms of the molecule (except H)
    j is an inner sum of atoms attached to the i atom [96]

The groups of compounds analyzed were esters, alcohols, and alkylbenzenes. Different stationary phases of different polarity were applied. The correlation between I_{SET} and RI as well as others parameters were analyzed. The GC indices were taken from the literature. In the case of esters, simple linear regression between the RIs and semi-empirical electrotopological indices were obtained [96,99]. Similar results were obtained in the case of alcohols, where the relationship between retention and electrotopological indices can be described with the use of simple linear regression [76,98]. In the case of different kinds of alkylbenzenes, linear regression can be used for the description of the relationship between gas RIs and semi-empirical electrotopological indices [97]. Many different indexes, including topological ones, taken from Dragon software were considered in relationships with Kovat's RI in the case of the terpenoids group [100]. MLR and ANN were utilized to settle these relationships. The validation model for all groups of compounds investigated showed that the final models are statistically significant and can be used for the prediction of RIs.

Applications of topological indices in modeling HPLC retention by topological indices are presented in Table 18.3.

18.5 TOPOLOGICAL INDICES IN DEVELOPMENT OF RETENTION BEHAVIOR OF SELECTED COMPOUNDS IN MIGRATION TECHNIQUES: CAPILLARY ZONE ELECTROPHORESIS AND MICELLAR ELECTROKINETIC CAPILLARY CHROMATOGRAPHY

Among various types of capillary electrophoresis, CZE and MEKC are the most popular techniques. There are only two papers so far concerning the search of a correlation between retention data in capillary chromatography and topological indices.

In 1996, Salo and coworkers confirmed that electrophoretic migration data of steroid hormones (t_M) obtained using sodium dodecyl sulfate (SDS)–borate system with a mixed micellar solution of SDS and sodium cholate can be predicted by means of linear correlation equation containing structural parameters of examined steroids, for example, sixth-order connectivity index (Randic index), $^6\chi$ [49]. It was stated that topological indices can be used in the modeling of migration of steroid hormones when the solutes form a congeneric series and stereochemical properties do not govern the separation process.

The aim of the second work by Liang et al. [101] was to develop correlations between the migration of 13 flavonoids in CZE and topological indices and to try to predict the migration of these flavonoids. Each of the developed models includes molecular connectivity index ($^m\chi_t^v$) and electrotopological state index (S_i). The results discussed in this paper demonstrate that $^m\chi_t^v$ and S_i can successfully be used to obtain a model of structure–electrophoretic mobility (μ_c) of flavonoids in CZE [101].

TABLE 18.3

Topological Indices Used in the Prediction of Chromatographic Retention in GC

Class of Compounds/Solutes	Topological Index	Retention Model	Stationary Phase/Column	Reference
Saturated and unsaturated hydrocarbons and halogenated hydrocarbons	Lu index	RI and Lu, linear	N/A	[69]
90 Saturated ester compounds	Lu index, Randic first-order connectivity index ($^1\chi$)	Lu and $^1\chi$ with RI, linear; Lu index and several DAIs with RI, MLR	GC-FID, polysiloxanes: SE-30, OV-7, DC-710, OV-25, XE-60, OV-225, Silar-5CP	[70]
132 Volatile organic compounds consisting of alkanes, alkenes, ethers, amines, alcohols, alkylbenzenes, and alkylhalides	Randic indexes: $^1\chi_v$, $^3\chi_v$, $^4\chi_v$, $^1\chi_{sol}$, $^2\chi_{sol}$, $^3\chi_{sol}$, and $^4\chi_{sol}$	Randic indices, multiparametric equations obtained by multivariate linear regression (MLR)	C_{67}, C_{103}, C_{78}, C_∞, POH, TTF, MTF, PCL, PBR, TMO, PSH, and PCN	[57]
Linear alkylbenzene isomers (LABs)	Wiener index (W), Balaban index (J), indices of molecular shape (different orders of kappa values χ)	Chromatographic retention index (I) of LABs and Wiener index, linear correlation	SE-54, DB-2 stationary phase	[72]
cis- and *trans*- Alkene isomers and alkanes	Semi-empirical topological index (I_{ET}), Wiener number ($^3W_{CH}$), molecular connectivity index of the first order ($^1\chi$)	Correlation between RI and topological indices (I_{ET}, $^3W_{CH}$, $^1\chi$) significant	N/A	[12]
Hydrocarbons, hydrocarbon derivatives containing oxygen, nitrogen, halogens Derivatives of benzene	Wiener index (W), Balaban index (J), molecular connectivity indices of Randić, valence connectivity indices of Lemont and Kier	Correlation between retention index (I) and $^1\chi^v$ for chemical benzene derivatives, linear; for halogenated derivatives, quadratic correlation	Different nonpolar and polar stationary phases TFPSI5, XF-1150, SE-30, Squalane, Apiezon L, PBD, PEA, PDMS	[74]
Aliphatic and aromatic amines	Connectivity index (^mQ)	RI and ^mQ, linear regression equation	Nonpolar and polar stationary phases OV-101, OV-225, and NGA	[75]
Saturated alcohols	Semi-empirical topological index (I_{ET})	Between semi-empirical topological index (I_{ET}), simple linear regression	OV-1, SE-30, OV-3, OV-7, OV-11, OV-17, and OV-25	[76]
PCBs	Balaban index (I_B), electrotopological index (TI^E)	Relationship between RI and Balaban index/ electrotopological index, linear	Silica capillary column PE-5MS, 5% diphenyl-95% dimethyl siloxane	[77]

(Continued)

TABLE 18.3 (*Continued*)
Topological Indices Used in the Prediction of Chromatographic Retention in GC

Class of Compounds/Solutes	Topological Index	Retention Model	Stationary Phase/Column	Reference
Several groups of isomeric organic compounds (heptanes, hexanes, hexynes, cycloalkanes, pentylbenzenes, mono- and disubstituted chloro- and bromobutanes)	Wiener (W), Hosoya (Z) indices	Between RI and topological indices (W, Z) linear correlation relation RI = a × Y + b where Y is the topological index	N/A	[78]
Volatile constituents of 96 essential oil components	Molecular connectivity indices, i.e., bonds, cluster, path and rotatable (Chi indices), Randić topological index, Balaban topological index, Wiener topological index, 13 nonzero E-state indexes	Two E-state indexes (ssCH and SOH) and five molecular connectivity indices ($^1\chi_B$, $^2\chi_p$, $^3\chi_C$, $^4\chi_C$, $^6\chi_p$) related to the GC t_R values, MLR	HP5-MS fused silica, 25 μm, 30 m × 0.25 mm	[79]
PAHs	Kier and Hall connectivity indices ($^n\chi$ where n = 0, 1, 2), Kier shape indices (K_n = 1, 2, 3)	Molecular polarizability (α), second-order molecular connectivity Kier and Hall index ($^2\chi$) showed significant correlation with RI, linear regression equation	Capillary columns coated with SE-52 (methylphenylsilicone) stationary phase	[80]
Phenol derivatives	Kier symmetry index (S0k), E-state topological parameter (Tie)	Between retention times and descriptors, MLR model	DB-5 nonpolar column chemically bonded with 95% dimethyl and 5% diphenyl-polysiloxane	[81]
Methylalkanes produced by insects	Balaban index (J)	Between Kovat's RI and fourth descriptor (i.e., Balaban index), significant correlation	Fused silica capillary column (DB-1)	[82]
Organic pesticides with diverse heteroatoms	Lu index, DAI, and χDAI	Between the retention time and the three topological indices, MLR	Four kinds of chromatographic columns with various polarity (DB-1, DB-5, DB-17, and DB1301)	[83]
Various oxygen-containing organic compounds (i.e., esters, aldehydes, ketones, alcohols)	Polarizability effect index (PEI index), modified polarizability effect index (MPEI$_m$), modified inner polarizability effect index (IMPEI$_m$)	Simple linear regression equations between the RI and the topological indices PEI, MPEI$_m$, and IMPEI$_m$ were established	OV-1 (dimethylpolysiloxane) and SE-54 (5% phenyl–95% dimethylpolysiloxane)	[84]

(*Continued*)

TABLE 18.3 (Continued)
Topological Indices Used in the Prediction of Chromatographic Retention in GC

Class of Compounds/Solutes	Topological Index	Retention Model	Stationary Phase/Column	Reference
Saturated esters	PEI, odd–even index (OEI), and steric effect index (SV)	Between Kovat's RI and topological indices PEI, OEI, and SV, MLR models were obtained	Seven stationary phases (SE-30, OV-7, DC-710, OV-25, XE-60, OV-225, Silar-5CP)	[85]
Mixed set of aldehydes and ketones	Xu index and atom-type-based AI topological indices	Gas chromatographic retention data correlated well with Xu index and atom-type-based AI topological indices, MLR	Different polar stationary phases (HP-1, HP-50, DB-210, HP-Innowax)	[86]
Saturated O-, N-, and S-heterocyclic compounds	Gutman's index, Balaban-type index, valence connectivity index chi-2, chi-0, and chi-1	Between gas chromatographic RIs and topological connectivity indices, linear-logarithmic equations were obtained	Polydimethyl siloxane stationary phase	[87]
PASHs	Different types of descriptors including 3D-Balaban index (J3D), sum of Kier–Hall electrotopological state (SS)	Between RIs and topological descriptors J3D (3D-Balaban index) and electrotopological state SS show good correlation (cross-validated $q^2 = 0.98$)	Bp × 5 (5% phenyl) stationary phase	[88]
Set of flavor compounds	Simple molecular connectivity Chi indices ($^4\chi^d$, $^1\chi^{a,b}$, $^0\chi^{a,b}$)	Multivariable linear regression (MLR) was used to map the topological indices to the gas chromatographic RI	Various polar and nonpolar columns (OV101, DB5, OV17, C20M)	[89]
Branched alkenes with C_4–C_9 carbons	Semi-empirical topological index (I_{ET})	Between topological index (I_{ET}) and chromatographic RI of branched alkenes, linear relation	Different stationary phases with squalane and 1-octadecane (Apiezon L, OV-1, DB-1)	[90]
Different alkanes and alcohols	Connectivity index $^1\chi^f$	Between gas chromatographic RIs and index $^1\chi^f$, linear correlation	N/A	[91]
Various alcohols and esters	Topological R_{AF} index and Δ_A index	Between retention times of examined alcohols and esters and topological indices R_{AF} and Δ_A, linear correlation	Six different capillary columns	[92]
Flavors and fragrances of volatile compounds	Various molecular descriptors	Good correlation between RI and solvation connectivity index of the first order	Nonpolar capillary column coated with methyl silicone OV-101	[93]

(Continued)

TABLE 18.3 (Continued)
Topological Indices Used in the Prediction of Chromatographic Retention in GC

Class of Compounds/Solutes	Topological Index	Retention Model	Stationary Phase/Column	Reference
PCBs	Different molecular descriptors including selected topological indices (i.e., Narumi topological index (HNar), second Mohar index (TI2), 3-path Kier alpha-modified-shape index	Between relative retention times and descriptors, multilinear regression model	Seventeen different capillary columns	[94]
Substituted coumarins	Wiener index (W), molecular connectivity indices of different orders (χ), indices of molecular shape, kappa values ($^1\kappa$, $^2\kappa$, $^3\kappa$), indices of electrotopological state (S_i), and topological state indices (Ti)	Between gas chromatographic RI and molecular descriptors, single or multiple linear regression equations	Fused silica capillary columns LM-1 and CBP5	[95]
Selected esters	Semi-empirical electrotopological index	Between RI and semi-empirical electrotopological index, linear relation	Silicone phases of different polarity level	[96]
Various esters	Semi-empirical electrotopological index	RI and semi-empirical electrotopological index, linear relation	Silicone phases of different polarity level	[99]
Alcohols	Semi-empirical electrotopological index	RI and semi-empirical electrotopological index, linear relation	Silicone phases of different polarity level	[98]
Various alkylbenzenes	Semi-empirical electrotopological index	Between RI and semi-empirical electrotopological index, linear relation	Silicone phases of different polarity level	[97]
Terpenoids	Different topological indices obtained using Dragon software	Between Kovat's RI and topological indices, MLR, ANN	Taken from literature	[100]

RI, gas chromatography retention index; FID, flame ionization detection; RI, Kovat's retention index; DAI, distance-based atom-type topological indices.

18.6   QSAR STUDY USING TOPOLOGICAL INDICES IN CHEMICAL AND PHARMACEUTICAL ANALYSIS

The intensive increase in the pharmaceutical industry and the necessity of developing new drugs for new diseases and disorders urgently demand the need to find an effective tool for pharmaceutical companies to speed up drug discovery processes, especially for the estimation of relevant properties and activities of potential drug candidates. In the past few decades, QSAR and QSPR have been the most prominent techniques applied in the fields of pharmaceutics in the discovery and development of new drugs. The main goal of these techniques is to develop mathematical models for the estimation of pharmaceutical properties of molecules, that is, pharmacological action, bioavailability, distribution to tissues, and others. This will minimize the need to conduct animal tests and reduce the costs of drug development in comparison with experimental studies.

In medicinal chemistry, there is interest in developing methods based on molecular descriptors for determining the n-octanol–water partition coefficient or its decimal logarithm (log P). This parameter is correlated with the pharmaceutical properties of molecules. Many QSAR/QSPR studies using topological indices have been developed for predicting log P. A large number of researchers have compared the predictive ability of log P calculation models. The compatibility of theoretical methods based on calculations using different molecular descriptors and experimental approaches (i.e., chromatographic data) in the determination of lipophilicity of biomolecules is still under investigation. The most used molecular descriptors in establishing QSAR and QSPR are topological indices.

Agrawal et al. [102,103] developed a number of mathematical models with the use of topological indices to obtain different classes of organic compounds. They proposed a novel method (multiparametric regression equations) of estimation of lipophilicity using distance-based topological indices of a heterogeneous set of 223 compounds. The results show that lipophilicity can be modeled with a multiparametric model where the W, 1Chi, B, J, and logRB indices are involved. Another paper prepared by Agrawal et al. [104] showed that a QSAR study on a set of carbonic anhydrase inhibitors can be well modeled using first-order valence connectivity index ($^1\chi^v$) in multiparametric regression [104].

A paper prepared by Lakshmi [105] demonstrates the utility of five new Kekule indices (K, K1, K2, K3, and K4) in the determination of log P values of polyacenes [105]. Excellent correlation between log P and K indices was obtained (r = 0.999).

Another work of Souza et al. [106] illustrates the applicability and efficiency of the I_{SET} semi-empirical electrotopological index in terms of predicting log P of a set of 131 aliphatic organic compounds from five different classes (hydrocarbons, alcohols, aldehydes, ketones, and esters) using QSAR. The I_{SET} model is linear and demonstrates good statistical quality (r > 0.99). The main advantage of the applied index is its simplicity.

Guo et al. [107] reported that in QSAR and QSPR studies of alkanes for the prediction of various properties of molecules containing heteroatoms, multiple bonds, and rings, the topological index EATI is useful. Satisfactory results have been obtained with the use of EATI to predict lipophilicity (log P) and biological activity of 44 examined alkanes.

The QSPR study on the octanol/air partition coefficient of polybrominated diphenyl ethers (PBDEs) illustrates the applicability of molecular distance-edge vector index as the structural descriptor for predicting the octanol/air partition coefficient of the 22 investigated PBDEs [108]. In this case, the MLR model and ANN model were established.

A research study of newly synthesized s-triazine derivatives reported by Jevric et al. [109] demonstrates that connectivity indices ($^0\chi$, $^1\chi$, $^2\chi$, $^3\chi$, $^4\chi$) were found to be important factors affecting the retention and also the lipophilicity, determined chromatographically by RP-HPTLC.

In the case of RP-TLC study of barbiturates under various chromatographic conditions, Kępczyńska et al. [110] present significant correlations between the chromatographic parameter R_{M0} and selected topological indices. The authors affirmed that the chromatographic parameter of lipophilicity (R_{M0}) of the examined compounds was best correlated with topological indices based on the adjacency matrix $^0\chi$, $^1\chi$, χ_{012}, $^0\chi^\nu$, $^1\chi^\nu$, and χ^ν_{012} as well as the distance matrix R, W, A, ^{1}B, and D ($r \geq 0.96$).

A paper by Pyka [111] focused on the application of topological indices for the prediction of the biological activity of selected alkoxyphenols affirmed that linear function describes very well the relationship between the toxicity on gram-positive bacteria and hydrophobic constant for examined meta- and para-alkoxyphenols. The best linear relationships were obtained for topological indices χ_{012}, ^{0}B, and D as well as Randic indices $^0\chi^\nu$, $^1\chi^\nu$, $^2\chi^\nu$, and $^3\chi^\nu_{p+c}$.

Selected structural descriptors including topological indices based on connectivity ($^0\chi^\nu$, $^1\chi^\nu$, and $^{02}\chi^\nu$, Randic), distance matrix (W, A, ^{0}B, ^{1}B, and ^{2}B, Pyka), information theory (I_{SA}, $\bar{I}_{SA}$), and electro-topological states (SaaCh, SaaN, SaasC, SdO, and SdssC) were used for QSAR analysis of nicotinic acid derivatives. Satisfactory linear correlations were obtained between chromatographic parameter lipophilicity determined by RP-TLC and some topological indices ($^0\chi^\nu$, $^1\chi^\nu$, ^{1}B, ^{2}B, I_{SA}, and $\bar{I}_{SA}$) in the case of the first set of compounds examined by Pyka [112]. The second step of this study confirmed the usefulness of $^0\chi^\nu$, $^1\chi^\nu$, $^2\chi^\nu$, ^{1}B, ^{2}B, I_{SA}, and $\bar{I}_{SA}$ in the prediction of chromatographic parameter of nicotinic acid derivatives. The electrotopological states SaaN and SdO correlate well with the lipophilicity parameter φ_0, which can also be the measure of lipophilicity [113].

The RP-HPTLC study of lipophilicity of selected higher fatty acids, hydroxyl acids, and their esters by Niestrój et al. [114] demonstrates that of all the applied topological indices the best correlations with partition coefficients are obtained with the biparametric equations relating to topological index A and electrotopological state E_{OH}.

The possibility of predicting the lipophilicity of a series of saturated fatty acids using topological indices is indicated in a paper prepared by Pyka and Bober [115]. The relationships between experimental partition coefficient (log P), partition coefficients obtained from an Internet database (IAlogP, ClogP™), and R_M values obtained by RP-HPTLC with topological indices based on the adjacency matrix (M, M^ν, $^0\chi$, $^1\chi$, $^2\chi$, $^0\chi^\nu$, $^1\chi^\nu$, $^2\chi^\nu$, F, N_2) and index of atomic composition ($\bar{I}_{SA}$) showed the best performance [115].

The use of selected topological indexes for the evaluation of lipophilicity of a series of steroid compounds determined using RP-HPTLC and theoretical methods was described by Pyka in 2009. It was found that lipophilicity determined chromatographically (as R_{MW} values) correlated best with topological indexes $^0\chi^\nu$, $^1\chi^\nu$, R, W, A, and ^{1}B. The theoretical log P values obtained from an Internet database correlated best with topological index ($\bar{I}_{SA}$) [116].

A lipophilicity study of steroid compounds by Pyka et al. [117] indicated that among different structural descriptors mentioned earlier, the lipophilicity parameters of examined free bile acids and conjugated with glycine (R_{MW}, φ_0 and log P) correlated best with topological indices Gutman (M^ν) and also Pyka (C).

One of the new papers prepared by Dołowy and Pyka [118] indicates the applicability of selected numerical values of topological indices in the prediction of log P of spironolactone, a member of the steroid group with diuretic property [118]. New methods of calculating the log P values by means of formulae containing topological indices W, ^{0}B, ^{1}B, I_B, and $^0\chi^\nu$ were proposed. The best agreement with experimental value of log P was indicated by the topological index developed by Pyka ^{0}B.

A significant role of topological index ^{0}B in the determination of the partition coefficient as a measure of lipophilicity of aliphatic compounds and tocopherols was confirmed in papers prepared by Niestrój [119,120].

New methods of log P calculations for naproxen using the retardation factor (R_F) and the numerical values of topological indices developed by Randic ($^1\chi^\nu$, $^2\chi$, $^1\chi^\nu$) and Pyka (^{0}B) were proposed to predict the lipophilicity of naproxen [121]. It was stated that the log P calculated using the procedure based on topological index ^{0}B value and R_F correlated best with experimental partition coefficient.

In further investigations, Dołowy and Pyka [122] used the numerical values of selected topological indices such as Gutman index M, Pyka ^{0}B, ^{1}B, Wiener W, and Balaban index I_B to predict the partition coefficient log P for two most popular anti-inflammatory agents, namely salicylic and ace-tylsalicylic acids. Partition coefficients of both compounds calculated by the new proposed methods based on topological indices ^0B, ^1B, M, and W correlated best with mean value of theoretical partition coefficient determined using various algorithms.

An assessment of the possibility of using topological indexes in the prediction of selected physicochemical properties such as partition coefficient (log P) and other pharmaceutically important compounds in TLC, HPLC, GC, and MEK includes the review paper prepared by Pyka [123].

In recent years, topological indices have emerged as useful molecular descriptors in the prediction of physicochemical properties of biomolecules from their structure, such as lipophilicity, with QSPR and QSRR studies. They have also been successfully applied in QSAR of various classes of pharmaceutically important compounds. Various mathematical models using topological indices have been described in the literature that aim at predicting the biological activity of different molecules and drug substances. It is important to find new lead drugs.

A QSAR study on antiulcer agents presented by Goel and Madan [124] indicates that the relationship of Wiener index (W) and molecular connectivity index developed by Randić (χ) with antiulcer activity of a series of 4-substituted-2-guanidino thiazole analogs determined by in vitro and in vivo tests is significant.

Another paper shows the utility of topological charge indices (TCIs), a combination of connectivity indices, in the selection of new potentially active cytostatic compounds [125].

A QSAR study on antidiabetic oral drugs reported by Popescu et al. [126] demonstrates that topological index I_P (Popescu index) allows predicting the activity of new compounds proposed as oral antidiabetics in the sulfonylureas, alfa-glucosidase inhibitors, metglinides, biguanides, and thiazolidinediones.

Another paper describes the relationships between the topological indices and human immunodeficiency virus (HIV) protease inhibitor activity. Agrawal et al. [127] proposed the tetra-parametric linear model containing topological indices Wiener (W), branching (B), first-connectivity ($^1\chi$), and log RB index, which allows predicting the inhibitory activity expressed as log IC50 of 5,6-dihydro-2-pyrones as HIV-1 protease inhibitors. Similar QSAR studies on the use of tetrahydropyrimidin-2-ones as HIV protease inhibitors have been performed by Lather and Madan [128]. Three topological indices, Wiener index (W), Zagreb group parameter proposed by Gutman (M_1), and eccentric connectivity index denoted by ξ^c, indicate significant correlations with HIV protease inhibitor activity of terahydropyrimidin-2-ones. Good predictability of diverse models based on various structural descriptors including topological index (Balaban-type index from Z-weighted distance matrix) has been observed by Khatri et al. for the prediction of anti-HIV activity of purine nucleoside analogs [129].

Another paper indicates the utility of six distance-based topological indices—Wiener (W), branching (B), first-order connectivity (χ), Balaban (J), Szeged (Sz), and log RB index—in modeling antihypertensive activity of 2-aryl-imino-imidazolidines [130].

The literature review shows applicability of topological descriptors also in quantitative structure–toxicity relationship (QSTR) studies. Nandan et al. [131] developed a QSTR multiparametric model based on topological parameters Wiener index and Balaban index to predict the toxicity of very toxic polychlorinated aromatic compounds that pollute the environment.

18.7 SUMMARY

Among various analytical methods, chromatography is still the most commonly used technique to separate and identify various compounds in a mixture, as well as to estimate the degree of purity of different chemically and biologically important compounds. The following chromatographic parameters, such as retardation factor in TLC, RT and relative RT or chromatographic RI (i.e., Kovat's index) in HPLC and GC, respectively, are used for the identification of compounds analyzed by these techniques.

Because chromatographic retention is a complex process that involves different interactions between the solute and the stationary phase, the correlation between chromatographic retention and molecular parameters can provide useful information not only on the molecular structure but also on RT and elution of chromatographed solutes. This fact has a huge impact on the development of theoretical models for estimating the retardation factor, RT, or RI, namely for QSRR analysis.* We showed that between various molecular descriptors, topological indices have become an important tool for predicting the retention behavior of different organic compounds in TLC, HPLC, and GC studies, as well as in electrochromatographic analysis by MEK and CZE. One of the main advantages of topological indices is that they can be rapidly calculated using proper computer software (i.e., Dragon).

We confirmed that many topological indices have been proposed to predict the most important chromatographic parameters such as retardation factor, RT, and RI of many well-described organic compounds in the literature and also new ones whose standards are not yet available. It may be observed that developing new topological indices allows increasing the predictive power of QSRR models. Moreover, multivariate statistical methods, such as MLR, PLS, and ANN, used in QSRR enable the selection of the most statistically significant QSRR models.

Another approach of topological indices observed in recent years is the use of topological indices to establish QSPR, QSAR, and QSTR in drug modeling. Many studies have evidenced that the mathematically calculated indices are correlated with lipophilicity, toxicity, or pharmacological activity of drugs (i.e., antidiabetic, analgesic, anti-HIV, anticancer). This fact emphasizes the significant role of topological indices in modeling not only chromatographic retention but also physicochemical properties and pharmacological activity of various compounds.

REFERENCES

1. Van de Waterbeemd, H., R.E. Carter, G. Grassy, H. Kuiny, Y.C. Martin, M.S. Tutte, and P. Willet. 1997. Glossary of terms used in computational drug design (IUPAC Recommendations 1997). *Pure Appl. Chem.,* 69(5):1137–1152.
2. Trinajstić, N. 1983. *Chemical Graph Theory.* CRC Press, Boca Raton, FL.
3. Gutman, I. 2013. Degree-based topological indices. *Croat. Chem. Acta,* 86(4):351–361.
4. Pyka, A. 1997. Topological indices and their significance in chromatographic investigations. *Wiad. Chem.,* 51(11–12):783–802.
5. Pyka, A. 1998. Topological indices and their significance in chromatographic investigations. Part II. *Wiad. Chem.,* 52(9–10):727–754.
6. Sabljič, A., and N. Trinajstic. 1981. Quantitative structure-activity relationships: The role of topological indices. *Acta Pharm. Jugosl.,* 31:189–214.
7. Balaban, A.T. 1979. Five new topological indices for the branching of tree-like graphs. *Theor. Chim. Acta,* 53:355–375.
8. Balaban, A.T., and I. Motoc. 1979. Chemical Graphs. XXXVI. Correlations between octane numbers and topological indices of alkanes. *MATCH Commun. Math. Comput. Chem.,* 5:197–218.
9. Mauri, A., V. Consonni, M. Pavan, and R. Todeshini. 2006. Dragon Software: An easy approach to molecular descriptor calculations. *MATCH Commun. Math. Comput. Chem.,* 56:237–248.
10. Nikmehr, M.J., N. Soleimani, and M. Veylaki. 2014. Topological indices based end-vertex degrees of edges on nanotubes. *Proc. IAM,* 3(1):89–97.
11. Khalifeh, M.H., H. Yousefi-Azari, and A.R. Ashrafi. 2008. The hyper-Wiener index of graph operations. *Comp. Math. Appl.,* 56:1402–1407.
12. Heinzen, V.E.F., M.F. Soares, and R.A. Yunes. 1999. Semi-empirical topological method for the prediction of the chromatographic retention of *cis-* and *trans-* alkene isomers and alkanes. *J. Chromatogr. A,* 849:495–506.
13. Śliwiok, J., B. Kocjan, B. Labe, A. Kozera, and J. Zalejska. 1993. Chromatographic studies of tocopherols. *J. Planar Chromatogr.,* 6:492–494.

---

* See also Chapter 17 in this book.

14. Pyka, A., and A. Niestrój. 2001. The application of topological indexes for prediction of the R_M values for tocopherols in RP-TLC. *J. Liq. Chromatogr. Relat. Technol.*, 24(16):2399–2413.
15. Pyka, A., and J. Śliwiok. 2001. Chromatographic separation of tocopherols. *J. Chromatogr. A*, 935:71–76.
16. Pyka, A., and M. Dołowy. 2004. Application of structural descriptors for the evaluation of some physicochemical properties of selected bile acids. *Acta Pol. Pharm. Drug Res.*, 61(6):407–413.
17. Dołowy, M. 2008. Application of selected topological indices to predict retention parameters of selected bile acids separated on modified TLC plates. *Acta Pol. Pharm. Drug Res.*, 65(1):51–57.
18. Pyka, A. 2009. Use of structural descriptors to QSRR analysis of selected bile acids separated by NP-TLC. *J. Liq. Chromatogr. Relat. Technol.*, 32:2739–2746.
19. Pyka, A. 2001. Chromatographic data-topological index dependence for selected steroids. *J. Liq. Chromatogr. Relat. Technol.*, 24(4):453–460.
20. Pyka, A., E. Kępczyńska, and J. Bojarski. 2003. Application of selected traditional structural descriptors to QSRR and QSAR analysis of barbiturates. *Indian J. Chem. Sec., A*, 42(A):1405–1413.
21. Pyka, A., K. Bober, D. Gurak, and A. Niestrój. 2002. Application of topological indices for evaluation of the TLC separation of selected essentials oil components. *Acta Pol. Pharm. Drug Res.*, 59:87–91.
22. Pyka, A., and J. Śliwiok. 2004. Use of traditional structural descriptors in QSRR analysis of nicotinic acid esters. *J. Liq. Chromatogr. Relat. Technol.*, 27(5):785–798.
23. Pyka, A. 1996. The use of selected topological indexes and R_M values to reveal the occurrence of hydrogen bonds in disubstituted benzene derivatives: Part XIII. *J. Planar Chromatogr.*, 9:288–290.
24. Pyka, A. 1999. Prediction of the TLC R_M values of selected hydrocarbons and their quinones. *J. Planar Chromatogr.*, 12:293–297.
25. Pyka, A. 2001. Investigations of the correlation between R_M values and selected topological indexes for higher alcohols, higher fatty acids and their methyl esters in RP-TLC. *J. Planar Chromatogr.*, 14:439–444.
26. Pyka, A., and K. Bober. 2002. Prediction of the R_M values of selected methyl esters of higher fatty acids in RPTLC. *J. Planar Chromatogr.*, 15:59–66.
27. Niestrój, A., A. Pyka, and J. Śliwiok. 2002. The use of topological indexes to predict the R_M values of higher fatty acids, hydroxyl fatty acids and their esters in RPTLC. *J. Planar Chromatogr.*, 15:177–182.
28. Pyka, A. 1994. The application of topological indexes (I_t) for prediction of the R_M values of isomeric alcohols: Part V. *J. Planar Chromatogr.*, 7:41–49.
29. Pyka, A. 1995. Topological indices and R_M values of isomeric phenol derivatives in structure–biological activity studies: Part VI. *J. Planar Chromatogr.*, 8:52–62.
30. Pyka, A. 2003. Use of structural descriptors to predict the R_M values of m- and p-alkoxyphenols in RP-TLC. *J. Planar Chromatogr.*, 16:131–135.
31. Garcia-March, F.J., G.M. Antos-Fos, F. Perez-Gimenez, M.T. Salabert-Salvador, R.A. Cercos-del-Pozo, and J.V. de Julian-Ortiz. 1996. Prediction of chromatographic properties for a group natural phenolic derivatives by molecular topology. *J. Chromatogr. A*, 719:45–51.
32. Pyka, A. 1994. The topological indices and R_M values of benzoic acid derivatives: A structure–activity investigations: Part IV. *J. Planar Chromatogr.*, 7:108–116.
33. Gutman, I., and A. Pyka. 1997. New topological indices for distinguishing between enantiomers and stereoisomers: A mathematical analysis. *J. Serb. Chem. Soc.*, 62:261–265.
34. Pyka, A. 1997. New topological indices for the study of isomeric compounds. *J. Serb. Chem. Soc.*, 62:251–259.
35. Pyka, A. 1993. A new optical topological index (I_{opt}) for predicting the separation of D and L optical isomers by TLC: Part III. *J. Planar Chromatogr.*, 6:282–288.
36. Pyka, A. 1994. A new stereoisomeric topological index (I_{STI}) for predicting the separation of stereoisomers in TLC. *J. Planar Chromatogr.*, 7:389–393.
37. Pyka, A. 1997. Dependence of R_M values on the topological index ^oB for selected isomers of fatty acids of cis and trans configuration. Part XIV. *J. Planar Chromatogr.*, 10:121–123.
38. Pyka, A., and K. Bober. 2001. Prediction of the TLC R_M values of selected isomers of higher fatty acids of cis and trans configuration. *J. Planar Chromatogr.*, 78:477–482.
39. Pyka, A. 1991. Correlation of topological indexes with the chromatographic separation of isomers. *J. Planar Chromatogr.*, 4:316–318.
40. Pyka, A. 2001. The application of topological indexes in TLC. *J. Planar Chromatogr.*, 14:152–159.
41. Pyka, A. 2010. Topological indices: TLC. In: *Encyclopedia of Chromatography*, 3rd edn., ed. J. Cazes, pp. 2340–2350, Taylor & Francis, Boca Raton, FL.
42. Ghosh, P., B. Chawla, P.V. Joshi, and S.B. Jaffe. 2006. Prediction of chromatographic retention times for aromatic hydrocarbons. *Energy & Fuels*, 20:609–619.

43. Amić, D., and D. Davidović-Amić. 1995. Calculation of retention times of anthocyanins with orthogonalized topological indices. *J. Chem. Inf. Comput. Sci.*, 35:136–139.

44. Pyka, A., and M. Miszczyk. 2005. Chromatographic evaluation of the lipophilic properties of selected pesticides. *Chromatographia*, 61(1/2):37–42.

45. Djaković-Sekulić, T., N. Perišić, and A. Pyka. 2003. Correlation of retention of anilides and some molecular descriptors. Application of topological indexes for prediction of log k values. *Chromatographia*, 58(1/2):47–51.

46. Lee, S.K., Y. Polyakova, and K.H. Row. 2003. Interrelation of retention factor of amino acids by QSPR and linear regression. *Bull. Korean Chem. Soc.*, 24(12):1757–1762.

47. Zheng, J., Y. Polyakova, and K.H. Row. 2006. Prediction of retention of nucleic compounds based on a QSPR model. *Chromatographia*, 64(3/4):129–137.

48. Hall, L.M., L.H. Hall, and L.B. Kier. 2003. Modeling drug albumin binding affinity with e-state topological structure representation. *J. Chem. Inf. Comput. Sci.*, 43:2120–2128.

49. Salo, M., H. Siren, P. Volin, S. Wiedmer, and H. Vuorela. 1996. Structure-retention relationships of steroid hormones in reversed-phase liquid chromatography and micellar electrokinetic capillary chromatography. *J. Chromatogr. A*, 728:83–88.

50. Tham, S.Y., and S. Agatonovic–Kustrin. 2002. Application of the artificial neural network in quantitative structure–gradient elution retention relationship of phenylthiocarbamyl amino acids derivatives. *J. Pharm. Biomed. Anal.*, 28:581–590.

51. Montana, M.P., N.B. Pappano, N.B. Debattista, J. Raba, and J.M. Luco. 2000. High-performance liquid chromatography of chalcones: Quantitative structure-retention relationships using partial least–squares (PLS) modeling. *Chromatographia*, 51(11/12):727–735.

52. Riahi, S., M.F. Mousavi, M.R. Ganjali, and P. Norouzi. 2008. Application of correlation ranking procedure and artificial neural networks in the modeling of liquid chromatographic retention times (t_R) of various pesticides. *Anal. Lett.*, 41:3364–3385.

53. D'Archivio, A.A., M.A. Maggi, P. Mazzeo, and F. Ruggieri. 2008. Quantitative structure–retention relationships of pesticides in reversed–phase high–performance liquid chromatography based on WHIM an GATEWAY molecular descriptors. *Anal. Chim. Acta*, 628:162–172.

54. Liu, T., I.A. Nicholls, and T. Oberg. 2011. Comparison of theoretical and experimental models for characterizing solvent properties using reversed phase liquid chromatography. *Anal. Chim. Acta*, 702:37–44.

55. Carlucci, G., A.A. D'Archivio, M.A. Maggi, P. Mazzeo, and F. Ruggieri. 2007. Investigation of retention behavior of non–steroidal anti–inflammatory drugs in high-performance liquid chromatography by using quantitative structure–retention relationships. *Anal. Chim. Acta*, 601(1):68–76.

56. Bączek, T., K. Macur, and L. Bober. 2009. Rapid HPLC method development of polynuclear aromatic hydrocarbons separation based on quantitative structure retention relationships. *J. Liq. Chromatogr. Relat. Technol.*, 32:668–679.

57. Ghavami, R., and Z. Rasouli. 2013. Investigation of retention behavior of anthraquinoids in RP-HPLC on 17 different C18 stationary phases by means of quantitative structure retention relationships. *Med. Chem. Res.*, 22:2677–2691.

58. Eugster, P.J., J. Boccard, B. Debrus, L. Breant, J.L. Wolfender, S. Martel, and P.A. Carrupt. 2014. Retention time prediction for dereplication of natural products ($C_xH_yO_z$) in LC-MS metabolite profiling. *Phytochemistry*, 108:196–207.

59. Dong, P.P., G.B. Ge, Y.Y. Zhang, C.Z. Ai, G.H. Li, L.L. Zhu, H.W. Luan, X.B. Liu, and L. Yang. 2009. Quantitative structure–retention relationships studies for taxanes including epimers and isomeric metabolites in ultrafast liquid chromatography. *J. Chromatogr. A*, 1216:7055–7062.

60. Lippa, K.A., L.C. Sander, and S.A. Wise. 2004. Chemometric studies of polycyclic aromatic hydrocarbon shape selectivity in reversed-phase liquid chromatography. *Anal. Bioanal. Chem.*, 378:365–377.

61. Liu, F., C. Cao, and B. Cheng. 2011. A Quantitative Structure–Property relationship (QSPR) study of aliphatic alcohols by the method of dividing the molecular structure into substructure. *Int. J. Mol. Sci.*, 12:2448–2462.

62. Albaugh, D.R., L.M. Hall, D.W. Hill, T.M. Kertesz, M. Parham, L.H. Hall, and D.F. Grant. 2009. Prediction of HPLC retention index using artificial neural networks and Igroup e-state indices. *J. Chem. Inf. Model.*, 49:788–799.

63. Talebi, M., G. Schuster, R.A. Shellie, and R. Szucs. 2015. Perform comparison of partial least squares-related variable selection methods for quantitative structure retention relationships modeling of retention times in reversed-phase liquid chromatography. *J. Chromatogr. A*, 1424:69–76.

64. Gou, W., Y. Lu, and X.M. Zheng. 2000. The prediction study for chromatographic retention index of saturated alcohols by MLR and ANN. *Talanta*, 51:479–488.

65. Pyka, A. 2010. Topological indices: Use in HPLC. In: *Encyclopedia of Chromatography*, ed. J. Cazes, pp. 2351–2360, Taylor & Francis, Boca Raton, FL.

66. Pyka, A. 2004. Topological indices: Use in HPLC. In: *Encyclopedia of Chromatography*, ed. J. Cazes, pp. 1–10, Marcel Dekker, New York.

67. Pyka, A. 2005. Topological indices: Use in HPLC. In: *Encyclopedia of Chromatography*, ed. J. Cazes, pp. 1705–1715, Taylor & Francis, Boca Raton, FL.

68. Polyakova, Y.L., and K.H. Row. 2005. Quantitative structure–retention relationships applied to reversed-phase high-performance liquid chromatography. *Med. Chem. Res.*, 14(8/9):488–522.

69. Lu, Ch., W. Guo, X. Hu, Y. Wang, and Ch. Yin. 2006. A Lu index for QSAR/QSPR studies. *Chem. Phys. Lett.*, 417:11–15.

70. Lu, Ch., W. Guo, and Ch. Yin. 2006. Quantitative structure-retention relationship study of the gas chromatographic retention indices of saturated esters on different phases using novel topological indices. *Anal. Chim. Acta*, 561:96–102.

71. Ghavami, R., and S. Faham. 2010. QSRR models for Kovat's retention indices of a variety of volatile organic compounds on polar and apolar GC stationary phases using molecular connectivity indexes. *Chromatographia*, 72:893–903.

72. Heinzen, V.E.F., and R.A. Yunes. 1996. Using topological indices in the prediction of gas chromatographic retention indices of linear alkylbenzene isomers. *J. Chromatogr. A*, 719:462–476.

73. Santiuste, J.M. 2000. Relationship between GLC retention data and topological indices for a wide variety of solutes on five stationary phases on different polarity. *Chromatographia*, 52:225–232.

74. Santiuste, J.M., and J.M. Takacs. 2003. Relationships between retention data of benzene and chlorobenzenes with their physic-chemical properties and topological indices. *Chromatographia*, 58:87–96.

75. Feng, C.J., and X.H. Du. 2001. Topological research of Kovat's indices for amines. *Se pu.* 19:124–127.

76. Junkes, B.S., R.D.M.C. Amboni, R.A. Yunes, and V.E.F. Heinzen. 2003. Prediction of the chromatographic retention of saturated alcohols on stationary phases of different applying the novel semi-empirical topological index. *Anal. Chim. Acta*, 477:29–39.

77. Krawczuk, A., A. Voelkel, J. Lulek, R. Urbaniak, and K. Szyrwińska. 2003. Use of topological indices of polychlorinated biphenyls in structure-retention relationships. *J. Chromatogr. A*, 1018:63–71.

78. Zenkevich, G., and A.N. Marinichev. 2001. Comparison of the topological and dynamic characteristics of molecules for calculating retention indices of organic compounds. *J. Struct. Chem.*, 42(5):747–754.

79. Quin, L.T., S.S. Liu, H.L. Liu, and J. Tong. 2009. Comparative multiple quantitative structure-retention relationships modeling of gas chromatographic retention time of essential oils using multiple linear regression, principal component regression, and partial least squares techniques. *J. Chromatogr. A*, 1216:5302–5312.

80. Drosos, J.C., M.V. Rhenals, and R. Vivas-Reyes. 2010. Quantitative structure-retention relationships of polycyclic aromatic hydrocarbons gas-chromatographic retention indices. *J. Chromatogr. A*, 1217:4411–4421.

81. Garkani-Nejad, Z. 2010. Quantitative structure-retention relationship study of some phenol derivatives in gas chromatography. *J. Chromatogr. Sci.*, 48:317–323.

82. Katritzky, A.K., and K. Chen. 2000. QSPR correlation and prediction of GC retention indexes for methyl-branched hydrocarbons produced by insects. *Anal. Chem.*, 72:101–109.

83. Hu, R., Ch. Yin, Y. Wang, Ch. Lu, and T. Ge. 2008. QSPR study on GC relative retention time of organic pesticides on different chromatographic columns. *J. Sep. Sci.*, 31:2434–2443.

84. Liu, F., Y. Liang, Ch. Cao, and N. Zhou. 2007. Theoretical prediction of the Kovat's retention index for oxygen-containing organic compounds using novel topological indices. *Anal. Chim. Acta*, 594:279–289.

85. Liu, F., Y. Liang, Ch. Cao, and N. Zhou. 2007. QSPR study of GC retention indices for saturated esters on seven stationary phases based on novel topological indices. *Talanta*, 72:1307–1315.

86. Ren, B. 2003. Atom-based AI topological descriptors for quantitative structure-retention index correlations of aldehydes and ketones. *Chem. Int. Lab. Syst.*, 66:29–39.

87. Farkas, O., K. Heberger, and I.G. Zenkevich. 2004. Quantitative structure relationships. IV. Prediction of gas chromatographic retention indices for saturated O-, N- and S-heterocyclic compounds. *Chem. Int. Lab. Syst.*, 72:173–184.

88. Can, H., A. Dimoglo, and V. Kovalishyn. 2005. Application of artificial neural networks for the prediction of sulfur polycyclic aromatic compounds retention indices. *J. Mol. Struct. Theochem.*, 723:183–188.

89. Yan, J., D.S. Cao, F.Q. Guo, L.X. Zhang, M. He, J.H. Huang, Q.S. Xu, and Y.Z. Liang. 2012. Comparison of quantitative structure–retention relationship models on four stationary phases with different polarity for a diverse set of flavor compounds. *J. Chromatogr. A*, 1223:118–125.

90. Junkes, B.S., R.D.M.C. Amboni, V.E.F. Heinzen, and R.A. Yunes. 2002. Use of semi-empirical topological method to predict the chromatographic retention of branched alkenes. *Chromatographia*, 55(1/2):75–81.

91. Randic, M., S.C. Basak, M. Pompe, and M. Novic. 2001. Prediction of gas chromatographic retention indices using variable connectivity index. *Acta Chim. Slov.*, 48:169–180.

92. Beteringhe, A., A.C. Radutiu, D.C. Culita, A. Mischie, and F. Spafiu. 2008. Quantitative-structure relationship (QSRR) study for predicting gas chromatographic retention times for some stationary phases. *QSAR Comb. Sci.*, 27(8):996–1005.

93. Rojas, C., P.R. Duchowicz, P. Tripaldi, and R.P. Diez. 1999. QSPR analysis for the retention index of flavors and fragrance on a OV-101 column. *Chemometr. Intell. Lab. Syst.*, 48:35–46.

94. D'Archivio, A.A., A. Incani, and F. Ruggieri. 2011. Cross-column prediction of gas-chromatographic retention of polychlorinated biphenyls by artificial neural networks. *J. Chromatogr. A*, 1218:8679–8690.

95. Soares, M.F., F.D. Monache, V.E.F. Heinzen, and R.A. Yunes. 1999. Prediction of gas chromatographic retention indices of coumarins. *J. Braz. Chem. Soc.*, 10(3):189–196.

96. Souza, E.S., C.A. Kuhnen, B.S. Junkes, R.A. Yunes, and V.E.F. Heinzen. 2008. Quantitative structure–retention relationships modeling of esters on stationary phases of different polarity. *J. Mol. Graph. Model.*, 28:20–27.

97. Porto, L.C., E.S. Souza, B.S. Junkes, R.A. Yunes, and V.E.F. Heinzen. 2008. Semi-empirical topological index: Development of QSPR/QSRR and optimization for alkylbenzenes. *Talanta*, 76:407–412.

98. Souza, E.S., C.A. Kuhnen, B.S. Junkes, R.A. Yunes, and V.E.F. Heinzen. 2010. Development of semi-empirical electrotopological index using the atomic charge in QSPR/QSRR models for alcohols. *J. Chemom.*, 24:149–157.

99. Amboni, R.D.M.C., B.S. Junkes, V.E.F. Heinzen, and R.A. Yunes. 2002. Semi-empirical topological method for prediction of the chromatographic retention of esters. *J. Mol. Struct.*, 579:53–62.

100. Hemmateenejad, B., K. Javadnia, and M. Elyasi. 2007. Quantitative structure–retention relationships for the Kovat's retention indices of a large set of terpenes: A combined data splitting-feature selection strategy. *Anal. Chim. Acta*, 592:72–81.

101. Liang, H.R., H. Vuorela, P. Vuorela, M.L. Riekkola, and R. Hiltunen. 1998. Prediction of migration behaviour of flavonoids in capillary zone electrophoresis by means of topological indices. *J. Chromatogr. A*, 798: 233–242.

102. Agrawal, V.K., S. Bano, and P.V. Khadikar. 2003. Topological approach to quantifying molecular lipophilicity of heterogeneous set of organic compounds. *Bioorg. Med. Chem.*, 11(18):4039–4047.

103. Agrawal, V.K., M. Gupta, J. Singh, and P.V. Khadikar. 2005. A novel method of estimation of lipophilicity using distance-based topological indices: Dominating role of equalized electronegativity. *Bioorg. Med. Chem.*, 13(6):2109–2120.

104. Agrawal, V.K., J. Singh, K.C. Mishra, P.V. Khadikar, Y.A. Jaliwala. 2006. QSAR studies on the use of 5,6-dihydro-2-pyrones as HIV-1 protease inhibitors. *ARKIVOC*, 2006(2):162–177.

105. Lakshmi, K.L. 2012. A highly correlated topological index for polyacenes. *J. Exp. Sci.*, 3(4):18–21.

106. Souza, E.S., L. Zaramello, C.A. Kuhnen, B.S. Junkes, R.A. Yunes, and V.E.F. Heinzen. 2011. Estimating the octanol/water partition coefficient for aliphatic organic compounds using semi-empirical electrotopological index. *Int. J. Mol. Sci.*, 12:7250–7264.

107. Guo, M., L. Xu, Ch.Y. Hu, and S.M. Yu. 1997. Study on structure-activity relationship of organic compounds—Applications of a new highly discriminating topological index. *MATCDY*, 35:185–197.

108. Jiao, L., M. Gao, X. Wang, and H. Lu. 2014. QSPR study on the octanol/air partition coefficient of polybrominated diphenyl ethers by using molecular distance-edge vector index. *Chem. Cent. J.*, 8:36.

109. Jevric, L.R.G., B. Koprivica, N.M. Misljenovic, A.N. Tepic, T.A. Kuljanin, and B.T. Jovanovic. 2011. Estimation of the correlation between the retention of s-triazine derivatives and some molecular descriptors. *APTEFF*, 42:231–239.

110. Kępczyńska, E., J. Bojarski, and A. Pyka. 2003. Lipophilicity of barbiturates determined by TLC. *J. Liq. Chromatogr. Relat. Technol.*, 26(19):3277–3287.

111. Pyka, A. 2002. Application of topological indices for prediction of the biological activity of selected alkoxyphenols. *Acta Pol. Pharm. Drug Res.*, 59(5):347–351.

112. Pyka, A. 2004. Study of lipophilicity and application of selected topological indexes in QSAR analysis of nicotinic acid derivatives. Part I. *J. Planar Chromatogr.*, 17:275–279.

113. Pyka, A., and W. Klimczok. 2005. Study of lipophilicity and application of selected structural descriptors in QSAR analysis of nicotinic acid derivatives. Investigations on RP18WF$_{254}$ plates. Part II. *J. Planar Chromatogr.*, 18:300–304.

114. Niestrój, A., A. Pyka, J. Klupsch, and J. Śliwiok. 2004. Use of RP-TLC and structural descriptors to predict the log P values of higher fatty acids, hydroxy acids, and their esters. *J. Liq. Chromatogr. Relat. Technol.*, 27(15):2449–2461.

115. Pyka, A., and K. Bober. 2006. Selected traditional structural descriptors and R_M values for estimation and prediction of lipophilicity of homologous series of saturated fatty acids. *JAOCS*, 83(9):747–751.

116. Pyka, A. 2009. Use of selected topological indexes for evaluation of lipophilicity of steroid compounds investigated by RP-HPTLC. *J. Liq. Chromatogr. Relat. Technol.*, 32:3056–3065.

117. Pyka, A., M. Dołowy, and D. Gurak. 2005. Use of selected structural descriptors for evaluation of the lipophilicity of bile acids investigated by RP-HPTLC. *J. Planar Chromatogr.*, 18:465–470.

118. Dołowy, M., and A. Pyka. 2015. Lipophilicity assessment of spironolactone by means of reversed chase liquid chromatography and by newly developed calculation procedures. *Acta Pol. Pharm. Drug Res.*, 72(2):235–244.

119. Niestrój, A. 2007. Comparison of methods for calculation of the partition coefficients of selected tocopherols. *J. Planar Chromatogr. Modern TLC*, 20(6):483–486.

120. Niestrój, A. 2010. Comparison of methods for calculation of the partition coefficients of selected aliphatic compounds. *J. Planar Chromatogr. Modern TLC*, 23(3):198–200.

121. Pyka A., A. Kazimierczak, and D. Gurak. 2013. Utilization of reversed-phase TLC and topological indices to the lipophilicity investigations of naproxen. *Pharm. Methods*, 4:16–20.

122. Dołowy, M., and A. Pyka. 2015. Lipophilicity study of salicylic and acetylsalicylic acids using both experimental and calculations methods. *J. Liq. Chromatogr. Relat. Technol.*, 38:485–491.

123. Pyka, A. 2012. Topological indexes and QSRR methodology in pharmaceutical and chemical analysis. *JAOCS*, 95(30):673–690.

124. Goel, A., and A.K. Madan. 1995. Structure-activity study on antiulcer agents using Wiener's topological index and molecular connectivity index. *J. Chem. Comput. Sci.*, 35(3):504–509.

125. Galvez, J., M.J. Gomez-Lechon, R. Garcia-Domenech, and J.V. Castell. 1996. New cytostatic agents obtained by molecular topology. *Bioorg. Med. Chem. Lett.*, 6(19):2301–2306.

126. Popescu, M., A. Velea, C. Michai, and S. Tivador. 2010. Quantitative structure-activity relationship in antidiabetic drugs by using topological descriptors. *Dig. J. Nanomater. Bios.*, 5(3):629–633.

127. Agrawal, V.K., J. Singh, P.V. Khadikar, and C.T. Supuran. 2006. QSAR study on topically acting sulfonamides incorporating GABA moieties: A molecular connectivity approach. *Bioorg. Med. Chem.*, 10(7):2044–2051.

128. Lather, V., and A.K. Madan. 2005. Topological models for the prediction of HIV-protease inhibitory activity of tetrahydropyrimidin-2-ones. *J. Mol. Graph. Model.*, 23:339–345.

129. Khatri, N., V. Lather, and A.K. Madan. 2015. Diverse models for anti-HIV activity of purine nucleoside analogs. *Chem. Centr. J.*, 9:29.

130. Agrawal V.K., S. Karmarkar, P.V. Khadikar, and S. Shrivastava. 2003. Use of distance-based topological indices in modeling antihypertensive activity: Case of 2-aryl-imino-imidazolidines. *Ind. J. Chem.*, 42A:1426–1435.

131. Nandan, K., K. Ranjan, Md.B. Ahmad, and B. Sah. 2013. QSAR studies on polychlorinated aromatic compounds using topological descriptors. *IJPSR*, 4(7):2691–2695.

Section VI

Application Overviews

19 Introduction to Herbal Fingerprinting by Chromatography

Johan Viaene and Yvan Vander Heyden

CONTENTS

19.1 INTRODUCTION

Global interest in traditional medicines, or more specifically in their active compounds, has risen to unprecedented levels. Although the importance of these medicines for specific ethnic groups is acknowledged, they have the potential to exceed this local level. In the light of current challenges in health sciences, for example, resistance to antibiotics and the constant quest for new pharmaceuticals, traditional herbal medicines have caught the attention of the scientific world. Since the active compounds in the majority of herbal medicines are unknown or not fully characterized, several research groups nowadays focus on studying traditional medicines, aiming to gain a deeper insight into their working mechanism to treat a given pathology. The compounds in a herbal medicine, responsible for their activity, can then be isolated for further study. An important condition for Western countries to adopt these medicines is that identification, standardization, and quality control of the herbal materials and their derived medicinal products are possible. Because of the often complex composition of herbal medicines, high requirements are set on both the chemical analysis techniques that are used to separate the compounds in these mixtures, as well as on the data analysis techniques that are used to extract the relevant information from the chemical analysis results. Both will be overviewed in this chapter, focusing mainly on the data-handling techniques.

19.1.1 Specific Characteristics of Herbal Medicines

Traditional herbal medicines are, in large parts of the world, very popular because of their high accessibility and low cost [1]. Herbal medicines exist in many forms, ranging from crude plant parts to more extensively purified or processed plant materials [2]. As the interest of the Western world in these traditional medicines increases, so does the demand for their reliable identification, standardization, and quality control [1]. Numerous guidelines were published that apply to herbal medicines [3–7], and the number of monographs on herbal medicines is steadily increasing not only in the Asian and Latin-American but also in the *European Pharmacopoeia* [8–11]. However, herbal medicines, as other natural products, are still a big challenge for the techniques that are currently applied in chemical analysis. A first difficulty is caused by the complex metabolic system in living organisms, which can render the composition of herbal medicines extremely complex [12–14]. Additionally, different samples of a species can, due to a multitude of factors, vary substantially both in qualitative and quantitative chemical composition, which is known as intraspecies variability [1]. A third important feature of herbal medicines (herb or preparation) is that their final effect is often caused by mutual interactions of different compounds. The combination of these features makes herbal analysis an extreme challenge for the current analysis and processing methods.

19.1.2 Analytical Strategies Used for Herbal Medicines

Analysis of herbal medicines may be required for diverse reasons. The applications discussed in this section are selected to illustrate the importance of herbal analysis and to explain what information is focused on in different fields.

In quality control of pharmaceuticals, pharmacopoeias describe a series of tests to demonstrate that a given sample conforms to the qualitative and quantitative requirements that are defined to guarantee drugs with sufficient safety and efficacy [10]. A first aspect included in a pharmacopoeial monograph is a check for the identity. For herbal medicines this means verifying gender and species. Thin-layer chromatography (TLC) is often the technique of choice for identity tests of herbal medicines thanks to its simplicity, inexpensiveness, versatility, speed, and selectivity [10]. TLC will be discussed in Section 19.2.1.1. It generates, for a sample, a pattern of bands or spots, reflecting the chemical composition. When a reference standard extract (or a certified genuine sample) is analyzed along with a sample, their patterns can be compared [8,10,11]. Besides TLC, profiles can be developed by many other separation techniques, for example, (ultra) high-performance liquid chromatography, resulting in a pattern of peaks. Traditionally, a number of marker compounds in the pattern are evaluated to decide on a sample's identity. The limited assessment is evidently a weakness in the process, as the evaluation of the markers is not exclusive enough for one species [10]. However, chemometric techniques that take the entire profile into consideration provide an objective way for identification purposes, and have already been shown beneficial in the context of identification for herbal material [15].

A second important step in quality control of drugs is demonstrating that the active substance is present in a sufficient amount, in order to guarantee its efficacy [8,10,11]. Various types of chromatographic separations allow quantification of (herbal) constituents, of which TLC [16], high-pressure liquid chromatography (HPLC) [17], and gas chromatography [18] are widely applied. The output of a chromatographic analysis is a set of consecutive detector measurements, where significant deviation of the noise (baseline) indicates the presence of detectable compounds. These data can be represented mathematically as a vector and graphically as a chromatogram.

Concentrations of compounds can be determined based on the comparison with the peak areas from compounds with known concentrations (see Chapter 6). Although the methodology of quantifying one or a limited number of marker compounds is applied in pharmacopoeial monographs on herbal medicines, it has some weaknesses. Usually one or a limited number of marker compounds that are important for the activity or that are characteristic for the herbal medicine are then quantified, for instance to identify samples [10]. Interactions with other compounds in the extract (for instance, synergistic effects) are thus completely neglected when marker information is linked to activity [12–14]. For identification purposes, the marker approach usually is insufficiently discriminative to identify only one species. Additionally, in most monographs on herbal medicines, different techniques are used for identity and assay tests [10]. The availability of one method serving both purposes would result in a decreased analysis time. From a fingerprint profile, the information relevant for identity testing for an assay or for evaluating another property of the sample (e.g., activity) can all be extracted applying the appropriate (multivariate) data analysis techniques.

Herbal analysis can also be applied in drug discovery. Worldwide, research groups screen herbal medicines with "long known" application in traditional medicine in order to scientifically study their secondary metabolites. The final aim is often to discover new compounds with mechanisms of action different from the currently applied drugs, which potentially could solve problems of therapy resistance, for instance, in the field of cancer or infection therapy [19–22]. This involves screening of samples from a herbal species for the pharmaceutical activity of interest, and identification of the potentially active compound(s), which requires their isolation. Usually, a multitude of purification steps (bioguided fractionation) is required to achieve this final goal [22]. After each step, the purified fraction is retested for its activity, in order to study in which fraction the active compound(s) are present. When an active compound is purified to a sufficient degree, techniques providing information on its chemical structure, that is, structure-elucidating techniques (nuclear magnetic resonance [NMR], mass spectrometry [MS], and infrared spectroscopy), are applied [19]. To estimate the biologic activity of a given herbal extract (crude extract or after purification steps), in vitro activity assays can be used [22]. Their advantage is that the global effect of all constituents present in the sample, which can mutually influence each other's activity, can be evaluated. Less advantageous is the fact that such tests may be rather time consuming. Additionally, determination of the compounds potentially responsible for the biological activity requires repeating the test on purified fractions until the active product is isolated, which is also time consuming.

Alternatively, linking the results of an activity test to fingerprint data using chemometric regression techniques (multivariate calibration) can give an insight into which compounds are potentially responsible for the activity, allowing researchers to proceed faster in drug discovery [22,23].

In other applications, fingerprints are the tools to study the stability of a herbal medicine, to optimize a process, for example, the extraction of a compound of interest, or to study the effects of harvest moments or of other factors on the chemical composition of a given herb [24–29]. Over the last years, experimental-design-based approaches were often used to optimize factors that influence a process, for instance, extraction parameters [24]. In this context, the use of fingerprints can help scientists to proceed faster and maximize the information on the herbal samples [30].

19.1.3 BENEFITS OF HERBAL FINGERPRINTING

A drawback of the currently applied quantitative (marker-based) pharmacopoeia assays is the inability to account for interactions (synergistic effects) between compounds. If such interactions are involved in the activity of the herb, conformity of the compound to the pharmacopoeial requirements will not be sufficient to guarantee that the sample meets the qualitative and quantitative requirements concerning its efficacy and safety, as interactions can moderate the activity of the sample [12–14]. Additionally, the specificity of the markers used in pharmacopoeia for identification is often insufficient to allow distinguishing a given species from all others.

A chromatogram can be developed to reflect the composition of a sample (or its extract) and can thus be considered as a characteristic profile or a fingerprint. In the fingerprinting approach, the

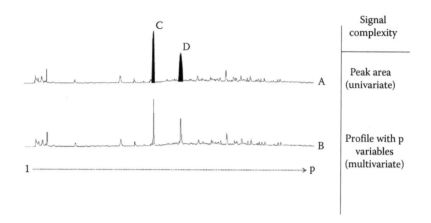

FIGURE 19.1 Traditional (A) peak-assay versus (B) fingerprint approach. Classically, individual compounds (C and D) are quantified.

entire profile (or occasionally a derived peak table) is considered to evaluate a sample's qualitative and/or quantitative properties. This is illustrated in Figure 19.1. Considering the entire profile of a sample has the advantage that all information on the compounds possibly influencing the activity or the class is included [12–14]. This can be beneficial in all aspects of herbal analysis mentioned in Section 19.1.2, as the information on all substances is considered simultaneously in order to find, for instance, groups of similar samples, to assign an unknown sample to a group (gender, species …) in the context of identification, or to link fingerprint data to quantitative properties of the sample, such as to biological activities [31]. The fingerprint methodology has become accepted by regulatory bodies and is slowly infusing into pharmacopoeial monographs, mainly for identification purposes [9,10]. The fact that fingerprints contain a high number of data points requires an appropriate data processing approach. To treat or visualize fingerprint data in a systematic way, multivariate techniques are the tools of choice [31]. These will be discussed in Section 19.3.

19.2 INSTRUMENTATION USED TO DEVELOP FINGERPRINTS

19.2.1 Chromatographic Techniques

Chromatographic methods are very popular for herbal analysis because of their ability to separate the compounds. Separation is induced by bringing the mixture in contact with two immiscible phases that are relatively moving in opposite directions. One of the phases, the stationary, is attached or coated to a carrier that often is silica-based and can have the properties of a solid or a liquid. The stationary-phase particles can be fixed on a plate (TLC) or immobilized in a column. The second phase, the mobile, moves over the stationary phase and can be a liquid, a gas, or a supercritical fluid. The stationary phase can have very diverse chemistries, ranging from high (e.g., bare silica) to low (e.g., C18 alkyl chains bound to silica) polarity. Since a good separation of the compounds in a mixture is obtained if their interactions with the two phases are sufficiently different, proper mobile and stationary phases are selected during method development. In the context of herbal fingerprinting, where one has very complex mixtures, method development often aims at obtaining a chromatogram where a maximal number of compounds are (partially) separated. However, this may depend on the purpose for which a fingerprint is developed [32].

19.2.1.1 Thin-Layer Chromatography

In TLC, the mobile phase is a liquid, while the stationary phase is composed of particles with liquid or solid properties attached to a plate. The sample is administered in small spots or bands on the plate, which is then put in contact with the mobile phase. The compounds are carried along by the mobile phase that usually mounts over the stationary phase through capillary forces. Compounds interacting more with the stationary phase will migrate rather slowly, while those that interact less with the stationary phase migrate faster. The result is a characteristic pattern of spots. After separation development, compounds are revealed using either reagents or by absorbance or fluorescence measurements, showing fingerprint patterns [16,33].

19.2.1.2   High-Performance Liquid Chromatography

In HPLC, the liquid mobile phase is pumped under high pressure at a constant flow rate through the stationary phase that is situated in a column. Sample (extract) is injected in the mobile phase flow and the compounds will move with different velocities through the column. Compounds eluting from the column are measured by a suitable detector [17]. The graphical representation of the detector signal as a function of time is a chromatogram, showing a peak for each detected compound. For a given chromatographic system, each compound has a characteristic retention time, which is the time between injection and detection of the maximal amount of the respective compound. Due to the complex composition of herbal extracts, with compounds that can largely differ in physicochemical properties, gradient elution is usually applied for fingerprinting. In this approach, the mobile phase composition is changed during analysis from low to high solvent strength. Isocratic conditions (i.e., fixed mobile phase used) would not allow this separation. Nonetheless, HPLC analyses taking around 60 min are not uncommon to develop fingerprints for herbal extracts. A mobile phase also may contain additives (e.g., acetic or formic acid or amines) to improve peak shapes [32].

Detectors play a key role in the compounds that appear in a chromatogram [17]. In herbal analysis, detection based on UV/Vis absorption is very often used, since the vast majority of compounds have structural properties allowing such absorption. Other detectors occasionally used are fluorescence [34], evaporative light scattering [35], electrochemical [34] and diode array (DAD) detectors. An extensively used detection technique for fingerprinting applications is mass spectrometry (MS) [36]. Since in fingerprint analysis the compounds are often unknown, occasionally several detectors are combined, either serially or in parallel, in order to get as much information as possible on a given extract's composition [34]. Evidently, the detector used will be decisive for the complexity of the fingerprints, as illustrated in Figure 19.2. Detectors that measure a single parameter (e.g., UV absorbance at one wavelength, fluorescence, or electrochemical properties) will generate fingerprints that exist of one row of measurements (a vector), while a fingerprint generated by a hyphenated detector (e.g., MS or DAD) will have an extra spectral dimension (fingerprint is a matrix).

19.2.1.3   Recent Evolutions

Since herbal samples contain a multitude of compounds, rather long fingerprints are often developed in herbal analysis [32]. Efforts to maintain the separations in shorter analysis times mainly focused on column properties. An important parameter, reflecting the separation power of an analytical column, is its efficiency [17], which is directly proportional to the column length, and inversely proportional to the height equivalent to a theoretical plate (HETP). Consequently, an increased efficiency can be obtained using longer chromatographic columns, or through reducing HETP. Several parameters influence HETP on a given column and their effect is described by means of the Van Deemter equation, and graphically represented in a Van Deemter plot [37]. The Van Deemter equation ($HETP = A + B/u + Cu$, where u is the linear velocity) expresses HETP as a function of the linear velocity, including an A term related to the Eddy diffusion, a B term related to the longitudinal diffusion, and a C term related to mass transfer [37]. The HETP for a classical HPLC column goes through a minimum at a given velocity. Consequently, the flow rate also goes through an optimum.

A first possibility to improve separations is to use longer columns. However, using a longer column leads to an increased back pressure. This limits the possibilities for classical columns.

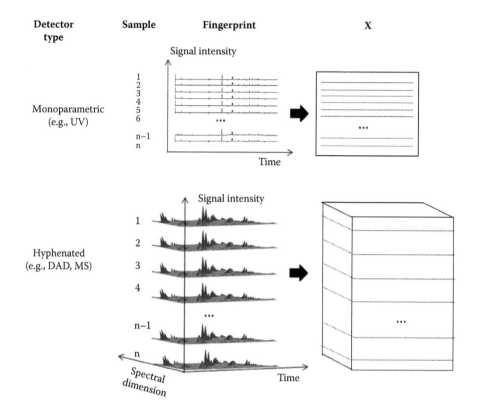

FIGURE 19.2 Output of monoparametric versus hyphenated detectors for fingerprint analysis.

However, this limitation can be overcome when monolithic (instead of particle-based) columns are used. Although the efficiency of these columns is less good than that of classical columns, their back pressure is much lower. Consequently, columns can be coupled, resulting in a sufficient efficiency without exceeding the workable back pressure limit. Therefore, these columns were recommended in a strategy to develop herbal fingerprints [32].

A second approach to improve the separation performance is using small particles, as this results in a decrease of both the A and the C terms, eventually leading to an increase in efficiency [38]. Classical particle sizes used in HPLC are in the range 2.5–5 μm, while in ultra-HPLC (UHPLC) applications, particles of 1.7 μm and smaller are used. However, pressures originating from these small particles are much higher than for classical particles. UHPLC nowadays is widely applied in fingerprint development.

A third possibility to improve the separation power is to use superficially porous, fused core, or core shell particles [39]. This results mainly in lower A and B terms [40,41].

19.2.2 OTHER TECHNIQUES

Other chromatographic techniques are also applied for herbal fingerprinting. When the compounds of interest are apolar, volatile compounds, gas chromatography is the preferred technique [42]. Evolutions in the field of supercritical fluid chromatography instrumentation also made this technique available for herbal analysis [43]. Non-chromatographic techniques were also successfully applied to develop herbal fingerprints, for example, capillary electrophoresis [44], direct infusion mass spectrometry [45], NMR [46] or selected ion flow tube mass spectrometry [47].

19.3 EXTRACTION OF RELEVANT INFORMATION

19.3.1 THE CRUCIAL ROLE OF DATA PRETREATMENT

A first step in the fingerprint data handling is to analyze the raw fingerprint data. However, these results can be suboptimal or difficult to interpret since experimental variability of different natures can have a major influence on the results of the multivariate techniques. Multivariate data analysis is based on matrix calculations. This data matrix is built by arranging the information (measurements of a signal at p time points) of n samples in an n × p matrix X. Ideally X would only contain the desired information on the samples, without being affected by noise or variability originating from other sources than those studied. However, in reality, undesired sources of variability often contribute to the fingerprint, and as a consequence adequate pretreatment of X can drastically improve the results, simplify models, and facilitate the interpretation of the data [15,48,49]. The fingerprint data pretreatment is dependent on the goal that is aimed. Thus, the best results are obtained when the preprocessing is optimized for a given data set and for a given aim. Data preprocessing is often a trial-and-error procedure, occasionally optimized by an experimental-design approach [50].

The objective of preprocessing is to minimize the influence of irrelevant information in the fingerprints, and to focus on or to maximize the variability of the information of interest. Chromatographic fingerprints are often recorded using gradient elution on different days, sometimes even over weeks or months. This situation often gives rise to undesired information in the fingerprints. The signal thus may change over time when the baseline signal changes due to mobile-phase composition changes. When measuring samples on different days, a solvent blank is often also measured daily. This allows subtracting the respective blank signal from each fingerprint, yielding only information on the compounds present in the extract [51]. Another example of undesired variability is the occurrence of retention time shifts, which result in information on one compound found in a given column of X for one sample that is shifted to another column for another sample. To correct for such shifts, alignment of the information over the samples is applied. Several alignment algorithms are available [31], for example, correlation optimized warping (Figure 19.3) [52,53], dynamic time warping [53],

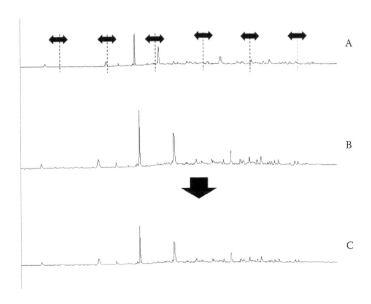

FIGURE 19.3 Correlation optimized warping. The chromatogram to be aligned (A) is divided in m parts of a given length, indicated by dashed lines. The double arrows represent a number of data points t that a chromatogram section can be stretched or compressed to correlate optimally to the target chromatogram (B). When both the segment length (determined by m) and the slack parameter t are optimized, the aligned chromatogram (C) is obtained.

parametric [54] and semi-parametric time warping [49], and automated peak alignment [55]. Other examples of often used pretreatments to highlight the information of interest in chromatographic fingerprint data are column centering, normalization by standard deviation, standard normal variate, and pareto scaling [48,56,57].

19.3.2 UNSUPERVISED DATA ANALYSIS

19.3.2.1 Exploratory Data Analysis

In exploratory data analysis, the information in **X** (raw data or after preprocessing) is used to reveal the global data structure. It concerns visualization techniques where similar objects will be close to each other or cluster, and dissimilar ones are far away. The data structure is thus revealed as groups of similar objects and outliers. The occurrence of outliers can have consequences for the subsequent data analysis [15,56,57].

Different techniques are used for this purpose [56–58]. Projection techniques, like principal component analysis (PCA) [56,57,59] or, more general, projection pursuit (PP) techniques, are often used and are illustrated in Figure 19.4. The common basic concept of PCA or PP is to define new axes, called latent variables (principal components or PCs), in the data space in such a way that every consecutive latent variable (LV) is in the direction of the maximal remaining variability in the data set. Each fingerprint is projected orthogonally on each LV, resulting in a score per principal component. Eventually, each fingerprint is represented by its scores on two (or three) LVs in a two (three)-dimensional score plot, which gives information on the relation among the fingerprints. As a consequence, groups of related samples can often be distinguished, while outlying samples can be identified as objects that are distant from the majority of the other fingerprints [15]. Scores are in fact weighted linear combinations of the variables in the **X** matrix. The weights used in these linear combinations are called loadings. The loading for a given PC can also be plotted in a loadings plot. They provide information on the variables. From such a loadings plot, one can observe to which extent the original variables contribute to the scores on a given PC and which variables provide similar information [56,57,59].

Clustering techniques are also applied for exploratory data analysis purposes [60]. Correlation or distance between fingerprints is considered as a measure of similarity, and the most similar fingerprints are clustered sequentially. The output of these techniques is illustrated in Figure 19.5 and is called a dendrogram. It provides also information on the structure of the data [57,60,61].

19.3.2.2 Similarity Analysis

For each pair of fingerprints, the similarity can be quantified using distance- or correlation-based parameters [62]. For distance-based parameters, the smaller the distance the more similar the fingerprints. When considering correlation-based parameters, high similarity between a pair of fingerprints is reflected by a value close to 1. When a set of n fingerprints is available, a similarity parameter

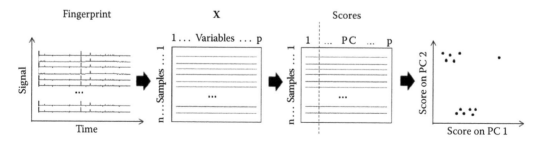

FIGURE 19.4 Principal component analysis. Fingerprints are converted to a data matrix **X**. Latent variables, here principal components or PCs, are calculated. The fingerprints are projected on each PC, resulting in a score per PC. Only a limited number of PCs are necessary to capture most (almost all) variability (for instance, left of the dashed line). A score plot displays the scores of each sample on two or three PCs.

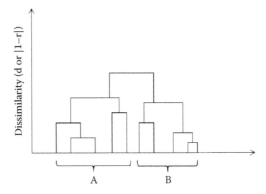

FIGURE 19.5 Hierarchical clustering analysis. Distance (d) or correlation (r) between each pair of finger-prints is calculated. Samples are linked based on similarity. The lower they are linked the more similar. Two main clusters (A) and (B) can be distinguished, each containing two subgroups.

can be calculated for all pairs of fingerprints. For a given parameter, the obtained values can be arranged in an n × n matrix, which shows symmetry around the main diagonal [62]. This matrix can be visualized in a color map, which allows the analyst to indicate fingerprints that are more distant or less correlated to the majority of the fingerprints. A statistical approach is to use reference samples and to estimate lower and upper warning and control limit values for each similarity parameter. For distance parameters, 97.5% of the values are below the upper warning limit and 99.85% below the upper control limit; for correlation parameters, 97.5% of the values are above the lower warning limit and 99.85% above the lower control limit. Alaerts et al. [62] applied this approach to check the authenticity of a set of green tea samples. Their approach is illustrated in Figure 19.6.

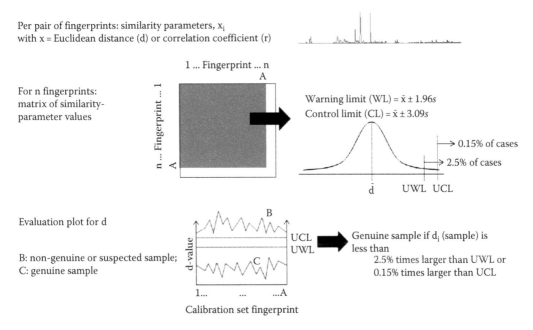

FIGURE 19.6 Similarity analysis. Per pair of fingerprints, a distance- or correlation-based similarity parameter is calculated. The obtained values are arranged in a matrix. When enough calibration samples (A) are available, warning and control limits can be calculated. For distance parameters, 97.5% of the values are expected below the UWL and 99.85% below the UCL. For a new fingerprint, its similarity values are compared to the warning and control limits in an evaluation plot.

The evaluation of the genuineness of test samples can be done either based on distance- or correlation-based similarity parameter values toward the fingerprints of a reference set. When a test sample is, for the distance parameter toward the reference samples, lower than the upper warning limit in at least 97.5% of the cases, the sample is considered genuine. The same is decided when its correlation with at least 97.5% of the reference samples is above the lower warning limit. Alternatively, the upper control limit for the distance or the lower for the correlation can be considered as more strict limits.

19.3.2.3 Fingerprint "Effect-Plots"

Klein-Júnior et al. [30] recently applied fingerprints and an experimental design approach to develop optimal extraction and fractionation conditions for multifunctional indole alkaloids from *Psychotria nemorosa* leaves. A part of the approach applied a screening design and is illustrated in Figure 19.7. The aim was determining which extraction-related factors significantly influenced a set of responses, including the UV absorptions at the different time points of a herbal fingerprint. The effects of each of the factors on the responses were calculated. Their significance was evaluated based on Dong's algorithm [63,64]. It involves an initial estimation of the standard error (s_0) based on the median of all effects. Finally, only the effects $|E_k| < 2.5s_0$ are used in the estimation of the final standard error $(SE)_e$ and the critical effect, E_{crit}. Plotting for a given factor, the effect per time point (for all responses) results in a fingerprint-effect plot. It allows observing how the entire fingerprint (given peaks or groups of peaks) is affected by changing a given factor from one level to another.

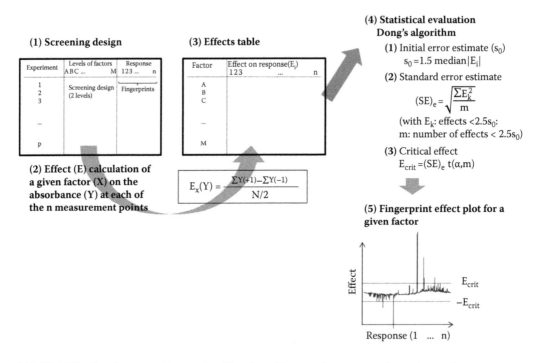

FIGURE 19.7 Development of fingerprint effect plots. The experiments are performed according to a screening design (1). The responses are LC-UV fingerprints. For each factor, its effect on the absorbance at each time point (n_i) of the fingerprint is estimated (2), put in a table (3), and statistically evaluated (4), using Dong's algorithm. This starts with an initial error estimate (s_0) based on the median of all effects (over n measurements). Then, the standard error $(SE)_e$ is estimated based on the effects below $2.5s_0$ (E_k). The critical effect is estimated as $(SE)_e$ multiplied by the student t-value with $\alpha = 5\%$ significance and m degrees of freedom. For a given factor, the fingerprint effect plot shows the n effects of a given factor.

19.3.3 Supervised Data Analysis

In supervised data analysis, besides the **X** matrix, containing the fingerprint data, at least one column vector **y** containing information on the fingerprints is required. The **y** information can be both continuous, for instance, a pharmaceutically interesting activity, or discrete, for example, class information (harvesting location or geographical origin, species, harvesting season). When the information is discrete, a discrimination/classification model, which is able to predict the class of a given new sample, can be built [56,57,65,66]. One may also focus on the variables that are important to distinguish the classes. When, on the other hand, the **y** information (property) is continuous (for instance, the antibiotic activity of a herbal sample), a regression model able to predict the property of a given new sample can be established [56,57,67]. From this type of model, one can also deduce which peaks (compounds) in a fingerprint are potentially involved in the observed activity. In both supervised multivariate modeling contexts, provided that the sample set is large enough, it will be split into a calibration set (around 70% or two-third of the samples) used to build the model and a test set (around 30% or one-third of the samples) used to externally validate the model. However, if the number of samples in the data set is limited, splitting into a calibration set and a test set is rather contraindicated since the quality of the developed model relies on a sufficient amount of expected variability contained in the calibration set used to build the model [15]. In that situation, a cross-validation (CV) approach is performed to validate the model.

19.3.3.1 Discrimination and Classification Modeling

In classification or discrimination modeling, the objective is to distinguish samples according to their class information **y**. The **X** and **y** data from the calibration set are used to build models that predict the class as a function of the fingerprint data in **X** [65]. Numerous techniques can be used in this context, but we will focus on some techniques that provide information both on the class and on the variables that are important for class prediction/distinction.

In a first group of techniques, the samples of all classes are used to develop a model distinguishing the classes. Techniques that use this principle are, for instance, linear and quadratic discriminant analysis (LDA and QDA) [15,57,65,66] and projection to latent structures discriminant analysis (PLS-DA) [56,66]. Both are projection techniques where the fingerprints are projected on LV axes, as in PCA (Figure 19.4). An important difference with PCA is that the axes are optimized to separate the fingerprints according to their class. For LDA and QDA, the samples are projected in LV space and a number of discriminant functions (DFs) that separate the sample projections according to their classes are defined (Figure 19.8). New samples are then assigned to the class their projection in LV space is located in (defined by the intersection of DFs).

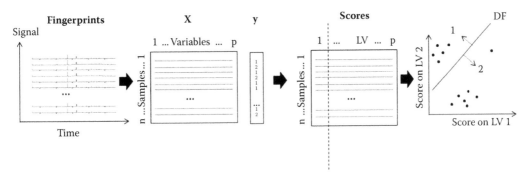

FIGURE 19.8 Linear or quadratic discriminant analysis. LVs are calculated that maximally and in decreasing order correlate to the class information in **y**. The fingerprints are projected on these LVs, resulting in a score for each fingerprint on each LV. The dashed line indicates the number of LVs necessary to get an optimal class separation (2 LVs in the example for graphical simplicity). A score plot displays the sample scores on both LVs. The discriminant function is indicated with DF. Samples situated in zone 1 are assigned to class 1, those in zone 2 to the other.

These methods are very useful, but one of their important drawbacks is that each sample is assigned to one of the classes that were considered in the model-building phase, meaning that samples that in reality do not belong to any of the classes will always be forced in one of the modeled classes. That is why these methods are called discrimination methods.

The principle of PLS-DA is illustrated in Figure 19.9. PLS-DA is a special form of the regression technique PLS (see Section 19.3.3.2), where the class information is transformed to as many columns as the number of classes in the data set. In fact, the members of one class are once opposed to all members of the other classes. For a given sample, its values in these columns are either 1 in the column where its class was opposed to the rest, and 0 in the other columns. The aim of PLS-DA is thus to build a model that predicts the values (around 0 or 1) in each column for a given representative new sample. In the end, the predicted values are converted to **coded y** values (0 and 1), which are then transformed to a class assignment (predicted class) [56,66].

Alternatively, models can be developed for each class individually, which is, for instance, the case for the technique soft independent modeling by class analogy [15,57,65]. It is illustrated in Figure 19.10. The models for each class are obtained by means of PCA, which does not create LVs optimizing class separation but rather capturing maximally the variability in each class. Each sample is then projected onto the LV spaces of the different classes. For each class, its members constitute a subspace with a centroid. For any sample, its distance to the centroid of a given class is converted to a probability of belonging to the considered class. A probability limit, often 95%, is used as a cutoff to consider a sample as a member of a given class. A new sample can be classified in one, more than one, or none of the calibration-set classes. Such methods are called classification methods.

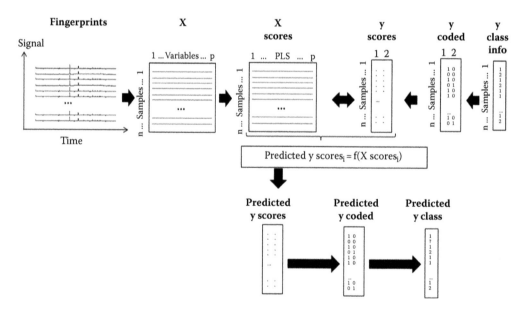

FIGURE 19.9 Projection to latent structures discriminant analysis (PLS-DA). The vector with class information (**y**) is converted to a matrix with as much columns as there are classes in the set (**y coded**), taking values zero (for non-members) and 1 (for members). PLS factors, or LVs that show maximal covariance, are computed for the **X** and the **y coded** matrices. Projection of the **X** data on the p LVs results in a matrix of **X scores**, while projection of the **y coded** data on the 2 LVs results in a matrix of **y scores**. A PLS model, linking **y scores** to **X scores** is built, which allows predicting **y scores** from the **X scores**. Through the predicted **y coded** values and **y class** labels, the classes are predicted. Conversion of the respective scores of new **X** into **predicted y class** data enables predicting the class of an unknown sample based on its fingerprint.

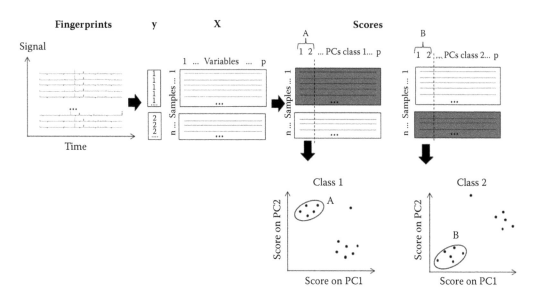

FIGURE 19.10 Soft independent modeling of class analogy. For each class, an individual PCA model is built (in this example, 2 PCs are used to simplify graphical representation, indicated with A and B for classes 1 and 2, respectively, marked in gray) based on its members data. The 95% probability limits of class membership (A and B on PC plots) are calculated for each class, based on the projections of its members. The non-members are also projected in the same PC space, and ideally fall outside the 95% confidence limit zone.

The number of LVs used to build the discrimination or classification model determines its complexity and the best model is usually determined using a CV procedure [56,57,67]. In this latter procedure, at different complexities, models are built based on a reduced calibration set, which is composed by all calibration samples with exclusion of a number of samples (often only one, leave-one-out CV). The classes of the excluded samples are predicted using the established models. This procedure is repeated until a prediction is obtained for each sample. The correct classification rate (CCR) of CV then expresses the ratio of the number of correctly assigned samples to the total number of samples in the calibration set. The number of LVs for the best model is selected as a compromise between a simple model (low number of LVs) and a high CCR. When the number of LVs to use is determined by means of CV, the final model is built based on the full calibration set used in the CV procedure. This model results in a CCR of calibration when applied to the calibration set, and in a CCR of prediction, when an independent test set is available. The CCRs quantify the discrimination/classification performance of a model and help in assessing whether or not a model is useful to predict the classes for future, real, and unknown samples. Approved models can then be used to classify or identify, for instance, different herbal species [15]. Classification techniques have been extensively successfully used on fingerprint data, and can also serve as identity tests for herbal medicines.

The main assumption in fingerprint analysis is that the variability within a group is smaller than the variability between groups. In general, the calibration and test sets used in modeling should be representative for the future samples. For instance, if relevant for future samples, the variability due to geographical origin or harvesting period, year-to-year variability, sample collection, and treatment procedures should be included in the calibration samples. When, over time, additional sources of variability occur, it will be necessary to update the model by incorporating calibration samples that cover these new sources.

19.3.3.2 Multivariate Calibration

19.3.3.2.1 Modeling and Prediction

Several multivariate regression techniques are available [67]. Again, in the context of this chapter, we will focus on a frequently used technique that allows not only modeling a property but also indicating the variables (compounds) probably contributing most. A widely applied technique, partial least squares or PLS regression existing in many variants (e.g., uninformative variable elimination PLS or orthogonal PLS), will be discussed to explain the principle of these projection techniques. Its principle is illustrated in Figure 19.11. PLS involves the creation of LVs, PLS factors, that explain as much as possible of the covariance between $\mathbf{X}$ and $\mathbf{y}$, where $\mathbf{y}$ can here be an activity measured for each sample. The fingerprints are projected on these LVs, resulting in scores on each PLS factor. The activity of a given sample can then be approximated by the weighted sum of its scores. This can be expressed as $\hat{y}_i = \sum_{i=1}^{q} w_q \cdot PLS_{iq}$, where $\hat{y}_i$ is the predicted activity of the ith sample, PLS_{iq} is the score of the ith sample on the qth PLS factor, and w_q the weight given to the qth PLS factor. The number of PLS factors in the best model (q) is usually determined by a CV procedure. One aims at a simple model with a minimal number of PLS factors with a minimal error on the predictions. For more detailed literature on multivariate regression, the reader is referred to [56,57,67].

When the activity (y-value) of each sample (in the calibration or test set) is predicted by the model, the correctness of the predictions can be evaluated, as illustrated in Figure 19.12. This is done by plotting the predicted activity as a function of the measured activity for all samples. The closer the samples are situated to a line through zero with unity slope (indicating equality), the better the model. Additionally, a root mean squared error (RMSE) can be calculated that quantitatively expresses the

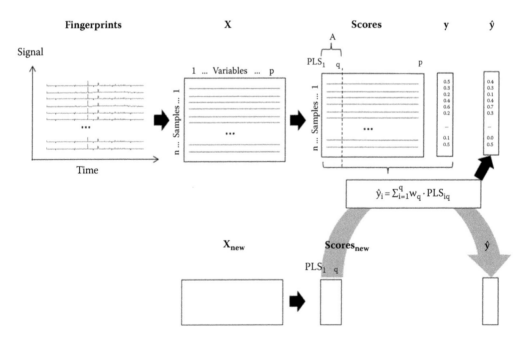

FIGURE 19.11 Projections to latent structures or partial least squares regression. Latent variables, PLS factors are calculated that maximally and in decreasing order explain as much as possible of the covariance between $\mathbf{X}$ and $\mathbf{y}$. A fingerprint is projected on each PLS factor, resulting in a score. The number of PLS factors (q) that sufficiently well models and predicts the property is indicated with the dashed line. This results in a model expressing y of a sample i as a function of its scores on the selected PLS factors (PLS_{iq}), taking a weight w_q into account for each PLS factor. Application of this model on (new) samples results in their predicted y values.

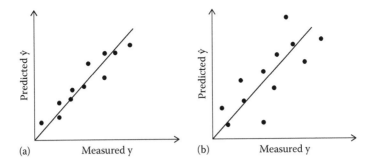

FIGURE 19.12 Evaluation of correspondence between measured and predicted activities. The predicted property is plotted as a function of the measured. (a) Example with good correspondence, that is, the points are close to the straight line indicating equality (i.e., going through zero and with slope one) and RMSE is low. (b) Example with less good prediction of property values.

quality of the predictions in the data set considered. $\text{RMSE} = \sqrt{\dfrac{\Sigma\left(y_i - \hat{y}_i\right)^2}{n}}$, with y_i and $\hat{y}_i$ the measured and predicted properties of the ith sample and n the number of samples taken into account. When the RMSE is estimated for the calibration-set samples, it is called RMSE of calibration (RMSEC); when it is obtained for the test set, RMSE of prediction (RMSEP), and when from a CV procedure, RMSE of cross validation (RMSECV). The smaller the RMSEP or RMSECV is, the better the predictive properties of the model. An example of the multivariate calibration of the antioxidant activity of green tea as a function of their fingerprints can be found in van Nederkassel et al. [68].

19.3.3.2.2 Indication of Interesting Compounds or Markers

Besides the prediction quality, in many studies it is even more interesting to determine which compounds can be linked to the observed activity. To obtain this goal, the model used to predict y is expressed as a function of the original variables in **X**, $y_i = \Sigma b_i \cdot x_i$, where x_i is the detector signal at the ith time point, and b_i the regression coefficient expressing to what extent x_i contributes to the approximation of y_i. Plotting the regression coefficients as a function of time along the fingerprints (illustrated in Figure 19.13)

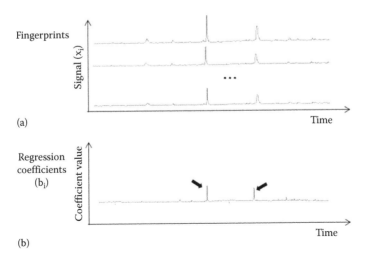

FIGURE 19.13 Regression coefficients plot (b) to indicate in the fingerprints (a) compounds (arrows) related to the studied activity.

results in a profile where positive regression-coefficient peaks indicate compounds behaving directly proportional with the activity (at least when y-value is higher for higher activity, else the opposite is true). Based on this information, potentially interesting peaks that require further study can be indicated. Examples of this approach can be found in References 69 and 70. Alternatively, as in [71], similar conclusions can be drawn based on plots of the weight vectors as a function of time points.

19.4   CONCLUSION

As discussed in this chapter, chromatographic fingerprints are used in many applications and for different purposes. In this chapter, we focused on their use for similarity analysis in the context of identification and quality control, for the classification of new samples to previously modeled groups, for the prediction of a property of new samples, and for the indication of interesting compounds related to a given property. It can be noticed that for given purposes, chromatographic fingerprints may be replaced by spectroscopic (e.g., properties modeling and prediction), but for others this is not possible (e.g., indication of interesting compounds) because information related to given compounds is needed and thus a separation technique is required.

REFERENCES

1. World Health Organisation. 2015. Essential medicines and health products. WHO. http://www.who.int/medicines/about/en/ (accessed February 17, 2017).
2. World Health Organisation. 2015. General guidelines for methodologies on research and evaluation of traditional medicine. WHO. http://www.who.int/medicines/areas/traditional/definitions/en/ (accessed February 17, 2017).
3. The European Parliament and the council of the European union. 2012. Annex I of Directive 2001/83/EC of the European Parliament and of the Council of 6 November 2001 on the Community code relating to medicinal products for human use. http://ec.europa.eu/health/files/ eudralex/vol-1/dir_2001_83_consol_2012/dir_2001_83_cons_2012_en.pdf (accessed February 17, 2017).
4. European Medicines Agency. Committee for Medicinal Products for Human Use (CHMP), Committee for Medicinal Products for Veterinary Use (CVMP), Committee on Herbal Medicinal Products (HMPC). 2011. Guideline on quality of herbal medicinal products/traditional herbal medicinal products. EMA/CPMP/QWP/2819/00 Rev. 2, EMA/CVMP/814/00 Rev. 2, EMA/ HMPC/201116/2005 Rev. 2. http://www.ema.europa.eu/docs/en_GB/document_library/Scientific_guideline/2011/09/WC500113209.pdf (accessed February 17, 2017).
5. European Medicines Agency. Committee for Medicinal Products for Human Use (CHMP), Committee for Medicinal Products for Veterinary Use (CVMP), Committee on Herbal Medicinal Products (HMPC). 2011. Guideline on specifications: Test procedures and acceptance criteria for herbal substances, herbal preparations and herbal medicinal products/traditional herbal medicinal products, EMA/CPMP/QWP/2820/00 Rev. 2, EMA/CVMP/815/00 Rev. 2 EMA/ HMPC/162241/ 2005 Rev. 2. http://www.ema.europa.eu/docs/en_GB/document_library/Scientific_guideline/2011/09/WC500113210.pdf (accessed February 17, 2017).
6. European Medicines Agency. Committee on Herbal Medicinal Products (HMPC), Committee for Medicinal Products for Human Use (CHMP), Committee for Medicinal Products for Veterinary Use (CVMP). 2008. Guideline on quality of combination herbal medicinal products/traditional herbal medicinal products, Doc. Ref. EMEA/HMPC/CHMP/CVMP/214869/2006. http://www.ema.europa.eu/docs/en_GB/document_library/Scientific_guideline/2009/09/WC500003286.pdf (accessed February 17, 2017).
7. European Medicines Agency. Committee on Herbal Medicinal Products (HMPC). 2012. Concept paper on non-pharmacopoeial reference standards for herbal substances, herbal preparations and herbal medicinal products/traditional herbal medicinal products, EMA/HMPC/312890/2012. http://www.ema.europa.eu/docs/en_GB/document_library/Scientific_guideline/2012/06/WC500129237.pdf (accessed February 17, 2017).

8. Zollner, T., and M. Schwarz. 2013. Herbal Reference Standards: Applications, definitions and regulatory requirements. *Braz J Pharmacog* 23: 1–21.
9. Chinese Pharmacopoeia Commission. 2005. *Pharmacopoeia of the People's Republic of China*, Vol. 1. Beijing, China: People Medical Publishing House.
10. European Directorate for the Quality of Medicines and Healthcare. 2017. *European Pharmacopoeia* edition 9.0. http://online.pheur.org/EN/entry.htm (accessed February 17, 2017).
11. Schwarz, M., Klier, B., and H. Sievers. 2009. Herbal Reference Standards. *Planta Med* 7: 689–703.
12. Wagner, H., and G. Ulrich-Merzenich. 2009. Synergy research: Approaching a new generation of phyto-pharmaceuticals. *Phytomedicine* 16: 97–110.
13. Lu, G.H., Chan, K., Liang, Y.Z., Leung, K., Chan, C.L., Jiang, Z.H., and Z.Z. Zhao. 2005. Development of high-performance liquid chromatographic fingerprints for distinguishing Chinese *Angelica* from related umbelliferae herbs. *J Chromatogr A* 1073: 383–392.
14. Li, Y., Wu, T., Zhu, J., Wan, L., Yu, Q., Li, X., Cheng, Z., and C. Guo. 2010. Combinative method using HPLC fingerprint and quantitative analyses for quality consistency evaluation of an herbal medicinal preparation produced by different manufacturers. *J Pharm Biomed Anal* 52: 597–602.
15. Viaene, J., Goodarzi, M., Dejaegher, B., Tistaert, C., Hoang Le Tuan, A., Nguyen Hoai, N., Chau Van, M., Quetin-Leclercq, J., and Y. Vander Heyden. 2015. Discrimination and classification techniques applied on *Mallotus* and *Phyllanthus* high performance liquid chromatography fingerprints. *Anal Chim Acta* 877: 41–50.
16. Sherma, J., and B. Fried (eds.). 2003. *Handbook of Thin-Layer Chromatography*. 3rd edn. New York: Marcel Dekker.
17. Lough, W.J., and I.W. Wainer. 1996. *High Performance Liquid Chromatography: Fundamental Principles and Practice*. 1st edn. Glasgow, Scotland: Blackie Academic and Professional.
18. Poole, C.F. 2012. *Gas Chromatography*. 1st edn. Oxford, U.K.: Elsevier.
19. Zhang, M., Zhao, H., Zhao, Z., Yan, H., Lv, R., Cui, L., Yuan, J. et al. 2016. Rapid screening, identification and purification of neuraminidase inhibitors from *Lithospermum erythrorhizon* Sieb.et Zucc. by ultrafiltration with HPLC-ESI-TOF-MS combined with semi-preparative HPLC. *J Sep Sci* 39: 2097–2104.
20. Wang, C., Hu, S., Chen, X., and X. Bai. 2016. Screening and quantification of anti-cancer compounds in traditional Chinese medicine by hollow fiber cell fishing and hollow fiber liquid/solid-phase microextraction. *J Sep Sci* 39: 1814–1824.
21. Arunachalam, K., Ascêncio, S.D., Soares, I.M., Souza Aguiar, R.W., da Silva, L.I., de Oliveira, R.G., Balogun, S.O., and D.T. de Oliveira Martins. 2016. *Gallesia integrifolia* (Spreng.) Harms: In vitro and in vivo antibacterial activities and mode of action. *J Ethnopharmacol* 184: 128–137.
22. Xu, J., Xu, Q.S., Chan, C.O., Mok, D.K., Yi, L.Z., and F.T. Chau. 2015. Identifying bioactive components in natural products through chromatographic fingerprint. *Anal Chim Acta* 870: 45–55.
23. Tistaert, C., Chataigné, G., Dejaegher, B., Rivière, C., Nguyen Hoai, H., Chau Van, M., Quetin-Leclercq, J., and Y. Vander Heyden. 2012. Multivariate data analysis to evaluate the fingerprint peaks responsible for the cytotoxic activity of *Mallotus* species. *J Chromatogr B* 910: 103–113.
24. Poojary, M.M., and P. Passamonti. 2015. Optimization of extraction of high purity all-trans-lycopene from tomato pulp waste. *Food Chem* 188: 84–91.
25. Alaerts, G., Merino-Arévalo, M., Dumarey, M., Dejaegher, B., Noppe, N., Matthijs, N., Smeyers-Verbeke, J., and Y. Vander Heyden. 2010. Exploratory analysis of chromatographic fingerprints to distinguish rhizoma *chianxiong* and rhizoma *ligustici*. *J Chromatogr A* 1217: 7706–7716.
26. Bueno, P.C.P., Pereira, F.M.V., Torres, R.B., and A.J. Cavalheiro. 2015. Development of a comprehensive method for analyzing clerodane-type diterpenes and phenolic compounds from *Casearia sylvestris* Swartz (Salicaceae) based on ultra high performance liquid chromatography combined with chemometric tools. *J Sep Sci* 38: 1649–1656.
27. Izadiyan, P., and B. Hemmateenejad. 2016. Multi-response optimization of factors affecting ultrasonic assisted extraction from Iranian basil using central composite design. *Food Chem* 190: 864–870.
28. Wu, J., Yu, D., Sun, H., Zhang, Y., Zhang, W., Meng, F., and X. Du. 2015. Optimizing the extraction of anti-tumor alkaloids from the stem of *Berberis amurensis* by response surface methodology. *Ind Crops Prod* 69: 68–75.
29. Ćujić, N., Šavikin, K., Janković, T., Pljevljakušić, D., Zdunić, G., and S. Ibrić. 2016. Optimization of polyphenols extraction from dried chokeberry using maceration as traditional technique. *Food Chem* 194: 135–142.

30. Klein-Júnior, L.C., Viaene, J., Salton, J., Koetz, M., Gasper, A.L., Henriques, A.T., and Y. Vander Heyden. 2016. The use of chemometrics to study multifunctional indole alkaloids from *Psychotria nemorosa* (Palicourea comb. nov.). Part I: Extraction and fractionation optimization based on metabolic profiling. *J Chromatogr A* 1463: 60–70.

31. Tistaert, C., Dejaegher, B., and Y. Vander Heyden. 2011. Chromatographic separation techniques and data handling methods for herbal fingerprints: A review. *Anal Chim Acta* 690: 148–161.

32. Dejaegher, B., Alaerts, G., and N. Matthijs. 2010. Methodology to develop liquid chromatographic fingerprints for the quality control of herbal medicines. *Acta Chromatogr* 22: 237–258.

33. Tang, T.X., Guo, W.Y., Xu, Y., Zhang, S.M., Xu, X.J., Wang, D.M., Zhao, Z.M., Zhu, L.P., and D.P. Yang. 2014. Thin-layer chromatographic identification of Chinese propolis using chemometric fingerprinting. *Phytochem Anal* 25: 266–272.

34. He, Y., Wu, Q., Hansen, S.H., Cornett, C., Møller, C., and P. Lai. 2013. Differentiation of tannin-containing herbal drugs by HPLC fingerprints. *Pharmazie* 68: 155–159.

35. Alaerts, G., Matthijs, N., Smeyers-Verbeke, J., and Y. Vander Heyden. 2007. Chromatographic fingerprint development for herbal extracts: A screening and optimization methodology on monolithic columns. *J Chromatogr A* 1172: 1–8.

36. Watson, J.T., and O.D. Sparkman. 2007. *Introduction to Mass Spectrometry: Instrumentation, Applications, and Strategies for Data Interpretation.* 4th edn. Chichester, U.K.: Wiley & Sons.

37. Dong, M.W. 2006. *Modern HPLC for Practicing Scientists.* Hoboken, NJ: Wiley & Sons.

38. Waksmundzka-Hajnos, M., and J. Sherma (eds.). 2011. *High Performance Liquid Chromatography in Phytochemical Analysis.* Boca Raton, FL: CRC Press.

39. Hayes, R., Ahmed, A., Edge, T., and H. Zhang. 2014. Core-shell particles: Preparation, fundamentals and applications in high performance liquid chromatography. *J Chromatogr A* 1357: 36–52.

40. Guiochon, G., and F. Gritti. 2011. Shell particles, trials, tribulations and triumphs. *J Chromatogr A* 1218: 1915–1938.

41. Griffin, C.T., Gosetto, F., Danaher, M., Sabatini, S., and A. Furey. 2014. Investigation of targeted pyrrolizidine alkaloids in traditional Chinese medicines and selected herbal teas sourced in Ireland using LC-ESI-MS/MS. *Food Addit Contam Part A Chem Anal Control Expo Risk Assess* 31: 940–961.

42. Liu, C.T., Zhang, M., Yan, P., Liu, H.C., Liu, X.Y., and R.T. Zhan. 2016. Qualitative and quantitative analysis of volatile components of Zhengtian pills using gas chromatography mass spectrometry and ultra-high performance liquid chromatography. *J Anal Methods Chem* doi: 10.1155/2016/1206391.

43. Li, J.R., Li, M., Xia, B., Ding, L.S., Xu, H.X., and Y. Zhou. 2013. Efficient optimization of ultra-high-performance supercritical fluid chromatographic separation of *Rosa sericea* by response surface methodology. *J Sep Sci* 36: 2114–2120.

44. Mazina, J., Vaher, M., Kuhtinskaja, M., Poryvkina, L., and M. Kaljurand. 2015. Fluorescence, electrophoretic and chromatographic fingerprints of herbal medicines and their comparative chemometric analysis. *Talanta* 139: 233–246.

45. Cabral, E.C., Sevart, L., Spindola, H.M., Coelho, M.B., Sousa, I.M., Queiroz, N.C., Foglio, M.A., Eberlin, M.N., and J.M. Riveros. 2013. *Pterodon pubescens* oil: Characterisation, certification of origin and quality control via mass spectrometry fingerprinting analysis. *Phytochem Anal* 24: 184–192.

46. Qin, H.L., Deng, A.J., Du, G.H., Wang, P., Zhang, J.L., and Z.H. Li. 2009. Fingerprinting analysis of Rhizoma chuanxiong of commercial types using 1H nuclear magnetic resonance spectroscopy and high performance liquid chromatography method. *J Integr Plant Biol* 51: 537–544.

47. Sovova, K., Dryahina, K., and P. Španel. 2011. Selected ion flow tube (SIFT) studies of the reactions of H_3O^+, NO^+ and O_2^+ with six volatile phytogenic esters. *Int J Mass Spectrom* 300: 31–38.

48. Zeaiter, M., and D. Rutledge. 2009. Preprocessing methods (Chapter 3.04). In *Comprehensive Chemometrics*, Vol. 3, eds. S.D. Brown, R. Tauler, B. Walczak, pp. 121–231. Amsterdam, the Netherlands: Elsevier.

49. Van Nederkassel, A.M., Xu, C.J., Lancelin, P., Sarraf, M., Mackenzie, D.A., Walton, N.J., Bensaid, F. et al. 2006. Chemometric treatment of vanillin fingerprint chromatograms. Effect of different signal alignments on principal component analysis plots. *J Chromatogr A* 1120: 291–298.

50. Gerretzen, J., Szymańska, E., Jansen, J.J., Bart, J., van Manen, H.J., van den Heuvel, E.R., and L.M.C. Buydens. 2015. Simple and effective way for data preprocessing selection based on design of experiments. *Anal Chem* 87: 12096–12103.

51. Alaerts, G., Pieters, S., Logie, H., Van Erps, J., Merino-Arévalo, M., Dejaegher, B., Smeyers-Verbeke, J., and Y. Vander Heyden. 2014. Exploration and classification of chromatographic fingerprints as additional tool for identification and quality control of several *Artemisia* species. *J Pharm Biomed Anal* 95: 34–46.

52. Vest Nielsen, N.P., Carstensen, J.M., and J. Smedsgaard. 1998. Aligning of single and multiple wavelength chromatographic profiles for chemometric data analysis using correlation optimised warping. *J Chromatogr A* 805: 17–35.

53. Pravdova, V., Walczak, B., and D.L. Massart. 2002. A comparison of two algorithms for warping of analytical signals. *Anal Chim Acta* 456: 77–92.

54. Eilers, P.H.C. 2004. Parametric time warping. *Anal Chem* 76: 404–411.

55. Daszykowski, M., Vander Heyden, Y., Boucon, C., and B. Walczak. 2010. Automated alignment of one-dimensional chromatographic fingerprints. *J Chromatogr A* 1217: 6127–6133.

56. Eriksson, L., Johansson, E., Kettaneh-Wold, N., Trygg, J., Wikstrom, C., and S. Wold. 2006. *Multi-and Megavariate Data Analysis. Part I: Basic Principles and Applications*. 2nd edn. Umea, Sweden: Umetrics Academy.

57. Vandeginste, B.G.M., Massart, D.L., Buydens, L.M.C., De Jong, S., Lewi, P.J., and J. Smeyers-Verbeke. 1998. *Handbook of Chemometrics and Qualimetrics—Part B*. Amsterdam, the Netherlands: Elsevier.

58. de Juan, A., and R. Tauler. 2009. Linear soft-modelling: Introduction (Chapter 2.12). In *Comprehensive Chemometrics*, Vol. 2, eds. S.D. Brown, R. Tauler, B. Walczak, pp. 207–210. Amsterdam, the Netherlands: Elsevier.

59. Esbensen, K.H., and P. Geladi. 2009. Principal component analysis: Concept, geometrical interpretation, mathematical background, algorithms, history, practice (Chapter 2.13). In *Comprehensive Chemometrics*, Vol. 2, eds. S.D. Brown, R. Tauler, B. Walczak, pp. 211–226. Amsterdam, the Netherlands: Elsevier.

60. Lee, I., and J. Yang. 2009. Common clustering algorithms (Chapter 2.27). In *Comprehensive Chemometrics*, Vol. 2, eds. S.D. Brown, R. Tauler, B. Walczak, pp. 557–618. Amsterdam, the Netherlands: Elsevier.

61. Shu, Z., Li, X., Rahman, K., Qin, L., and C. Zheng. 2016. Chemical fingerprint and quantitative analysis for the quality evaluation of *Vitex negundo* seeds by reversed-phase high-performance liquid chromatography coupled with hierarchical clustering analysis. *J Sep Sci* 39: 279–286.

62. Alaerts, G., Van Erps, J., Pieters, S., Dumarey, M., van Nederkassel, A.M., Goodarzi, M., Smeyers-Verbeke, J., and Y. Vander Heyden. 2012. Similarity analyses of chromatographic fingerprints as tools for identification and quality control of green tea. *J Chromatogr B* 910: 61–70.

63. F. Dong. 1993. On the identification of active contrasts in unreplicated fractional factorials. *Stat Sin* 3: 209–217.

64. Vander Heyden, Y., Nijhuis, A., Smeyers-Verbeke, J., Vandeginste, B.G.M., and D.L. Massart. 2001. Guidance for robustness/ruggedness tests in method validation. *J Pharm Biomed Anal* 24: 723–753.

65. Lavine, B.K., and W.S. Rayens. 2009. Classification: Basic concepts (Chapter 3.15). In *Comprehensive Chemometrics*, Vol. 3, eds. S.D. Brown, R. Tauler, B. Walczak, pp. 507–515. Amsterdam, the Netherlands: Elsevier.

66. Lavine, B.K., and W.S. Rayens. 2009. Statistical discriminant analysis (Chapter 3.16). In *Comprehensive Chemometrics*, Vol. 3, eds. S.D. Brown, R. Tauler, B. Walczak, pp. 517–540. Amsterdam, the Netherlands: Elsevier.

67. Kalivas, J.H.. 2009. Calibration methodologies (Chapter 3.01). In *Comprehensive Chemometrics*, Vol. 3, eds. S.D. Brown, R. Tauler, B. Walczak, pp. 1–32. Amsterdam, the Netherlands: Elsevier.

68. van Nederkassel, A.M., Daszykowski, M., Massart, D.L., and Y. Vander Heyden. 2005. Prediction of total green tea antioxidant capacity from chromatograms by multivariate modelling. *J Chromatogr A* 1096: 177–186.

69. Ben Ahmed, Z., Yousfi, M., Viaene, J., Dejaegher, B., Demeyer, K., Mangelings, D., and Y. Vander Heyden. 2016. Antioxidant activities of *Pistacia atlantica* extracts modelled as a function of chromatographic fingerprints in order to identify antioxidant markers. *Microchem J* 128: 208–217.

70. Klein-Junior, L.C., Viaene, J., Tuenter, E., Salton, J., Gasper, A.L., Apers, S., Andries, J.P., Pieters, L., Henriques, A.T., and Y. Vander Heyden. 2016. The use of chemometrics to study multifunctional indole alkaloids from *Psychotria nemorosa* (Palicourea comb. nov.). Part II: Indication of peaks related to the inhibition of butyrylcholinesterase and monoamine oxidase-A. *J Chromatogr A* 1463: 71–80.

71. Felipe-Sotelo, M., Tauler, R., Vives, I., and J.O. Grimalt. 2008. Assessment of the environmental and physiological processes determining the accumulation of organochlorine compounds in European mountain lake fish through multivariate analysis (PCA and PLS). *Sci Total Environ* 404: 148–161.

20 Chemometric Strategies in Analysis of Chromatographic–Mass Spectrometry Data

Samantha Riccadonna and Pietro Franceschi

CONTENTS

20.1 INTRODUCTION

The coupling of chromatography and mass spectrometry (MS) provides a powerful tool for the chemical analysis of complex matrices [1–3]. The most common analytical techniques used in the separation step are gas chromatography (GC), liquid chromatography (LC), and capillary electrophoresis (CE) [4–12]. GC-MS allows the separation of volatile and semivolatile compounds, LC-MS is used for nonvolatile compounds with higher molecular weight, both polar and nonpolar, while CE is mainly used for polar, charged analytes. Nowadays, these techniques are routinely used in many fields such as clinical biochemistry [13–15], pharmacy [16], food science [17], environmental analyses [6,18], toxicology [19,20], and forensics [21,22], to list a few.

MS and chromatography complement each other, and this is the reason for the success of hyphenated techniques. MS is sensitive and "universal," but it measures only mass to charge ratios, with the additional drawback that ionization is extremely sensitive to matrix effects, that is, it highly depends on the matrix composition. A chromatographic separation before MS makes the analytical platform also sensitive to the chemical characteristics of the analytes, with the additional bonus that the components of a complex matrix do not reach the ionization source at the same time, thus reducing the impact of matrix effects. Even in presence of excellent separation, however, the problems arising from co-elution cannot be fully solved [23,24] and they can be a subtle confounding factor leading to biased results.

The analytical platforms available nowadays allow the fast measurement of large numbers of samples and can be used to perform targeted and untargeted analyses [25–28]. Targeted experiments aim to measure the concentration of a set of selected compounds/analytes, which is chosen in advance. From the data analysis point of view, then, their output is a matrix where each row represents one sample and each column one of the selected metabolites. Untargeted experiments, instead, are designed with the ideal objective of analyzing all the analytes present in the samples, including chemical unknowns. Not surprisingly, these type of experiments are more "powerful", but the price to pay is a more complex data analysis workflow, which can more easily lead to false discoveries. It is important to point out that in untargeted experiments what is actually measured is not the concentration of specific metabolites, but instead the intensity of the ions generated in the ionization source. The common output of this type of analysis is not anymore a two-dimensional matrix of metabolites and samples, but a three-dimensional data "cube," where each variable (ion) is characterized by its *m/z* value, its retention time (*rt*), and its intensity. The data cube can still be organized in two-dimensional matrix form: the samples will still be the rows, but now each column will represent an ion and not a metabolite. These variables are commonly indicated by the term "features."

Different chemometric strategies should be applied for the analysis of the outcomes of targeted and untargeted experiments, in particular because the first ones are usually employed in hypothesis testing and model validation, while the latter are more suitable in the hypothesis generation phase. Untargeted experiments also require tailored chemometric strategies to optimize the initial processing of the "raw" data and to associate the features to specific metabolites.

Regardless of its origin, each data matrix can be analyzed from a univariate or multivariate point of view. With the first perspective, each variable is treated independently, while multivariate approaches try to benefit from the relations among the variables to highlight the overall structure of the dataset. The univariate approach is appealing because the results can be easily interpreted on the bases of the measured variables and this is the natural choice, for example, if one is interested to test if the difference in concentration of a metabolite in a two-class problem is significant or not. Univariate approaches, however, suffer from known statistical limitation when they are used "sequentially" to tackle more global scientific questions like "are these two classes significantly different?" In this case, indeed, it is necessary to take into account the well-known "multiple testing" problem [29]. Multivariate tools concentrate on the full data matrix, with the objective of exploiting the "correlations" among the variables. This choice is particularly appealing when the interactions among the variables are an inherent aspect of the problem under study because they are, for example, the result of chemical or biological processes. Going multivariate, however, it is not as straightforward as can be naively expected. Due to the recent technological improvements in the analytical platforms, indeed, the general trend is to measure more variables than samples and in this case the associations among the variables will also be present by chance, and will not mirror real chemical or biological relations. Spurious associations, however, will be picked up by multivariate analyses with obvious implication on the interpretability of the results. In this regard, it is mandatory to perform a robust validation of the results: statistical models fitting the experimental data too closely are not describing well the chemical or biological phenomena, since they are biased by experimental variability and by the presence of spurious associations. An exhaustive treatment of all these aspects goes beyond the scope of this chapter and the interested reader can refer to the specific literature [30–33]. As a general comment, we would like just to remark that the path from data to knowledge can be intricate and require feedback loops between the design of the experiment, the data processing, the data analysis, and the validation.

In the remainder of the chapter, we will focus on some characteristics of chromatographic–mass spectrometry data that have to be taken into account when performing univariate and multivariate analyses. We will then introduce two widely used multivariate tools that are popular in metabolomics and analytical chemistry: principal component analysis (PCA) and partial least squares (PLS) regression, also known as projection on latent structures. Both tools are based on the idea that the information encoded in a dataset can be effectively summarized by a limited number of "new" variables resulting from the combination of the ones actually measured. PCA and PLS are already well described in literature and implemented in many software packages, thus we will discuss them from a methodological point of view. A most comprehensive introduction to multivariate chemometrics tools can be found in [34,35], both providing detailed discussions of worked-out examples in the popular R programming language.

20.2   DATA TRANSFORMATION

The data analyst must be aware of the characteristic of chromatography–MS data in order to be able to highlight the interesting biological variability, controlling as much as possible the impact of unwanted analytical variability. We have already mentioned that the MS ion source itself can be a source of variability as well as matrix effects. Moreover, sample-to-sample differences can be introduced in the sample handling and data generation phases [36]. A normalization step is often necessary to limit the impact of these phenomena. In targeted approaches, when the number of measured compounds is limited, a widely adopted strategy involves the use of standards, which are known compounds with known concentrations [36,37]. In untargeted approaches, instead, the use of standards is possible but not straightforward, because a limited number of chemicals is not able to fully capture the complexity of the real samples. In such cases, it is common to rely on quality control (QC) samples in order to both control and calibrate the system in terms of performances and stability [38]. The use of QCs for normalization purposes, also called model-based normalization, is based on the assumption that the model of variation extrapolated from the QCs holds for the whole dataset, which of course has its own limits [39]. In addition to model-based approaches, a number of data-driven normalization algorithms have been developed in order to account for the systematic variation affecting the measurements both across different runs of the same datasets and across different related datasets [37,38]. They can vary from a simple scaling factor for all the spectra to more complex transformations (based, for instance, on LOESS or mixed linear models) [39,40].

A second critical aspect of these data is known in statistics as heteroscedasticity [37]. The variance (both technical and biological) affecting each variable is not constant but is larger for more intense signals [37,41]. This could be an issue in the statistical analysis or modeling phase, since a homogenous structure of the variance is often an assumption of the statistical models, for example, in linear models or generalized additive models. Van den Berg and colleagues discuss a list of transformations (and their pros and cons) that could be applied to correct for unwanted characteristics of the data [41]. As an example, log transformation or power transformation can be used to correct for data heteroscedasticity. It is worth noting here that this brings as a consequence that the differences between small and large values are reduced. This is of particular interest for chromatography–MS data considering the high dynamic range of these techniques. Of course, the choice of the transformations to be applied is constrained by the statistical analysis and modeling tools adopted, which in turn depends on the given task (such as classification, feature selection, etc.), and vice versa.

Finally, it has to be taken into account that chromatography–MS datasets are also characterized by the presence of missing values, that is, the concentration or intensity is missing for a subset of the metabolites or features, respectively. In targeted approaches, this happens when a compound is under the detection threshold and thus a quite simple strategy for data imputing can be adopted. The easiest choice would be to replace missing values with the same (small) constant value. However, this choice does not take into consideration the natural variability in concentration, thus impacting in following data analysis phase. Imputing missing values with any value between zero and the instrument detection limit could be a better choice, even if zero is a critical value in case of log-transformation. In untargeted experiments, the picture is more complex: we have to consider that a peak could be missed by the peak picking algorithm or it could simply be absent in a specific sample. In such cases, double check of the peak picking procedure is needed. Eventually, the same imputing strategy discussed in the targeted approach can be adopted.

20.3 PRINCIPAL COMPONENT ANALYSIS*

Before spending a lot of time building, tuning, and interpreting the results of complex models, it is a good practice to check how the measured data look like and if there are abnormalities or "weird" points. The best tool for exploring multidimensional data produced with analytical techniques such as LC-MS or GC-MS is certainly PCA.

The main ingredients of PCA are the matrices represented in Figure 20.1. The matrix X stores the information about N samples (on the rows) for which K variables were measured. The N samples

---

* See also Chapter 8.

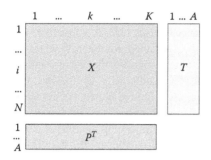
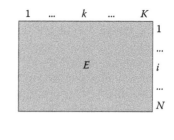

FIGURE 20.1   Schematic representation of the key elements of the PCA: the $N \times K$ data matrix X; the matrix of the loadings P^T, which are the new A dimensions obtained as a linear combination of the original K ones; the matrix of the scores T, which are the new coordinates of the points; the matrix of the residuals E, containing the part of X not explained by PCA.

can be represented as a cloud of points in the geometric space defined by the K variables. In this geometric view, PCA can be seen as an algorithm looking for a projection of the data in a lower dimensional space, which captures most of the dataset variability. Here, the underlying assumption is that variability is equivalent to information. When the X matrix has the mean of the columns equal to zero, or, in other words, when all the variables have zero mean, the projection will be immediately related to the covariance of the initial variables. Software implementations of PCA* usually provide the mean-centered PCA as the default choice, but also the noncentered version is allowed. In this last case, if the centroid of the cloud of samples is quite distant from the origin, the first PC will not be determined by the variance of the variables but, instead, by their means.

PCA represents the matrix X as a linear combination of the *scores T* and the *loadings P* (where both the matrices have as many rows as X and A columns): this decomposition can be written in matrix form as $X = TP^T + E$.[†] The presence of the error matrix E indicates that the representation obtained by PCA is an approximation of the original data X. The residuals can be interpreted as the divergence between the original points and their projection onto the new space: such difference can be expressed as sum of squared residuals, also written as $\|X - TP^T\|^2$, which should be as small as possible.

The loading vectors p_a^T (which are the rows in P^T) define the contribution of the original variables to each principal component, while the score vectors t_a (which are the columns in T) are the coordinates of the samples in the new space with A dimensions. In this "geometric" representation, PCA is commonly calculated by the singular value decomposition matrix factorization technique [35,42].

It is interesting to point out that PCA decomposition can be seen in a more "statistical" framework. Each sample is indeed described by measuring several variables, which will be characterized by their distribution and their covariance structure. The principal components can then be expressed as linear combinations of the K variables under the constraints that each one of them must be uncorrelated with the ones previously calculated and it has to have the maximum possible variance. In this form, PCA reduces to an optimization problem, which can be solved by the Lagrange multiplier technique [42,43].

An example of such projection is provided in Figure 20.2, which shows the "score plot" obtained performing PCA on a public LC-MS untargeted dataset on yellow and red raspberries (which is available from MetaboLights with the access number MTBLS333 [44]). In this case, X is a matrix with the intensity of $K = 6365$ features (couples *m/z*, *rt*) measured in $N = 26$ samples (13 red, 12 yellow, and 1 pink), which are extracts of raspberries of different varieties collected in different places and in different years. As can be seen from Figure 20.2, the first two components are enough

---

* For example, the `prcomp` R function.
[†] P^T is the matrix transpose of P.

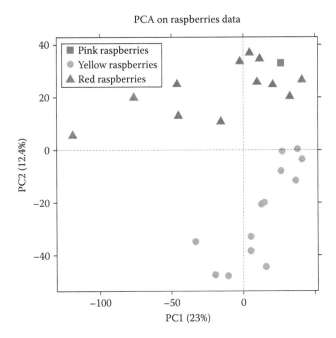

FIGURE 20.2   Score plot of the raspberries data, where the sample membership is highlighted using different colors and shapes for the points. The data have been centered and scaled before performing PCA.

to capture the most striking characteristic of the data, the color of the fruits, which subdivides the sample in two distinct clusters. To avoid misleading interpretation based only on low variance components, it is recommended to explicitly declare the percentage of variance captured by the plotted components. In the example in Figure 20.2, the first two components account for more than 30% of the total variance. It is possible to show that the variance explained by each component is quantified by the eigenvalues of the covariance matrix of X [42,43].

The information about the captured variance and the number of components can be efficiently represented in the "Scree plot." The Scree plot can be used to decide how many components are needed to have a good representation of our original dataset X by using the following rule-of-thumb: if there is a clear drop in variance, then only the previous components could be retained. In reality, the trend is usually quite smooth, as can be noticed in the raspberries example reported in Figure 20.3. In such cases, the choice of a threshold is largely subjective.

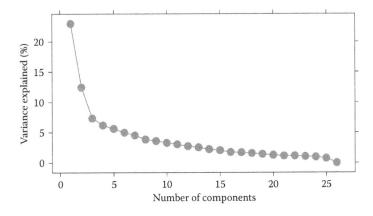

FIGURE 20.3   Scree plot on the raspberries data.

More formally, the problem of the optimal number of components can be stated as a model selection problem, since there is the need to balance the model complexity, in this case the number of components, and the accuracy of the representation of the data. In the context of explorative analysis, the previously mentioned heuristics can be enough, but more formal decision criterion can be found in literature [42,43].

As previously discussed, to take full advantage of PCA for the inspection of chromatography–MS data, it is necessary to take into account their specific characteristics. First of all, the different variables measured with these technologies can show extreme differences in intensity (due to the high dynamic range of MS-based technologies), but since the PCA projection is looking for the directions of maximal variance, the most intense variables will most likely dominate the picture. This is absolutely fair if one is interested in the "strong" signals but can hide interesting phenomena occurring in the low-intensity range. To counterbalance this effect, it is possible to rely on different forms of variable scaling [41]. One of the most popular form of scaling is "autoscaling," that is, each variable is divided by its variance. With this choice all the variables will have the same impact on PCA. This is obviously only one of the possible choices (logarithmic transformation of the intensities can be an alternative as discussed in the previous section) and it has the side effect of enhancing the importance of low-intensity variables, which are usually more difficult to measure and can be affected by the presence of missing values. The choice of the type of scaling depends on the aim of the analysis and to the type of effects we are interested to explore.

Due to its characteristics, PCA can be effectively used to identify outliers and to highlight the presence of obvious clusters in the data. Outliers can be roughly defined as points that are not consistent with the rest of the data and in a PCA score plot show up as isolated points that lie quite far away from the "cloud" formed by the other samples. In literature, formal approaches exist for the identification of outliers [42,43], but from an explorative point of view it is probably more useful to identify the critical samples from the score plot and go back to check the raw data and their metadata.

Given that PCA maximizes the variation along the principal components, this approach can be successfully applied also for checking the impact of potential confounding factors. As an example, in Figure 20.4 we show the possible confounding factors in the raspberries dataset (storage temperature, origin of the samples, and sampling year). The plot does not show any well-defined cluster that can be linked to experimental factors, thus indicating that the variance associated to the difference in color dominates the dataset (which is shown in Figure 20.2).

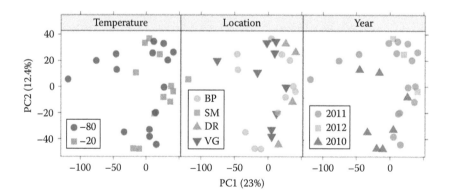

FIGURE 20.4 PCA on raspberries data: the three panels show the impact of the main possible confounding factors (storage temperature, origin of the samples, and sampling year), which were identified from the design of the experiments.

20.4 PARTIAL LEAST SQUARES OR PROJECTION TO LATENT STRUCTURE*

PLS is a multivariate analysis approach quite popular in chemometrics, because it works well also when many collinear variables and even missing values are present in the data [45]. PLS shares with PCA the idea of representing the data in a low dimensional space described by "latent variables." As before, these new variables are obtained as a linear combination of the original ones, but now they should both explain the variance in the dataset and have "predictive" potential toward some property of the samples. We are thus moving into a regression context and have to deal with two data matrices: the X matrix of independent or predictor variables (where X has essentially the same meaning as for PCA) and the Y matrix of dependent or response variables. For example, one could be interested in predicting the systolic blood pressure (Y) from the metabolic profile of the serum (X). Actually, several variants of the PLS algorithm exist: for example, when Y is multivariate ($M > 1$) it is known as PLS2, in case of Y equal to a simple vector ($M = 1$) it is known as PLS1 [46]. The approach is also used for classification purposes (e.g., when Y contains the class membership of the samples), defining Y as a vector of numeric classes or combining PLS with discriminant analysis (PLS-DA) [47,48]. The original idea of PLS has also been extended with the aim of enhancing the model interpretability by imposing orthogonal constraints or transformations to the scores, such as in orthogonal PLS (OPLS) [49].

Here we briefly present the main ingredients and an intuitive idea of the geometrical interpretation of the technique, with the help of Figure 20.5 for the description of the involved matrices. As for PCA, the use of mean-centered data allows us to discard the intercept vector. The X matrix can be described again as $X = TP^T + E$, where T is the score matrix, P^T is the loading matrix, and E is the matrix of the residuals. Equivalently, the same reasoning can be applied to Y, but now the aim is not to focus only on Y, but to link the two decompositions. In order to accomplish this task, one could think that X and Y, which contain different variables, can be both represented in the geometric space defined by the samples, instead of the usual representation in the space defined by the variables. In this geometrical space, it is possible to describe the relationship between X and Y, identifying a set of common latent variables that allow us to formally write $Y = TC^T + F$, where F is the residual matrix of the model. In other words, the scores T now will be both good summary of X and also good predictors of Y. The solution is usually

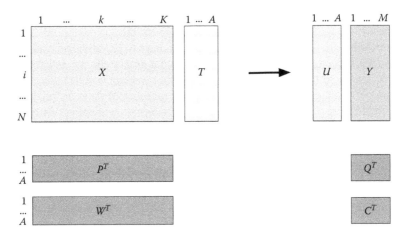

FIGURE 20.5   Schematic representation of the key elements of the NIPALS algorithm used to compute PLS. (From Wold, S. et al., *Chemometrics Intellig. Lab Syst.*, 58, 109, 2001; Höskuldsson, A., *J. Chemom.*, 2, 211, 1988.)

---

* See also Chapter 14.

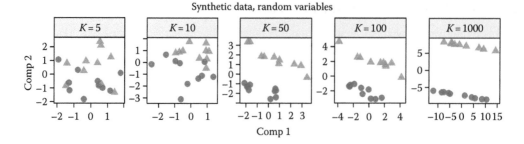

FIGURE 20.6   Score plot on random data obtained with a nonvalidated PLS. The *K* variables are computed as randomly sampled from uniform distribution in 0, 1, and then centered and scaled before computing PLS. The point color is assigned according to the sample class.

built iteratively, as for example in the Nonlinear Iterative PArtial Least Squares (NIPALS) algorithm, solving at each step a simple (bivariate) regression problem [45,50].

The *X*-score plot provides a representation of *X* in the projected space, as for PCA. The scores will again explain a portion of the variation, but now we have to consider the variation of both *X* and *Y*. In particular, the scores of *X* and *Y* could be represented either in separate plots or together, the *X*-scores in the abscissa and *Y*-scores in the ordinate, so that we can inspect the correlation between the two datasets. The similarities between PCA and PLS also affect the loadings: they give us information about the components and their relationships with respect to the original variables (e.g., if a variable has a large *X*-loading it is important for the modeling of *X*). The weights and the regression coefficients instead are new elements, typical of the PLS framework (as can be seen comparing Figures 20.1 and 20.5). The first ones encode the information about the relationships between *X* and *Y* variables. The latter ones quantify instead the predictive power of the original variables, given the number of latent variables considered in the model (*A*), which is usually computed by cross-validation [35]. We will come back later to this need to generate "robust" estimates of the model parameter. Finally, to summarize the variable importance for both modeling *X* and predicting *Y*, the variable importance for the projection (VIP) statistic has been defined as the weighted sum of squares of the weights and is used for variable selection purposes [51]. An extended discussion of the interpretation of the PLS elements can be found in [33,35,45].

Here it is important to stress that the outcomes of PLS have to be validated, exactly as for all statistical models [33]. Indeed, good predictive power does not necessarily mean that the results found in the dataset at hand hold for all the population. This phenomenon is called "overfitting." In addition, without proper validation, PLS will detect also the presence of random "relations" among the variables, which are more likely when for a given set of samples the number of variables increases. To illustrate this effect, we run the PLS algorithm (in its discriminant flavor) on a "random" dataset that was obtained by sampling 5, 10, 50, 100, and 1000 variables from a uniform distribution for 20 samples, assigned to two groups. In these data, then, there is no "real" difference between the two classes. Data have been centered and scaled before computing PLS, and the score plots representing the first two "components" are shown in Figure 20.6. Moving from left to right in Figure 20.6, the points start separating as the number of variables increases. The point here is that all the variables are totally uninformative. The observed separation, then, is the result of the presence of random association in the dataset, which is more likely as the number of variables increases.

REFERENCES

1. Gross JH. *Mass Spectrometry: A Textbook*. Springer Science & Business Media, Berlin, Germany; 2011.
2. Wilson ID, Brinkman UAT. Hyphenation and hypernation the practice and prospects of multiple hyphenation. *J Chromatogr A*. 2003;1000: 325–356.

3. Andrew Shalliker R. *Hyphenated and Alternative Methods of Detection in Chromatography*. CRC Press, Boca Raton, FL; 2011.

4. David Sparkman O, Penton Z, Kitson FG. *Gas Chromatography and Mass Spectrometry: A Practical Guide*. Academic Press, Oxford, U.K.; 2011.

5. Chauhan A. GC-MS technique and its analytical applications in science and technology. *J Anal Bioanal Tech*. 2014;5: 222. doi:10.4172/2155-9872.1000222.

6. Sneddon J, Masuram S, Richert JC. Gas chromatography-mass spectrometry-basic principles, instrumentation and selected applications for detection of organic compounds. *Anal Lett*. 2007;40: 1003–1012.

7. Ferrer I, Thurman EM, eds. *Liquid Chromatography Time-of-Flight Mass Spectrometry*. Hoboken, NJ: John Wiley & Sons, Inc.; 2009.

8. Zhang Y, Yuan Z, Dewald HD, Chen H. Coupling of liquid chromatography with mass spectrometry by desorption electrospray ionization (DESI). *Chem Commun*. 2011;47: 4171.

9. Theodoridis GA, Gika HG, Want EJ, Wilson ID. Liquid chromatography-mass spectrometry based global metabolite profiling: A review. *Anal Chim Acta*. 2012;711: 7–16.

10. Cai J, Henion J. Capillary electrophoresis-mass spectrometry. *J Chromatogr A*. 1995;703: 667–692.

11. Dakna M, He Z, Yu WC, Mischak H, Kolch W. Technical, bioinformatical and statistical aspects of liquid chromatography-mass spectrometry (LC-MS) and capillary electrophoresis-mass spectrometry (CE-MS) based clinical proteomics: A critical assessment. *J Chromatogr B Analyt Technol Biomed Life Sci*. 2009;877: 1250–1258.

12. Nesbitt CA, Zhang H, Yeung KKC. Recent applications of capillary electrophoresis-mass spectrometry (CE-MS): CE performing functions beyond separation. *Anal Chim Acta*. 2008;627: 3–24.

13. Pitt JJ. Principles and applications of liquid chromatography-mass spectrometry in clinical biochemistry. *Clin Biochem Rev*. 2009;30: 19–34.

14. Kolch W, Neusüss C, Pelzing M, Mischak H. Capillary electrophoresis-mass spectrometry as a powerful tool in clinical diagnosis and biomarker discovery. *Mass Spectrom Rev*. 2005;24: 959–977.

15. Desiderio C, Rossetti DV, Iavarone F, Messana I, Castagnola M. Capillary electrophoresis—Mass spectrometry: Recent trends in clinical proteomics. *J Pharm Biomed Anal*. 2010;53: 1161–1169.

16. Patel KN, Patel JK, Patel MP, Rajput GC, Patel HA. Introduction to hyphenated techniques and their applications in pharmacy. *Pharm Methods*. 2010;1: 2–13.

17. Picó Y. Mass spectrometry in food quality and safety. In: *Advanced Mass Spectrometry for Food Safety and Quality*. Elsevier, Amsterdam, the Netherlands; 2015. pp. 3–76.

18. Petrović M, Hernando MD, Díaz-Cruz MS, Barceló D. Liquid chromatography–tandem mass spectrometry for the analysis of pharmaceutical residues in environmental samples: A review. *J Chromatogr A*. 2005;1067: 1–14.

19. Kintz P, Mangin P. Simultaneous determination of opiates, cocaine and major metabolites of cocaine in human hair by gas chromotography/mass spectrometry (GC/MS). *Forensic Sci Int*. 1995;73: 93–100.

20. Inoue S, Saito T, Mase H, Suzuki Y, Takazawa K, Yamamoto I et al. Rapid simultaneous determination for organophosphorus pesticides in human serum by LC-MS. *J Pharm Biomed Anal*. 2007;44: 258–264.

21. Bogusz MJ. Hyphenated liquid chromatographic techniques in forensic toxicology. *J Chromatogr B Biomed Sci Appl*. 1999;733: 65–91.

22. Eckenrode BA. Environmental and forensic applications of field-portable GC-MS: An overview. *J Am Soc Mass Spectrom*. 2001;12: 683–693.

23. Trufelli H, Palma P, Famiglini G, Cappiello A. An overview of matrix effects in liquid chromatography-mass spectrometry. *Mass Spectrom Rev*. 2011;30: 491–509.

24. Silvestro L, Tarcomnicu I, Rizea S. Matrix effects in mass spectrometry combined with separation methods—Comparison HPLC, GC and discussion on methods to control these effects. In: Coelho AV, ed. *Tandem Mass Spectrometry—Molecular Characterization*. InTech, Rijeka, Croatia; 2013; 4–37.

25. Roberts LD, Souza AL, Gerszten RE, Clish CB. Targeted metabolomics. *Curr Protoc Mol Biol*. 2012;Chapter 30: Unit 30.2: 1–24.

26. Vinayavekhin N, Saghatelian A. Untargeted metabolomics. *Curr Protoc Mol Biol*. 2010; Chapter 30: Unit 30.1: 1–24.

27. Cajka T, Fiehn O. Toward Merging untargeted and targeted methods in mass spectrometry-based metabolomics and lipidomics. *Anal Chem*. 2016;88: 524–545.

28. Vrhovsek U, Masuero D, Gasperotti M, Franceschi P, Caputi L, Viola R et al. A versatile targeted metabolomics method for the rapid quantification of multiple classes of phenolics in fruits and beverages. *J Agric Food Chem*. 2012;60: 8831–8840.

29. Franceschi P, Giordan M, Wehrens R. Multiple comparisons in mass-spectrometry-based-omics technologies. *Trends Analyt Chem*. 2013;50: 11–21.

30. Hastie T, Tibshirani R, Friedman J. *The Elements of Statistical Learning: Data Mining, Inference, and Prediction*, 2nd edn. Springer Science & Business Media, New York; 2009.
31. Good PI. *Permutation, Parametric, and Bootstrap Tests of Hypotheses*. Springer Science & Business Media, New York; 2006.
32. Alonso A, Marsal S, Julià A. Analytical methods in untargeted metabolomics: State of the art in 2015. *Front Bioeng Biotechnol*. 2015;3: 23.
33. Kjeldahl K, Bro R. Some common misunderstandings in chemometrics. *J Chemom*. 2010;24: 558–564.
34. Varmuza K, Kurt V, Peter F. *Introduction to Multivariate Statistical Analysis in Chemometrics*, CRC Press, Boca Raton, FL; 2009.
35. Wehrens R. *Chemometrics with R: Multivariate Data Analysis in the Natural Sciences and Life Sciences*. Springer Science & Business Media, Berlin, Germany; 2011.
36. Chen M, Rao RSP, Zhang Y, Zhong CX, Thelen JJ. A modified data normalization method for GC-MS-based metabolomics to minimize batch variation. *Springerplus*. 2014;3: 439.
37. Sysi-Aho M, Katajamaa M, Yetukuri L, Oresic M. Normalization method for metabolomics data using optimal selection of multiple internal standards. *BMC Bioinformatics*. 2007;8: 93.
38. Veselkov KA, Vingara LK, Masson P, Robinette SL, Want E, Li JV et al. Optimized preprocessing of ultra-performance liquid chromatography/mass spectrometry urinary metabolic profiles for improved information recovery. *Anal Chem*. 2011;83: 5864–5872.
39. Ejigu BA, Valkenborg D, Baggerman G, Vanaerschot M, Witters E, Dujardin J-C et al. Evaluation of normalization methods to pave the way towards large-scale LC-MS-based metabolomics profiling experiments. *OMICS*. 2013;17: 473–485.
40. Kultima K, Nilsson A, Scholz B, Rossbach UL, Fälth M, Andrén PE. Development and evaluation of normalization methods for label-free relative quantification of endogenous peptides. *Mol Cell Proteomics*. 2009;8: 2285–2295.
41. van den Berg RA, Hoefsloot HCJ, Westerhuis JA, Smilde AK, van der Werf MJ. Centering, scaling, and transformations: Improving the biological information content of metabolomics data. *BMC Genomics*. 2006;7: 142.
42. Vidal R, Ma Y, Sastry SS. *Generalized Principal Component Analysis*. Springer, New York; 2003.
43. Jolliffe IT. *Principal Component Analysis*. Springer, New York; 2002.
44. Carvalho E, Franceschi P, Feller A, Herrera L, Palmieri L, Arapitsas P et al. Discovery of A-type procyanidin dimers in yellow raspberries by untargeted metabolomics and correlation based data analysis. *Metabolomics*. 2016;12: 144. doi:10.1007/s11306-016-1090-x.
45. Wold S, Sjöström M, Eriksson L. PLS-regression: A basic tool of chemometrics. *Chemometrics Intellig Lab Syst*. 2001;58: 109–130.
46. Brereton RG. *Chemometrics for Pattern Recognition*. John Wiley & Sons, Chichester, U.K.; 2009.
47. Barker M, Rayens W. Partial least squares for discrimination. *J Chemom*. 2003;17: 166–173.
48. Brereton RG, Lloyd GR. Partial least squares discriminant analysis: Taking the magic away. *J Chemom*. 2014;28: 213–225.
49. Trygg J, Wold S. Orthogonal projections to latent structures (O-PLS). *J Chemom*. 2002;16: 119–128.
50. Höskuldsson A. PLS regression methods. *J Chemom*. 1988;2: 211–228.
51. Farrés M, Platikanov S, Tsakovski S, Tauler R. Comparison of the variable importance in projection (VIP) and of the selectivity ratio (SR) methods for variable selection and interpretation: Comparison of variable selection methods. *J Chemom*. 2015;29: 528–536.

21 Chemometric Strategies in Chromatographic Analysis of Pharmaceuticals

Erdal Dinç

CONTENTS

21.1 INTRODUCTION

The advancements in mathematical and statistical applications in chemistry (and related fields of science and industry) and the rapid developments of optical and microelectronic technologies used in chemical instruments with computers for controlling devices, managing system operation, data acquisition, and reporting experimental results gave rise to the emergence of a new scientific discipline known as chemometrics for the qualitative and quantitative resolution of complex systems containing chemicals or pharmaceuticals. Over the years, chemometrics has become a promising interdisciplinary intersection with a significant impact on analytical chemistry and its neighboring branches. In the applications of the basic theory and methodology in research and development projects, a chemometric process covers two different options including development of new theories and algorithms for processing chemical data, and new applications of the chemometric techniques to different fields of chemistry, for example, analytical chemistry, medicinal chemistry, environmental chemistry, food chemistry, agricultural chemistry, and chemical engineering. In general, chemometrics is used in the mentioned areas to solve the problem of qualitative and quantitative analysis and others. Obviously, chemometrics provides researchers efficient ways to solve complex chemical problems for the desired goal. The major aim of chemometrics is to extract more useful information from signals such as spectra, voltammogram, chromatogram, and other several formats obtained from chemical instruments.

Use of chemometric techniques allows spectral resolution of overlapping signals (or co-elution of peaks) in spectrophotometric, spectrofluorimetric, and chromatographic analyses, and noise removal (or noise elimination), baseline correction or background correction in chromatographic multiway analysis differentiation, data smoothing and filtering, multivariate calibration, pattern recognition, classification, and design and optimization of operational experimental parameters in chemical processes and separation science, especially in chromatography and others. In the applications of other analytical methodologies (e.g., near-infrared [IR] spectroscopy), chemometrics has been applied to poor resolution problems and data processing.

In practice, chromatographic separation of chemicals and pharmaceuticals may occasionally result in poor chromatographic performance and poor peak resolution or co-elution of peaks in a chromatogram due to similarity of chemical and physical features of analyzed substances and other chromatographic operational conditions. Therefore, chromatographic analyses in most research and development projects require the use of chemometric tools for the optimization of experimental operational conditions giving better chromatographic performance.

Due to the similar chemical and physical characteristics of analyzed compounds and nonoptimal chromatographic operational conditions (or other experimental and apparatus factors), chromatographic separation of analytes in samples may result in poor chromatographic performance and poor peak resolution or co-elution of peaks in a chromatogram. Therefore, chromatographic analyses in most research and development projects can require the use of chemometric tools for the optimization of experimental operational conditions, giving better chromatographic performance. For example, the surface response methodologies can be used for the optimization of chromatographic methods, in order to obtain the optimal operational experimental conditions.

The most common chemometric approaches used for the optimization of chromatographic systems are full and fractional factorial, central composite design (CCD), Box–Behnken, Doehlert, and mixture design [1–5].*

In some cases, conventional chemical evaluation of the signals obtained by the chemical instrumentation techniques may not always provide successful results for chemical and pharmaceutical analysis. Hence, the conventional analysis techniques require the use of chemical prior separation, derivation, and extraction process to reach desirable analysis results. All these give rise to sometimes time-consuming and tedious chemical procedures for the multicomponent analysis of coexisting compounds in samples. Hence, conventional analysis techniques coupled with chemometric signal processing can be necessary for the analysis of complex systems and in order to enhance their resolution ability in analyses.

Nowadays, two-dimensional or two-way data obtained by "hyphenated instruments" such as high-performance liquid chromatography with a diode array detector (HPLC-DAD), gas chromatography with a mass spectroscopic detector (GC-MS), gas chromatography with an infrared spectroscopic detector (GC-IR), and HPLC with a mass spectroscopic detector (HPLC-MS) introduced to chemical laboratories are processed by multivariate curve resolution–alternating least squares (MCR-ALS)† to perform the qualitative and quantitative analyses of commercial and natural complex samples. Multiway analysis methods (e.g., parallel factor analysis [PARAFAC] [85,86], generalized rank annihilation method [GRAM] [89], N-PLS [96], and Tucker [88]) are applied to the decomposition of three-way data array or N-way data array obtained by arranging the two-way data matrices into pure signal profiles (e.g., spectral profile and time profile in HPLC or UPLC) and relative concentration profiles.

Several reviews including chemometrics and its applications from optimization to classification approaches using different analytical methods or instruments from spectroscopy to chromatography were reported [6–17].

The objective of this chapter is to review basic chemometric strategies used in the chromatographic analysis of pharmaceuticals, namely, raw drug material, drug and its metabolite, commercial drug dosage form, herbal medicine and drug–food supplements, and active compounds in biological fluids.

---

* See Chapter 1.
† See Chapter 11.

21.1.1 Brief History of Chemometrics

In the 1970s, chemometrics appeared as a subdiscipline of analytical chemistry. The term "chemometrics" was introduced by Svante Wold and Bruce R. Kowalski, considered as the founders of chemometrics. A popular definition was given by Massart: "Chemometrics is the chemical discipline that uses mathematical, statistical and other methods employing formal logic to design or select optimal measurement procedures and experiments, and to provide maximum relevant chemical information by analyzing chemical data" [18]. Another definition for the term "chemometrics" was given by Wold [19]. Several definitions of chemometrics were presented by experts working in the field and relevant journals. It can be said that all definitions, which are complementary to each other, are acceptable to many scientists. The International Chemometrics Society was established in 1974. Since then, workshops and courses related to chemometrics have been organized regularly at conferences by the scientists working in chemometrics and applications. During the 1980s, three different journals, namely, *Journal of Chemometrics*, *Chemometrics and Intelligent Laboratory Systems*, and *Journal of Chemical Information and Modeling*, were published in chemometrics. The scopes of these journals have been both fundamental and methodological research in chemometrics. Today, papers related to current chemometric methods and their applications are frequently published in common journals, such as *Talanta*, *Analytical Chemistry*, *Applied Spectroscopy*, and *Analytica Chimica Acta*. In chemometrics, some useful books were published in the 1980s, that is, the first edition of Malinowski's *Factor Analysis in Chemistry* [20], Sharaf, Illman, and Kowalski's *Chemometrics* [21], Massart et al.'s *Chemometrics: A Textbook* [18], *Multivariate Calibration* by Martens and Naes [22], and *Chemometrics: Statistics and Computer Applications in Analytical Chemistry* by Otto [23]. The history of chemometrics began with the review and publishing of a series of interviews by Geladi and Esbensen [24,25]. In chemometrics, typical application areas today are molecular modeling and QSAR; cheminformatics; the "-omics" fields of genomics, proteomics, metabonomics, and metabolomics; process modeling; and process analytical technology.

21.1.2 Experimental Difficulties in Chromatographic Analysis

The main problems of chromatographic analysis of pharmaceuticals consisting of raw material–containing drug substances, drug preparation, medicinal herbal plants, active compounds in biological fluids, and pharmaceutical ingredients in different samples are co-elution or incomplete separation, peak asymmetry, baseline drift, background and time shift of peak, the presence of unexpected compounds in samples, matrix effect, and so on. Several chemometric strategies can be used for the elimination of the mentioned difficulties in chromatographic analysis of pharmaceutical substances.

In chromatography, the similarity of physical and chemical features of analyzed pharmaceuticals gives rise to co-eluted peaks or overlapping peaks. This problem can be resolved using two different chemometric strategies. In the first strategy, developing a method generally requires the multivariate optimization of the experimental conditions in order to provide complete separation of all sample components. The second strategy is the second-order calibration methods and others (for) to achieve individual spectral profile, individual chromatographic profile, and concentration profile of analyzed drug materials in samples.

In addition, the characterization and classification of samples using pharmaceutical fingerprints are some of the difficulties in qualitative chromatographic analysis. In these cases, pattern recognition techniques are the most commonly used strategies applied to chromatographic data sets in order to characterize and classify samples.

As well as in the spectroscopic studies, chemometric pretreatments in chromatographic data analyses have a very important role in improving the quality of investigated data. Data pretreatments are other chemometric strategies in chromatographic analyses of samples containing pharmaceuticals. Some of them are mean-centering, noise elimination, background correction, and data smoothing.

21.2   CHEMOMETRIC STRATEGIES AND APPLICATIONS

Chemometrics provides valuable tools for enhancing the chromatographic resolution in the analysis of pharmaceuticals with less laboratory work. In the chromatographic analysis of drug substances, chemometric methodologies have been used for processing and interpretation of chromatographic data, and optimization of experimental separation conditions to get better elution of analyzed drug substances in a chromatogram. Other uses of chromatography for the analysis of pharmaceuticals are exploration, classification, and discrimination. We will discuss the extended applications based on the use of chemometric strategies in the chromatographic analysis of pharmaceuticals.

21.2.1   EXPERIMENTAL DESIGN AND OPTIMIZATION METHODOLOGIES*

Chromatographic analysis usually involves three steps: sample preparation, compound separation, and compound quantification. The steps of sample preparation and compound separation have been frequently optimized employing multivariate statistical techniques [6]. In this chapter, compound separation is one of the chemometric strategies in chromatographic analysis of pharmaceuticals. Finding the optimal chromatographic conditions for the separation of the analyzed active compound require the use of design and optimization techniques.

Experimental design and optimization is a widespread and powerful methodology for the method development processes to reach optimal experimental conditions and instrumental settings. This methodology provides us rich and rational information to make better decisions for optimal factor settings and experimental domain that produces reliable and accurate analysis results. In the development of new analytical methods or techniques, the general strategy of experimental design and optimization is to expose factor effects and their interactions to obtain optimal experimental conditions that produce real and reliable results. If there are not any effects from some factors on the analytical response, the experimental design and optimization allow us to eliminate ineffective factors from the experimental design model. A review of statistical designs and response surface techniques for the optimization of chromatographic systems has been reported in the literature [6].

The optimization of the operational experimental parameters providing better chromatographic separation is one of the most important chemometric strategies in chromatographic analysis of pharmaceuticals. In most research and development projects in academic and industrial sectors, the principal goal of the chemometric optimization of a chromatographic response as a function of several factor variables is to uncover the experimental conditions that give the best output.

In the chemical and pharmaceutical researches, the principal goal of the chemometric optimization of a chromatographic response (dependent variable) as a function of several factor variables (independent variables) is to uncover the experimental conditions that give the best output. In most research and development studies in the academic and industrial sectors, the classical optimization strategy is accomplished by exploring the influence of one factor variable on an experimental response variable when other related factors are constant. This "one factor at a time" optimization strategy may lead to time consumption and incorrect interpretation of results, and even wrong decisions for discovering optimal experimental conditions, if there are significant interactions between factors in experimental domain. In this case, it is difficult or even impossible to find rational optimal operational conditions with one factor optimization approach.

In the applications of chemometrics in chromatographic analysis, the multivariate optimization methodologies are very efficient chemometric tools for the identification of main factor effects and their interactions on experimental response in order to overcome the disadvantages of "one factor at a time" optimization. Usually, they are very useful tools for the optimization of chromatographic systems. For instance, in chromatographic analysis, the advantages of multivariate optimization

* See also Chapter 1.

techniques are less laboratory work, short analysis time, and to find real optimal chromatographic conditions in the presence of the factor interactions on a response function practically.

Experimental design and response surface methodology (RSM) considering all the effects of chemical factors and their interactions on analytical response is a very useful strategy tool for chromatographic method development to reach optimal operational conditions or instrumental settings.

The surface response methodologies such as full and fractional factorial designs [26–28], CCD [1,2,29], Doehlert matrix [30,31], and Box–Behnken design [32] are used for the chemometric optimization of chromatographic separation and quantitation of pharmaceuticals. Another optimization approach is the simplex design and its modified versions known as sequential designs. On the other hand, several types of mixture designs were used for the optimization of the proportional of the solvents in a mobile-phase system.

In the optimization of chromatographic operational conditions, the usual three steps are screening, modeling, and quantification approaches. The screening step is related to finding the most important factors and their interactions that have a significant influence on the result. The modeling step is related to the regression of the factors on the response (to perform a response surface experiment to produce a prediction model to determine curvature, and detect interactions among the factors).

The final step is related to optimizing the process or optimal operational conditions. For instance, the most important thing in HPLC is to obtain the optimum resolution in the minimum time.

In chromatography, a number of factors such as flow rate, pH, temperature, solvent composition, concentration, and buffer solution % in the mobile phase have a great influence on chromatographic performance or chromatographic response (resolution (R), retention factor (k), selectivity factor (α), and peak symmetry).

Some typical applications of multivariate approaches for the optimization of chromatographic methods for the quantitation of pharmaceuticals are reviewed in the following sequence. The core details of the review papers are listed in Table 21.1.

Karlsson and Hermansson [33] used a fractional factorial design with center points for the optimization of chromatographic system in a separation of the enantiomers of omeprazole and its main metabolite (hydroxyl-omeprazole) in a patient plasma sample. In this design, mobile-phase pH, acetonitrile (ACN) % in mobile phase, ionic strength, and column temperature were tested as the variables, and capacity factor (k), enantioselective retention (α), asymmetry factor (As), and column efficiency expressed by the number of theoretical plates (N) were dependent factors. In this study, evaluation of the experimental design showed that column temperature and ACN were variables most important for the enantioseparations. Ionic strength and mobile-phase pH were not very important for enantioselectivity, but for this application not negligible, as the α1-glycoprotein column has a longer lifetime at pH = 5.7 than at higher pH. The retention order of (R)-1 and (s)-2 could be controlled by column temperature and the content of ACN. The chromatographic system was used for the enantioresolution of 1 and its metabolite, 2, in human plasma samples.

An HPLC method was developed by Ho et al. [34] for the simultaneous separation and determination of famotidine, ranitidine hydrochloride, cimetidine, and nizatidine in commercial products. In HPLC analysis, a two-level, full factorial design and computer program were used to choose an acceptable HPLC condition. It is experimentally demonstrated that the application of factorial design to the rapid selection of an acceptable HPLC condition for the simultaneous separation and quantitation of four different drugs in commercial products is very useful.

Wsól and Fell [35] applied the Box–Wilson CCD to find the optimum resolution of enantiomers of rac-11-dihydrooracin (DHO), the principal metabolite of a potential cytostatic drug oracin. The optimum factor space was defined by three parameters consisting of temperature, buffer concentration, and modifier concentration. Chromatographic response functions (CRFs), resolution R_s, and retention time of the last component eluted t_{RL} were used to evaluate the resolution with regard to quality and quantity. In the chromatographic separation of DHO enantiomers, all four different CRFs give comparable results. As a result, the following optimum mobile phase and conditions were selected: 0.33 M $NaClO_4$ (pH = 3.00) and ACN (70.8 and 29.2, v/v) at temperature 25°C.

TABLE 21.1
Core Details in the Applications of the Chemometric Approaches to the Method Optimization of Chromatographic Analysis of Pharmaceuticals

Pharmaceuticals	Optimization Method	Chromatographic Factor	Number of Experiments	Response	Reference
Omeprazole, hydroxyomeprazole	FrFFD with center points	pH, ACN %, ionic strength, column temperature	28 and then 24	k, α, N, As	[33]
Famotidine, ranitidine hydrochloride, cimetidine, nizatidine	FFD with a two-level	MeOH %, TEA %, phosphate buffer concentration	8	t_R of the analytes	[34]
Rac-11-dihydrooracin	Box–Wilson CCD	Column temperature, ACN %, buffer concentration	23	CRF	[35]
Oxytetracycline and impurities	CCD	Column temperature, molarity of TBA and EDTA, ACN %, pH	29	R_s	[36]
R-timolol and related substances	Box–Behnken three-level design	Column temperature, added DEA %, 2-propanol % in hexane	15	DDF	[37]
Nine anthraquinones, bianthrones	FrFFD (2^{4-1}) and then CCF	Buffer pH, buffer concentration, SDC concentration, voltage then ACN %, buffer pH, SDC concentration	8 and then 17	Overall R_s, analysis time (T), As	[38]
Abacavir, lamivudine, and zidovudine	FFD	MeOH %, pH of the aqueous of the mobile phase	14 + 5 replications	Relative t_r	[39]
Carbamazepine, clonazepam, diazepam, chlordiazepoxide, lorazepam, alprazolam, oxazepam, flurazepam	Face-centered cube response surface design	Cm %, N, SDS (M) and Brij-35 (M), pH and temperature	34	t_r	[40]
Fosinopril sodium and its degradation product (fosinoprilat)	2^3 FFD with a zero level	Temperature, pH, MeOH %	9	k, α	[41]
Fosinopril sodium and its degradation product (fosinoprilat)	2^3 FFD with a zero level and then CCD	Sodium dodecyl sulfate, n-butanol, pH of the mobile phase	9 and then 18	k_1, k_2	[42]
Protocatechuic acid, 4-hydroxybenzoic acid, syringic acid, ferulic acid, p-coumaric acid, chlorpropamide, tolbutamide, indoprofen	OAD and then CCD	Type and concentration of organic modifier, pH and concentration of buffer, temperature, flow rate then MeOH %, pH, and concentration of buffer	16 and then 16	R_s, t_r, COF	[43]

(Continued)

TABLE 21.1 (*Continued*)
Core Details in the Applications of the Chemometric Approaches to the Method Optimization of Chromatographic Analysis of Pharmaceuticals

Pharmaceuticals	Optimization Method	Chromatographic Factor	Number of Experiments	Response	Reference
Risperidone, clozapine, venlafaxine, 9-hydroxy-risperidone, N-desmethylclozapine, O-desmethylvenlafaxine, levomepromazine, carbamazepine, N-desmethyllevomepromazine	Mixture design	Buffer %, ACN %, MeOH %, buffer molarity, pH	36	DDF	[44]
Caffeic, ferulic, vanillic, ellagic, gallic, benzoic, cinnamic, and hydrocinnamic acids	CCF	Initial ACN %, ACN % at the end of gradient elution, duration of the gradient elution, flow rate	23	CRF	[45]
Valsartan and its metabolite valeryl-4-hydroxy-valsartan	FrFD and then CCD	Flow rate, column temperature, TFA %, initial ACN %, ACN steepness and then flow rate, pH, initial ACN %, gradient steepness	18 and then 27	Peak areas and R_s	[46]
Amlodipine, atorvastatin	2^{4-1} FrFD and then CCD	MeOH %, pH and concentration of buffer, flow rate	12 and then 20	DDF	[47]
Haloperidol and its degradation products	FFD	Organic phase variation, flow rate and gradient rise time	11	CRF	[48]
Lansoprazole and its impurities	2^3 FFD and then RSM	pH of mobile phase, TEA % and ACN %	8	CRF	[49]
Nimodipine and its impurities	2^5 FFD and then CCD	Type and concentration of organic modifier, column temperature, flow rate, pH and then ACN %, column temperature, flow rate	32 and then 20	R_s, t_r, COF	[50]
Oxcarbazepine	3^2 and 2^3 FFD	Methanol %, flow rate, pH	9 and then 8	Peak area ratio	[51]
Betaxolol, propranolol, prilocaine, metoprolol, econazole, miconazole, sotalol, bupivacaine, mepivacaine, oxprenolol	FrFD and then CCF	Hexane %, temperature of column, nature and % of the acidic additive, nature and % of the basic additive	12 and then 11	R_s	[52]

(Continued)

TABLE 21.1 (Continued)
Core Details in the Applications of the Chemometric Approaches to the Method Optimization of Chromatographic Analysis of Pharmaceuticals

Pharmaceuticals	Optimization Method	Chromatographic Factor	Number of Experiments	Response	Reference
Cholecalciferol, retinol palmitate, phylloquinone, α-tocopherol acetate	CCF	Surfactant concentration, cosurfactant %, organic oily solvent %, temperature, and pH	32	MCEF	[53]
Raloxifen and its impurities	CCD	ACN %, pH of the mobile phase, SDS %, temperature	30	CRF, CEF and Duarte's CRF	[54]
Darifenacin hydrobromide and its degradation product	2^5 FFD and then CCD	Organic modifier type and content, mobile-phase pH, flow rate, column oven temperature	32 and then 20	k' and R then DDF	[55]
Selegiline, mianserin, sertraline, moclobemide, fluoxetine, and maprotiline	2^3 FFD	ACN %, pH of the mobile phase, concentration of ammonium acetate	30	N_{CRF}*	[56]
Desonide and its degradation product, sorbic acid, methylparaben, propyl gallate	FrFD and then CCF	Buffer pH and %, ACN %, column temperature, flow rate	11 and 11	k, ATR and then DDF	[57]
Dicentrine and neolitsine	FFD	pH, initial MeOH %, gradient slope	43	$\log(k_{iR})$, $\log(w_i)$, $\log(w_t)$	[58]
Irbesartan and hydrochlorothiazide	2^3 FFD	Methanol %, pH, column temperature	8	DDF	[59]
Itraconazole and its impurities	Box–Behnken design and FrFD	ACN %, pH, column temperature and then ACN %, pH, column temperature, flow rate	15 and 11	Separation function and peak area	[60]
Losartan and hydrochlorothiazide	2^3 FFD	Methanol %, pH, flow rate	8 + 6	t_r, As, R_s, DDF	[61]
Metaxalone and its impurities	Box–Behnken design	Column temperature, pH, and concentration of buffer	17	N, R_s, k, tailing factor, analysis time	[62]
Moxifloxacin hydrochloride and ketorolac tromethamine	CCF	MeOH %, buffer pH, and flow rate	15	DDF	[63]
Emtricitabine	CCD	MeOH %, pH	13	t_r, R_s, As	[64]
Ofloxacin and nimorazole	CCD	THF %, flow rate, buffer molarity	20	DDF	[65]

(Continued)

TABLE 21.1 (*Continued*)
Core Details in the Applications of the Chemometric Approaches to the Method Optimization of Chromatographic Analysis of Pharmaceuticals

Pharmaceuticals	Optimization Method	Chromatographic Factor	Number of Experiments	Response	Reference
Bisoprolol hemifumarate and hydrochlorothiazide	Full factorial design	Column temperature, flow rate, H_3PO_4 %	27	N_{CRF}*	[66]
Benzalkonium chloride	2^{7-3} FrFD	Buffer pH, TEA %, ACN %, start % mobile phase A, gradient time point, flow rate, column temperature	19	t_r, tailing factor, retention difference, R_s	[67]
Valsartan	FFD	Flow rate, detection wavelength, pH of buffer	27	Peak area, tailing factor, N	[68]
Piroxicam	Box–Behnken design	ACN %, flow rate, column temperature	17	DDF	[69]
Paracetamol and zaltaprofen	CCD	MeOH %, pH, flow rate	20	DDF	[70]
Azithromycin, secnidazole, fluconazole	CCD	Flow rate, MeOH %	9	Tailing factor, R_s, analysis time	[71]

Notes: ACN, acetonitrile; MeOH, methanol; k, capacity factor; R_s, resolution; α, selectivity factor; N, number of theoretical plates; As, asymmetry factor; t_R, retention time; DEA, diethylamine; DDF, Derringer's desirability function; CCD, central composite design; FFD, full factorial design; FrFD, fractional factorial design; CCF, central composite face-centered; SDS, sodium dodecyl sulfate; ANOVA, analysis of variance; RSM, response surface methodology; CSP, chiral stationary phase; CRF, chromatographic response function; CEF, chromatographic exponential function; MCEF, modified chromatographic exponential function; N_{CRF}*, improved chromatographic response function; $\log(k_{tR})$, logarithm of retention factor; $\log(w_l)$, $\log(w_r)$, logarithm of the half-widths of the peaks; OAD, orthogonal array design; COF, chromatographic optimization function; ATR, arc tangents resolution function; THF, tetrahydrofuran; N, number of theoretical plates.

It was described that the use of Box–Wilson CCD provided optimum separation of DHO enantiomers based on a relatively small number of experiments.

Diana and coworkers [36] developed an HPLC method for separation of oxytetracycline and its impurities with the use of CCD. The selection of stationary and mobile phases was based on a previous study performed by the group. The mobile phase was a mixture of ACN, tetrabutylammonium hydrogen sulfate (TBA) solution, ethylenediaminetetraacetic acid (EDTA) solution, and water. Three levels were determined for five chromatographic parameters: column temperature, molarity of TBA and EDTA, ACN content, and pH of the mobile phase. CCD consisted of 29 (2^{5-1} + 2.5 + 3) experiments. The dependent variable was resolution between the selected chromatographic peak pairs. Response surface plots of responses as a function of independent variables were created to select a chromatographic condition suitable for the purpose. The method was validated and showed good linearity, repeatability, and robustness.

A liquid chromatography (LC) method based on cellulose tris(3,5-dimethylphenylcarbamate) (5 μm) as chiral stationary phase and a mixture of hexane, 2-propanol, and diethylamine as mobile phase was developed for the simultaneous determination of R-timolol and other potential related substances in S-timolol maleate samples [37]. In the optimization of the LC method, an experimental design was elaborated in order to test the effects of the different selected factors. In order to get optimal operational LC conditions, three factors, namely, the concentrations of 2-propanol, diethylamine, and the column temperature, were studied simultaneously using the Derringer's desirability function for the responses of interest. In the optimized LC conditions obtained by application of Box–Behnken experimental design, the separation of S-timolol, R-timolol, isotimolol, and dimer maleate as well as dimorpholinothiadiazole was performed in less than 20 min. The developed LC method was compared to the European pharmacopoeia method and was applied to the determination of R-timolol and other related substances in several samples of S-timolol maleate from different sources.

Kuo and Sun [38] have developed an efficient micellar electrokinetic chromatography (MEKC) for the analysis of nine anthraquinones and bianthrones in rhubarb, which has laxative, antibacterial, hemostatic, and antispasmodic properties. In the development of the MEKC technique, two experimental designs were used. The first one is the screening design to find factors showing significant effect. Screening design was carried out through a two-level fractional factorial design. As a result, sodium dodecyl sulfate (SDS) concentration, buffer pH, and ACN % were selected as suitable factors, which have a significant effect, to be considered in the optimization process. In the next step, a central composite face-centered (CCF) design was used for the determination of the optimum conditions for the selected factors for the analysis of major active principals in rhubarb crude drugs.

A reversed-phase high-performance liquid chromatographic (RP-HPLC) method was developed and validated by Djurdjevic and coworkers [39] for the simultaneous analysis of abacavir, lamivudine, and zidovudine in tablets. The RSM was used for the optimization process. In the development of the RP-HPLC method, pH and the type of organic modifier were tested to achieve the successful separation of three drugs. In order to determine the optimal RP-HPLC conditions, two factors, methanol % (v/v) in mobile phase and pH (adjusted with H_3PO_4 in the presence of 0.2% trimethylamine, TEA), were investigated in the range of 20%–50% v/v and pH: 2.5–4.0, respectively, and the relative retention was considered as the response variable. The optimal mobile phase was found to be a mixture of water and methanol (60: 40, v/v + % 2 TEA) with a pH of 3.20 to get desirable separation of components in samples with acceptable elution times. Optimized and validated RP-HPLC was applied for the analysis of drugs in pharmaceutical forms.

Hadjmohammadi and Ebrahimi [40] applied the chemometric approach for the optimization of separation of eight anticonvulsant agents in mixed micellar liquid chromatography (MLC) by experimental design and regression models. The optimization of the experimental conditions is a complicated process in MLC due to a large number of the factors (or variables), which must be simultaneously treated. These factors include type and concentration of surfactants, type and volume fraction of organic modifier, temperature, and pH of mobile phases. Quantitative variation of the mentioned factors in MLC analyses affected both the retention of analytes and the extent of the surfactant monomers adsorbed onto the stationary phase. In order to optimize the separation of anticonvulsant agents, the effect of six experimental parameters, namely, the organic modifier (Cm) in the mobile phase, the length of the alkyl chain of the organic modifier (N), the concentration of SDS, and Brij-35 concentration, pH of mobile phase, and temperature on retention time, was considered. Experiments were performed according to a face-centered cube response surface experimental design, and then Pareto-optimality method was used for the optimization of chromatographic separation.

Safa et al. [41] have applied the experimental design to identify the optimum chromatographic conditions for the separation of fosinopril sodium and its degradation product, fosinoprilat. A 2^3 full factorial design with two levels (low and high) with the zero level has been used for the experimental screening process. In this paper, methanol % in mobile phase, pH of the mobile phase, and column temperature have been considered as independent variables or factors and capacity and selectivity factors were used as dependent variables. The estimation of coefficients of the model has been performed by regression of the independent variables on the dependent variables (capacity and selectivity factors). After experimental screening processing, RSM has been applied for optimization. Optimum conditions have been X-Terra™ 150 mm × 4.6 mm, 5 μm particle column at 45°C, and methanol–water (75:25 v/v) at pH 3.1 as a mobile phase, with a flow rate of mL/min. The proposed method has been used for the quantitative analysis of fosinopril sodium and its degradation product in tablets. Applying the chemometric approach enables a relatively limited number of experiments to define factors that affect the chromatographic behavior of investigated substances and obtain optimum conditions for their fosinoprilat, named SQ 27519, a specific competitive inhibitor of angiotensin-converting enzyme.

A microemulsion liquid chromatographic method was developed by Jančić and coworkers [42] for the separation of fosinopril sodium and its degradation product, fosinoprilat, using chemometric support. In the liquid chromatographic separation, a microemulsion was used as the mobile phase. The modifications of the mobile phase included the changes to the type of the lipophilic phase, the type and concentration of cosurfactant and surfactant, as well as the pH of the mobile phase. In this study, the screening process using a 2^3 full factorial design was applied for selecting factors that had an influence on separation. In the next step, optimization was done by a CCD. An appropriate resolution with reasonable retention times was obtained with a microemulsion containing 0.9% w/w of cyclohexane, 2.2% w/w of SDS, 8.0% w/w of n-butanol, and 88.9% of aqueous 25 mM disodium phosphate, the pH of which was adjusted to 2.8 with 85% orthophosphoric acid. Separations were performed on an X-Terra 50 mm × 4.6 mm, 3.5 μm particle size column at 30°C. UV detection was performed at 220 nm and with a flow rate of 0.3 mL/min. The established method was validated and applied for analysis of appropriate tablets.

Wanga et al. [43] made a study on optimizing reversed-phase liquid chromatographic separation of an acidic mixture on a monolithic stationary phase with the aid of RSM and experimental design. An orthogonal array design was used for the selection of the significant parameters for the optimization. The significant factors were optimized by using a CCD, and then the quadratic models based on the relationships between the independent and dependent parameters were constructed and the optimal conditions for separation were determined. The optimized method was successfully applied for the analysis of the acidic drugs, protocatechuic acid, 4-hydroxybenzoic acid, syringic acid, ferulic acid, p-coumaric acid, chlorpropamide, tolbutamide, and indoprofen investigated in this study on the monolithic packing.

Cutroneoa and coworkers [44] described a chemometric procedure for the optimization of the separation of some psychotropic drugs. The choice of experimental factors and responses were investigated. Briefly, a series of designs and optimizations were applied to visualize the effects of three solvents (percentage of buffer, percentage of ACN, and percentage of methanol) in mobile phase on selectivity and retention. In this study, a chemometric approach was applied for the selection of the optimum separation conditions by using Derringer's desirability function in a combined experimental design.

Kiendrebeogo and coworkers [45] used CCF design for HPLC optimization of screening procedure of phenolic acids in several folk medicines in Burkina Faso. A suitable HPLC separation of caffeic, ferulic, vanillic, ellagic, gallic, benzoic, cinnamic, and hydrocinnamic acids in their mixtures was proposed with the use of experimental design. In the gradient elution program, four different factors were chosen as independent factors: initial ACN percentage, ACN percentage at the end of gradient elution, duration of the gradient elution, and flow rate. The effect of these factors on the CRF was investigated by CCF design. Using the optimized HPLC conditions, they performed the screening of eight phenolic acids in *S. hermonthica*, *G. senegalensis*, and honey extracts.

Iriarte and coworkers [46] used experimental design and optimization approach for a solid-phase extraction-HPLC-UV-fluorescence method in order to quantify valsartan and its metabolite, valeryl-4-hydroxy-valsartan, in human plasma samples. The analytical column, organic modifier of the mobile phase, elution mode, internal standard, and excitation and emission wavelengths were chosen one variable at a time. Then, two different experimental designs were used: fractional factorial design to choose significant variables and CCD to obtain the optimal values for the significant variables. Both experimental design procedures had the same response, peak areas, and resolution. After optimizing the significant factors, they applied this chromatographic method to simultaneously determine valsartan and its plasma in human plasma samples. They also used experimental design and optimization approach to extraction procedure of valsartan in human plasma samples.

Sivakumar et al. [47] applied statistical experimental design and Derringer's desirability function to develop an RP-HPLC method for the analysis of amlodipine and atorvastatin in pharmaceutical formulations. In the screening step, a 2^{4-1} fractional design was used to identify the significant factors affecting the analysis time response. The significance of the investigated factors was determined using the analysis of variance (ANOVA) test. According to the results of the fractional factorial design, the key factors were selected. Then, a CCD was performed to optimize the selected factors in order to get the optimal chromatographic conditions. The responses such as retention factors, resolutions, and retention times for amlodipine and atorvastatin with the internal standard (propylparaben) were measured. Derringer's desirability function based on the use of multiple responses was considered as a response function for both screening and optimization processes. The experimental results of the CCD were fitted with a second-order polynomial function and then the optimum condition for separation of the related compounds was located.

Petkovska and Dimitrovska [48] have described the development and validation of an RP-HPLC method for simultaneous determination of haloperidol and its degradation products using a design of experiments (DOE). A 2^3 full factorial design was applied to optimize the chromatographic operational conditions for the separation of haloperidol and its degradation impurities. The process is based on the chemometric optimization of the organic phase variation, flow rate, and gradient rise time for the gradient elution of analytes. Eleven experiments were carried out, and the resolution (R_s) values of all consecutive peak pairs were computed. The CRF calculated from R_s values was used. The best result, which corresponds to high CRF values, was obtained using the mobile phase containing ACN as organic modifier and phosphate buffer pH 6.5, with organic phase variation from 20:80% to 72:28% v/v, gradient rise time of 7 min, and flow rate of 1.5 mL/min at 25°C.

Lansoprazole in the presence of its five related compounds was determined by using the RP-HPLC method [49]. A full factorial design at two levels was used for the screening step to determine the influence of a number of effects on a response and to eliminate those that are not significant. Nine experiments with a center point were performed. In the estimation of the system response, the resolution (R_s) between peak pairs was selected and then the total number of detected peak pairs was five. CRF values were calculated for 30 experiments. In order to investigate the chromatographic behavior of the analyzed substances for the given experimental range, and to find the optimum separation conditions, optimization of the method was performed using RSM.

Experimental design methodologies were used by Barmpalexis et al. [50] for developing and optimizing a validated isocratic RP-HPLC separation of nimodipine and impurities in tablets. In this study, a 2^5 full factorial design was applied to screen five independent factors, while the optimum chromatographic conditions were evaluated by a CCD using both a graphical (overlay contour plots) and a mathematical (Derringer's desirability function) global optimization approach. Optimum separation conditions were reported as ACN/H$_2$O (67.5/32.5, v/v) as a mobile phase at a flow rate of 0.9 mL/min at the column temperature of 40°C.

Rao [51] developed a high-performance liquid chromatographic method for the determination of oxcarbazepine in pharmaceutical formulations. Optimization of significant factors and robustness study was performed by factorial design. The effect of methanol percentage, flow rate, and pH of mobile phase on the peak area ratio was investigated. A linear model was postulated, and a 2^3 full factorial design was employed to estimate the model coefficients. RSM gave the optimum chromatographic conditions, which are a mobile phase consisting of ACN and phosphoric acid adjusted to pH 3 (50:50, v/v) at a flow rate of 1 mL/min.

Dossou and coworkers [52] used experimental design methodology for the optimization of the LC enantiomeric separation of chiral pharmaceuticals using cellulose tris(4-chloro-3-ethylphenylcarbamate) as chiral selector and polar nonaqueous mobile phases. A fractional factorial design was applied to screen the effect of factors on the LC enantioresolution. Then, a face-centered CCD has been used to get optimal LC conditions for the enantioresolution of the investigated 10 chiral basic drugs. In this study, the effect of the formic acid proportion on enantioresolution was confirmed and high-resolution values obtained for almost all compounds tested in relatively reasonable analysis times.

A chromatographic method for separation and determination of fat-soluble vitamins was developed by Momenbeik [53] using isocratic microemulsion liquid chromatography. Optimization of the parameters affecting the separation (surfactant concentration, percentage of cosurfactant, percentage of organic oily solvent, temperature, and pH) was performed using a genetic algorithm method with face-centered CCD. A new software was developed to apply genetic algorithm to find the operational conditions that provide the minimized chromatographic responses for all evaluated analytes simultaneously. The optimum conditions were as follows: 73.6 mM sodium dodecyl sulfate, 13.64% (v/v) 1-butanol, 0.48% (v/v) diethyl ether, column temperature of 32.5°C, and 0.02 M phosphate buffer of pH 6.99. After calibration and validation studies, the optimum chromatographic method was applied to determine the amount of vitamins in multivitamin syrup and a sample of fish oil capsule.

A new CRF for chromatographic optimization strategies was designed by Stojanović et al. [54]. The performance of this new CRF was compared to the previously developed chromatographic exponential function and Duarte's CRF. In this study, four important factors were recognized. In the next step, CCD was applied. Then global optimization conditions were computed for the chromatographic separation and analysis of raloxifen and its impurities.

A set of chemometric tools including in silica simulation and DOE was used by Meneghini and coworkers [55] to develop and validate an RP-LC/UV method for the analysis of darifenacin hydrobromide (DF) and its degradation product. An exploratory two-level factorial design was applied for the identification of the effects of the studied factors on the DF retention factor and resolution (k' and R). In the optimization procedure, Derringer's desirability was used. In addition, a Plackett–Burman design was used to test robustness in the validation stage.

Rakić et al. [56] presented exploration of chromatographic behavior in system by 3^3 experimental design and improved CRF, denoted as improved CRF ($N_{CRF}*$). The main aim of the study was to use experimental design and improved CRF in the evaluation of retention behavior of antidepressant mixtures in hydrophilic interaction chromatography (HILIC) system. This $N_{CRF}*$ provides the simultaneous estimation of the separation of analytes. The applied function allowed identification of experimental regions where examined mixture showed optimal behavior in HILIC system.

Lopes and coworkers [57] have developed a reversed-phase liquid chromatographic method for the separation and simultaneous quantitation of desonide, sorbic acid, methylparaben, propyl gallate, and degradation product of desonide in a hydrophilic cream with the use of experimental design, resolution loss functions, and chemometrics methods. In this study, a screening phase was used for identification of the most influent chromatographic variables (pH and organic solvent content). Then, these factor variables were further optimized using a CCD.

Rafamantanana et al. [58] used DOE and design space methodology for HPLC separation of 13 alkaloids from leaves of *Spirospermum penduliflorum* Thouars. Full factorial design was used in order to investigate the effects of a selected variable on logarithm of the retention factor and logarithms of peaks' half-width. The optimal separation was achieved with a mobile phase consisting of methanol and pH 3 ammonium formate buffer with gradient elution, starting at 32% of methanol and a gradient slope of 0.42%/min. The developed analytical method was used for the quantification of dicentrine, the major vasorelaxing alkaloid, in two plant samples.

An HPLC method was developed with the aid of experimental design and optimization technique by Vujić and coworkers [59] for the estimation of irbesartan and hydrochlorothiazide in pharmaceutical preparations. Independent variables were chosen as methanol content, pH of the mobile phase, and column temperature at two levels in full factorial experimental design. Their influence on four criteria was investigated: resolution and symmetry of the irbesartan peak, resolution and symmetry of the hydrochlorothiazide peak, retention factor of irbesartan, and retention factor of hydrochlorothiazide. All five responses were included in Derringer's desirability function and its value was used as chromatographic response. After optimizing the experimental conditions, the separation was conducted with a mobile phase consisting of methanol-tetrahydrofuran-acetate buffer (47:10:43 v/v/v), pH 6.5 at a column temperature of 25°C. The method was applied to the simultaneous determination of irbesartan and hydrochlorothiazide in their commercial dosage forms.

Chemometric optimization and validation of RP-HPLC method was performed by Kasagić Vujanović and coworkers [60] for the analysis of itraconazole and its impurities. In order to reach the desired chromatographic resolution with a limited number of experiments in minimum time, Box–Behnken design was used to simultaneously optimize the selected chromatographic parameters. Fractional factorial 2^{4-1} design was used for robustness testing, and the flow rate of the mobile phase was identified as a factor that influences the method significantly.

An HPLC method for simultaneous analysis of losartan potassium and hydrochlorothiazide was developed by Smajić et al. [61] with the aid of an experimental design strategy. Three independent variables, flow rate, methanol content, and pH value of mobile phase, were selected as input. Their influence on the responses, namely, resolution, symmetry of irbesartan peak, symmetry of hydrochlorothiazide peak, retention factor of irbesartan, and retention factor of hydrochlorothiazide, was investigated by a 2^3 full factorial design with six replicates at zero level. All independent variables were found to be significant. Then, Derringer's desirability function was used as a response to optimize these parameters in the experimental domain. The optimal conditions were found to be 45% for methanol content, mobile-phase pH 4.8, and 0.82 mL/min for flow rate. The optimized method was validated and applied to the simultaneous analysis of losartan potassium and hydrochlorothiazide in tablets.

Sahu and Patro [62] applied chemometric response methodology for the development and optimization of an RP-HPLC method for the separation of metaxalone and its two base hydrolytic impurities. The paper describes the use of the Box–Behnken experimental design to identify the significant variables' effects influencing the separation of the related substances in a stability indicating RP-HPLC method. They concluded that the use of experimental design and RSM is an economic procedure, able to perform the analysis efficiently in a very short time period and with a small number of experiments for the desired separation of metaxalone and its impurities.

Experimental design and response surface technique was applied by Kalariya and coworkers [63] to choose the optimum RP-HPLC conditions for the determination of moxifloxacin HCl and ketorolac tromethamine in eye drops. Three independent factors, methanol content, buffer pH, and flow rate, were used to design mathematical models. Derringer's desirability function was applied to simultaneously optimize the retention time of last eluting peak (ketorolac tromethamine) and tailing factor of moxifloxacin. The use of experimental design and response surface technique provided a better insight into the sensitivity of chromatographic factors and their interaction effects on the attributes of separation.

Singh and Pai [64] used CCD experiments to optimize an HPLC method for the determination of emtricitabine in nanoparticles. The effects of methanol concentration and pH of the mobile phase on the retention time, peak resolution, and peak asymmetry were investigated. Thirteen experiments were conducted using three levels of the selected independent factors. The optimized method was successfully applied for the determination of in vitro and in vivo release studies of emtricitabine nanoparticles.

Giriraj and Sivakkumar [65] described a rapid chemometrics-assisted RP-HPLC method with photodiode array (PDA) detection for the simultaneous estimation of ofloxacin and nimorazole in pharmaceutical formulation. The proposed method was optimized by using CCD in RSM. Derringer's desirability function was used to concurrently optimize the selected two responses, retention time and resolution.

Dinç et al. [66] developed a new reversed-phase ultra-performance liquid chromatography (UPLC) method for the simultaneous determination of bisoprolol hemifumarate and hydrochlorothiazide. Optimization of the method was done by using three-factor, three-level full factorial design in order to obtain better peak resolution with a short runtime.

Zakrajšek and coworkers [67] developed an HPLC method to quantify benzalkonium chloride, which is a mixture of different preservative homologs, in nasal formulations. They used design of experiments methodology for the separation of all interfering peaks, active compound and benzalkonium chloride homologs. This approach also provided a more robust analytical method for the determination of benzalkonium chloride. The gradient HPLC elution condition was optimized by a randomized 2^{7-3} fractional factorial design. Nineteen experiments were carried out and six retention factors, tailing factor, retention difference, and resolution of benzalkonium chloride homologs were measured as responses for each experiment. The method was successfully applied for the determination of the assay of benzalkonium chloride (BKC) in two different nasal spray formulations.

Kumar et al. [68] used experimental design and optimization method to develop an RP-HPLC method for the quantitative determination of valsartan in nanoparticle formulations. Full factorial design with three factors and three levels was used to optimize the effect of variable factors, which were flow rate, detection wavelength, and pH of buffer. The responses were peak area, tailing factor, and number of theoretical plates. Optimized chromatographic parameters for the determination of valsartan were reported as 1.0 min/mL flow rate, UV detection at 250 nm, and buffer pH 3.0. The developed method was applied to the quantitation of valsartan in nanoparticle formulations.

Dragomiroiu and coworkers [69] developed an RP HPLC-UV method for the quantitative determination of piroxicam in bulk materials and pharmaceutical formulations using a DOE approach. Chromatographic variables were chosen as ACN percentage in mobile phase, flow rate, and column temperature, and their influence on the chromatographic behavior was evaluated by Box–Behnken design with 17 experiments and Derringer's desirability function. Evaluated responses in Derringer's desirability function were retention time, peak symmetry, resolution of the chromatographic separation, number of theoretical plates, and capacity factor. The mobile phase consisting of trifluoroacetic acid/ACN mixture (60:40, v/v), a flow rate of 1.1 mL/min, and a column temperature of 40°C were found to be the optimum experimental conditions for the separation.

A new RP-HPLC method for the estimation of paracetamol and zaltaprofen in commercial formulations was developed by Sathiyasundar and Valliappan [70] by using of CCD and Derringer's desirability function. The independent variables were chosen as methanol content of mobile phase, pH of aqueous component of the mobile phase, and flow rate. Twenty experiments were performed according to CCD, and values of capacity factor of the first peak, selectivity factor, and retention time of the last peak were included in Derringer's desirability function. The optimal chromatographic conditions were found to be 60% of methanol and 40% water (pH 3.5) for the mobile-phase composition and 0.98 mL/min for the flow rate.

Sahoo and Sahu [71] described a chemometric approach to RP-HPLC determination of azithromycin, secnidazole, and fluconazole using RSM. In this study, CCD was used for the optimization of flow rate and methanol affecting three different chromatographic factors: tailing factor, resolution, and analysis time. The optimal conditions were found as MeOH and water (37:63; v/v) at a flow rate of 1.20 mL/min by using RSM.

21.2.2 Multivariate Calibration Approaches and Their Applications

Hyphenated chromatographic systems, for example, HPLC (or UPLC) coupled with PDA detection (HPLC-PDA or UPLC-PDA) and mass spectrometry (HPLC-MS or UPLC-MS), are the most commonly used methods for the analysis of complex systems containing pharmaceuticals. However, successful applications of these hyphenated methods have been limited in consequence of co-elution problems, chromatographic baseline drift, fluctuations in column system, and so on. Particularly, in the analysis of more complex samples, the chromatographic profiles of analyzed drugs (or with interferences in samples) may be partially or totally superimposed because of the co-eluting peaks in a chromatogram. Multivariate calibration algorithms are powerful chemometric tools to overcome these drawbacks in chromatographic analysis of pharmaceutical substances.

In the literature, some interesting applications of two-way and three-way (or N-way) calibration models in the quantitation of drug materials in samples have been reported. For the quantitative resolution of chromatographic data sets, two-way calibration models are the most used ordinary PCR and partial least squares (PLS) algorithms for analyzing first-order data obtained by chromatographic instrument systems [72–84].

In three-way calibration applications, second-order data produced by hyphenated instruments, for example, HPLC-DAD and HPLC-MS, have been processed by chemometric methods with second-order advantage to resolve the overlapped chromatographic peaks of analyzed drugs in a given sample in a short experimental period. Various chemometric algorithms have been applied for the decomposition of a chromatographic data set into its individual original contributions. Some of the most popular algorithms are PARAFAC [85–87], Tucker [88], the GRAM [89–91], direct trilinear decomposition [92], MCR-ALS [93,94], alternating trilinear decomposition (ATLD) [95], and trilinear PLS [96].* Some typical applications of the mentioned chemometric calibration methods will be briefly explained in the following.

Zissis and coworkers [97] applied two-way, unfolded three-way, and three-mode PLS calibration to HPLC-DAD data for the determination of a low-level pharmaceutical impurity. Two-way diode array chromatograms of 2-hydroxipridine and 3-hydroxipridine were recorded at two levels of resolution. 3-hydroxipridine was treated as a minor impurity and samples were prepared with two different percentages (0.1%–0.5% and 1%–5%). Two-way PLS was applied separately in spectral profile and elution profile. Relative errors for impurity predictions were tabulated for different calibration methods for different chromatographic separations and different impurity levels. All methods were found to be effective; however, the quality of the predictions was better in good chromatographic resolution and with high content of impurity. Two-way spectral method gave better results because of the inconsistency problems in elution time.

Wiberg and Jacobsson [98] used PARAFAC method to three-dimensional HPCL-DAD data for binary mixtures containing different amounts of prilocaine with a constant amount of lidocaine. A set of 17 samples were analyzed by HPLC-DAD at three different levels of separation, and PARAFAC decomposition was applied to these three data sets. The PARAFAC results of the first case, which had no chromatographic separation at all, gave a good estimate of spectral and concentration profiles of the data. When PARAFAC method was applied to the second and third cases, which had a chromatographic resolution of 0.7 and 1.0, respectively, underlying chromatographic, spectral, and concentration profiles were resolved more accurately than the first case. The loadings describing the prilocaine content were regressed against the true prilocaine concentration, and good predication results were obtained.

A second-order calibration method based on the ATLD algorithm was developed for simultaneous determination of cortisol and prednisolone in body fluids by Zhang and coworkers [99]. Three-way data of urine and plasma samples were recorded by HPLC-DAD. Although heavily overlapping peaks of the analytes, ATDL successfully resolved the chromatographic, spectral, and concentration profiles even in the presence of unknown co-eluting interferences.

---

* See also Chapter 23.

Schmidt and coworkers [100] used PARAFAC algorithm in combination with HPLC-DAD method for the characterization of commercial preparations of St. John's wort. Twenty-four different commercial samples including tablets and capsules were analyzed by HPLC-DAD after an extraction procedure. The obtained data were preprocessed by correlation optimized warping (COW) to eliminate time-shifting caused by matrix effects. The relative concentration, elution profiles, and UV spectra of the individual metabolites were obtained as loadings from the PARAFAC analysis. Principal component analysis (PCA) was applied to the new matrix, which was constituted from the relative concentration score vectors or matrices of PARAFAC analysis in order to compare samples.

García and colleagues [101] improved the algorithm unfolded-PLS followed by residual bilinearization (U-PLC/RBL) for the simultaneous determination of eight tetracyclines in wastewaters and compared it with MCR-ALS. The HPLC analysis provided well-separated peaks of the drugs of interest (tetracycline), but because the sample matrix is highly complex, univariate calibration had not been an option for the study. In order to reduce the total interferences retained after the pre-concentration of wastewaters, baseline correction and piecewise direct standardization (PDS) were applied. Models were simplified by partitioning HPLC-DAD data in eight regions, each region corresponding to one analyte. The combination of U-PLS with RBL was reported to be an alternative to MCR-ALS for processing HPLC-DAD data while preserving the second-order advantage.

Cañada-Cañada et al. [102] applied PARAFAC, N-PLS, and MCR to HPLC-FLR data to quantify several fluoroquinolones in tap and water samples as well as human urine samples. The substances of interest were pipemidic acid, marbofloxacin, ofloxacin, norfloxacin, ciprofloxacin, enrofloxacin, lomefloxacin, and danofloxacin. Data matrices containing the fluorescence intensity as a function of retention time and emission wavelength were recorded with the use of HPLC method with spectrofluorimetric detection. The chromatographic method provided complete separation of pipemidic acid, marbofloxacin, ciprofloxacin, and lomefloxacin, so their quantitation was done by univariate calibration. Second-order calibration methods were used for the quantitative determination of co-eluted drugs, namely, ofloxacin, norfloxacin, enrofloxacin, and danofloxacin. Determination of marbofloxacin in urine samples was also performed by second-order calibration methods because the peak of marbofloxacin overlapped with the interferences (salicylic acid and metabolites) in urine. The combination of emission fluorescence–retention time matrices with selected second-order algorithms allowed the successful determination of fluoroquinolones in samples with and without interferences.

The simultaneous determination of naproxen, ketoprofen, diclofenac, piroxicam, indomethacin, sulindac, diflunisal, and carbamazepine in river and wastewaters was performed by Gil Garcia and coworkers [103] with the use of different analytical and chemometric methods. The analysis of river water was performed by following various steps: pre-concentration, solid-phase microextraction, HPLC, and PDS. However, the analysis of the wastewaters was not possible with this approach due to high content of interference. The team used MCR-ALS algorithm after background correction.

Galera et al. [104] proposed an HPLC-DAD method for simultaneous determination of nine beta-blockers and two analgesics in river water. The method includes a pre-concentration step with the use of precolumn switching and MCR-ALS approach along with standard addition technique. Moreover, Eilers methodology was used to reduce the matrix background in three-way data. The use of MCR-ALS algorithm enabled overcoming problems such as overlapping peaks and additive and matrix effect errors. The method was successfully applied for the determination of eleven pharmaceuticals in river water with acceptable recoveries and precision values despite the complexity of the samples.

Li and coworkers [105] used a second-order calibration method based on the ATLD algorithm for the quantitative HPLC analysis of levodopa, carbidopa, and methyldopa in human plasma samples. Before applying ATLD algorithm, three chromatographic regions were selected for each analyte as their own elution time domain to avoid collinearity. The use of ATLD allowed correct estimation of each analyte even in the presence of uncalibrated interferences. A three-component model was built for each analyte. Chromatographic, spectral, and concentration profiles of the drugs were resolved from the decomposition of three-way data. This method provided a satisfactory prediction and accuracy to quantitatively determine the target drugs in human plasma.

Ouyang et al. [106] proposed an HPLC-DAD method coupled with second-order calibration based on alternating penalty trilinear decomposition algorithm to analyze metronidazole and tinidazole in plasma samples. Overlapping chromatographic peaks of the analytes and the plasma interferences were obtained with simple and green chromatographic conditions, which resulted in short analysis time. Chromatographic and spectral profiles of the analytes were resolved accurately with ATDL and for the quantitative analysis, good recoveries were obtained for test and plasma samples.

Yu and colleagues [107] used the ATLD method combined with HPLC-DAD to mathematically separate the co-eluted peaks and to quantify quinolones in honey samples. Overlapping peaks were successfully resolved and the developed method was applied to determine 12 quinolones at the same time in the presence of uncalibrated interfering components in complex background. This method was reported to be more effective and simpler compared with traditional chromatographic methods.

Hu and coworkers [108] developed an HPLC-DAD method coupled with second-order calibration algorithms for the determination of vancomycin and cephalexin in human plasma. They applied PARAFAC and self-weight-alternative-trilinear-decomposition methods to a set of HPLC-DAD data lacking an efficient separation of vancomycin and cephalexin. These two methods provided a good agreement of added and found drug content in the presence of uncalibrated interferences caused by plasma samples.

Arroyo and coworkers [109] used PARAFAC and PARAFAC2 algorithms to determine seven nonsteroidal anti-inflammatory drugs in bovine milk by GC-MS. One of the aims of the study was to optimize derivatization reaction and solid-phase microextraction procedures in order to reach best results. Quantitative results were gathered with the use of PARAFAC and PARAFAC2 algorithms. For the selection of solid phase microextraction (SPME) fiber type, loadings of PARAFAC and PARAFAC2 decomposition were used. The optimization of SPME and derivatization procedures with eight experimental factors was evaluated with D-optimal design. The extracted quantity of each analyte was used as the experimental responses that were determined by PARAFAC2 calibration. Although GC-MS analysis resulted in overlapping peaks of the drugs, the second-order advantage of the calibration methods overcame this problem and successful determination results along with the suitable figures of merit were obtained.

Grisales and colleagues [110] used HPLC-DAD method in combination with U-PLS regression in order to determine enantiomeric composition of ibuprofen in tablets. The use of U-PLS enabled the quantitation of enantiomers in their co-eluted chromatograms. Because the time-shifting was observed in the chromatographic analyses, a time alignment procedure was performed. The quantitative estimation of enantiomeric purity was achieved below 0.1% levels even in the presence of strongly overlapped peaks with the use of U-PLS algorithm.

Culzoni and colleagues [111] developed a method for simultaneous determination of galantamine and its major metabolites in plasma with the use of second-order calibration. Matrices of fluorescence intensity were recorded as a function of retention time and emission wavelength after a chromatographic run of 6 min. After baseline correction with asymmetric least squares methodology, the matrices were divided in five time regions, each one including a single analyte due to the significant spectral overlapping of the analytes. Then MCR-ALS algorithm was applied to the second-order data, and good results of predictions were obtained. The use of MCR-ALS provided sensitive and selective determination of galantamine and its metabolites with a relatively simple analytical procedure, including only a simple serum deproteinization step.

Yajuan and coworkers [112] used ATLD algorithm for the determination of costunolide and dehydrocostuslactone in human plasma and in traditional Chinese medicine by HPLC-DAD.

Both samples gave heavily overlapped chromatographic peaks among the analytes and interferents. But second-order calibration method based on ATLD algorithm enabled resolution and quantification of the analytes. The determination results were also compared with those obtained by HPLC-MS/MS, and no significant difference was reported.

Yu et al. [113] used PARAFAC algorithm for the qualitative and quantitative determination of 11 antibiotic drugs in tap water by HPLC-DAD. The antibiotics of interest were sulfacetamide, sulfamerazine, sulfamethoxazole, tetracycline, pipemidic acid, pefloxacin, danofloxacin, lomefloxacin, metronidazole, ornidazole, and oxytetracycline. HPLC analysis of the samples gave overlapping peaks of the analytes, as well as background drift problems. After solving background drift problems by orthogonal spectral signal projection (OSSP), PARAFAC algorithm was applied. With the background correction, the use of excessive factors to model the effect of background drift was eliminated and significant improvement in the quantitative capability of PARAFAC was reported. Eleven antibiotics in tap water were accurately quantified from overlapped chromatographic peaks.

Vosough and Esfahani [114] developed an HPLC-DAD method for quantification of five antibiotics in wastewaters with the use of MCR-ALS algorithm. Quantitative determination of sulfamethoxazole, metronidazole, chloramphenicol, sulfadiazine, and sulfamerazine was carried out after a solid-phase extraction procedure. Standard addition calibration in combination with MCR-ALS algorithm was applied due to the matrix interferences and sensitivity changes. Application of MCR/ALS algorithm was able to resolve highly drifted background constituents and strongly overlapped peaks of the analytes and matrix interferences.

Chen and colleagues [115] developed a new strategy for fast chiral screening by combining HPLC-DAD with a MCR-ALS algorithm. The solutions containing the racemates of althiazide, benzoin, N-(3,5-dinitrobenzoyl)-DL-leucine, propranolol hydrochloride, and verapamil were analyzed by HPLC-DAD using a chiral stationary phase and resulted in a chromatogram with overlapping parts. By using MCR-ALS, the data matrix was deconvoluted into resolved chromatograms and spectra. This approach provided a chiral screening with one-fifth of the conventional analysis time.

Akvan and Parastar [116] developed an HPLC-DAD analysis method coupled with a new combination of second-order calibration methods for simultaneous determination of carbamazepine, naproxen, diclofenac, gemfibrozil, and mefenamic acid in water samples. A pre-concentration step of solid-phase extraction procedure was used after optimization using face-centered CCD. New combination of multivariate curve resolution–correlation optimized warping–parallel factor analysis (MCR–COW–PARAFAC) and multivariate curve resolution–alternating least squares with trilinearity constraint (MCR–COW–MCR) was used due to the complexity of water matrices and other chromatographic issues. MCR was followed by COW for efficient chromatographic peak alignment. Analytical figures of merit, sensitivity, and limit of detection (LOD) and limit of quantitation (LOQ) values were reported to be improved for both methods compared to univariate techniques. The developed method was suitable for efficient fast, simple, and cost-effective quantification of pharmaceuticals in river and well waters.

An HPLC-DAD method was proposed by Yin and coworkers [117] for simultaneous determination of eight co-eluted compounds by using second-order calibration method based on the ATLD algorithm. Gallic acid, caffeine, and six catechins in tea infusions were analyzed, and data matrices were recorded as a function of elution time and wavelength. The ATLD algorithm successfully predicted the quantitative amounts of the analytes in samples in the presence of co-eluted peaks, uncalibrated interferences, and baseline drift. Figures of merit were calculated to validate the proposed method. In order to confirm the reliability of the method, liquid chromatography tandem-mass spectrometry (LC-MS/MS) with multiple reaction modes was employed and gave comparable results with the proposed HPLC-DAD method. Furthermore, PCA was used for cluster analysis of investigated teas, based on the quantitative results obtained from the ATLD algorithm.

Vosough and coworkers [118] proposed a strategy based on solid-phase extraction followed by fast HPLC with diode array detection coupled with MCR-ALS algorithm for the determination of methamphetamine and pseudoephedrine in river water. The sample preparation step included a solid-phase extraction procedure. Analysis of methamphetamine and pseudoephedrine in groundwater was performed by the use of classical univariate calibration methods. Analysis of the river water sample, which contained uncalibrated interferences and uncorrected background signals, required the use of MCR-ALS algorithm.

Padró and colleagues [119] used U-PLS algorithm to determine ketoprofen enantiomeric composition from strongly overlapped HPLC chromatograms. Three different chromatographic methods were developed in order to evaluate the power of U-PLS, with different resolution values ranging from 0.17 to 0.81. U-PLS algorithm was applied to these three different chromatographic three-way data. The analytical figures of merit were compared with each other, as well as to those obtained by a classical univariate calibration method. The classical method was obtained by a fully chromatographic separation of enantiomers. The quantification of enantiomeric composition from strongly overlapped profiles was successfully achieved with an enantiomeric purity well below 1% with excellent precision even from highly overlapped signals.

Alcaráz and coworkers [120] developed a new model to determine fluoroquinolones in water samples. Analysis of ofloxacin, ciprofloxacin, and danofloxacin in tap, underground, and mineral waters was carried out in the presence of enoxacin and marbofloxacin as interferents. They obtained a third-order data by recording the excitation–emission matrices of each fraction collected every 2 s during an HPLC run. These third-order chromatographic-excitation emission matrix (EEM) data were processed by augmented PARAFAC and MCR-ALS algorithms and the models were compared in terms of sensitivity, selectivity, relative error predictions, LOD, and LOQ. The use of augmented PARAFAC gave better analytical predictions as well as better figures of merit. Both methods saved cost and time, following green chemistry principles.

Gu et al. [121] developed a second-order calibration method based on the ATLD algorithm to be used with three-way data obtained by liquid chromatography coupled with mass spectrometry (LC-MS). This strategy was employed to simultaneously determine 10 β-blockers in human urine and plasma samples. This strategy was reported to solve the problems of co-eluting peaks and uncalibrated interferences in quantitative LC–MS technique and enhanced the selectivity and sensitivity of LC–MS in full-scan mode. The average recoveries were reported between 90% and 110% with standard deviations and average relative prediction errors less than 10%. In order to confirm the reliability of the method, the samples were analyzed by multiple reaction monitoring (MRM) method. T-test demonstrated that there are no significant differences between the prediction results of the two methods.

Dinç and Ertekin [122] used PARAFAC and trilinear PLS models for quantification of olmesartan medoxomil and hydrochlorothiazide in tablets. Co-eluted chromatographic analysis of samples by UPLC-PDA provided a three-dimensional data array with a dimension of $662 \times 120 \times 85$. This data array did not have time shift of background drift, hence was directly subjected to two different three-way calibration methods. PARAFAC model gave three loadings, namely, chromatographic, spectral, and relative concentration profiles that ease the interpretation of the results. In order to prove the ability of proposed three-way models, a conventional UPLC analysis was also performed by separation of the peaks chromatographically. The results were compared by ANOVA and no significant difference was reported between the methods.

Dinç and Büker [123] developed a UPLC-PDA method coupled with PARAFAC and trilinear PLS to quantify two active substances in eye drop formulations. The samples containing the compounds of interest, brimonidine tartrate and timolol maleate, as well as the internal standard ornidazole were analyzed by a chromatographic method that resulted in overlapping peaks. The three-way data obtained by UPLC-PDA were decomposed by PARAFAC and trilinear PLS models. The assay results of tablet samples were statistically compared with a traditional UPLC method, and no difference was reported.

Dinç and coworkers [124] developed three different chemometric approaches to simultaneously determine ciprofloxacin and ornidazole in tablets using the three-dimensional data obtained by UPLC-PDA instrument. These methods were PARAFAC, trilinear PLS, and unfolded PLS models.

The three-way data were of overlapping peaks of ciprofloxacin, ornidazole, and internal standard, but all three methods were able to quantify the drug amounts in tablets in spite of co-elution in chromatographic region.

21.2.3 Data Processing, Signal Processing, and Assessment of Peak Purity

21.2.3.1 Data Processing

In chemometrics, preliminary data processing is a very important treatment in order to improve the quality of results. These data processing approaches are centering, scaling and auto-scaling mathematical treatments, and so on. Others are Savitzky–Golay method [125,126] for reducing the effect of systematic errors, that is, removal of the systematic shift of the data, multiplicative scattering corrections [22,127], and orthogonal signal correction [128], which is a method for data processing in solving problems of quantitative analysis.

In instrumental analysis, significant background signal and interferences coming from unknown constituents in analyzed samples are the more difficult problems affecting the analysis of chemicals and pharmaceuticals, especially baseline correction, background correction, noise removal, data smoothing and filtering, and resolution of peak. These examples of the mentioned treatments are termed as data processing or signal processing. For example, Daszykowski and Walczak [129] stressed that chromatographic performance can be enhanced by the elimination of noise and background components. This is a crucial step for reducing both the complexity and the number of unexpected components. Pretreatments of data in chemometrics improve the quality of spectroscopic and chromatographic measurements [12].

Xu and coworkers [130] used two different chromatographic pretreatment procedures for the quality evaluation of the herbal medicine *Houttuynia cordata*. Target peak alignment and multiplicative signal correction were performed in order to correct the retention time shifts and for response correction in 18 different GC-MS chromatograms. Effect of pretreatments was investigated by the quality control of the samples with and without pretreatment. Quality control was conducted by correlation and congruence coefficients as well as PCA of chromatographic fingerprints. It was reported that data correction reduced the variations from the experimental procedure and provided more accurate results.

Zhang et al. [131] developed a technique to remove three-dimensional background drift from three-dimensional chromatographic data of herbal medicine *Rhizoma chuanxiong*. The three-dimensional data were obtained by an LC × LC analyzer that couples a first separation column and a second separation column with a multi-wavelength UV absorbance detector. These data were decomposed using the ATLD algorithm. In the decomposition step, background drift was modeled as one component just like the analytes of interest. Then, the background component was subtracted from the raw data. The technique was reported to yield a good removal of background drift, without the need to perform a blank chromatographic run, and required no prior knowledge about the sample composition.

Fredriksson and colleagues [132] compared different preprocessing methods for enhancement of a liquid chromatography–mass spectrometry method to analyze a pharmaceutical drug and its degradation products. The compared methods were component detection algorithm and three kinds of digital filters—matched filtration, Gaussian second derivative, and Savitzky–Golay. The preprocessing methods were applied with different settings to data sets. The performance and robustness of these methods were evaluated for extracted ion chromatogram, total ion chromatogram, and base peak chromatogram in the presence of different types of noise. The best improvements in signal-to-noise ratio of the extracted ion chromatograms were obtained with matched filtration under the ideal case with random white noise. Gaussian second derivative and component detection algorithm improved the signal-to-noise ratio for both total ion and base peak chromatograms and reduced the background in the spectral domain. Data reduction ability of these two methods was also reported in the study.

De Zan and coworkers [133] used PDS and baseline correction approaches for MCR-ALS resolution of eight tetracycline antibiotics in wastewaters. Because the sample matrix was highly complex, a large baseline drift as well as additive interferences were present in real samples. These problems were solved by using asymmetric least squares method for baseline correction and PDS to correct changes in instrumental response due to sample pretreatment procedures. MCR-ALS was applied to the corrected and uncorrected data sets. The quality of the spectral and chromatographic profiles obtained from corrected data sets was better, and the calibration model was simpler.

Yu and coworkers [113] used OSSP to correct background drift for the determination of 11 antibiotics in tap water. Several data matrices were obtained by the analysis of pre-concentrated tap water samples by HPLC-DAD. Background drift resulting from matrix effect was corrected by OSSP. In order to evaluate the performance of the correction method, PARAFAC was applied to simulated chromatographic data set, with and without correction. A significant improvement was reported in quantitative results. This strategy was successfully performed to quantify 11 antibiotics in tap water.

Yu et al. [134] developed a new chemometric strategy for automatic chromatographic peak detection and background drift correction to be used in liquid and gas chromatography. A statistical method was used for estimation of instrumental noise level coupled with first-order derivative of chromatographic signal to automatically extract chromatographic peaks in the data.

Then, a local curve-fitting strategy was employed for background drift correction. In order to verify the performance of the proposed strategy, simulated and real liquid chromatographic data were designed with various kinds of background drift and degree of overlapped chromatographic peaks of 11 antibiotics in tap water samples. The underlying chromatographic peaks were automatically detected and reasonably integrated by this strategy. The proposed method was used to analyze a complex data set of plant extracts with GC coupled with flame ionization detector in order to monitor quality changes during storage procedure.

21.2.3.2 Signal Processing

In practice, the traditional estimation of analytical signals, for example, spectra, chromatogram, voltammogram, and other data formats obtained from chemical instruments, may not always provide the desired results for chemical and pharmaceutical analysis with a higher complexity of analyzed samples. Hence, combined use of conventional chromatographic methods and some chemometric signal processing methods can be necessary for the analysis of complex systems. As a consequence of the mentioned approach, conventional HPLC (or UPLC) technique coupled with signal processing tools enhances their ability of resolution, separation, and analysis of pharmaceuticals tremendously. In this regard, several signal processing techniques have been developed for many application areas from data analysis to data compression. One of the newest additions has been wavelets [135–137] and their applications in overlapping chromatograms [138,139]. Wavelet transform approach is a powerful signal processing tool for data reduction, de-noising, baseline correction, and resolution of overlapping spectra. In our previous studies, the wavelet transform signal processing tools combined with conventional spectral and chromatographic techniques were applied to the analysis of pharmaceuticals in multicomponent samples. In this study, some typical applications of the wavelet transforms to chromatographic signals for pharmaceutical analysis will be described as follows.

Dinç and Büker [140] applied continuous wavelet transform to overlapping chromatograms of amiloride hydrochloride and hydrochlorothiazide to quantify them in tablets. The UPLC analysis of samples gave overlapping peaks of the compounds of interest. The overlapping UPLC data vectors were processed by two continuous wavelet transform methods, namely, Bior 1.3 (a = 48) and Mexh (a = 14). The minimum and maximum amplitudes were selected for the quantitative determination of amiloride hydrochloride and hydrochlorothiazide. The proposed methods were validated and successfully applied to the tablet samples. The proposed methods showed that continuous wavelet transform methods can be used effectively in the analysis of active compounds without the need of chromatographic separation.

21.2.3.3   Peak Purity*

Qualitative and quantitative analyses are the main objectives of scientists who work on chromatography. For qualitative analysis of unknown samples, one might be interested in some retention regions and want to find what chemical components are in the corresponding chromatographic peaks. Evaluation of the peak purity is a necessary step to make the qualitative analysis reliable. Some aspects of peak purity evaluation might include whether impurities or minor components are present in the peaks, the elution distribution of impurities, or the relative concentrations of different components. A regular task for quantitative analysis of a compound of interest is calibrating peak areas with standard samples. Impurities in the objective peak will bring bias and error, so peak purity evaluation might be one of the key factors affecting the quality of the analysis. [10] Some typical examples of chemometric approaches for peak purity assessment are given in the following.

Hu et al. [141] proposed an artificial neural networks approach for the evaluation of chromatographic peak purity. The selected active substances were caffeic acid and salicylic acid with the impurities 3,4-dihydroxybenzoic acid and 4-hydroxybenzoic acid, respectively. The chromatographic data were predicted by a nonlinear transformation function with a back-propagation algorithm. The purity of the chromatographic peaks was assessed by Mann–Whitney U-test. The performance of this approach was compared with the results from PCA.

De Braekeleer and colleagues [142] evaluated the purity of tetracycline hydrochloride by means of HPLC-DAD analysis along with orthogonal projection approach (OPA) and the fixed size moving window evolving factor analysis approach (FSW-EFA). Detection of four impurities was successfully done; even retention time of some impurities coincided with the main peak. While USP XXIII univariate method failed to detect the impurity 4-epianhydrotetracycline, it was successfully detected by OPA and FSW-EFA methods. After the presence of impurities was confirmed, their concentration profiles and spectra were modeled by MCR-ALS resolution method.

Wiberg and coworkers [143] evaluated the peak purity of lidocaine by PCA and HPLC-DAD method. Prilocaine was modeled as an impurity with a minor amount in samples. Different separation levels of two drugs were used to test the abilities and strength of the method. The peak purity determination was made by examination of relative observation residuals, scores, and loadings from the PCA decomposition of the data over a chromatographic peak. The peak purity function for the software of the HPLC system was used as a reference method. The PCA method showed good results outperforming the reference method by a factor of 10 and was suitable to determine whether or not a chromatographic peak is pure.

Van Zomeren and coworkers [144] used augmented iterative target transformation factor analysis to resolve overlapping peaks of nitrazepam, clonazepam, and lorazepam in HPLC and MEKC. Augmentation, which means the combination of multiple data matrices, was applied to data that originate from the same mode of detection but different separation methods. Alignment between the wavelength scales of instruments was performed by wavelength shift eigenstructure tracking (WET), and this step also enabled correction of wavelength shift between the HPLC-DAD and MEKC-DAD. After WET, which is based on singular value decomposition and cubic spline interpolation, the data were augmented and decomposed by iterative target transformation factor analysis. Augmented curve resolution was used to estimate chromatographic, electrophoretic, and spectral profiles of three different drugs from a single peak.

Wiberg [145] reported the use of PCA decomposition of HPLC-DAD data for relative impurity profiling of prilocaine. Prilocaine and its six different impurities were separated by an isocratic HPLC method. Since the molar absorption coefficients of different peaks are different from each other, exact estimation of impurity is not possible from a chromatogram recorded at a single wavelength. Instead, the score vectors obtained by PCA decomposition of HPLC-DAD data matrix contain more accurate quantitative information of the main peak and impurities. A good

---

* See also Chapter 11.

estimation of impurity quantities by integration of score chromatograms was obtained by the proposed method without calibration. MCR-ALS was used for comparison and no significant difference was reported.

21.2.4 EXPLORATION, CLASSIFICATION, AND DISCRIMINATION*

Multivariate data analysis is a very important tool for the exploration analysis of chemical data or pharmaceutical data, particularly spectroscopic and chromatographic ones. In this context, PCA is the most commonly used for pattern recognition, classification, modeling, and other aspects of data evaluation with chromatographic analysis or chromatographic fingerprint in order to assess the quality and consistency of botanical products or other samples containing medicinal plants. For the same purpose, other chemometric approaches are soft independent modeling of class analogy (SIMCA), hierarchical cluster analysis (HCA), linear discriminate analysis, and so on. Related papers were reviewed in the following.

Detroyer and coworkers [146] performed an exploratory chemometric analysis of 83 drug substances from 8 pharmacological groups based on their chromatographic behaviors. The data set included molecular weight, log P values, and log k values of the drug substances obtained by eight different chromatographic systems. PCA, clustering, and sequential projection pursuit were applied to the data as complementary techniques. The first three principal components gave interpretable results, and auto-scaled data provide a better understanding. The families were assigned to a big cluster, and dendrogram was obtained by normalized scores of the substances. Sequential projection pursuit was able to find some contrast more easily than PCA. This study was able to uncover the information present in data and predicted the most useful chromatographic systems useful for the purpose.

Yan and coworkers [147] used PCA and cluster analysis for the classification of the medicinal plant *Cnidium monnieri*. HPLC analysis was performed to determine 11 coumarin constituents in methanol extracts of the plant, collected from different regions of China. Fifty-three HPLC data sets were studied by PCA and cluster analysis to reveal the relationship between the samples and their geographic distribution, which resulted from their different coumarin content. PCA alone was not useful for the classification of the samples, but successful results were obtained when further study of PCA and cluster analysis were performed.

Xu and colleagues [130] evaluated the similarity of GC-MS fingerprints of the herbal medicine *Houttuynia cordata* by correlation coefficient and congruence coefficient as well as kernel PCA. Before evaluation of quality, the HPLC fingerprints were preprocessed to reduce time shifts and for response correction. With the use of kernel PCA, clustering of the fingerprints was explored quickly and four clusters of the geographic origin were revealed.

Quality evaluation of *Radix Salvia miltiorrhiza* was performed by Ma et al. [148] by HPLC combined with PCA analysis. Simultaneous determination of seven components was performed, and the HPLC-UV data were used to carry out PCA. Scores and loading plots for PC1 versus PC2 were used for the quality assessment. Thirty plant samples were divided into three groups in the scores plot. The groups represented the genetic relationship between the samples. PCA was used to differentiate the quality of samples from different origins and the method provided a rapid screening for the quality evaluation of *Radix Salvia miltiorrhiza*.

Chen and coworkers [149] used HPLC fingerprints combined with chemometric methods to establish an objective pattern recognition system to discriminate *Ganoderma lucidum* samples from different origins and to control the quality of the plant. The relative peak areas of 19 characteristic compounds were calculated for 60 samples. Different pattern recognition procedures were applied to classify the samples according to their cultivated origins. The discrimination abilities of the HCA, PCA, SIMCA, and PLS–discrimination analysis (PLS-DA) were compared. Four marker

---

* See also Chapter 8.

constituents were found out to be the most discriminant variables. They efficiently differentiated the two species and classified the plant from three main planting areas.

Ni and colleagues [150] used PCA to discriminate a traditional Chinese medicine, which was the processed rhizome of *Atractylis chinensis*. The molecular and metal fingerprint profiles of the medicine were obtained by HPLC and inductively coupled plasma atomic emission spectroscopy techniques. The data matrices obtained from the two different techniques provided two principal component biplots, which showed that the HPLC fingerprint data were discriminated on the basis of the processing methods of the raw plant, while the metal analysis grouped according to the geographical origin. When the two data matrices were combined into a one two-way matrix, the resulting biplot showed a clear separation on the basis of the HPLC fingerprints. Within each group, the objects were separated according to their geographical origin. It is reported that improved characterization of the complex traditional medicine material was possible by using such an approach. In addition, K-nearest neighbor method and linear discriminant analysis were successfully applied to the individual data matrices and supported PCA approach.

Peng and colleagues [151] used similarity analysis, HCA, and PCA for classification of *Artemisia selengensis* Turcz. The information for matching and discrimination of HPLC fingerprints was reported to be suitable for quality assurance of the medicinal plant. The developed HPLC method enabled the quantification of rutin, which is the main active component, and obtaining detailed HPLC fingerprint data. The common fingerprint at a fixed wavelength and enhanced fingerprint with additional spectral data were compared, and the enhanced fingerprint was found to be more informative for the data analysis.

Deconinck and coworkers [152] used chromatographic fingerprints of genuine and counterfeit samples of Viagra® and Cialis® to differentiate between genuine and counterfeit samples and to classify counterfeit samples. Several exploratory chemometric techniques were applied to reveal structures in the data sets as well as differences among the samples. Then, the differences between the samples were modeled to obtain a predictive model for both the differentiation between genuine and counterfeit samples as well as the classification of the counterfeit samples.

Projection pursuit and hierarchical clustering were used for the exploratory analysis, and differences among the genuine and the counterfeit samples as well as differences among the different groups of counterfeits were clearly revealed. Predictive models were obtained perfectly for the differentiation between genuine and counterfeit medicines. High correct classification rates for the classification in the different classes of counterfeit medicines were also obtained. The best performing models were obtained with least squares support vector machines and SIMCA.

21.3 CONCLUSIONS

This chapter provides an overview on the applications of chemometric strategies in chromatographic analysis of pharmaceuticals. Multivariate chemometric techniques applied to chromatographic analysis of samples containing pharmaceuticals have been used for the optimization of chromatographic system. Their applications were reviewed and summarized in Table 21.1. In analyses of pharmaceuticals, two-way multivariate calibration methods were reported (see Table 21.2). As it is known very well overlapped chromatographic peaks (or co-eluted chromatographic peaks) with unknown interferences are very common phenomena in qualitative and quantitative chromatographic analysis of pharmaceuticals. In this context, a related topic reviewed here was the three-way (or N-way) analysis methods, which have been used as a chemometric strategy for the analysis of drug substances in samples. For the similar purposes, MCR-ALS approach has been implemented. These methods offer the benefit of short time analysis due to the avoidance of long and tedious pretreatment of samples to remove interferences. Applications of the reviewed papers were presented in Table 21.2. Finally, other chemometric strategies in chromatography to obtain better analysis of pharmaceuticals are data processing, signal processing, and assessment of peak purity. Their practical implementations in chromatographic analysis were reviewed.

TABLE 21.2

Three-Way Chemometric Methods and Their Applications to the Chromatographic Analysis of Pharmaceuticals

Analyzed Pharmaceuticals	Analytical Method	Chemometric Method	Reference
2-hydroxipridine and 3-hydroxipridine	HPLC-DAD	Two-way PLS, unfolded PLS, three-way PLS	[97]
Prilocaine	HPLC-DAD	PARAFAC	[98]
Cortisol and prednisolone	HPLC-DAD	ATLD	[99]
Commercial samples of St. John's wort	HPLC-DAD	PARAFAC	[100]
Tetracycline, oxytetracycline, meclocycline, minocycline, metacycline, chlortetracycline, demeclocycline, doxycycline	HPLC-DAD	U-PLS/RBL and MCR-ALS	[101]
Pipemidic acid, marbofloxacin, ofloxacin, norfloxacin, ciprofloxacin, enrofloxacin, lomefloxacin	HPLC-FLR	PARAFAC, N-PLS, MCR-ALS	[102]
Naproxen, ketoprofen, diclofenac, piroxicam, indomethacin, sulindac, diflunisal, carbamazepine	HPLC-DAD	MCR-ALS	[103]
Sotalol, atenolol, nadolol, pindolol, metoprolol, timolol, bisoprolol, propranolol, betaxolol, paracetamol, phenazone	HPLC-DAD	MCR-ALS	[104]
Levodopa, carbidopa, methyldopa	HPLC-DAD	ATLD	[105]
Metronidazole and tinidazole	HPLC-DAD	ATLD	[106]
Ciprofloxacin, danofloxacin, difloxacin, enoxacin, enrofloxacin, fleroxacin, lomefloxacin, marbofloxacin, ofloxacin, orbifloxacin, pefloxacin, sarafloxacin	HPLC-DAD	ATLD	[107]
Vancomycin and cephalexin	HPLC-DAD	PARAFAC-ALS and SWATLD	[108]
Ibuprofen, naproxen, ketoprofen, diclofenac, flufenamic acid, tolfenamic acid, meclofenamic acid	GC-MS	PARAFAC and PARAFAC2	[109]
R-(−)-ibuprofen and S-(+)-ibuprofen	HPLC-DAD	U-PLS	[110]
Galantamine and its major metabolites	HPLC-FLR	MCR-ALS	[111]
Costunolide and dehydrocostuslactone	HPLC-DAD	ATLD	[112]
Eleven antibiotics	HPLC-DAD	PARAFAC	[113]
Sulfamethoxazole, metronidazole, chloramphenicol, sulfadiazine, and sulfamerazine	HPLC-DAD	MCR-ALS	[114]
Five racemic pairs	HPLC-DAD	MCR-ALS	[115]
Carbamazepine, naproxen, diclofenac, gemfibrozil, and mefenamic acid	HPLC-DAD	MCR-COW and PARAFAC	[116]
Gallic acid, caffeine epicatechin gallate, catechin, epicatechin, epigallocatechin, epigallocatechin gallate, gallocatechin gallate	HPLC-DAD	ATLD	[117]
Pseudoephedrine and methamphetamine	HPLC-DAD	MCR-ALS	[118]
Racemic ketoprofen	HPLC-DAD	U-PLS	[119]
Ofloxacin, ciprofloxacin, danofloxacin	HPLC-EEM	Augmented PARAFAC and MCR-ALS	[120]
Ten beta-blockers	LC-MS	ATLD	[121]
Hydrochlorothiazide and olmesartan medoxomil	UPLC-PDA	PARAFAC and trilinear PLS	[122]
Brimonidine tartrate and timolol maleate	UPLC-PDA	PARAFAC and trilinear PLS	[123]
Ciprofloxacin and ornidazole	UPLC-PDA	PARAFAC, trilinear PLS, U-PLS	[124]

Notes: HPLC, high-performance liquid chromatography; DAD, diode array detector; UPLC, ultra-performance liquid chromatography; PDA, photodiode array; LC-MS, liquid chromatography–mass spectrometry; SWATLD, self-weight-alternative-trilinear-decomposition; PARAFAC, parallel factor analysis; U-PLS, unfolded partial least squares; MCR-ALS, multivariate curve resolution–alternating least squares; MCR-COW, multivariate curve resolution–correlation optimized warping; ATLD, alternating trilinear decomposition.

REFERENCES

1. Bruns, R.E., I.S. Scarminio, and B. de Barros Neto. 2006. *Statistical Design—Chemometrics*. Amsterdam, the Netherlands: Elsevier.
2. Box, G.E.P., J. Stuart Hunter, and W.G. Hunter. 2005. *Statistics for Experimenters: Design, Innovation, and Discovery*. Hoboken, NJ: Wiley-Interscience.
3. Montgomery, D.C. 2013. *Design and Analysis of Experiments*. Hoboken, NJ: John Wiley & Sons, Inc.
4. Vander Heyden, Y., C. Perrin, and D.L. Massart. 2000. Optimization strategies for HPLC and CZE. In *Handbook of Analytical Separations, 1*, ed. Valkó, K., pp. 163–212. Amsterdam, the Netherlands: Elsevier.
5. Siouffi, A.M. and R. Phan-Tan-Luu. 2000. Optimization methods in chromatography and capillary electrophoresis. *Journal of Chromatography A* 892 (1–2): 75–106. doi:10.1016/s0021-9673(00)00247-8.
6. Ferreira, S.L.C., R.E. Bruns, E.G.P. da Silva, W.N.L. dos Santos, C.M. Quintella, J.M. David, J.B. de Andrade, M.C. Breitkreitz, I.C.S.F. Jardim, and B.B. Neto. 2007. Statistical designs and response surface techniques for the optimization of chromatographic systems. *Journal of Chromatography A* 1158 (1): 2–14.
7. Roggo, Y., P. Chalus, L. Maurer, C. Lema-Martinez, A. Edmond, and N. Jent. 2007. A review of near infrared spectroscopy and chemometrics in pharmaceutical technologies. *Journal of Pharmaceutical and Biomedical Analysis* 44 (3): 683–700. doi:10.1016/j.jpba.2007.03.023.
8. Gómez, V. and M.P. Callao. 2008. Analytical applications of second-order calibration methods. *Analytica Chimica Acta* 627 (2): 169–183. doi:10.1016/j.aca.2008.07.054.
9. Zhang, Y.-J., W.-J. Gong, J.-M. Zhang, Y.-P. Zhang, S.-M. Wang, L. Wang, and H.-Y. Xue. 2008. Optimization strategies using response surface methodologies in high performance liquid chromatography. *Journal of Liquid Chromatography & Related Technologies* 31 (19): 2893–2916. doi:10.1080/10826070802424493.
10. Xu, L., L.-J. Tang, C.-B. Cai, H.-L. Wu, G.-L. Shen, R.-Q. Yu, and J.-H. Jiang. 2008. Chemometric methods for evaluation of chromatographic separation quality from two-way data—A review. *Analytica Chimica Acta* 613 (2): 121–134. doi:10.1016/j.aca.2008.02.061.
11. Daszykowski, M. and B. Walczak. 2011. Methods for the exploratory analysis of two-dimensional chromatographic signals. *Talanta* 83 (4): 1088–1097. doi:10.1016/j.talanta.2010.08.032.
12. Goicoechea, H.C., M.J. Culzoni, M.D. Gil García, and M. Martínez Galera. 2011. Chemometric strategies for enhancing the chromatographic methodologies with second-order data analysis of compounds when peaks are overlapped. *Talanta* 83 (4): 1098–1107. doi:10.1016/j.talanta.2010.07.057.
13. Arancibia, J.A., P.C. Damiani, G.M. Escandar, G.A. Ibañez, and A.C. Olivieri. 2012. A review on second- and third-order multivariate calibration applied to chromatographic data. *Journal of Chromatography B* 910: 22–30. doi:10.1016/j.jchromb.2012.02.004.
14. Gindy, A.E. and G.M. Hadad. 2012. Chemometrics in pharmaceutical analysis: An introduction, review, and future perspectives. *Journal of AOAC International* 95 (3): 609–623. doi:10.5740/jaoacint.sge_el-gindy.
15. Ruckebusch, C. and L. Blanchet. 2013. Multivariate curve resolution: A review of advanced and tailored applications and challenges. *Analytica Chimica Acta* 765: 28–36. doi:10.1016/j.aca.2012.12.028.
16. Murphy, K.R., C.A. Stedmon, D. Graeber, and R. Bro. 2013. Fluorescence spectroscopy and multi-way techniques. PARAFAC. *Analytical Methods* 5 (23): 6557. doi:10.1039/c3ay41160e.
17. Kumar, N., A. Bansal, G.S. Sarma, and R.K. Rawal. 2014. Chemometrics tools used in analytical chemistry: An overview. *Talanta* 123: 186–199. doi:10.1016/j.talanta.2014.02.003.
18. Massart, D.L., B.G.M. Vandeginste, S.N. Deming, Y. Michotte, and L. Kaufman. 1988. *Chemometrics: A Textbook*. Amsterdam, the Netherlands: Elsevier.
19. Wold, S. 1995. Chemometrics; what do we mean with it, and what do we want from it? *Chemometrics and Intelligent Laboratory Systems* 30 (1): 109–115. doi:10.1016/0169-7439(95)00042-9.
20. Malinowski, E.R. 2002. *Factor Analysis in Chemistry*. New York: Wiley.
21. Sharaf, M.A., D.L. Illman, and B.R. Kowalski. 1986. *Chemometrics*. New York: Wiley.
22. Martens H. and T. Naes. 1989. *Multivariate Calibration*. Chichester, U.K.: Wiley.
23. Otto, M. 1999. *Chemometrics: Statistics and Computer Applications in Analytical Chemistry*. New York: Wiley.
24. Geladi, P. and K. Esbensen. 1990. The start and early history of chemometrics: Selected interviews. Part 1. *Journal of Chemometrics* 4 (5): 337–354. doi:10.1002/cem.1180040503.
25. Esbensen, K. and P. Geladi. 1990. The start and early history of chemometrics: Selected interviews. Part 2. *Journal of Chemometrics* 4 (6): 389–412. doi:10.1002/cem.1180040604.

26. Box, G.E.P. and J.S. Hunter. 1957. Multi-factor experimental designs for exploring response surfaces. *The Annals of Mathematical Statistics* 28 (1): 195–241. doi:10.1214/aoms/1177707047.

27. Box, G.E.P. and J.S. Hunter. 1961. The 2^{k-p} fractional factorial designs: Part I. *Technometrics* 3 (3): 311–352.

28. Box, G.E.P. and J.S. Hunter. 1961. The 2^{k-p} fractional factorial designs: Part II. *Technometrics* 3 (3): 449–458.

29. Box, G.E.P. and K.G. Wilson. 1951. On the experimental attainment of optimum conditions. *Journal of the Royal Statistical Society B* 13: 1–45.

30. Doehlert, D.H. 1970. Uniform shell designs. *Applied Statistics* 19: 231–239.

31. Doehlert, D.H. and V.L. Klee. 1972. Experimental designs through level reduction of the d-dimensional cuboctahedron. *Discrete Mathematics* 2 (4): 309–334.

32. Box, G.E.P. and D.W. Behnken. 1960. Some new three level designs for the study of quantitative variables. *Technometrics* 2 (4): 455–475.

33. Karlsson, A. and S. Hermansson. 1997. Optimisation of chiral separation of omeprazole and one of its metabolites on immobilized A1-acid glycoprotein using chemometricsglycoprotein using chemometrics. *Chromatographia* 44 (1–2): 10–18. doi:10.1007/bf02466509.

34. Ho, C., H.-M. Huang, S.-Y. Hsu, C.-Y. Shaw, and B.-L. Chang. 1999. Simultaneous high-performance liquid chromatographic analysis for famotidine, ranitidine HCl, cimetidine, and nizatidine in commercial products. *Drug Development and Industrial Pharmacy* 25 (3): 379–385. doi:10.1081/ddc-100102186.

35. Wsól, V. and A.F. Fell. 2002. Central composite design as a powerful optimisation technique for enantioresolution of the Rac-11-dihydrooracin—The principal metabolite of the potential cytostatic drug oracin. *Journal of Biochemical and Biophysical Methods* 54 (1–3): 377–390. doi:10.1016/s0165-022x(02)00138-0.

36. Diana, J., G. Ping, E. Roets, and J. Hoogmartens. 2002. Development and validation of an improved liquid chromatographic method for the analysis of oxytetracycline. *Chromatographia* 56 (5–6): 313–318. doi:10.1007/bf02491938.

37. Marini, R.D., P. Chiap, B. Boulanger, W. Dewe, P. Hubert, and J. Crommen. 2003. LC method for the simultaneous determination R-timolol and other closely related impurities S-timolol maleate: Optimization by use of an experimental design. *Journal of Separation Science* 26 (9–10): 809–817. doi:10.1002/jssc.200301367.

38. Kuo, C.-H. and S.-W. Sun. 2003. Analysis of nine rhubarb anthraquinones and bianthrones by micellar electrokinetic chromatography using experimental design. *Analytica Chimica Acta* 482 (1): 47–58. doi:10.1016/s0003-2670(03)00169-7.

39. Djurdjevic, P., A. Laban, S. Markovic, and M. Jelikic-Stankov. 2004. Chemometric optimization of a RP-HPLC method for the simultaneous analysis of abacavir, lamivudine, and zidovudine in tablets. *Analytical Letters* 37 (13): 2649–2667. doi:10.1081/al-200031946.

40. Hadjmohammadi, M.R. and P. Ebrahimi. 2004. Optimization of the separation of anticonvulsant agents in mixed micellar liquid chromatography by experimental design and regression models. *Analytica Chimica Acta* 516 (1–2): 141–148. doi:10.1016/j.aca.2004.04.019.

41. Ivanovic, D., M. Medenica, B. Jancic, A. Malenovic, and S. Markovic. 2004. Chemometrical approach in fosinopril-sodium and its degradation product fosinoprilat analysis. *Chromatographia* 60 (S1): S87–S92. doi:10.1365/s10337-004-0324-7.

42. Jančić, B., M. Medenica, D. Ivanović, A. Malenović, and S. Marković. 2005. Microemulsion liquid chromatographic method for characterisation of fosinopril sodium and fosinoprilat separation with chemometrical support. *Analytical and Bioanalytical Chemistry* 383 (4): 687–694. doi:10.1007/s00216-005-0074-x.

43. Wang, Y., M. Harrison, and B.J. Clark. 2006. Optimising reversed-phase liquid chromatographic separation of an acidic mixture on a monolithic stationary phase with the aid of response surface methodology and experimental design. *Journal of Chromatography A* 1105 (1–2): 199–207. doi:10.1016/j.chroma.2005.11.101.

44. Cutroneo, P., M. Beljean, R. Phan Tan Luu, and A.-M. Siouffi. 2006. Optimization of the separation of some psychotropic drugs and their respective metabolites by liquid chromatography. *Journal of Pharmaceutical and Biomedical Analysis* 41 (2): 333–340. doi:10.1016/j.jpba.2005.10.050.

45. Kiendrebeogo, M., L. Choisnard, C.E. Lamien, A. Meda, D. Wouessidjewe, and O.G. Nacoulma. 2005. Experimental design optimization for screening relevant free phenolic acids from various preparations used in Burkina Faso folk medicine. *African Journal of Traditional, Complementary, and Alternative Medicines* 3 (1): 115–128. doi:10.4314/ajtcam.v3i1.31146.

46. Iriarte, G., N. Ferreirós, I. Ibarrondo, R.M. Alonso, M.I. Maguregi, L. Gonzalez, and R.M. Jiménez. 2006. Optimization via experimental design of an SPE-HPLC-UV-fluorescence method for the determination of valsartan and its metabolite in human plasma samples. *Journal of Separation Science* 29 (15): 2265–2283. doi:10.1002/jssc.200600093.

47. Sivakumar, T., R. Manavalan, C. Muralidharan, and K. Valliappan. 2007. An improved HPLC method with the aid of a chemometric protocol: Simultaneous analysis of amlodipine and atorvastatin in pharmaceutical formulations. *Journal of Separation Science* 30 (18): 3143–3153. doi:10.1002/jssc.200700148.

48. Petkovska, R. and A. Dimitrovska. 2008. Use of chemometrics for development and validation of an RP-HPLC method for simultaneous determination of haloperidol and related compounds. *Acta Pharmaceutica* 58 (3): 243–256. doi:10.2478/v10007-008-0019-y.

49. Petkovska, R., C. Cornett, and A. Dimitrovska. 2008. Chemometrical approach in lansoprazole and its related compounds analysis by rapid resolution RP-HPLC method. *Journal of Liquid Chromatography & Related Technologies* 31 (14): 2159–2173. doi:10.1080/10826070802225478.

50. Barmpalexis, P., F.I. Kanaze, and E. Georgarakis. 2009. Developing and optimizing a validated isocratic reversed-phase high-performance liquid chromatography separation of nimodipine and impurities in tablets using experimental design methodology. *Journal of Pharmaceutical and Biomedical Analysis* 49 (5): 1192–1202. doi:10.1016/j.jpba.2009.03.003.

51. Rao, A.A. 2008. Chromatographic method development: Computer simulated statistical design approach. *African Journal of Pure and Applied Chemistry* 2 (1): 001–005.

52. Dossou, K.S.S., P. Chiap, B. Chankvetadze, A.-C. Servais, M. Fillet, and J. Crommen. 2010. Optimization of the LC enantioseparation of chiral pharmaceuticals using cellulose tris(4-chloro-3-methylphenylcarbamate) as chiral selector and polar non-aqueous mobile phases. *Journal of Separation Science* 33 (12): 1699–1707. doi:10.1002/jssc.201000049.

53. Momenbeik, F., M. Roosta, and A.A. Nikoukar. 2010. Simultaneous microemulsion liquid chromatographic analysis of fat-soluble vitamins in pharmaceutical formulations: Optimization using genetic algorithm. *Journal of Chromatography A* 1217 (24): 3770–3773. doi:10.1016/j.chroma.2010.04.012.

54. Jančić-Stojanović, B., T. Rakić, N. Kostić, A. Vemić, A. Malenović, D. Ivanović, and M. Medenica. 2011. Advancement in optimization tactic achieved by newly developed chromatographic response function: Application to LC separation of raloxifene and its impurities. *Talanta* 85 (3): 1453–1460. doi:10.1016/j.talanta.2011.06.029.

55. Meneghini, L.Z., C. Junqueira, A.S. Andrade, F.R. Salazar, C.F. Codevilla, P.E. Fröehlich, and A.M. Bergold. 2011. Chemometric evaluation of darifenacin hydrobromide using a stability-indicating reversed-phase LC method. *Journal of Liquid Chromatography & Related Technologies* 34 (18): 2169–2184. doi:10.1080/10826076.2011.585486.

56. Rakić, T., B.J. Stojanović, A. Malenović, D. Ivanović, and M. Medenica. 2012. Improved chromatographic response function in HILIC analysis: Application to mixture of antidepressants. *Talanta* 98: 54–61. doi:10.1016/j.talanta.2012.06.040.

57. Lopes, S.B., J.M. Sarraguça, J.A.V. Prior, and J.A. Lopes. 2012. Development of an HPLC assay methodology for a desonide cream with chemometrics assisted optimization. *Analytical Letters* 45 (11): 1390–1400. doi:10.1080/00032719.2012.675494.

58. Rafamantanana, M.H., B. Debrus, G.E. Raoelison, E. Rozet, P. Lebrun, S. Uverg-Ratsimamanga, P. Hubert, and J. Quetin-Leclercq. 2012. Application of design of experiments and design space methodology for the HPLC-UV separation optimization of aporphine alkaloids from leaves of *Spirospermum pendulliflorum* thouars. *Journal of Pharmaceutical and Biomedical Analysis* 62: 23–32. doi:10.1016/j.jpba.2011.12.028.

59. Vujić, Z., N. Mulavdić, M. Smajić, J. Brborić, and P. Stankovic. 2012. Simultaneous analysis of irbesartan and hydrochlorothiazide: An improved HPLC method with the aid of a chemometric protocol. *Molecules* 17 (12): 3461–3474. doi:10.3390/molecules17033461.

60. Kasagić, I., A. Malenović, M. Jovanović, T. Rakić, B.J. Stojanović, and D. Ivanović. 2013. Chemometrically assisted optimization and validation of RP-HPLC method for the analysis of itraconazole and its impurities. *Acta Pharmaceutica* 63 (2): 159–173. doi:10.2478/acph-2013-0015.

61. Smajić, M., Z. Vujić, N. Mulavdić, and J. Brborić. 2013. An improved HPLC method for simultaneous analysis of losartan potassium and hydrochlorothiazide with the aid of a chemometric protocol. *Chromatographia* 76 (7–8): 419–425. doi:10.1007/s10337-013-2388-8.

62. Sahu, P.K. and C.S. Patro. 2014. Application of chemometric response surface methodology in development and optimization of a RP-HPLC method for the separation of metaxalone and its base hydrolytic impurities. *Journal of Liquid Chromatography & Related Technologies* 37 (17): 2444–2464. doi:10.1080/10826076.2013.840841.

63. Kalariya, P.D., D. Namdev, R. Srinivas, and S. Gananadhamu. 2014. Application of experimental design and response surface technique for selecting the optimum RP-HPLC conditions for the determination of moxifloxacin HCl and ketorolac tromethamine in eye drops. *Journal of Saudi Chemical Society* 21 (1): S373–S382. doi:10.1016/j.jscs.2014.04.004.

64. Singh, G. and R.S. Pai. 2014. Optimization (central composite design) and validation of HPLC method for investigation of emtricitabine loaded poly(lactic-*co*-glycolic acid) nanoparticles: In vitro drug release and in vivo pharmacokinetic studies. *The Scientific World Journal* 2014: 1–12. doi:10.1155/2014/583090.

65. Giriraj, P. and T. Sivakkumar. 2014. A rapid-chemometrics assisted RP-HPLC method with PDA detection for the simultaneous estimation of ofloxacin and nimorazole in pharmaceutical formulation. *Journal of Liquid Chromatography & Related Technologies* 38 (8): 904–910. doi:10.1080/10826076.2014.991870.

66. Dinç, E., Z.C. Ertekin, and G. Rouhani. 2015. A new RP-UPLC method for simultaneous quantitative estimation of bisoprolol hemifumarate and hydrochlorothiazide in tablets using experimental design and optimization. *Journal of Liquid Chromatography & Related Technologies* 38 (9): 970–976. doi:10.1080/10826076.2014.999200.

67. Zakrajšek, J., V. Stojić, S. Bohanec, and U. Urleb. 2015. Quality by design based optimization of a high performance liquid chromatography method for assay determination of low concentration preservatives in complex nasal formulations. *Acta Chimica Slovenica* 62 (1): 72–82. doi:10.17344/acsi.2014.718.

68. Kumar, L., M. Sreenivasa Reddy, R.S. Managuli, and G.K. Pai. 2015. Full factorial design for optimization, development and validation of HPLC method to determine valsartan in nanoparticles. *Saudi Pharmaceutical Journal* 23 (5): 549–555. doi:10.1016/j.jsps.2015.02.001.

69. Dragomiroiu, G.T.A.B., A. Cimpoieşu, O. Ginghina, C. Baloescu, M. Barca, D.E. Popa, A.M. Ciobanu, and V. Anuta. 2015. The development and validation of a rapid HPLC method for determination of piroxicam. *Farmácia* 63: 123–131.

70. Sathiyasundar, R. and K. Valliappan. 2015. Experimental design approach to optimization of the new commercial RP-HPLC discrimination conditions for the estimation of paracetamol and zaltaprofen in pharmaceutical formulation. *International Journal of Pharmaceutical Sciences and Research* 6 (1): 183–189. doi:10.13040/ijpsr.0975-8232.

71. Sahoo, D.K. and P.K. Sahu. 2015. Chemometric approach for RP-HPLC determination of azithromycin, secnidazole, and fluconazole using response surface methodology. *Journal of Liquid Chromatography & Related Technologies* 38 (6): 750–758. doi:10.1080/10826076.2014.968664.

72. Dinç, E. and Ö. Üstündağ. 2005. Application of multivariate calibration techniques to HPLC data for quantitative analysis of a binary mixture of hydrochlorathiazide and losartan in tablets. *Chromatographia* 61 (5–6): 237–244. doi:10.1365/s10337-005-0511-1.

73. Dinç, E., Ö. Üstündağ, A. Özdemir, and D. Baleanu. 2005. A new application of chemometric techniques to HPLC data for the simultaneous analysis of a two-component mixture. *Journal of Liquid Chromatography & Related Technologies* 28 (14): 2179–2194. doi:10.1081/jlc-200064041.

74. Dinç, E., A. Özdemir, H. Aksoy, Ö. Üstündağ, and D. Baleanu. 2006. Chemometric determination of naproxen sodium and pseudoephedrine hydrochloride in tablets by HPLC. *Chemical & Pharmaceutical Bulletin* 54 (4): 415–421. doi:10.1248/cpb.54.415.

75. Dinç, E., A. Ozdemir, H. Aksoy, and D. Baleanu. 2006. Chemometric approach to simultaneous chromatographic determination of paracetamol and chlorzoxazone in tablets and spiked human plasma. *Journal of Liquid Chromatography & Related Technologies* 29 (12): 1803–1822. doi:10.1080/10826070600717023.

76. Dinc, E., A. Hakan Aktas, D. Baleanu, and O. Ustundag. 2006. Simultaneous determination of tartrazine and allura red in commercial preparation by chemometric HPLC method. *Journal of Food and Drug Analysis* 14 (3): 284–291.

77. Tosun, A., Ö. Bahadır, and E. Dinç. 2007. Determination of anomalin and deltoin in *Seseli resinosum* by LC combined with chemometric methods. *Chromatographia* 66 (9–10): 677–683. doi:10.1365/s10337-007-0392-6.

78. Dinç, E., A. Bilgili, and B. Hanedan. 2007. Simultaneous determination of trimethoprim and sulphamethoxazole in veterinary formulations by chromatographic multivariate methods. *Die Pharmazie—An International Journal of Pharmaceutical Sciences* 62 (3): 179–184.

79. Küçükboyacı, N., A. Güvenç, E. Dinç, N. Adıgüzel, and B. Bani. 2010. New HPLC-chemometric approaches to the analysis of isoflavones in *Trifolium lucanicum* gasp.. *Journal of Separation Science* 33 (17–18): 2558–2567. doi:10.1002/jssc.201000273.

80. Aşçı, B., Ö.A. Dönmez, A. Bozdoğan, and S. Sungur. 2011. Simultaneous determination of paracetamol, pseudoephedrine hydrochloride, and dextromethorphan hydrobromide in tablets using multivariate calibration methods coupled with HPLC-DAD. *Journal of Liquid Chromatography & Related Technologies* 34 (16): 1686–1698. doi:10.1080/10826076.2011.578321.

81. Dinç, E. and D. Baleanu. 2012. Ultra-performance liquid chromatography for the multicomponent analysis of a ternary mixture containing thiamine, pyridoxine, and lidocaine in ampules. *Journal of AOAC International* 95 (3): 903–912. doi:10.5740/jaoacint.11-199.

82. Kamal, A.H., M.M. Mabrouk, H.M. El-Fatatry, and S.F. Hammad. 2015. Simultaneous determination of esomeprazole magnesium trihydrate and naproxen by combined HPLC-chemometric techniques and RP-HPLC method. *International Journal of Pharmacy*, 5 (1): 107–121.

83. Kendir, G., E. Dinç, and A.K. Güvenç. 2015. Ultra-performance liquid chromatography for the simultaneous quantification of rutin and chlorogenic acid in leaves of *Ribes* L. species by conventional and chemometric calibration approaches. *Journal of Chromatographic Science* 53 (9): 1577–1587. doi:10.1093/chromsci/bmv060.

84. Kendir, G., E. Dinç, and A. Köroğlu. 2016. Quantitative analysis of *Melissa officinalis* L. samples by chromatographic multivariate calibration methods. *Chromatographia* 79 (3–4): 189–198. doi:10.1007/s10337-015-3005-9.

85. Harshman, R.A. 1970. Foundations of the PARAFAC procedure: Models and conditions for an "explanatory" multimodal factor analysis. *UCLA Working Papers in Phonetics*, 16: 1–84.

86. Carroll, J.D. and J.-J. Chang. 1970. Analysis of individual differences in multidimensional scaling via an N-way generalization of "Eckart-Young" decomposition. *Psychometrika* 35 (3): 283–319. doi:10.1007/bf02310791.

87. Bro, R. 1997. PARAFAC. Tutorial and applications. *Chemometrics and Intelligent Laboratory Systems* 38 (2): 149–171. doi:10.1016/s0169-7439(97)00032-4.

88. Tucker, L.R. 1966. Some mathematical notes on three-mode factor analysis. *Psychometrika* 31 (3): 279–311. doi:10.1007/bf02289464.

89. Sanchez, E. and B.R. Kowalski. 1986. Generalized rank annihilation factor analysis. *Analytical Chemistry* 58 (2): 496–499. doi:10.1021/ac00293a054.

90. Sanchez, E., L. Scott Ramos, and B.R. Kowalski. 1987. Generalized rank annihilation method: I. Application to liquid chromatography—Diode array ultraviolet detection data. *Journal of Chromatography A* 385: 151–164. doi:10.1016/s0021-9673(01)94629-1.

91. Ramos, L.S., E. Sanchez, and B. R. Kowalski. 1987. Generalized rank annihilation method. II. Analysis of bimodal chromatographic data. *Journal of Chromatography A* 385: 165–180. doi:10.1016/s0021-9673(01)94630-8.

92. Sanchez, E. and B.R. Kowalski. 1990. Tensorial resolution: A direct trilinear decomposition. *Journal of Chemometrics* 4 (1): 29–45. doi:10.1002/cem.1180040105.

93. Tauler, R. 1995. Multivariate curve resolution applied to second order data. *Chemometrics and Intelligent Laboratory Systems* 30 (1): 133–146. doi:10.1016/0169-7439(95)00047-x.

94. De Juan, A., E. Casassas, and R. Tauler. 2002. Soft modeling of analytical data. In *Encyclopedia of Analytical Chemistry. Applications, Theory and Instrumentation*, ed. R.A. Myers, pp. 9800–9837. Chichester, U.K.: John Wiley & Sons.

95. Wu, H.-L., M. Shibukawa, and K. Oguma. 1998. An alternating trilinear decomposition algorithm with application to calibration of HPLC-DAD for simultaneous determination of overlapped chlorinated aromatic hydrocarbons. *Journal of Chemometrics* 12 (1): 1–26. doi:10.1002/(sici)1099-128x(199801/02)12:1<1::aid-cem492>3.0.co;2-4.

96. Bro, R. 1996. Multiway calibration. Multilinear PLS. *Journal of Chemometrics* 10 (1): 47–61. doi:10.1002/(sici)1099-128x(199601)10:1<47::aid-cem400>3.0.co;2-c.

97. Zissis, K.D., R.G. Brereton, S. Dunkerley, and R.E.A. Escott. 1999. Two-way, unfolded three-way and three-mode partial least squares calibration of diode array HPLC chromatograms for the quantitation of low-level pharmaceutical impurities. *Analytica Chimica Acta* 384 (1): 71–81. doi:10.1016/s0003-2670(98)00844-7.

98. Wiberg, K. and S.P. Jacobsson. 2004. Parallel factor analysis of HPLC-DAD data for binary mixtures of lidocaine and prilocaine with different levels of chromatographic separation. *Analytica Chimica Acta* 514 (2): 203–209. doi:10.1016/j.aca.2004.03.062.

99. Zhang, Y., H.-L. Wu, Y.-J. Ding, A.-L. Xia, H. Cui, and R.-Q. Yu. 2006. Simultaneous determination of cortisol and prednisolone in body fluids by using HPLC–DAD coupled with second-order calibration based on alternating trilinear decomposition. *Journal of Chromatography B* 840 (2): 116–123. doi:10.1016/j.jchromb.2006.04.043.

100. Schmidt, B., J.W. Jaroszewski, R. Bro, and M. Witt. 2008. Combining PARAFAC analysis of HPLC-PDA profiles and structural characterization using HPLC-PDA-SPE-NMR-MS experiments: Commercial preparations of St. John's Wort. *Analytical Chemistry* 80 (6): 1978–1987. doi:10.1021/ac702064p.

101. García, M.D.G., M.J. Culzoni, M.M. De Zan, R. Santiago Valverde, M. Martínez Galera, and H.C. Goicoechea. 2008. Solving matrix effects exploiting the second-order advantage in the resolution and determination of eight tetracycline antibiotics in effluent wastewater by modelling liquid chromatography data with multivariate curve resolution-alternating least squares and unfolded-partial least squares followed by residual bilinearization algorithms: II. Prediction and figures of merit. *Journal of Chromatography A* 1179 (2): 115–124. doi:10.1016/j.chroma.2007.11.049.

102. Cañada-Cañada, F., J.A. Arancibia, G.M. Escandar, G.A. Ibañez, A. Espinosa Mansilla, A. Muñoz de la Peña, and A.C. Olivieri. 2009. Second-order multivariate calibration procedures applied to high-performance liquid chromatography coupled to fast-scanning fluorescence detection for the determination of fluoroquinolones. *Journal of Chromatography A* 1216 (24): 4868–4876. doi:10.1016/j.chroma.2009.04.033.

103. Gil García, M.D. F. Cañada Cañada, M.J. Culzoni, L. Vera-Candioti, G.G. Siano, H.C. Goicoechea, and M. Martínez Galera. 2009. Chemometric tools improving the determination of anti-inflammatory and antiepileptic drugs in river and wastewater by solid-phase microextraction and liquid chromatography diode array detection. *Journal of Chromatography A* 1216 (29): 5489–5496. doi:10.1016/j.chroma.2009.05.073.

104. Martínez Galera, M., M.D. Gil García, M.J. Culzoni, and H.C. Goicoechea. 2010. Determination of pharmaceuticals in river water by column switching of large sample volumes and liquid chromatography–diode array detection, assisted by chemometrics: An integrated approach to green analytical methodologies. *Journal of Chromatography A* 1217 (13): 2042–2049. doi:10.1016/j.chroma.2010.01.082.

105. Li, S.-F., H.-L. Wu, Y.-J. Yu, Y.-N. Li, J.-F. Nie, H.-Y. Fu, and R.-Q. Yu. 2010. Quantitative analysis of levodopa, carbidopa and methyldopa in human plasma samples using HPLC-DAD combined with second-order calibration based on alternating trilinear decomposition algorithm. *Talanta* 81 (3): 805–812. doi:10.1016/j.talanta.2010.01.019.

106. Ouyang, L.Q., H.L. Wu, Y.J. Liu, J.Y. Wang, Y.J. Yu, H.Y. Zou, and R.Q. Yu. 2010. Simultaneous determination of metronidazole and tinidazole in plasma by using HPLC-DAD coupled with second-order calibration. *Chinese Chemical Letters* 21 (10): 1223–1226. doi:10.1016/j.cclet.2010.04.016.

107. Yu, Y.-J., H.-L. Wu, S.-Z. Shao, C. Kang, J. Zhao, Y. Wang, S.-H. Zhu, and R.-Q. Yu. 2011. Using second-order calibration method based on trilinear decomposition algorithms coupled with high performance liquid chromatography with diode array detector for determination of quinolones in honey samples. *Talanta* 85 (3): 1549–1559. doi:10.1016/j.talanta.2011.06.044.

108. Hu, L.-Q., C.-L. Yin, Y.-H. Du, and Z.-P. Zeng. 2012. Simultaneous and direct determination of vancomycin and cephalexin in human plasma by using HPLC-DAD coupled with second-order calibration algorithms. *Journal of Analytical Methods in Chemistry* 2012: 1–8. doi:10.1155/2012/256963.

109. Arroyo, D., M. Cruz Ortiz, and L.A. Sarabia. 2011. Optimization of the derivatization reaction and the solid-phase microextraction conditions using a D-optimal design and three-way calibration in the determination of non-steroidal anti-inflammatory drugs in bovine milk by gas chromatography–mass spectrometry. *Journal of Chromatography A* 1218 (28): 4487–4497. doi:10.1016/j.chroma.2011.05.010.

110. Grisales, J.O., J.A. Arancibia, C.B. Castells, and A.C. Olivieri. 2012. Determination of enantiomeric composition of ibuprofen in pharmaceutical formulations by partial least-squares regression of strongly overlapped chromatographic profiles. *Journal of Chromatography B* 910: 78–83. doi:10.1016/j.jchromb.2012.04.018.

111. Culzoni, M.J., R.Q. Aucelio, and G.M. Escandar. 2012. High-performance liquid chromatography with fast-scanning fluorescence detection and multivariate curve resolution for the efficient determination of galantamine and its main metabolites in serum. *Analytica Chimica Acta* 740: 27–35. doi:10.1016/j.aca.2012.06.034.

112. Liu, Y., H. Wu, S. Zhu, C. Kang, H. Xu, Z. Su, H. Gu, and R. Yu. 2012. Rapid determination of costunolide and dehydrocostuslactone in human plasma sample and Chinese patent medicine Xiang Sha Yang Wei capsule using HPLC-DAD coupled with second-order calibration. *Chinese Journal of Chemistry* 30 (5): 1137–1143. doi:10.1002/cjoc.201100677.

113. Yu, Y.-J., H.-L. Wu, H.-Y. Fu, J. Zhao, Y.-N. Li, S.-F. Li, C. Kang, and R.-Q. Yu. 2013. Chromatographic background drift correction coupled with parallel factor analysis to resolve coelution problems in three-dimensional chromatographic data: Quantification of eleven antibiotics in tap water samples by high-performance liquid chromatography coupled with a diode array detector. *Journal of Chromatography A* 1302: 72–80. doi:10.1016/j.chroma.2013.06.009.

114. Vosough, M. and H.M. Esfahani. 2013. Fast HPLC-DAD quantification procedure for selected sulfonamids, metronidazole and chloramphenicol in wastewaters using second-order calibration based on MCR-ALS. *Talanta* 113: 68–75. doi:10.1016/j.talanta.2013.03.049.

115. Chen, K., F. Lynen, L. Hitzel, M. Hanna-Brown, R. Szucs, and P. Sandra. 2013. A new strategy for fast chiral screening by combining HPLC-DAD with a multivariate curve resolution–alternating least squares algorithm. *Chromatographia* 76 (17–18): 1055–1066. doi:10.1007/s10337-013-2520-9.

116. Akvan, N. and H. Parastar. 2014. Second-order calibration for simultaneous determination of pharmaceuticals in water samples by solid-phase extraction and fast high-performance liquid chromatography with diode array detector. *Chemometrics and Intelligent Laboratory Systems* 137: 146–154. doi:10.1016/j.chemolab.2014.07.004.

117. Yin, X.-L., H.-L. Wu, H.-W. Gu, X.-H. Zhang, Y.-M. Sun, Y. Hu, L. Liu, Q.-M. Rong, and R.-Q. Yu. 2014. Chemometrics-enhanced high performance liquid chromatography-diode array detection strategy for simultaneous determination of eight co-eluted compounds in ten kinds of Chinese teas using second-order calibration method based on alternating trilinear decomposition algorithm. *Journal of Chromatography A* 1364: 151–162. doi:10.1016/j.chroma.2014.08.068.

118. Vosough, M., H. Mohamedian, A. Salemi, and T. Baheri. 2014. Multivariate curve resolution-assisted determination of pseudoephedrine and methamphetamine by HPLC-DAD in water samples. *Journal of Chromatographic Science* 53 (2): 233–239. doi:10.1093/chromsci/bmu046.

119. Padró, J.M., J. Osorio-Grisales, J.A. Arancibia, A.C. Olivieri, and C.B. Castells. 2015. Scope of partial least-squares regression applied to the enantiomeric composition determination of ketoprofen from strongly overlapped chromatographic profiles. *Journal of Separation Science* 38 (14): 2423–2430. doi:10.1002/jssc.201500217.

120. Alcaráz, M.R., S.A. Bortolato, H.C. Goicoechea, and A.C. Olivieri. 2015. A new modeling strategy for third-order fast high-performance liquid chromatographic data with fluorescence detection. Quantitation of fluoroquinolones in water samples. *Analytical and Bioanalytical Chemistry* 407 (7): 1999–2011. doi:10.1007/s00216-014-8442-z.

121. Gu, H.-W., H.-L. Wu, X.-L. Yin, Y. Li, Y.-J. Liu, H. Xia, S.-R. Zhang et al. 2014. Multi-targeted interference-free determination of ten β-blockers in human urine and plasma samples by alternating trilinear decomposition algorithm-assisted liquid chromatography–mass spectrometry in full scan mode: Comparison with multiple reaction monitoring. *Analytica Chimica Acta* 848: 10–24. doi:10.1016/j.aca.2014.08.052.

122. Dinç, E. and Z.C. Ertekin. 2016. Three-way analysis of the UPLC–PDA dataset for the multicomponent quantitation of hydrochlorothiazide and olmesartan medoxomil in tablets by parallel factor analysis and three-way partial least squares. *Talanta* 148: 144–152. doi:10.1016/j.talanta.2015.10.074.

123. Dinç, E. and E. Büker. 2016. Parallel factor analysis and trilinear partial least squares applied to the UPLC-PDA data array for the quantification of brimonidine tartrate and timolol maleate in an eye drop formulation. *Journal of Liquid Chromatography & Related Technologies* 39 (7): 374–383. doi:10.1080/10826076.2016.1174941.

124. Dinç, E., Z.C. Ertekin, and E. Büker. 2016. Two-way and three-way approaches to ultra-performance liquid chromatography-photodiode array dataset for the quantitative resolution of a two-component mixture containing ciprofloxacin and ornidazole. *Journal of Separation Science* doi:10.1002/jssc.201600382.

125. Brereton, R.G. 2003. *Chemometrics: Data Analysis for the Laboratory and Chemical Plant.* Chichester, U.K.: Wiley.

126. Savitzky, A. and M.J.E. Golay. 1964. Smoothing and differentiation of data by simplified least squares procedures. *Analytical Chemistry* 36 (8): 1627–1639. doi:10.1021/ac60214a047.

127. Geladi, P., D. MacDougall, and H. Martens. 1985. Linearization and scatter-correction for near-infrared reflectance spectra of meat. *Applied Spectroscopy* 39 (3): 491–500. doi:10.1366/0003702854248656.

128. Trygg, J. and S. Wold. 2003. O2-PLS, a two-block (X-Y) latent variable regression (LVR) method with an integral OSC filter. *Journal of Chemometrics* 17 (1): 53–64. doi:10.1002/cem.775.

129. Daszykowski, M. and B. Walczak. 2006. Use and abuse of chemometrics in chromatography. *Trends in Analytical Chemistry* 25 (11): 1081–1096. doi:10.1016/j.trac.2006.09.001.

130. Xu, C.-J., Y.-Z. Liang, F.-T. Chau, and Y.V. Heyden. 2006. Pretreatments of chromatographic fingerprints for quality control of herbal medicines. *Journal of Chromatography A* 1134 (1–2): 253–259. doi:10.1016/j.chroma.2006.08.060.

131. Zhang, Y., H.-L. Wu, A.-L. Xia, L.-H. Hu, H.-F. Zou, and R.-Q. Yu. 2007. Trilinear decomposition method applied to removal of three-dimensional background drift in comprehensive two-dimensional separation data. *Journal of Chromatography A* 1167 (2): 178–183. doi:10.1016/j.chroma.2007.08.055.

132. Fredriksson, M., P. Petersson, M. Jörntén-Karlsson, B.-O. Axelsson, and D. Bylund. 2007. An objective comparison of pre-processing methods for enhancement of liquid chromatography–mass spectrometry data. *Journal of Chromatography A* 1172 (2): 135–150. doi:10.1016/j.chroma.2007.09.077.

133. De Zan, M.M., M.D. Gil García, M.J. Culzoni, R.G. Siano, H.C. Goicoechea, and M. Martínez Galera. 2008. Solving matrix-effects exploiting the second order advantage in the resolution and determination of eight tetracycline antibiotics in effluent wastewater by modelling liquid chromatography data with multivariate curve resolution-alternating least squares and unfolded-partial least squares followed by residual bilinearization algorithms. *Journal of Chromatography A* 1179 (2): 106–114. doi:10.1016/j.chroma.2007.11.091.

134. Yu, Y.-J., Q.-L. Xia, S. Wang, B. Wang, F.-W. Xie, X.-B. Zhang, Y.-M. Ma, and H.-L. Wu. 2014. Chemometric strategy for automatic chromatographic peak detection and background drift correction in chromatographic data. *Journal of Chromatography A* 1359: 262–270. doi:10.1016/j.chroma.2014.07.053.

135. Daubechies, I. 1992. *Ten Lectures on Wavelets*. Philadelphia, PA: Society for Industrial and Applied Mathematics.

136. Walczak, B. 2000. *Wavelets in Chemistry*. Amsterdam, the Netherlands: Elsevier Science B.V.

137. Dinç, E. and D. Baleanu. 2007. A review on the wavelet transform applications in analytical chemistry. In *Mathematical Methods in Engineering*, eds. K. Taş, J.A. Tenreiro Machado, and D. Baleanu, pp. 265–284. Dordrecht, the Netherlands: Springer.

138. Shao, X., W. Cai, P. Sun, M. Zhang, and G. Zhao. 1997. Quantitative determination of the components in overlapping chromatographic peaks using wavelet transform. *Analytical Chemistry* 69 (9): 1722–1725. doi:10.1021/ac9608679.

139. Shao, X., W. Cai, and P. Sun. 1998. Determination of the component number in overlapping multicomponent chromatogram using wavelet transform. *Chemometrics and Intelligent Laboratory Systems* 43 (1–2): 147–155. doi:10.1016/s0169-7439(98)00066-5.

140. Dinç, E. and E. Büker. 2012. A new application of continuous wavelet transform to overlapping chromatograms for the quantitative analysis of amiloride hydrochloride and hydrochlorothiazide in tablets by ultra-performance liquid chromatography. *Journal of AOAC International* 95 (3): 751–756. doi:10.5740/jaoacint.sge_dinc.

141. Hu, Y., G. Zhou, J. Kang, Y. Du, F. Huang, and J. Ge. 1996. Assessment of chromatographic peak purity by means of artificial neural networks. *Journal of Chromatography A* 734 (2): 259–270. doi:10.1016/0021-9673(95)01303-2.

142. De Braekeleer, K., A. de Juan, and D.L. Massart. 1999. Purity assessment and resolution of tetracycline hydrochloride samples analysed using high-performance liquid chromatography with diode array detection. *Journal of Chromatography A* 832 (1–2): 67–86. doi:10.1016/s0021-9673(98)00985-6.

143. Wiberg, K., M. Andersson, A. Hagman, and S.P. Jacobsson. 2004. Peak purity determination with principal component analysis of high-performance liquid chromatography–diode array detection data. *Journal of Chromatography A* 1029 (1–2): 13–20. doi:10.1016/j.chroma.2003.12.052.

144. van Zomeren, P.V., H.J. Metting, P.M.J. Coenegracht, and G.J. de Jong. 2005. Simultaneous resolution of overlapping peaks in high-performance liquid chromatography and micellar electrokinetic chromatography with diode array detection using augmented iterative target transformation factor analysis. *Journal of Chromatography A* 1096 (1–2): 165–176. doi:10.1016/j.chroma.2005.08.047.

145. Wiberg, K. 2006. Quantitative impurity profiling by principal component analysis of high-performance liquid chromatography–diode array detection data. *Journal of Chromatography A* 1108 (1): 50–67. doi:10.1016/j.chroma.2005.12.077.

146. Detroyer, A., V. Schoonjans, F. Questier, Y. Vander Heyden, A.P. Borosy, Q. Guo, and D.L. Massart. 2000. Exploratory chemometric analysis of the classification of pharmaceutical substances based on chromatographic data. *Journal of Chromatography A* 897 (1–2): 23–36. doi:10.1016/s0021-9673(00)00803-7.

147. Yan, F. 2001. Analysis of *Cnidium monnieri* fruits in different regions of China. *Talanta* 53 (6): 1155–1162. doi:10.1016/s0039-9140(00)00594-4.

148. Ma, H.-L., M.-J. Qin, L.-W. Qi, G. Wu, and P. Shu. 2007. Improved quality evaluation of *Radix Salvia miltiorrhiza* through simultaneous quantification of seven major active components by high-performance liquid chromatography and principal component analysis. *Biomedical Chromatography* 21 (9): 931–939. doi:10.1002/bmc.836.

149. Chen, Y., S.-B. Zhu, M.-Y. Xie, S.-P. Nie, W. Liu, C. Li, X.-F. Gong, and Y.-X. Wang. 2008. Quality control and original discrimination of *Ganoderma lucidum* based on high-performance liquid chromatographic fingerprints and combined chemometrics methods. *Analytica Chimica Acta* 623 (2): 146–156. doi:10.1016/j.aca.2008.06.018.

150. Ni, Y., Y. Peng, and S. Kokot. 2008. Fingerprinting of complex mixtures with the use of high performance liquid chromatography, inductively coupled plasma atomic emission spectroscopy and chemometrics. *Analytica Chimica Acta* 616 (1): 19–27. doi:10.1016/j.aca.2008.04.015.

151. Peng, L., Y. Wang, H. Zhu, and Q. Chen. 2011. Fingerprint profile of active components for Artemisia selengensis Turcz by HPLC–PAD combined with chemometrics. *Food Chemistry* 125 (3): 1064–1071. doi:10.1016/j.foodchem.2010.09.079.

152. Deconinck, E., P.Y. Sacré, P. Courselle, and J.O. De Beer. 2012. Chemometrics and chromatographic fingerprints to discriminate and classify counterfeit medicines containing PDE-5 inhibitors. *Talanta* 100: 123–133. doi:10.1016/j.talanta.2012.08.029.

Section VII

Miscellaneous

22 Bayesian Methods in Chromatographic Science

David Brynn Hibbert

CONTENTS

22.1   INTRODUCTION

Thomas Bayes (1702–1761) has given his name to a method of inferential probability that is most suited to the modern scientific method. Starting with a hypothesis (H), the likelihood of data obtained from experiment is used to update the initial probability (prior) to give a new probability (posterior). Thus, P(H|data) ∝ P(data|H) × P(H). This simple yet very powerful equation (derived by Laplace, not in fact Bayes) can be applied to all scientific conclusions based on experimental data.

Chromatography, and all its hyphenated variants, produces data that can be treated by Bayesian methodology. This chapter reviews Bayes Theorem* and illustrates different applications of it through examples from chromatography.

22.2   HISTORY OF BAYES THEOREM

A most readable general history of Bayes Theorem is written by Sharon McGrayne.[1] Known as "inverse probability" for many years, the approach that considers data as given and parameters random variables was only called after Thomas Bayes in the early years of the twentieth century—and then often in derogatory terms.[2] It could be said that the availability of fast computers allowing numerical solutions (often by Markov Chain Monte Carlo, MCMC) has given a boost to solving the complex mathematics that arise in Bayesian calculations. There has also been an acceptance of subjective priors as a valid approach, although empirical Bayes (using the data to establish priors) is also popular. The description below is a very brief introduction to the Bayesian methodology. Useful introductory books are References 3 and 4, and in chemistry, the reviews by the author and Armstrong[5,6] give a more detailed description of the different formalisms and examples in chemistry.

22.2.1   Use in Chemistry and Chromatography

Bayes Theorem is a statistical tool. It was not made for chemistry *per se*, and certainly not for chromatography. However, the Bayesian paradigm is a powerful approach that can be applied in classification (pattern recognition), model selection, and for parameter estimation. Chromatography tends to produce multivariate data (detector response by time), which can also be multidimensional (when coupled with mass spectrometry, or multiwavelength spectroscopy). Therefore, these data have distributions and can be modeled in a Bayesian way. Parameters such as numbers of peaks, peak positions, and peak broadening can be estimated by Bayes Theorem. There are many uses in general classification where the data just happens to be from a detector sitting at the end of a chromatographic column. These will also be treated in this chapter, although less comprehensively. Once the methodology is established, it can be applied to any chromatographic data. This review will concentrate on chemical applications, although some reference is made to the larger fields of medicine and bioinformatics.

22.2.2   Theory (Without Too Many Equations)

22.2.2.1   Bayes for a Hypothesis with Evidence

Bayes Theorem unites two basic rules of probability, the product rule and the addition rule, and tells us how to treat conditional probabilities. In its simplest form (see Reference 5 for the derivation), Bayes Theorem for values of a hypothesis (H, the white powder contains cocaine) in the light of some evidence E (say chromatographic data of a white powder) is

$$p(H|E,I) = \frac{p(E|H,I) \times p(H|I)}{p(E|I)}$$

(22.1)

---

* Following rules of grammar this should be written Bayes' Theorem, but the possessive apostrophe is often dropped. Bayes Theorem is also known as Bayes Law, or Bayes Rule.

The terminology used here is $p(A|B)$ that means "the probability of A given B." $p(E|H)$ is also called the "likelihood" of the evidence given the truth of the hypothesis (likelihood has a more restricted meaning in statistics, but the term is widely used to cover any measure of the probability of the evidence under the hypothesis). The appearance of "I" for information in each term is to remind us that there is often background knowledge about a problem that will influence the way we treat it. For example, parameters in models are often constrained to be positive and nonzero. The appearance of the same "I" in Equation 22.1 also implies that this information does not change as we go from the prior probability to the posterior probability. However, in the body of this chapter, "I" will be dropped from general equations for clarity. The denominator in Equation 22.1 is often seen as a normalizing factor, leaving the simple relation that the probability of H given some data is proportional to the likelihood of the data if the hypothesis is true multiplied by the prior probability of the hypothesis being true:

$$p(H|E,I) \propto p(E|H,I) \times p(H,I) \tag{22.2}$$

Where there is just a single hypothesis, which is either true (H) or it is not ($\bar{H}$), Equation 22.1 becomes

$$p(H|E,I) = \frac{p(E|H,I) \times p(H|I)}{p(E|H,I) \times p(H|I) + p(E|\bar{H},I) \times p(\bar{H}|I)} \tag{22.3}$$

and Equation 22.3 can be manipulated further knowing $p(H) + p(\bar{H}) = 1$. When the evidence E is a test result that can only be positive (E ≡ T = P) or negative (E ≡ T = N), the probabilities in Equation 22.3 can be identified with false positive and negative probabilities in terms of a contingency table (Table 22.1).

Bayes Theorem is popular and useful because it gives us what we want to know (the probability of our hypothesis) rather than requiring us to nominate a hypothesis (e.g., the null hypothesis of significance testing) and then to determine the probability of the data, given the assumption of truth of the hypothesis. It is considered more natural that Bayes Theorem is applied to parameters having probability distributions that are determined from fixed experimental data than the approach that treats data as having a probability under a given set of parameters.

Finally, to offer another view of Bayes Theorem, Bayes tells us how to update our knowledge on the receipt of some new information (evidence, data), that is, to go from a prior probability $p(H)$ before we have done our experiment, to a posterior probability given the data obtained from the experiment $p(H|E)$—by multiplying by the likelihood and normalizing by the probability of the evidence.

TABLE 22.1

Contingency Table for the Outcomes of Testing with Binary (Positive/Negative) Result

		Hypothesis Is in Reality	
		True (H)	False ($\bar{H}$)
Test result	Positive (P)	$p(T=P|H,I)$ true positive probability	$p(T=P|\bar{H},I)$ false positive probability
	Negative (N)	$p(T=N|H,I)$ false negative probability	$p(T=N|\bar{H},I)$ true negative probability

22.2.2.2 Odds and Likelihood Ratios

Equation 22.1 for the positive and negative hypotheses can be written as

$$\frac{p(H|E,I)}{p(\bar{H}|E,I)} = \frac{p(E|H,I)}{p(E|\bar{H},I)} \times \frac{p(H|I)}{p(\bar{H}|I)}$$

or   (22.4)

$$O(H|E,I) = LR \times O(H|I)$$

where the ratio of probabilities of H and $\bar{H}$ is known as the odds of H, prior odds before the evidence is considered, and posterior odds afterward. The odds are updated by the likelihood ratio (*LR*). Probability is related to odds by

$$p = \frac{O}{1+O}$$ (22.5)

For example, even odds ($O = 1$) becomes a probability of H of 0.5.

22.2.2.3 Parameter Estimation

Bayes Theorem can also be written in terms of probability density functions (pdf), and this is seen when discussing continuous parameters that can be described by a distribution

$$\pi(\theta|x,I) = \frac{f(x|\theta,I) \times \pi(\theta|I)}{f(x|I)}$$ (22.6)

where
   θ is a parameter of interest (e.g., a peak retention time)
   x is data (a digitized chromatogram)
   $\pi(\cdot)$ is the density function that may or may not be Gaussian
   $f(x|\theta,I)$ is the likelihood of x given θ and I
   $f(x|I)$ is the marginal density of the data, which as discussed before is usually a normalizing factor

The use of Bayes Theorem to estimate parameter values is possible because the set of parameter values that maximize their posterior probability are, *ipso facto*, the best estimates. Consider the simple case of n independent measurements of mass concentration from which an estimate of the true value with a standard uncertainty is required. It is possible to show[5] that if the distributions of data and parameters are Gaussian, and there is no information about the prior probabilities, Bayes Theorem gives the usual result, namely, the estimate of μ is the average of the data and the standard deviation is the sample standard deviation of the mean, following a distribution close to the Student-t distribution for small n and tending to Gaussian for larger n. However, the advantage of Bayes Theorem is that prior knowledge about the problem is naturally incorporated. For example, mass concentration cannot be negative, and we may know an upper bound on the quantity. The prior distribution may also be known. If not, we take a uniform distribution between lower and upper bounds for the value and a scale invariant uniform prior for the standard deviation (Jeffrey's prior).

   Apart from the simplest cases, numerical methods are required to obtain values and credibility intervals.

   Having the distribution of estimated parameters allows comparison of credibility intervals (the equivalent of confidence intervals of frequentist approaches) leading to probabilities obtained directly that two sets of data, for example, come from a common population. This is the Bayesian equivalent of significance testing, but having the distribution of the parameters the comparison is, in the eyes of many statisticians, a more natural process.

22.2.2.4 Markov Chain Monte Carlo

MCMC methods are particularly important in Bayesian analysis for numerically sampling the posterior pdf and determining posterior probabilities. MCMC methods draw samples from the posterior pdf by performing a random walk in the parameter space of the model.[7] The length of a step in the random walk is proportional to the posterior pdf. A common approach uses the Gibbs sampler, a special case of the Metropolis–Hastings algorithm.[8]

In general, the raw MCMC data is correlated and requires resampling at the correlation length scale or lag length, which is determined from autocorrelation function and the sampling efficiency of the MCMC process. Descriptive statistics (e.g., mean, median, and standard deviation) can be obtained from the resampled data. The resampled data can also be binned to produce a histogram of the data.

So Bayes Theorem is a powerful statistical method that can be used to obtain chromatographic parameters and then help with interpretation of the results of chromatography experiments. The remainder of the chapter will illustrate the use of Bayes Theorem in chromatography.

22.3 CHROMATOGRAPHY PROBLEMS TREATED BY BAYES

A Bayesian method of obtaining chromatographic peak parameters was described as long ago as 1971,[9] using the maximum of the posterior probability of parameters of the Fourier transform of a peak. Since then, the problem of locating and describing peaks has been revisited on occasion.

22.3.1 Number of Peaks in a Chromatogram

One of the fundamental tasks of a chromatographer is to decide where her peaks are located in a chromatogram. Easy for small numbers of not-overlapping peaks, but for complex mixtures of similar molecules (e.g., petroleum, metabolites), how many peaks, and where they are becomes a problem that can be helped by mathematics. Algorithmic methods of peak detection usually rely on changes in the derivatives of the detector response, or by the application of a filter to obtain peaks that are determined to be above a given threshold. Lopatka et al. argue that the traditional methods yield a binary response to the existence of a peak, it is either there or not, and this response depends on tunable parameters. In their paper of 2014, a probabilistic method is described that may be classed as Bayesian and illustrates nicely possible advantages of the Bayesian approach.[10]

The data are a series of discrete detector responses with time. With knowledge of the normalized peak width (σ_{peak}), a window is described containing $2n + 1$ points such that n is about $4\sigma_{peak}$. Each point in the chromatogram is considered to be influenced by a peak (hypothesis i, H_i*) or not (hypothesis $\bar{H}_i$), where i is a running index. The response is modeled for Gaussian peaks covering a window centered on k as

$$y_k = a + bk + c_{i-n}g_{i-n}(k)\cdots c_{i+n}g_{i+n}(k) + \varepsilon_k \tag{22.7}$$

where $a + bk$ is a linear term to account for baseline drift and

$$g_j(k) = \begin{cases} \dfrac{1}{\sigma_{peak}\sqrt{2\pi}}\exp\left(-\dfrac{(k-j)^2}{2\sigma_{peak}^2}\right), & \text{if there is a peak centered at } j \\ 0, & \text{otherwise} \end{cases} \tag{22.8}$$

---

* Throughout this chapter, symbols have been changed from those in the papers referenced to achieve some consistency across all examples.

Every possible combination of a given number of peaks (about 3) is considered in a window of $2n + 1$. The matrix of these possibilities is denoted V_m.

The prior probability for a configuration of t coeluting compounds is

$$p_t = \alpha \exp(-2\alpha)(1 - \exp(-\alpha))^{t-1} \tag{22.9}$$

where α is a saturation parameter (= number of compounds/peak capacity). Responses are calculated under the competing hypotheses and residuals calculated. As all possibilities are calculated, Bayes Theorem can be written in the discrete form

$$p(H_i|x) = \frac{p(x|H_i)p(H_i)}{p(x|H_i)p(H_i) + P(x|\bar{H}_i)p(\bar{H}_i)} \tag{22.10}$$

The likelihood is given in terms of V_m

$$p(x|H_i) = \sum_{V_m \in H_i} p(x|V_m) \times p(V_m|H_i) \tag{22.11}$$

and similarly for H_i. As H_i and $\bar{H}_i$ are mutually exclusive $\bar{H}_i = 1 - H_i$. The result of these calculations is the probability of a peak centered on every point in a chromatogram. By choosing a suitably high probability threshold, peaks can be identified. Assignments appear to tolerate noise and coelution, although a conservatively low saturation parameter is recommended.

Recently, the use of comprehensive two-dimensional chromatography coupled with mass spectrometry detection has allowed further refinement.[11] The proposed normal-exponential-Bernoulli model performs simultaneous baseline correction and removal of noise from total ion chromatograms (TICs), followed by finding candidate peak regions using a statistical test based on conditional Bayes factors. Peaks are then picked by fitting experimental data with probabilistic mixture models. The conditional Bayes factor is expressed as a posterior odds ratio for a response above baseline. For the ith TIC with data x_i and true signal Θ_i $x_i \sim N(\Theta_i + \mu, \sigma^2)$ and $\Theta_i \sim \exp(\phi)$ for a peak and $x_i \sim N(\mu, \sigma^2)$, $\Theta_i = 0$ for the baseline. The authors use expectation maximization to obtain the parameters μ, σ, ϕ, and r. The odds ratio is

$$w_i = \frac{p_1(x_i)}{p_0(x_i)} \times \frac{\hat{r}}{1 - \hat{r}} \tag{22.12}$$

where
r is the fraction of TICs for which $\Theta_i \neq 0$
p_0 is the normal probability distribution function

$$p_1(x_i) = \frac{1}{\phi} \exp\left(\frac{\sigma^2}{2\phi^2} - \frac{x_i - \mu}{\phi}\right) \Phi\left(\frac{x_i - \mu - \sigma^2/\phi}{\sigma}\right) \tag{22.13}$$

where $\Phi(\cdot)$ is the cumulative distribution function of the standard normal distribution. The priors are unity. An odds ratio of >10 is taken to indicate the presence of a peak.

Befekadu et al.[12] consider the related problem of peak alignment given LC-MS data (retention time and total ion current) using a Bayesian hierarchical model. A fully Bayesian mixed-effects model is reported that effectively accounts for population homogeneous behavior across biological groups (i.e., fixed systematic changes) and for heterogeneity within groups (random effects).

Bayesian inference of unknown parameters is carried out via MCMC using the Gibbs sampling technique with conjugate priors.

22.3.2 BAND BROADENING IN CHROMATOGRAPHY

Armstrong[13] has shown a nice example of using Bayes Theorem to choose the most appropriate model for describing band broadening in chromatography. We, of course, know that chromatography peaks are not perfect tall thin rectangles, but often neither are they perfect Gaussians/Lorentzian shapes. The parameter in question is H, the height of a theoretical plate (or N, the number of theoretical plates, $= L/H$), which is obtained from the length of the column (L), retention time (t_R), and width of the peak at baseline (W): $H = LW^2/16t_R^2$. In 1956, Van Deempter described the variation of H with the velocity of the mobile phase u, which passes through a minimum at the optimum flow for that peak[14]:

$$\hat{H} = a_1/u + a_2u + a_3u^2 \tag{22.14}$$

In the equations in this section, $a_1, \ldots, a_n$ are constants of the model. Following the Van Deempter equation, some seven more models were assessed by Armstrong, each having up to five constants. The point of the study was to show how Bayesian analysis could identify the generating model with up to 6% added noise more reliably than simple statistical measures and other information-based approaches (e.g., Akaike and Schwartz Information Criteria).

Starting with the discrete model ($\hat{H}$, with j parameters) for $i = 1, 2, 3, \ldots, M$ data points representing the response of the chromatographic detector with time

$$H\left(u_i\right) = \hat{H}\left(u_i; a_{1\ldots j}\right) + \varepsilon\left(u_i\right) \tag{22.15}$$

It is assumed each point is affected by Gaussian noise with mean zero and standard deviation σ proportional to H ($\sigma_i = \beta H_i$). Bayes Theorem for the parameters $\mathbf{a} = a_{1\ldots j}$ and β given experimental data $\mathbf{H} = H_{1\ldots i}$, and model

$$p\left(\mathbf{a}, \beta | \mathbf{H}, M_q, I\right) = \frac{p\left(\mathbf{H} | \mathbf{a}, \beta, M_q, I\right) p\left(\beta | \mathbf{H}, I\right) p\left(\mathbf{a} | M_q, I\right)}{p\left(\mathbf{H} | M_q, I\right)} \tag{22.16}$$

$p(\beta | \mathbf{H}, I)p(\mathbf{a} | M_q, I)$ are the independent prior probabilities for the noise parameter and the model parameters, and are taken as uniform. "I" is background information of the experimental conditions and ensures that all models are being tested under the same conditions. The likelihood of $\mathbf{H}$ given a set of parameters $\mathbf{a}$, β, and qth model Mq is $p(\mathbf{H} | \mathbf{a}, \beta, M_q, I)$. The denominator of (22.16) is known as the marginalized likelihood or "evidence." It is the likelihood of $\mathbf{H}$ given the model and is evaluated by integrating out the systematic influence of the parameters over suitable spaces $\int_\Delta d\beta \int_A D\mathbf{a}\, p(\mathbf{a}, \beta, \mathbf{H} | M_q, I)$. Once the parameters (the reader is directed to the paper[13] for details of the mathematics) are obtained for each model, the models are compared by the discrete Bayes Theorem applied to the N models

$$p\left(M_q | \mathbf{H}, I\right) = \frac{p\left(\mathbf{H} | M_q, I\right) p\left(M_q | I\right)}{\sum_{q=1}^{q=N} p\left(\mathbf{H} | M_q, I\right) p\left(M_q | I\right)} \tag{22.17}$$

The prior for the models can be taken as uniform $p(M_q | I) = 1/N$ in that before any experiments no model is favored over any other.

On real data for propylparaben of Kirkup[15], the model

$$\hat{H} = a_1/u + a_2 e^{a_3 u} \tag{22.18}$$

was clearly the best (by orders of magnitude of p) by the full Bayesian treatment, whereas comparing weighted regression sums of squares preferred a five-parameter model with (22.18) coming second. It was noted that different approximate Bayesian approaches also did not choose the best model.

22.4 CLASSIFICATION

22.4.1 Bayes' Classifiers Explained

If we have $j = 1\ldots K$ sets of data $\mathbf{X}_j$ that are known to belong to classes y_j, then for some new data $\mathbf{X}_u$ the Bayes classification is

$$C_{\text{Bayes}}(x) = \underset{j \in \{1,2,\ldots,K\}}{\arg\max} \, p(y_j | \mathbf{X}_u) \tag{22.19}$$

which, put simply, is the classification that minimizes the probability of misclassification. The problem, of course, is finding the right form of the distributions.

22.4.1.1 Naïve Bayes Classifier

The most used Bayes classifier is the "naïve" (also "simple" or "independence") Bayes classifier, which covers different approaches, some of which are not strictly Bayes (relying on maximum likelihood methods), but all assume independence between features. An advantage of naïve Bayes is that it can estimate the parameters necessary for classification from minimum training data. Applying the assumption that the $x_{1\ldots N}$ are independent

$$\hat{y} = \underset{k \in \{1,2,\ldots,K\}}{\arg\max} \, p(y_k) \prod_{i=1}^{i=N_k} p(x_i | y_k) \tag{22.20}$$

The class priors can be equal ($=1/K$) or can be weighted by the number of data describing each class $\left(= N_k \Big/ \sum_{r=1}^{r=K} N_r \right)$. Typical distributions are multinomial or Bernoulli.

22.4.1.2 Bayesian Interpretation of the Lasso Method

A method of regression that chooses variables from a large number of candidates and has become popular recently is called the Lasso method. It has been used for biomarker identification,[16] and seems to improve on traditional methods, such as stepwise regression and ridge regression. The Bayesian element in this method, which constrains the sum of all coefficients in the model to a maximum value, is that the method can be shown to be Bayesian with a Laplacian prior. The probability density function of the Laplace distribution is $f(x|\mu,b) = \dfrac{1}{2b} \exp\left(-\dfrac{|x-\mu|}{b} \right)$, which peaks at $x = 0$, falling away exponentially on either side.

The Bayesian Lasso allows for the full posterior of the model coefficients to be explored rather than just a point estimate and so can give more instructive information regarding variable selection.[17] The Lasso minimizes $\left\{ \dfrac{1}{N} \sum_{i=1}^{i=N} (y_i - \beta_0 - \beta \mathbf{X}_i)^2 \right\}$ subject to $\sum_{j=1}^{j=p} |\beta_j| \le t$, for data $y_{1\ldots N}$ described by a model in p variables $x_{1\ldots N, 1\ldots p}$. The coefficients of the model $\boldsymbol{\beta}$ are constrained to sum to a

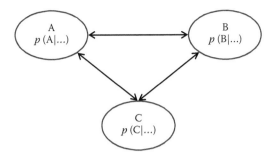

FIGURE 22.1   A simple Bayesian network with mutual implications.

maximum of t, itself a parameter of the problem. During the minimization, values of β can be forced to zero, thus creating a simpler model.

22.4.1.3   Bayesian Networks

A Bayesian network is not a classifier *per se*, but as decision trees are used as classifiers and a Bayesian network is typified by probability densities at nodes, we shall briefly describe them. Suppose two peaks in a mass spectrum (A, B) are associated with a particular compound C. They mutually imply each other (if compound C is present then we expect A and B, if we see B then this lends support for the presence of C, more so if A is also seen) (Figure 22.1).

The joint probability of finding A, B, and C given some data is written $p(A,B,C|Data)$. This can be decomposed to give

$$p(A,B,C|Data) = p(C|A,B,Data)\,p(A|B,Data)\,p(B|Data) \tag{22.21}$$

The network can be interrogated with questions such as "what is the probability of C if we observe A and don't know about B." This is, dropping "|Data" in each probability expression for clarity

$$p(C|A) = \frac{p(C,A)}{p(A)} = \frac{p(C,A,B) + p(C,A,\bar{B})}{p(A,C,B) + p(A,\bar{C},B) + p(A,C,\bar{B}) + p(A,\bar{C},\bar{B})} \tag{22.22}$$

where $\bar{B}$ is "not B" (i.e., B is not true). Bayesian networks satisfy the local Markov property, namely, each variable is conditionally independent of its non-descendants given its parent variables.

22.4.2   Bayesian Models for Oil Spills with Chromatographic Data

The problem of determining the origin of an oil spill is made hard by the weathering that inevitably occurs as soon as the oil enters the environment. Using artificially weathered samples for comparison can facilitate identification, but the process of providing sufficiently similar conditions (light, bacterial action, temperature) can be difficult and will certainly be time consuming. However, as long as an oil is not converted to another oil during weathering, classification techniques can be used to discriminate among them. An early example was by Clark and Jurs[18] using 13 GC-peak areas of n-C_{16} to n-C_{25} plus pristines, phytanes, and unresolved background to discriminate among two crude oils, two mixtures, and then weathered samples. The data were 80 chromatograms with 10 from each class.

In a recent paper in which the oil classification problem is posed from a forensic perspective, well suited to a Bayesian approach, Blomstedt et al.[19] compared Bayesian and frequentist (t-test) approaches. They assumed that peak height ratios followed a normal relative standard deviation model characterized by mean µ and standard deviation $c\mu$, where c was a predetermined constant. The Bayes factor B in favor of the hypothesis H that two samples $(\mathbf{x}, \mathbf{y})$ are drawn from populations having the same µ, with the alternative hypothesis that they are from populations with different µ, is given by

$$B = \frac{p(\mathbf{x},\mathbf{y}|H)}{p(\mathbf{x},\mathbf{y}|\bar{H})} \tag{22.23}$$

where

$$p(\mathbf{x},\mathbf{y}|H) = \int_{\Theta} f(\mathbf{x},\mathbf{y}|\mu)\pi(\mu)d\mu \tag{22.24}$$

$$p(\mathbf{x},\mathbf{y}|\bar{H}) = \int_{\Theta} f(\mathbf{x}|\mu_1)\pi(\mu_1)d\mu_1 \int_{\Theta} f(\mathbf{y}|\mu_2)\pi(\mu_2)d\mu_2 \tag{22.25}$$

where
   Θ is the parameter space
   $\pi(\cdot)$ is an inverse Gamma function representing the prior

22.4.3 FORENSIC APPLICATIONS

Perhaps because the forensic world has embraced the notion of the probability of guilt or innocence of an accused person, and Bayes has found his way into analysis of DNA data, we have many examples of the use of Bayes Theorem applied to chromatographic data for legal purposes. LRs are often given in evidence in the form of "It is 100 times more likely to have found the shoe print if it came from the defendant's shoes, than if it came from a shoe worn by a random person," that is, $LR = p(E|H)/p(E|\bar{H}) = 100$, where the evidence E is the shoe print found at the crime scene and a database of shoe prints, and H = "the defendant's shoe made the print." Forensic experts giving evidence are trained not to succumb to the "Prosecutors Fallacy"[20] and say "It is 100 times more likely that the shoe print came from the defendant than someone else" (i.e., $p(H|E)/p(\bar{H}|E) = 100$).

22.4.3.1 Accelerants in Fire Residues

Williams et al.[21] show how chromatographic data, in the form of total ion spectra (TIS) from GC-MS, can be classified, even in the presence of confounding data. The system was fire residues, used with target factor analysis (TFA). A library of TIS from reference ignitable liquids with assigned ASTM classification was used as the target factors in TFA. A Bayesian decision rule was used, expressed in terms of correlation coefficients between the TIS (or TFA vectors) of a test sample and the M reference classes.

$$p(y_i|r) = \frac{p(r|y_i)p(y_i)}{\sum_{j=1}^{j=M} p(r|y_j)p(y_j)} \tag{22.26}$$

Uniform priors on the classes y_i were used, and the kernel-approximated distribution of the correlation coefficients (r) was

$$p(r|y_i) = \frac{1}{n_i} \sum_{j=1}^{j=n_i} \frac{1}{\sqrt{2\pi h_i^2}} \exp\left(-\frac{(r-r_i)^2}{2h_i^2}\right) \tag{22.27}$$

where

  y_i is the ith class
  h_i is an adjustable bandwidth
  n_i is the number of samples in each reference class

In this study, classification is assessed by identifying the class with the highest probability of having correlations that exceed a specified lower limit, $r > r_{LL}$. Expressed in terms of the integrated area of the kernel distribution between -1 and $r_{LL}(I_i)$

$$p(y_i | I_i) = \frac{(1-I_i)p(y_i)}{\sum_{j=1}^{j=M}(1-I_i)p(y_j)} \tag{22.28}$$

This stratagem helps with the problem that $p(y_i | r)$ does not necessarily go to 0 at the boundary $r = 1$, because of the finite bandwidth h_i. The procedure is called "soft" Bayesian classification, because the probabilities of all classes are considered, not just the greatest when assigning membership. In the case of fire debris, it is quite possible that multiple accelerants were used and so two classes will have mutually high probabilities compared with other candidates.

22.4.3.2 Types of Cannabis

A straightforward application of Bayes Theorem is made by Bossa et al.[22] to carry out a two-state classification using GC-MS analysis of cannabis seedlings to determine whether they are the drug-producing strains or those for legal production of hemp fiber. Normalized and standardized peak areas, each sample measured three times, of cannabidiol, tetrahydrocannabinol, and cannabinol, from 132 drug plants and 158 legal plants, made the data on which the classification was built. The Bayes factor, expressed as the ratio of the posterior odds and the prior odds, can be shown simply to be the LR of the two hypotheses. Here the Bayes factor ($B_{1,0}$ where "1" = classification as drug H_{drug}, "0" = classification as fiber H_{fiber}) tells us how much the hypothesis that the sample is a drug is improved when the available data (evidence) is considered compared with our prior knowledge of the odds:

$$B_{1,0} = \frac{p(H_{drug}|D)/p(H_{fiber}|D)}{p(H_{drug})/p(H_{fiber})} = \frac{p(D|H_{drug})}{p(D|H_{fiber})} \tag{22.29}$$

where D is data. This form of $B_{1,0}$ can only be used if the distribution of the data is known. For GC peak areas of samples of cannabis, we can assume a Gaussian distribution, but this must be an approximation. (Peak areas cannot be negative and have some upper bound limited by the instrumentation and amount injected.) The authors of the paper[22] consider a Gaussian distribution, and, alternatively, a kernel density estimate of the distribution (i.e., a numerical approach based on the empirical data). This paper is recommended for study because the theory is laid out clearly in detail and an Appendix gives a worked example that can be easily followed.

22.4.3.3 Links between Samples of Heroin

A method of deciding whether samples of seized heroin can be determined to be linked, by virtue of their GC profiles was the subject of a study by Dujourdy et al.[23] Taking a database of samples known to be independent, the correlation coefficients, which ranged from 0.22 to 1.00, with a large number above 0.85, were fitted to two beta distributions, above and below $r = 0.95$. Given samples from two seizures, correlation coefficients are computed for all pairs of samples and the median ($\tilde{x}$) is used as the test statistic. The LR for the two hypotheses that the samples are linked or not linked is

$$LR = \frac{f_{\text{sample}}\left(\tilde{x}|H_{\text{linked}}\right)}{f_{\text{database}}\left(\tilde{x}|H_{\text{not linked}}\right)} \tag{22.30}$$

where f is the relevant beta distribution. The authors show that it can be concluded that the samples are linked for $LR > 24.5$, and not linked $LR < 0.95$, showing a clear discriminating power of the approach.

Two papers using Bayesian inference in sports drug testing are interesting reading, although they are not specifically directed to chromatography. They consider carbon dioxide concentration in equine plasma,[24] and testosterone abuse in human athletes detected by stable isotope analysis.[25] The former shows how modeling both undoped and doped populations allows a proper determination of false positive and false negative rates for a given threshold, and the latter paper argues that priors should include a reasonable probability of administration.

22.4.3.4 Workplace Drug Testing

Ellison et al. give a detailed discussion of the use of the discrete Bayes (see, e.g., Equation 22.10) for qualitative drug testing with known error rates.[26] The example given in the paper uses HPLC and enzyme multiplied immunoassay technique (EMIT), methods which give just a positive/negative result, to detect benzoylecgonine (a cocaine abuse indicator). The false positive and false negative rates for each test are known from validation studies, and because the tests are based on very different chemistries it is possible to assume independence and combine probabilities in a straightforward way. Assume the hypothesis H is that a drug is actually present, and $\bar{H}$ is that it is absent, and T is a test result. Note that $p(H) + p(\bar{H}) = 1$. Applying Bayes Theorem for the probability of H

$$p\left(H|T_{\text{HPLC}}, T_{\text{EMIT}}\right) = \frac{p\left(T_{\text{HPLC}}|H\right)p\left(H|T_{\text{EMIT}}\right)}{p\left(T_{\text{HPLC}}|H\right)p\left(H|T_{\text{EMIT}}\right) + p\left(T_{\text{HPLC}}|\bar{H}\right)p\left(\bar{H}|T_{\text{EMIT}}\right)} \tag{22.31}$$

and $p(H_1|T_{\text{EMIT}})$, the prior for Equation 22.31, is then given by the usual Bayes expression, Equation 22.3.

22.4.4 BIOMARKERS BY BAYESIAN MODELS

Hernandez et al.[16] have written a comprehensive review on Bayes in proteomics biomarker development. Data used in proteomics are from LC-MS experiments and are typically characterized by many more variables than samples (called "small n large p" in the jargon). The motivation for using Bayes is the ability to input other information about the problem by manipulating priors and, indeed, the distribution models. For example, peaks near the detection limit and with known greater variability can have less informative priors assigned. The use of the Lasso method (see above) is described, and priors other than Laplacian discussed. The issue is, how much to penalize parameters with a double Pareto, normal-gamma, and so-called Zellner's g-prior offering different degrees of shrinkage. Nonparametric Bayes applied to classification and regression decision trees and different modifications are discussed and suggested to be tried. Modeling data at a node of a decision tree as a multinomial distribution allows a probability to be calculated on a classification. The review includes pieces of R code, and packages containing different algorithms are identified.

22.4.4.1 Nonparametric Bayesian Models

Suvitaival et al.[27] used a sophisticated methodology using UPLC-MS data in which peak shapes and correlated peaks were used to analyze metabolomics data. Peaks were clustered to identify compounds by a nonparametric Bayesian Dirichlet process model, having redefined the prior distributions from a normal distribution to a beta distribution to improve the match to the peak shape similarity observations. A Bayesian multiway model was then used to infer responses to covariates from clusters obtained in the first part of the method.

22.4.4.2 Peptide Identification from a Database

A peptide matching algorithm, Annotated Peptide Matching Algorithm, uses a probability from a distribution of LC-MS data, m/z peaks, and corresponding peak reverse-phase elution concentrations (% acetonitrile, ACN) to obtain a score for a test profile against the database.[28] The probability of a peptide coordinate (x, a m/z or its ACN) being part of a database peak with mean $\mu_{database}$ and standard deviation $\sigma_{database}$ is given by

$$p = 2 - \left[1 + erf\left(\frac{|x - \mu_{database}|}{\sqrt{2}\sigma_{database}}\right)\right]$$

(22.32)

The contribution to the score is the product of the m/z and ACN probabilities for a coordinate. The authors claim that the approach, using only LC-MS data, outperforms LC-MS/MS searches.

22.4.4.3 Bayesian Networks for Peptide Identification

Building a library of MS/MS fragments associated with known peptides allows matching of unknown spectra by a dynamic Bayesian network (DBN).[29] Given ion fragments from LC-MS/MS, a database of known peptide-spectrum matches (PSM) is used to train two DBNs, one for positive matches and one for negative matches. A vector of log likelihood values for match to no match is then made the input to a support vector machine (SVM) to effect the final classification.

$$LLR = \log\left(\frac{p(I,P|M)}{p(I,P|\bar{M})}\right)$$

(22.33)

where $p(I,P|M)$ is the probability that a given ion sequence (I) is associated with a peptide (P), given a match (M). The network has nodes such as "Ion detectable," "Ion detected," "Intensity of peak," "C-terminal residue," "N-terminal residue," and "Peptide residue."

22.4.5 OTHER APPLICATIONS USING CHROMATOGRAPHIC DATA

22.4.5.1 Cholera Diagnosis Using Bayesian Latent Class Models

The performance of a field test kit for cholera using a dipstick immunochromatographic test, also known as a rapid diagnostic test (RDT), was evaluated without a "gold standard" by Bayesian latent class modeling.[30] The RDT and a culture were used together to obtain information about a discrete latent class—the true disease status—knowing that each method was imperfect. Details of the method are given by Branscum.[31] Briefly, the sensitivity of a test j, $Se = p(T_j = P|H)$, the specificity $Sp = p(T_j = N|\bar{H})$, and the prevalence of the disease π (the symbols from Table 22.1 are used here) are the desired parameters. A single test on a population does not allow evaluation of these parameters unless extra information is available, but two independent tests can be analyzed for the parameters. Beta prior distributions were used to model the uncertainty about the disease and tests, using published data or elicitation of opinions from experts.

22.4.6 COMPARISONS BETWEEN BAYES AND OTHER CLASSIFIERS

Bayesian approaches tend to compare well against other classification algorithms, especially if the distribution and priors are well known. When no particular attention is paid to the features of the model, then Bayes is less outstanding. A comprehensive comparison of several methods to pick wound biomarkers from UPLC-TOF/MS data for a model plant used Naïve Bayes with a Gaussian distribution of parameters.[32] Naïve Bayes appeared mid-table, being outperformed by SVMs, random forest decision trees, and a multilayer perceptron artificial neural network.

Random forest decision trees were also reported to be superior to Naïve Bayes in classifying three kinds of cut tobacco from GC-FID data.[33] In each of these cases, the assumption of independence of factors is almost certainly not correct.

An earlier study[34] looked at the ability of univariate, multivariate (SIMCA, PLS regression, KNN, LDA), and probabilistic methods to distinguish ultrafiltration from microfiltration of apple juice based on HPLC–fluorescence data of amino acids and riboflavin. Although a method is described as Bayesian, it is not clear whether it is a full Bayesian model with priors, or relies on simply fitting Gaussian distributions providing a covariance matrix. The problem is sufficiently easy that all multivariate methods gave satisfactory performance.

A study using high-performance thin-layer chromatography to discriminate among Chinese herbal products (*Bupleuri Radix*) employed Naïve Bayes, SVM, KNN, radial basis function, network NN, and logistic modeling to construct prediction models.[35] In this work, the Naïve Bayes performed below average, but no details were given of distributions, priors etc.

22.4.7 METHOD VALIDATION AND METROLOGY

The validation and use of chromatography to solve real-world problems inevitably focuses on the reliability of results obtained and the ability of the client to make useful inferences from the results obtained. We now understand the need for proper assessment of measurement uncertainty[36] and the importance of metrological traceability of those results.[37] Statistics can go beyond simple repetition and application of Student-t tests,[38] in particular the proper use of Experimental Design,[39] and probabilistic approaches to treating validation data.

22.4.7.1 Bayes Theorem Applied to Reliability Studies of HPLC-UV

Dharuman and Vasudevan[40], in a paper on method validation of HPLC-UV using a probabilistic methodology to achieving reliability, discuss the perceived shortfalls of the present approaches to validation of analytical methods. They argue for a reliability approach that models the analytical results and provides a probability distribution for results that can then be judged relative to predetermined acceptance limits. The validation result, $X_{i,j,k}$ is modeled for $i = 1\ldots I$ concentration levels of validation standards, $j = 1\ldots J$ runs, and $k = 1\ldots K$ replicates.

$$X_{i,j,k} = \left(\beta_0 + \beta_i \mu_i\right) + \left(u_{0,j} + u_{1,j}\mu_i\right) + \varepsilon_{i,j,k} \tag{22.34}$$

where
  β_0 and β_1 are constant and proportional bias terms
  $u_{0,j}$ and $u_{1,j}$ are random effects for the jth run
  $\varepsilon_{i,j,k}$ is a residual normally distributed error term

The so-called reliability probability π is obtained as a function of μ_i. This can be multiply sampled by MCMC to obtain the posterior probability of results lying within the specified acceptance limits $[-\lambda, \lambda]$. The paper is somewhat light on mathematical details (see Rozet et al. for more[41]), but an example of validating an HPLC-UV method for the antibiotics cefepime and tazobactam is given. The paper stresses the recurring theme with Bayesian analysis, that Bayes delivers what the user wants to know (here the reliability of the method to give results within some tolerance range with a certain probability), rather than information from which further inference is needed.

22.4.7.2   Robustness Testing

An interesting use of a Bayesian approach comes from Oca et al. who studied results from a highly fractional factorial design of the analysis of drugs in animal muscle by LC-MS. In this robustness study of sample preparation, seven factors (analyst, SPE method, time between preparation of elution solution and analysis, time between preparation of buffer and analysis, volume of buffer, volume of eluent, volume of SPE rinsing solution) were investigated in eight runs of a two-level fractionated design. The model is

$$y = \beta_0 + \sum_{i=1}^{i=7} \beta_i x_i + \varepsilon \tag{22.35}$$

where

y is the analytical result
$x_{1...7}$ are coded levels (-1, $+1$) of the factors
β_0 is the mean result
$\beta_{1...7}$ are the coefficients from analysis of the design
ε is random error (repeatability)

A robust method will show no effects of the investigated factors, that is, estimates of the confidence (credible) intervals of β will include zero. Traditional methods use replicate runs to estimate ε against which the values of β can be tested, or a probability plot of β may be assessed for extreme values being significantly away from a normal variability. The Bayesian approach is to calculate a posterior probability that the effect is not zero under the assumption that an unknown proportion α of the factors are active ($\beta > 0$), and the β's are normally distributed $\sim \left(0, \sigma_{\text{inactive}}^2\right)$ for inactive factors and $\sim \left(\bar{\beta}_{\text{active}}, \sigma_{\text{active}}^2\right)$. The ratio of the variances for active and inactive factors $k = \sigma_{\text{active}}^2 / \sigma_{\text{inactive}}^2$, which is expected to be greater than 5, is also modeled.

22.5   CONCLUSIONS

As stated at the outset, Bayes Theorem is not written for chromatography or even chemistry, but it is a powerful tool to properly answer questions in our field. The popularity of Bayes has caused authors to append the sobriquet "Bayes" to any method that calculates a probability distribution, whether or not it is done in the framework of priors, likelihood, marginalization, and posterior probabilities that characterize Bayes Theorem. Bayes Theorem is conceptually simple, but algorithmically hard for all but the simplest problems. However, the bottom line of Bayes is that it answers questions we want to know—what is the best estimate and how sure am I of the answer?, are my results reliable, and how can I demonstrate that reliability?, what is the risk if I make the wrong decision? It is this that makes Bayes worth studying and spending time with your favorite statistician. This chapter has scratched the surface, but together with the references it may stimulate the reader to try Bayes for her particular problem.

GLOSSARY OF ACRONYMS AND INITIALISMS

ACN   % acetonitrile
APMA   Annotated peptide matching algorithm
ASTM   ASTM International
CART   Classification and regression decision trees
DBN   Dynamic Bayesian network
DNA   Deoxyribonucleic acid

DoE   Design of experiments
EM   Expectation maximization
EMIT   Enzyme multiplied immunoassay technique
FID   Flame ionization detector
GC   Gas chromatography
H Height of a theoretical plate
HPLC   High-performance liquid chromatography
HPTLC   High-performance thin-layer chromatography
KNN k-nearest neighbors
LC   Liquid chromatography
LDA   Linear discriminant analysis
MCMC   Markov chain Monte Carlo
MS   Mass spectrometry
NEB   Normal exponential Bernoulli
NN   Neural network
NRSD   Normal relative standard deviation
PLS   Partial least squares
PSM   Peptide-spectrum matches
RDT   Rapid diagnostic test
SIMCA   Soft independent modeling of class analogy
SPE   Solid-phase extraction
TFA   Target factor analysis
TIC   Total ion chromatogram
TIS   Total ion spectra
TOFMS   Time-of-flight mass spectrometry
UPLC   Ultrahigh-performance liquid chromatography
UV   Ultraviolet

REFERENCES

1. McGrayne SB. *The Theory That Would Not Die: How Bayes' Rule Cracked the Enigma Code, Hunted Down Russian Submarines, & Emerged Triumphant from Two Centuries of Controversy* 2011. Boston, MA: Yale University Press.
2. Fienberg SE. When did Bayesian inference become "Bayesian"? *Bayesian Analysis* 2006:1(1): 1–40.
3. Sivia DS, Skilling J. *Data Analysis: A Bayesian Tutorial* 2006, 2nd ed. Oxford, U.K.: Oxford University Press.
4. Bolstad WM. *Introduction to Bayesian Statistics* 2004, 1st ed. Hoboken, NJ: John Wiley & Sons.
5. Armstrong N, Hibbert DB. An introduction to Bayesian methods for analyzing chemistry data. Part 1: An introduction to Bayesian theory and methods. *Chemometrics and Intelligent Laboratory Systems* 2009:97(2): 194–210.
6. Hibbert DB, Armstrong N. An introduction to Bayesian methods for analyzing chemistry data. Part II: A review of applications of Bayesian methods in chemistry. *Chemometrics and Intelligent Laboratory Systems* 2009:97(2): 211–220.
7. Gregory PC. *Bayesian Logical Data Analysis for the Physical Sciences* 2005. Cambridge, U.K.: Cambridge University Press.
8. Ó Ruanaidh JJK, Fitzgerald WJ. *Numerical Bayesian Methods Applied to Signal Processing* 1996. Statistics and Computing. New York: Springer-Verlag.
9. Kelly PC, Harris WE. Application of method of maximum posterior probability to estimation of gas-chromatographic peak parameters. *Analytical Chemistry* 1971:43(10): 1184–1195.
10. Lopatka M, Vivó-Truyols G, Sjerps MJ. Probabilistic peak detection for first-order chromatographic data. *Analytica Chimica Acta* 2014:817: 9–16.
11. Kim S, Ouyang M, Jeong J, Shen C, Zhang X. A new method of peak detection for analysis of comprehensive two-dimensional gas chromatography mass spectrometry data. *Annals of Applied Statistics* 2014:8(2): 1209–1231.

12. Befekadu GK, Tadesse MG, Ressom HW. A Bayesian based functional mixed-effects model for analysis of lc-ms data. *Proceedings of the Annual International Conference of the IEEE Engineering in Medicine and Biology Society*, Minneapolis, MN, 2009, pp. 6743–6746.

13. Armstrong N. Bayesian analysis of band-broadening models used in high performance liquid chromatography. *Chemometrics and Intelligent Laboratory Systems* 2006:81: 188–201.

14. Van Deemter JJ, Zuiderweg F, Klinkenberg AV. Longitudinal diffusion and resistance to mass transfer as causes of nonideality in chromatography. *Chemical Engineering Science* 1956:5(6): 271–289.

15. Kirkup L, Foot M, Mulholland M. Comparison of equations describing band broadening in high-performance liquid chromatography. *Journal of Chromatography A* 2004:1030(1): 25–31.

16. Hernández B, Pennington SR, Parnell AC. Bayesian methods for proteomic biomarker development. *EuPA Open Proteomics* 2015:9: 54–64.

17. Park T, Casella G. The Bayesian lasso. *Journal of the American Statistical Association* 2008:103(482): 681–686.

18. Clark HA, Jurs PC. Classification of crude oil gas chromatograms by pattern recognition techniques. *Analytical Chemistry* 1979:51(6): 616–623.

19. Blomstedt P, Gauriot R, Viitala N, Reinikainen T, Corander J. Bayesian predictive modeling and comparison of oil samples. *Journal of Chemometrics* 2014:28(1): 52–59.

20. Leung WC. The prosecutor's fallacy—A pitfall in interpreting probabilities in forensic evidence. *Medicine Science and the Law* 2002:42(1): 44–50.

21. Williams MR, Sigman ME, Lewis J, Pitan KM. Combined target factor analysis and Bayesian soft-classification of interference-contaminated samples: Forensic fire debris analysis. *Forensic Science International* 2012:222(1–3): 373–386.

22. Bozza S, Broséus J, Esseiva P, Taroni F. Bayesian classification criterion for forensic multivariate data. *Forensic Science International* 2014:244: 295–301.

23. Dujourdy L, Barbati G, Taroni F, Guéniat O, Esseiva P, Anglada F, Margot P. Evaluation of links in heroin seizures. *Forensic Science International* 2003:131(2–3): 171–183.

24. Hibbert D, Armstrong N, Vine J. Total CO_2 measurements in horses: Where to draw the line. *Accreditation and Quality Assurance* 2011:16(7): 339–345.

25. Flenker U, Geppert LN, Ickstadt K. Validity of stable isotope data in doping control: Perspectives and proposals. *Drug Testing and Analysis* 2012:4(12): 934–941.

26. Ellison SLR, Gregory S, Hardcastle WA. Quantifying uncertainty in qualitative analysis. *The Analyst* 1998:123: 1155–1161.

27. Suvitaival T, Rogers S, Kaski S. Stronger findings from mass spectral data through multi-peak modeling. *BMC Bioinformatics* 2014:15(1): 1–11.

28. Chen SS, Deutsch EW, Yi EC, Li X-J, Goodlett DR, Aebersold R. Improving mass and liquid chromatography based identification of proteins using Bayesian scoring. *Journal of Proteome Research* 2005:4: 2174–2184.

29. Klammer AA, Reynolds SM, Bilmes JA, Maccoss MJ, Noble WS. Modeling peptide fragmentation with dynamic Bayesian networks for peptide identification. *Bioinformatics* 2008:24(13): i348–i356.

30. Page AL, Alberti KP, Mondonge V, Rauzier J, Quilici ML, Guerin PJ. Evaluation of a rapid test for the diagnosis of cholera in the absence of a gold standard. *PLoS One* 2012:7(5): e37360.

31. Branscum AJ, Gardner IA, Johnson WO. Estimation of diagnostic-test sensitivity and specificity through Bayesian modeling. *Preventive Veterinary Medicine* 2005:68(2–4): 145–163.

32. Boccard J, Kalousis A, Hilario M, Lantéri P, Hanafi M, Mazerolles G, Wolfender J-L, Carrupt P-A, Rudaz S. Standard machine learning algorithms applied to uplc-tof/ms metabolic fingerprinting for the discovery of wound biomarkers in arabidopsis thaliana. *Chemometrics and Intelligent Laboratory Systems* 2010:104(1): 20–27.

33. Lin X, Sun L, Li Y, Guo Z, Li Y, Zhong K, Wang Q, Lu X, Yang Y, Xu G. A random forest of combined features in the classification of cut tobacco based on gas chromatography fingerprinting. *Talanta* 2010:82(4): 1571–1575.

34. Blanco-Gomis D, Fernandez-Rubio P, Guiterrez-Alvarez MD, Mangas-Alonso JJ. Use of high-performance liquid chromatographic-chemometric techniques to differentiate apple juices clarified by micro-filtration and ultrafiltration. *The Analyst* 1998:123(1): 125–129.

35. Cheng X, Cai H, He P, Zhang Y, Tian R. Combination of effective machine learning techniques and chemometric analysis for evaluation of bupleuri radix through high-performance thin-layer chromatography. *Analytical Methods* 2013:5(22): 6325–6330.

36. Joint Committee for Guides in Metrology. *Evaluation of Measurement Data—Guide to the Expression of Uncertainty in Measurement*. Genève, Switzerland: JCGM 100; Sèvres, http://www.bipm.org/utils/common/documents/jcgm/JCGM_100_2008_E.pdf: BIPM, 2008.

37. De Bièvre P, Dybkaer R, Fajgelj A, Hibbert DB. Metrological traceability of measurement results in chemistry: Concepts and implementation (iupac technical report). *Pure and Applied Chemistry* 2011:83(10): 1873–1935.

38. Rozet E, Lebrun P, Debrus B, Hubert P. New methodology for the development of chromatographic methods with bioanalytical application. *Bioanalysis* 2012:4(7): 755–758.

39. Hibbert DB. Experimental design in chromatography: A tutorial review. *Journal of Chromatography B* 2012:910: 2–13.

40. Dharuman JG, Vasudevan M. Reliability-targeted hplc-uv method validation—A protocol enrichment perspective. *Journal of Separation Science* 2014:37(3): 228–236.

41. Rozet E, Govaerts B, Lebrun P, Michail K, Ziemons E, Wintersteiger R, Rudaz S, Boulanger B, Hubert P. Evaluating the reliability of analytical results using a probability criterion: A Bayesian perspective. *Analytica Chimica Acta* 2011:705(1–2): 193–206.

23 Multiway Methods in Chromatography

Łukasz Komsta and Yvan Vander Heyden

CONTENTS

23.1 DEFINITION OF MULTIWAY DATASETS

A multivariate dataset, arranged as a matrix, can be interpreted as an example of *two-way* dataset. In this case, each value in the matrix is characterized by two factors (one factor is an *object* and the other factor is a *property*), so each cell contains the corresponding property of the corresponding object. Each row contains properties for the same object and each column the same property for various objects. For instance, chromatograms have to be recorded exactly in the same way (the same number of points, the same time sampling frequency, etc.) to fit all in one matrix.

A two-way dataset is the simplest example of a *multiway* dataset. For *three-way* data, a dataset can be arranged as a tensor (cube), which is a three-dimensional analog of a matrix. The cell (entry) of this cube is a value related to the unique combination of three factors. Some examples of three-way datasets are

1. Retention times of 50 compounds, measured with 30 mobile phases, each on 10 columns ($50 \times 30 \times 10$)
2. Retention times of 50 compounds, measured with three modifiers, each in the same 10 modifier fraction values of mobile phase ($50 \times 3 \times 10$)
3. R_F values of 50 compounds in TLC, registered with 10 mobile phases, each mobile phase on 5 plate types ($50 \times 10 \times 5$)
4. R_F values of 50 compounds in TLC, registered with 20 mobile phases, each mobile phase developed on 3 different plate distances ($50 \times 20 \times 3$)

There are also situations where the chromatographer obtains *four-way* datasets, and theoretically the number of "ways" is infinite. In general, we speak about *N-way* datasets. For example, all the above experiments can be extended to four-way research by performing *each full* experiment set on several distinct temperatures. Then, a four-way dataset can be perceived as a "vector of cubes," a five-way one as a "matrix of cubes," a six-way one as a "cube of cubes," etc. The general name is *hypercube*, of which visualization is not possible.

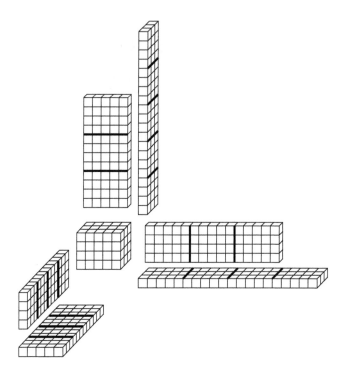

FIGURE 23.1 The six ways of unfolding a three-way array into a two-dimensional matrix.

The above examples can be analyzed after rearranging to two-way matrices, with dimensions 50×300, 50×300, 50×50, and 50×600, respectively. This can be done in several ways (Figure 23.1) and is called *unfolding*. Although unfolding is a simple workaround, it does not take into account the *trilinear* structure of the data. For example, principal component analysis* treats each column separately and "does not know" that they are interconnected with an additional factor. Another approach to avoid the three-way data analysis is called *flattening*. It works by splitting the cube into individual *slices* (matrices) and analyzes each of them separately. In general, this workaround is also not recommended, as it does not respect the multiway design of the experiment, nor interconnections between elements of the (hyper)cube. The best method for exploration of the multiway datasets is an appropriate multiway method.

23.2 MULTIWAY ANALYSIS

The principal component analysis[†] treats the dataset as a *bilinear* object. The simplest one-component PCA model represents the matrix as a product of two vectors, one being called *scores* and another called *loadings*. Each entry in the matrix is a product of the corresponding score and loading. Therefore, if a one-component PCA model is sufficient to describe the dataset, each entry in the matrix can be represented as the product of two independent coefficients—one representing a contribution of the object and the second representing a contribution of the property (variable). The whole matrix has then rank equal to one, as it represents the points lying on the line in multivariate space (in one-dimensional subspace[‡]). By adding another principal component to the model, the second

---

* See Chapter 8.
[†] See Chapter 8.
[‡] See Chapter 7.

score and loading vectors are introduced. Their product contains the information explained by the second principal component. The dataset is then described as a sum of several matrices, each being a product of the corresponding score and loading vectors.

By analogy, three-way analysis of the data cube assumes that the data inside this cube are *trilinear* [1–3]. If the cube has rank equal to one, each entry in this cube can be represented as a product of three coefficients attributed to three factors. When doing a typical PCA on a matrix, the result obtained on the transposed matrix yields different values because of centering along the column dimension. In the case of a trilinear approach, the tensor is usually not centered, so the factors (dimensions) are equally treated and there is no difference in which is a "score" and in which is a "loading."

The main and basic methods of three-way analysis are parallel factor analysis (PARAFAC, sometimes referred also as CANDECOMP; Figure 23.2) and Tucker3 (Figure 23.3). They have smaller numbers of degrees of freedom than PCA on an unfolded bivariate matrix (as they must preserve the trilinearity), so the explained variance is almost always smaller (but the obtained results are more meaningful) and there is a number of better visualization approaches [4]. PARAFAC can be perceived as a constrained version of Tucker3, so if the dataset is well modeled by PARAFAC, there is no need to try Tucker3, because PARAFAC results are easier to interpret (take a look at [5], see also [6] for extended discussion of intercollinearities in such datasets). Tucker3 cannot hold datasets with dimensions higher than three. Therefore, the multilinear datasets must always be modeled by PARAFAC (there is a Tucker4 approach, but it is very rarely used).

In the last years, the importance of other methods for multilinear decomposition increased, mainly for multiway calibration. The reader interested in them could start the additional study on the following approaches: self-weighted alternating trilinear decomposition [7], alternating slice-wise diagonalization [8], alternating coupled matrices resolution [9], "structuration des tableaux à trois indices de la statistique" (STATIS) [10], alternating asymmetric trilinear decomposition [11], alternating penalty quadrilinear decomposition [12], alternating weighted residue constraint quadrilinear decomposition [13], regularized self-weighted alternating quadrilinear decomposition [14], three-way nonnegative matrix approximation [15], and bilinear least squares [16]. Two comparative

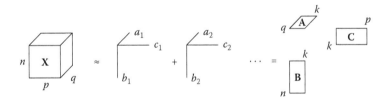

FIGURE 23.2 The idea of PARAFAC decomposition.

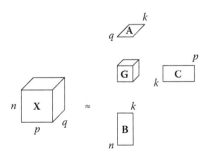

FIGURE 23.3 The idea of Tucker3 decomposition.

reviews of the above (and other less important) methods are also worth reading [17,18]. An interesting approach is also the generalized rank annihilation method (GRAM) [19], which extracts features from exactly *two* matrices, which can be treated as a special case of a trilinear dataset.

23.2.1 PARAFAC

PARAFAC [5] is a basic method of decomposition of a three-way (or multiway) array to several sets of "scores" (Figure 23.2). One-component PARAFAC treats each value in the array as a product of several independent coefficients associated with the factors. Therefore, the impact of each (there are three or more of them) factor can be separated and interpreted as independent values. For instance, a three-way array X is modeled as a product of three independent coefficients a, b, and c (forming a triad) in the following way:

$$x_{ijk} = \sum_{f=1}^{F} a_{if} b_{jf} c_{kf} + e_{ijk}$$

where
   F is the number of PARAFAC factors (their optimal number can be chosen during analysis)
   e is an error unexplained by the model

One of the main advantages of PARAFAC is that the solution is unique and there is no rotational freedom as with PCA. To be more specific, the result can be rotated, but then the fit will decrease. PCA, contrary to PARAFAC, finds only one of many possible solutions, optimized for maximal explained variance in first components (in other words, the solution can be rotated in an infinite number of ways). Although PARAFAC has the above advantage, the result is meaningful only if the dataset is really trilinear (or multilinear; for a discussion on the possible deviations [20]. The rotational ambiguity can occur rarely for datasets with special properties [21]. The sign ambiguity is still present just like in PCA; however, there are some approaches to deal with this problem [22]. It must also be emphasized that the value of each score differs when changing the total number of decomposition factors.

The scores of the PARAFAC model are not forced to be orthogonal and, in many cases, they are intercorrelated. To improve the interpretability, an orthogonality constraint can be applied to force some score pairs to be orthogonal. It is possible only for a subset of factors, that is, there is no possibility to force orthogonality between all possible pairs. A factor can be also constrained (forced) to be nonnegative or unimodal [23]. This is often useful when modeling spectroscopic data (e.g., DAD spectrum along one dimension).

The data preprocessing for PARAFAC is not a straightforward task, and in most cases, the dataset is centered *across* one chosen mode and optionally scaled *within* this mode (i.e., whole slices are scaled). This is equal to unfolding the array to a wide matrix with the rows corresponding to the chosen mode, then center across the columns and scale across the rows. Any other method of scaling and centering can easily destroy the information inside a dataset and should be avoided. Moreover, contrary to PCA, centering and scaling is not a standard way, and the first trials should be done on the original array without any preprocessing.

The optimal number of PARAFAC factors can be chosen by visual analysis of the results, values of explained variance, or some diagnostic values, such as core consistency diagnostics (CORCONDIA). PARAFAC analysis is possible with large (up to 70%) numbers of missing values [24,25].*

---

* See also Chapter 13.

Several extensions were elaborated to PARAFAC: a penalty diagonalization error approach, which can deal with some datasets causing very slow convergence of the algorithm [26], a connection with ASCA (ANOVA—simultaneous component analysis), called PARAFASCA [27], a weighted version of the algorithm [28], automated peak extraction and quantification [29], a combination with MCR-based peak alignment [30], and augmented PARAFAC version useful in multiway calibration [31]. The reader should be also aware of PARAFAC2 [32], which is a modification of PARAFAC to deal with datasets of variable length along one dimension. Every sample can have its own distinct set of elution-time loadings, so this technique can deal with deviations from trilinearity caused by time shifts in chromatograms.

PARAFAC is most often fitted with alternating least squares (ALS) algorithm; however, some other methods also exist [5,33]. Fortunately, a practitioner does not really need to dig deeply into the computational details.

23.2.2 Tucker3

Tucker3 (Figure 23.3) differs from PARAFAC by the presence of an additional tensor $\mathbf{G}$ with weights, which is a cube with dimensions equal to decomposition factors. It incorporates additional degrees of freedom, as the final tensor is a weighted sum of score vectors. Therefore, there is always a rotational ambiguity and an infinite number of possible solutions. PARAFAC can be seen as a special case of Tucker3, where $\mathbf{G}$ is forced to be the tensor with ones on the diagonal and zeros in the other cells [34].

There are also Tucker2 and Tucker1 versions of the algorithm. Tucker2 can be understood as Tucker3, but the number of scores along one chosen mode is set to the dimension of this mode. In other words, Tucker2 does not do any compression of one of the modes, so the explained variance (and fitting quality) is always higher than for Tucker3. Further increase of complexity leads to Tucker1 model, when only one dimension is compressed. Similar to PARAFAC, the properties of the obtained scores can be forced to fulfill some requirements (i.e., they are constrained) [35].

23.3 MULTIWAY CALIBRATION

The classical approaches for building a model to predict a sample property from a chromatogram (treated as a signal) are based on multivariate regression methods.* In these cases, the property is predicted from a vector (univariate dataset), consisting—in the case of chromatogram—of consecutive response values. However, when using multiresponse detectors (being able to record the entire spectrum on each time point), a chromatogram becomes a matrix—it is a two-way dataset, having time as the one factor/dimension and some spectral or detection property along the second dimension.

The use of classical regression methods (as PLS) is still possible in such cases after unfolding this matrix to a vector. However, similar to the PCA-PARAFAC difference, the model does not assume any interconnection between these factors. This may lead to less transparent models with less good prediction ability; moreover, such models are difficult to interpret. The alternative is to use N-way calibration methods [36–39].

The simplest approach is a modification of classical PLS (on unfolded matrix) to force at each fitting step the residuals to be bilinear or multilinear [40]. This is called residual bilinearization (RBL) [41], residual trilinearization (RTL) [42–44], or residual quadrilinearization [45]. Another approach is to make a PARAFAC analysis and then use the scores to build the model [46]. This approach is analogous to Principal Component Regression† for two-way datasets.

---

* See Chapter 14.
† See Chapter 14.

The method of choice is N-way PLS [47,48]. It is the modification of PLS, which works with three-way or multiway datasets, preserving the multilinear structure [49]. In general, the use of N-way PLS is very similar to the classical PLS approach [50]. The validation of the predictive power of multiway models is slightly different—in most cases, the leave-bar-out validation approach is used [51].

23.4   SOFTWARE

There are two widely used toolboxes for MATLAB®, providing all algorithms for multiway data analysis and calibration: the MVC3 toolbox [52] and N-way toolbox [53]. They are freely available to download and come with extensive documentation and examples. PARAFAC and Tucker models can be also fitted in R using "multiway" package from CRAN repository.

23.5   EXAMPLES OF APPLICATIONS

23.5.1   TLC Retention Analysis with PARAFAC

PARAFAC was applied to analyze trilinear datasets containing retention coefficients [54,55]. In the first example [54], 35 model compounds were chromatographed on silica and RP18 phases using 9 concentrations of 6 modifiers (diluted with hexane or water). The R_F values obtained in this study were arranged as two tensors with dimensions 35 (compounds) × 9 (concentrations) × 6 (modifiers). The first decision was to choose the optimal number of PARAFAC components. The one-component model explained more than 90% of the variance for both adsorbents, while the CORCONDIA was 100%. Adding the second component did not increase the explained variance significantly, and CORCONDIA decreased to an unacceptable value. Therefore, this dataset was finally modeled with a one-component PARAFAC as the optimal model.

Each entry in the tensor (an R_F value) is modeled as a product of three scores. Therefore, the score of a compound can be interpreted as a coefficient expressing its "contribution" to retention, that is, some experimental chromatographic descriptor. Analogously, the score of a solvent can be treated as an experimental coefficient describing its elution strength. The score of the modifier concentration has no cognitive meaning, but it can be used to see if the dependence between the retention and the modifier concentration is generally linear.

Figure 23.4a presents compound scores on silica and RP18, compared in one graph. The compound list is not needed to see a general tendency, the main conclusion is that the scores represent the average affinity to the adsorbent, and the chromatographic behavior on RP18 is not clearly correlated with that of silica gel. In both cases (Figure 23.4b), one can see the linear dependence between concentration and the obtained score of the concentration. The modifier scores (Figure 23.4b and c) express very well the elution strength from ethyl acetate (weakest) to isopropanol (strongest) on silica, and from methanol (weakest) to acetone (strongest) on RP18. It is worth underlining again here that this simple one-component multiplicative model explains more than 90% of the total variance caused by differences in retention.

However, there are cases where a one-component model is not enough to explain and interpret the dataset and a two-component model has to be created and interpreted [55]. Here the same compounds were chromatographed with 20 pure solvents as mobile phases, each solvent on 9 adsorbents. Therefore, the obtained R_F values were arranged as a tensor with dimensions 35 (compounds) × 20 (solvents) × 9 (adsorbents). Due to the presence of various adsorbents, more phenomena were responsible for the retention, and this dataset could not be modeled properly with one-component PARAFAC. Although the one-component model explained about 80% of the total variance, the results were almost impossible to interpret. Adding the second component increased the variance to

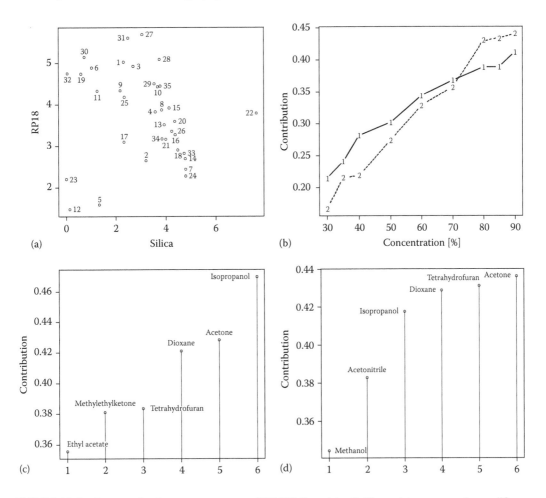

FIGURE 23.4 An example of two one-component PARAFAC models of trilinear data, compound × modifier × concentration: (a) compound scores, (b) modifier concentration scores, (c) modifier scores on silica, and (d) modifier scores on RP18. See text for discussion. (From Komsta, Ł., Radoń, M., Radkiewicz, B., and Skibiński, R.: Trilinear multiplicative modeling of thin layer chromatography retention as a function of solute, organic modifier and its concentration. *J. Sep. Sci.* 2011. 34. 59–63. Copyright Wiley-VCH Verlag GmbH & Co. KGaA. Reproduced with permission.)

90%, still with good CORCONDIA. The insufficiency of the one-factor model can be easily seen in Figure 23.5a, where three clusters are formed by the used solvents and the differences are also expressed by the second score. The first cluster contains apolar hydrocarbons (CLH, cyclohexane; OCT, octanol; HEX, hexane; and HEP, heptane). The second contains chlorinated solvents (CBU, chlorobutane; CFM, chloroform; DCM, dichloromethane; DCE, dichloroethane; and TOL, toluene). The last consists of highly polar solvents (ACT, acetone; ACN, acetonitrile; BUT, 1-butanol; DIO, dioxane; ETH, ethanol; ETA, ethyl acetate; MET, methanol; PRO, propan-1-ol; IPA, propan-2ol; and THF, tetrahydrofuran). While the first PARAFAC score explains an average solvent strength, the second contains the difference between chlorinated solvents (and their effect on the retention) and the other solvents.

The compound scores (Figure 23.5b) are also not intercorrelated, so each score represents a different trend in the chromatographic behavior of a particular compound. The scores of the adsorbents (Figure 23.5c) show the different properties of NH$_2$, alumina, and cellulose, while the other

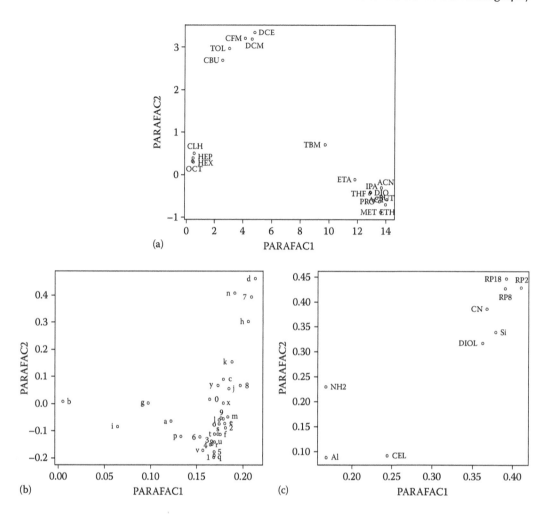

FIGURE 23.5 An example of a two-component PARAFAC model of trilinear data, compound × solvent × adsorbent: (a) solvent scores, (b) compound scores, and (c) adsorbent scores. See text for discussion. (From Komsta, Ł., Skibiński, R., Bezpalko, N., Mielniczek, A., and Stępkowska, B.: Trilinear analysis of thin-layer chromatography retention of 35 model compounds chromatographed on nine adsorbents with 20 pure solvents. *J. Sep. Sci.* 2016. 39. 4258–4262. Copyright Wiley-VCH Verlag GmbH & Co. KGaA. Reproduced with permission.)

adsorbents are placed in accordance with their polarity (DIOL close to silica, RP adsorbents in upper right corner of the plot).

The above examples clearly show the possibility to separate the effects of each factor (compound, concentration and modifier or compound, solvent and adsorbent) and to investigate separately their properties.

23.5.2 OTHER APPLICATIONS

Because of the large number of papers presenting other application of multiway methods in chromatography, this chapter contains only carefully selected examples with well-defined theoretical background (Table 23.1). They can be considered as a recommended further literature.*

---

* For the introduction about MCR-ALS methods, connected frequently with PARAFAC, see Chapter 11.

TABLE 23.1

Examples of Multiway Method Applications in Chromatography

Reference	Technique	Analytes	Methods
Akvan et al. [56]	HPLC-DAD	Pharmaceuticals in water samples	MCR-COW-PARAFAC
Arancibia et al. [42]	HPLC with fluorimetric detector	Urea herbicides	MCR-ALS, PLS/RBL,
Arroyo et al. [57]	GC-MS	Steroid hormones	PARAFAC
Arroyo et al. [58]	LC-MS-MS	Malachite green	PARAFAC
Baunsgaard et al. [59]	HPLC with fluorescence detector	Fluorescent colorants and color precursors	PARAFAC
Bechmann [60]	FIA	Blue dextran, potassium hexacyanoferrate, and heparin	PARAFAC, tri-PLS
Beltran et al. [61]	HPLC with fast-scanning fluorescence detector	Polycyclic aromatic hydrocarbons	PARAFAC
Cañada-Cañada et al. [62]	HPLC with fast-scanning fluorescence	Fluoroquinolones	PARAFAC, N-PLS, MCR-ALS
Culzoni et al. [63]	HPLC-DAD	Dyes	PLS/RBL, MCR-ALS
De Godoy et al. [64]	GCxGC with FID	Kerosene in gasoline	PARAFAC, PARAFAC2, N-PLS
De Llanos et al. [65]	HPLC with fluorimetric detector	Urine markers	MCR-ALS
Durante et al. [66]	GC with FID	Vinegar samples	N-PLS
Fraga and Corley [67]	LCxLC with UV detection (one wavelength)	Test mixture of several organic acids	GRAM, PARAFAC
Hoggard and Synovec [68]	GCxGC with TOF-MS	Various samples	PARAFAC with modifications
Komsta et al. [69–71]	TLC	Retention values of 35 compounds (lipophilicity modeling)	PARAFAC
Komsta et al. [72]	TLC	Retention values of 35 compounds (interpretation)	PARAFAC
Lozano et al. [73]	HPLC with fluorimetric detector	Olive oils	PARAFAC, PLS/RTL
Porter et al. [74]	LCxLC with DAD	Maize seedling digests	PARAFAC
Pravdova et al. [75]	GC-FID	Maillard reaction of various amino acids and sugars	PARAFAC, Tucker3
Qing et al. [76]	HPLC-DAD	Hydrolysis of naptalam	AQQLD
Schmidt et al. [77]	HPLC-DAD-NMR-MS	St. John's wort samples	PARAFAC
Sinha et al. [78]	GCxGC with QTOF-MS	Environmental samples	TLD, PARAFAC
Van Mispelaar et al. [79]	GCxGC with FID	Perfume substances	PARAFAC
Vosough et al. [80]	HPLC-DAD	Antibiotics in wastewater	MCR-ALS, PLS/RBL
Walker et al. [81]	Several methods (combined results)	Water fingerprints	PARAFAC
Watson et al. [82]	GCxGCxGC	Test mixture (26 compounds)	PARAFAC
Wiberg et al. [83]	HPLC-DAD	Lidocaine and prilocaine (partially separated)	PARAFAC
Wu et al. [84]	HPLC-DAD	p-chlorotoluene and o-chlorotoluene	PARAFAC, ATLD

Note: AQQLD, alternating quinquelinear decomposition; ATLD, alternating trilinear decomposition; TLD, trilinear decomposition.

REFERENCES

1. Bro, R. 2006. Review on multiway analysis in chemistry—2000–2005, *Critical Reviews in Analytical Chemistry*, 36(3–4): 279–293.
2. Daszykowski, M. and B. Walczak. 2011. Methods for the exploratory analysis of two-dimensional chromatographic signals, *Talanta*, 83(4): 1088–1097.
3. Smilde, A.K. 1992. Three-way analyses problems and prospects, *Chemometrics and Intelligent Laboratory Systems*, 15(2–3): 143–157.
4. Omidikia, N., H. Abdollahi, and M. Kompany-Zareh. 2013. Visualization and establishment of partial uniqueness and uniqueness rules in parallel factor analysis, *Journal of Chemometrics*, 27(10): 330–340.
5. Bro, R. 1997. PARAFAC. Tutorial and applications, *Chemometrics and Intelligent Laboratory Systems*, 38(2): 149–171.
6. Lorho, G., F. Westad, and R. Bro. 2006. Generalized correlation loadings. Extending correlation loadings to congruence and to multi-way models, *Chemometrics and Intelligent Laboratory Systems*, 84(1–2): 119–125.
7. Chen, Z.-P., H.-L. Wu, J.-H. Jiang, Y. Li, and R.-Q. Yu. 2000. A novel trilinear decomposition algorithm for second-order linear calibration, *Chemometrics and Intelligent Laboratory Systems*, 52(1): 75–86.
8. Jiang, J.-H., H.-L. Wu, Y. Li, and R.-Q. Yu. 2000. Three-way data resolution by alternating slice-wise diagonalization (ASD) method, *Journal of Chemometrics*, 14(1): 15–36.
9. Li, Y., J.-H. Jiang, H.-L. Wu, Z.-P. Chen, and R.-Q. Yu. 2000. Alternating coupled matrices resolution method for three-way arrays analysis, *Chemometrics and Intelligent Laboratory Systems*, 52(1): 33–43.
10. Stanimirova, I., B. Walczak, D.L. Massart, V. Simeonov, C.A. Saby, and E. Di Crescenzo. 2004. STATIS, a three-way method for data analysis. Application to environmental data, *Chemometrics and Intelligent Laboratory Systems*, 73(2): 219–233.
11. Hu, L.-Q., H.-L. Wu, Y.-J. Ding, D.-M. Fang, A.-L. Xia, and R.-Q. Yu. 2006. Alternating asymmetric trilinear decomposition for three-way data arrays analysis, *Chemometrics and Intelligent Laboratory Systems*, 82(1–2): 145–153.
12. Xia, A.-L., H.-L. Wu, S.-F. Li, S.-H. Zhu, L.-Q. Hu, and R.-Q. Yu. 2007. Alternating penalty quadrilinear decomposition algorithm for an analysis of four-way data arrays, *Journal of Chemometrics*, 21(3–4): 133–144.
13. Fu, H.-Y., H.-L. Wu, Y.-J. Yu, L.-L. Yu, S.-R. Zhang, J.-F. Nie, S.-F. Li, and R.-Q. Yu. 2011. A new third-order calibration method with application for analysis of four-way data arrays, *Journal of Chemometrics*, 25(8): 408–429.
14. Kang, C., H.-L. Wu, Y.-J. Yu, Y.-J. Liu, S.-R. Zhang, X.-H. Zhang, and R.-Q. Yu. 2013. An alternative quadrilinear decomposition algorithm for four-way calibration with application to analysis of four-way fluorescence excitation-emission-pH data array, *Analytica Chimica Acta*, 758: 45–57.
15. Sun, J., T. Li, P. Cong, W. Xiong, S. Tang, and L. Zhu. 2010. Direct decomposition of three-way arrays using a non-negative approximation, *Talanta*, 83(2): 541–548.
16. Braga, J.W.B., C.B.G. Bottoli, I.C.S.F. Jardim, H.C. Goicoechea, A.C. Olivieri, and R.J. Poppi. 2007. Determination of pesticides and metabolites in wine by high performance liquid chromatography and second-order calibration methods, *Journal of Chromatography A*, 1148(2): 200–210.
17. Faber, N.M., R. Bro, and P.K. Hopke. 2003. Recent developments in CANDECOMP/PARAFAC algorithms: A critical review, *Chemometrics and Intelligent Laboratory Systems*, 65(1): 119–137.
18. Kolda, T.G.A. and B.W.B. Bader. 2009. Tensor decompositions and applications, *SIAM Review*, 51(3): 455–500.
19. Ortiz, M.C., L.A. Sarabia, I. García, D. Giménez, and E. Meléndez. 2006. Capability of detection and three-way data, *Analytica Chimica Acta*, 559(1): 124–136.
20. De Juan, A. and R. Tauler. 2001. Comparison of three-way resolution methods for non-trilinear chemical data sets, *Journal of Chemometrics*, 15(10): 749–772.
21. Abdollahi, H. and S.M. Sajjadi. 2010. On rotational ambiguity in parallel factor analysis, *Chemometrics and Intelligent Laboratory Systems*, 103(2): 144–151.
22. Bro, R., R. Leardi, and L.G. Johnsen. 2013. Solving the sign indeterminacy for multiway models, *Journal of Chemometrics*, 27(3–4): 70–75.
23. Bro, R. and N.D. Sidiropoulos. 1998. Least squares algorithms under unimodality and non-negativity constraints, *Journal of Chemometrics*, 12(4): 223–247.
24. Smoliński, A., B. Walczak, and J.W. Einax. 2002. Exploratory analysis of data sets with missing elements and outliers, *Chemosphere*, 49(3): 233–245.

25. Tomasi, G. and R. Bro. 2005. PARAFAC and missing values, *Chemometrics and Intelligent Laboratory Systems*, 75(2): 163–180.
26. Cao, Y.-Z., Z.-P. Chen, C.-Y. Mo, H.-L. Wu, and R.-Q. Yu. 2000. A PARAFAC algorithm using penalty diagonalization error (PDE) for three-way data array resolution, *Analyst*, 125(12): 2303–2310.
27. Jansen, J.J., R. Bro, H.C.J. Hoefsloot, F.W.J. Van Den Berg, J.A. Westerhuis, and A.K. Smilde. 2008. PARAFASCA: ASCA combined with PARAFAC for the analysis of metabolic fingerprinting data, *Journal of Chemometrics*, 22(2): 114–121.
28. Andersson, G.G., B.K. Dable, and K.S. Booksh. 1999. Weighted parallel factor analysis for calibration of HPLC-UV/Vis spectrometers in the presence of Beer's law deviations, *Chemometrics and Intelligent Laboratory Systems*, 49(2): 195–213.
29. Furbo, S. and J.H. Christensen. 2012. Automated peak extraction and quantification in chromatography with multichannel detectors, *Analytical Chemistry*, 84(5): 2211–2218.
30. Parastar, H. and N. Akvan. 2014. Multivariate curve resolution based chromatographic peak alignment combined with parallel factor analysis to exploit second-order advantage in complex chromatographic measurements, *Analytica Chimica Acta*, 816: 18–27.
31. Bortolato, S.A., V.A. Lozano, A.M. de la Peña, and A.C. Olivieri. 2015. Novel augmented parallel factor model for four-way calibration of high-performance liquid chromatography-fluorescence excitation-emission data, *Chemometrics and Intelligent Laboratory Systems*, 141: 1–11.
32. Amigo, J.M., T. Skov, R. Bro, J. Coello, and S. Maspoch. 2008. Solving GC-MS problems with PARAFAC2, *Trends in Analytical Chemistry*, 27(8): 714–725.
33. Van Benthem, M.H. and M.R. Keenan. 2008. Tucker1 model algorithms for fast solutions to large PARAFAC problems, *Journal of Chemometrics*, 22(5): 345–354.
34. Andersson, C.A. and R. Bro. 1998. Improving the speed of multi-way algorithms: Part I. Tucker3, *Chemometrics and Intelligent Laboratory Systems*, 42(1–2): 93–103.
35. Ten Berge, J.M.F. and A.K. Smilde. 2002. Non-triviality and identification of a constrained Tucker3 analysis, *Journal of Chemometrics*, 16(12): 609–612.
36. Escandar, G.M., H.C. Goicoechea, A. Muñoz de la Peña, and A.C. Olivieri. 2014. Second- and higher-order data generation and calibration: A tutorial, *Analytica Chimica Acta*, 806: 8–26.
37. Gómez, V. and M.P. Callao. 2008. Analytical applications of second-order calibration methods, *Analytica Chimica Acta*, 627(2): 169–183.
38. Olivieri, A.C., G.M. Escandar, and A.M.D.L. Peña. 2011. Second-order and higher-order multivariate calibration methods applied to non-multilinear data using different algorithms, *Trends in Analytical Chemistry*, 30(4): 607–617.
39. Ortiz, M.C. and L. Sarabia. 2007. Quantitative determination in chromatographic analysis based on N-way calibration strategies, *Journal of Chromatography A*, 1158(1–2): 94–110.
40. Olivieri, A.C., G.M. Escandar, H.C. Goicoechea, and A.M. de la Peña. 2015. Unfolded and multiway partial least-squares with residual multilinearization: Applications, *Data Handling in Science and Technology*, 29: 365–397.
41. Olivieri, A.C. 2005. On a versatile second-order multivariate calibration method based on partial least-squares and residual bilinearization: Second-order advantage and precision properties, *Journal of Chemometrics*, 19(4): 253–265.
42. Arancibia, J.A. and G.M. Escandar. 2014. Second-order chromatographic photochemically induced fluorescence emission data coupled to chemometric analysis for the simultaneous determination of urea herbicides in the presence of matrix co-eluting compounds, *Analytical Methods*, 6(15): 5503–5511.
43. Muñoz De La Peña, A., I. Durán Merás, and A. Jiménez Girón. 2006. Four-way calibration applied to the simultaneous determination of folic acid and methotrexate in urine samples, *Analytical and Bioanalytical Chemistry*, 385(7): 1289–1297.
44. Muñoz de la Peña, A., I.D. Merás, A. Jiménez Girón, and H.C. Goicoechea. 2007. Evaluation of unfolded-partial least-squares coupled to residual trilinearization for four-way calibration of folic acid and methotrexate in human serum samples, *Talanta*, 72(4): 1261–1268.
45. Maggio, R.M., A. Muñoz De La Peña, and A.C. Olivieri. 2011. Unfolded partial least-squares with residual quadrilinearization: A new multivariate algorithm for processing five-way data achieving the second-order advantage. Application to fourth-order excitation-emission-kinetic-pH fluorescence analytical data, *Chemometrics and Intelligent Laboratory Systems*, 109(2): 178–185.
46. Ortiz, M.C., L.A. Sarabia, M.S. Sánchez, A.A. Herrero, S.A. Sanllorente, and C.A. Reguera. 2015. Usefulness of PARAFAC for the quantification, identification, and description of analytical data, *Data Handling in Science and Technology*, 29: 37–81.

47. Cantwell, M.T., S.E.G. Porter, and S.C. Rutan. 2007. Evaluation of the multivariate selectivity of multi-way liquid chromatography methods, *Journal of Chemometrics*, 21(7–9): 335–345.

48. Smilde, A.K., R. Tauler, J. Saurina, and R. Bro. 1999. Calibration methods for complex second-order data, *Analytica Chimica Acta*, 398(2–3): 237–251.

49. Yu, Y.-J., H.-L. Wu, J.-F. Nie, S.-R. Zhang, S.-F. Li, Y.-N. Li, S.-H. Zhu, and R.-Q. Yu. 2011. A comparison of several trilinear second-order calibration algorithms, *Chemometrics and Intelligent Laboratory Systems*, 106(1): 93–107.

50. Smilde, A.K. 1997. Comments on multilinear PLS, *Journal of Chemometrics*, 11(5): 367–377.

51. Louwerse, D.J., A.K. Smilde, and H.A.L. Kiers. 1999. Cross-validation of multiway component models, *Journal of Chemometrics*, 13(5): 491–510.

52. Olivieri, A.C., H.-L. Wu, and R.-Q. Yu. 2012. MVC3: A MATLAB graphical interface toolbox for third-order multivariate calibration, *Chemometrics and Intelligent Laboratory Systems*, 116: 9–16.

53. Andersson, C.A. and R. Bro. 2000. The N-way toolbox for MATLAB, *Chemometrics and Intelligent Laboratory Systems*, 52(1): 1–4.

54. Komsta, Ł., M. Radoń, B. Radkiewicz, and R. Skibiński. 2011. Trilinear multiplicative modelling of thin layer chromatography retention as a function of solute, organic modifier and its concentration, *Journal of Separation Science*, 34(1): 59–63.

55. Komsta, Ł., R. Skibiński, N. Bezpalko, A. Mielniczek, and B. Stępkowska. 2016. Trilinear analysis of thin-layer chromatography retention of 35 model compounds chromatographed on nine adsorbents with 20 pure solvents, *Journal of Separation Science*, 39(21): 4258–4262.

56. Akvan, N. and H. Parastar. 2014. Second-order calibration for simultaneous determination of pharmaceuticals in water samples by solid-phase extraction and fast high-performance liquid chromatography with diode array detector, *Chemometrics and Intelligent Laboratory Systems*, 137: 146–154.

57. Arroyo, D., M.C. Ortiz, and L.A. Sarabia. 2007. Multiresponse optimization and parallel factor analysis, useful tools in the determination of estrogens by gas chromatography-mass spectrometry, *Journal of Chromatography A*, 1157(1–2): 358–368.

58. Arroyo, D., M.C. Ortiz, L.A. Sarabia, and F. Palacios. 2008. Advantages of PARAFAC calibration in the determination of malachite green and its metabolite in fish by liquid chromatography-tandem mass spectrometry, *Journal of Chromatography A*, 1187(1–2): 1–10.

59. Baunsgaard, D., C.A. Andersson, A. Arndal, and L. Munck. 2000. Multi-way chemometrics for mathematical separation of fluorescent colorants and colour precursors from spectrofluorimetry of beet sugar and beet sugar thick juice as validated by HPLC analysis, *Food Chemistry*, 70(1): 113–121.

60. Bechmann, I.E. 1997. Second-order data by flow injection analysis with spectrophotometric diode-array detection and incorporated gel-filtration chromatographic column, *Talanta*, 44(4): 585–591.

61. Beltrán, J.L., J. Guiteras, and R. Ferrer. 1998. Parallel factor analysis of partially resolved chromatographic data. Determination of polycyclic aromatic hydrocarbons in water samples, *Journal of Chromatography A*, 802(2): 263–275.

62. Cañada-Cañada, F., J.A. Arancibia, G.M. Escandar, G.A. Ibañez, A. Espinosa Mansilla, A. Muñoz de la Peña, and A.C. Olivieri. 2009. Second-order multivariate calibration procedures applied to high-performance liquid chromatography coupled to fast-scanning fluorescence detection for the determination of fluoroquinolones, *Journal of Chromatography A*, 1216(24): 4868–4876.

63. Culzoni, M.J., A.V. Schenone, N.E. Llamas, M. Garrido, M.S. Di Nezio, B.S. Fernández Band, and H.C. Goicoechea. 2009. Fast chromatographic method for the determination of dyes in beverages by using high performance liquid chromatography-Diode array detection data and second order algorithms, *Journal of Chromatography A*, 1216(42): 7063–7070.

64. Godoy, L.A.F. De, E.C. Ferreira, M.P. Pedroso, C.H.D.V. Fidélis, F. Augusto, and R.J. Poppi. 2008. Quantification of kerosene in gasoline by comprehensive two-dimensional gas chromatography and N-way multivariate analysis, *Analytical Letters*, 41(9): 1603–1614.

65. Llanos, A.M. De, M.M. De Zan, M.J. Culzoni, A. Espinosa-Mansilla, F. Cañada-Cañada, A.M. De La Peña, and H.C. Goicoechea. 2011. Determination of marker pteridines in urine by HPLC with fluorimetric detection and second-order multivariate calibration using MCR-ALS, *Analytical and Bioanalytical Chemistry*, 399(6): 2123–2135.

66. Durante, C., M. Cocchi, M. Grandi, A. Marchetti, and R. Bro. 2006. Application of N-PLS to gas chromatographic and sensory data of traditional balsamic vinegars of modena, *Chemometrics and Intelligent Laboratory Systems*, 83(1): 54–65.

67. Fraga, C.G. and C.A. Corley. 2005. The chemometric resolution and quantification of overlapped peaks form comprehensive two-dimensional liquid chromatography, *Journal of Chromatography A*, 1096(1–2): 40–49.

68. Hoggard, J.C. and R.E. Synovec. 2008. Automated resolution of nontarget analyte signals in GC × GC-TOFMS data using parallel factor analysis, *Analytical Chemistry*, 80(17): 6677–6688.
69. Gowin, E. and Ł. Komsta. 2012. Revisiting thin-layer chromatography as a lipophilicity determination tool. Part III—A study on CN adsorbent layers, *Journal of Planar Chromatography—Modern TLC*, 25(5): 471–474.
70. Gowin, E. and Ł. Komsta. 2012. Revisiting thin-layer chromatography as a lipophilicity determination tool. Part II—Is silica gel a reliable adsorbent for lipophilicity estimation?, *Journal of Planar Chromatography—Modern TLC*, 25(1): 5–9.
71. Komsta, Ł., R. Skibiński, A. Berecka, A. Gumieniczek, B. Radkiewicz, and M. Radoń. 2010. Revisiting thin-layer chromatography as a lipophilicity determination tool-A comparative study on several techniques with a model solute set, *Journal of Pharmaceutical and Biomedical Analysis*, 53(4): 911–918.
72. Komsta, Ł., R. Skibiński, A. Gumieniczek, and A. Wojnar. 2010. Multi-way analysis of retention of model compounds in thin-layer chromatography, *Acta Chromatographica*, 22(1): 27–36.
73. Lozano, V.A., A. Muñoz de la Peña, I. Durán-Merás, A. Espinosa Mansilla, and G.M. Escandar. 2013. Four-way multivariate calibration using ultra-fast high-performance liquid chromatography with fluorescence excitation-emission detection. Application to the direct analysis of chlorophylls a and b and pheophytins a and b in olive oils, *Chemometrics and Intelligent Laboratory Systems*, 125: 121–131.
74. Porter, S.E.G., D.R. Stoll, S.C. Rutan, P.W. Carr, and J.D. Cohen. 2006. Analysis of four-way two-dimensional liquid chromatography-diode array data: Application to metabolomics, *Analytical Chemistry*, 78(15): 5559–5569.
75. Pravdova, V., C. Boucon, S. De Jong, B.A. Walczak, and D.L.A. Massart. 2002. Three-way principal component analysis applied to food analysis: An example, *Analytica Chimica Acta*, 462(2): 133–148.
76. Qing, X.-D., H.-L. Wu, X.-H. Zhang, Y. Li, H.-W. Gu, and R.-Q. Yu. 2015. A novel fourth-order calibration method based on alternating quinquelinear decomposition algorithm for processing high performance liquid chromatography-diode array detection- kinetic-pH data of naptalam hydrolysis, *Analytica Chimica Acta*, 861: 12–24.
77. Schmidt, B., J.W. Jaroszewski, R. Bro, M. Witt, and D. Stærk. 2008. Combining PARAFAC analysis of HPLC-PDA profiles and structural characterization using HPLC-PDA-SPE-NMR-MS experiments: Commercial preparations of St. John's wort, *Analytical Chemistry*, 80(6): 1978–1987.
78. Sinha, A.E., C.G. Fraga, B.J. Prazen, and R.E. Synovec. 2004. Trilinear chemometric analysis of two-dimensional comprehensive gas chromatography-time-of-flight mass spectrometry data, *Journal of Chromatography A*, 1027(1–2): 269–277.
79. Van Mispelaar, V.G., A.C. Tas, A.K. Smilde, P.J. Schoenmakers, and A.C. Van Asten. 2003. Quantitative analysis of target components by comprehensive two-dimensional gas chromatography, *Journal of Chromatography A*, 1019(1–2): 15–29.
80. Vosough, M., M. Rashvand, H.M. Esfahani, K. Kargosha, and A. Salemi. 2015. Direct analysis of six antibiotics in wastewater samples using rapid high-performance liquid chromatography coupled with diode array detector: A chemometric study towards green analytical chemistry, *Talanta*, 135: 7–17.
81. Walker, S.A., R.M.W. Amon, C. Stedmon, S. Duan, and P. Louchouarn. 2009. The use of PARAFAC modeling to trace terrestrial dissolved organic matter and fingerprint water masses in coastal Canadian Arctic surface waters, *Journal of Geophysical Research: Biogeosciences*, 114(4): 1–12.
82. Watson, N.E., W.C. Siegler, J.C. Hoggard, and R.E. Synovec. 2007. Comprehensive three-dimensional gas chromatography with parallel factor analysis, *Analytical Chemistry*, 79(21): 8270–8280.
83. Wiberg, K. and S.P. Jacobsson. 2004. Parallel factor analysis of HPLC-DAD data for binary mixtures of lidocaine and prilocaine with different levels of chromatographic separation, *Analytica Chimica Acta*, 514(2): 203–209.
84. Wu, H.-L., M. Shibukawa, and K. Oguma. 1997. Second-order calibration based on alternating trilinear decomposition: A comparison with the traditional PARAFAC algorithm, *Analytical Sciences*, 13(Suppl): 53–58.

24 Recurrent Relationships in Separation Science

Igor Zenkevich

CONTENTS

24.1 IMPORTANT CHROMATOGRAPHIC REGULARITIES: INTRODUCTION TO RECURRENCES

Principal predestinations of different kinds of chromatography and related (hyphenated) separation methods are identification of analytes in multicomponent samples by non-spectroscopic methods and their quantitation. The first problem is solved using retention parameters of analytes, while the second is based on processing their peak areas. These tasks determine the set of mathematical relationships preferably used in chromatography*:

1. The dependence of so-called adjusted retention times $(t_R - t_0)$ of topologically related (e.g., normal linear) homologues of any series versus the number of carbon atoms (n_C) in their molecules:

$$\log\left(t_R - t_0\right) = a'n_C + b' \tag{24.1}$$

    where
        t_R is net retention time
        t_0 is holdup (void, dead, etc.) time (retention time of the compound not retained in
            chromatographic system)
        coefficients "a'" and "b'" of all regression equations are calculated using the least
            squares method (LSM)

---

* Here and hereinafter the coefficients (a', b', etc.) of nonrecurrent equations are indicated with asterisks, while the coefficients of recurrent equations without them.

Equation 24.1 approximates the retention times of analytes in isothermal gas chromatography (GC) and isocratic reversed-phase (RP) high-performance liquid chromatography (HPLC).

2. Temperature dependence of retention times of any analytes in GC described by the following linear regression:

$$\log\left(t_R - t_0\right) = a'/T + b' \tag{24.2}$$

where T is absolute temperature, K.

The analogue of this equation in RP HPLC is the dependence of the logarithms of retention factors $k' = (t_R - t_0)/t_0$ versus the volume fraction of organic modifier in an eluent, C:

$$\log k' = \log k_1' - SC \tag{24.3}$$

where

$\log k_1'$ is the conventional value of the logarithm of the retention factor for eluents containing no organic modifier

S is the coefficient interpreted as a characteristic of the elution ability of the organic solvent

Since $k' = (t_R - t_0)/t_0$, Equation 24.3 can be rewritten in the form similar to that of Equation 24.1, that is:

$$\log\left(t_R - t_0\right) = a'C + b' \tag{24.4}$$

3. Equations 24.1 through 24.4 include the values of holdup time, t_0, which can be measured experimentally, but often it seems better to evaluate them by different mathematical relations using data for other analytes.

4. Quantitative determinations are carried out using various methods of quantitation. Different modifications of these methods often require application of chemometric approaches.

All problems indicated here can be solved using various known mathematical approaches and algorithms. However, the search for new (more preferable) approaches remains an actual problem in separation science. Thus, such an unusual class of mathematical dependencies as *recurrent relations* was recommended for use in chemistry and, in particular, in chromatography. This type of function is well known in mathematics, but, surprisingly, rarely used in chemistry. These relations were proposed for the first time in 2005–2006 for approximation of various physicochemical properties (A) of organic compounds within homologous series. Most of the monotonous variations of these properties versus number of carbon atoms in the molecule (n) can be described by simple linear (first-order) recurrent equations [1–9]:

$$A\left(n+1\right) = aA\left(n\right) + b \tag{24.5}$$

The correlation coefficients (r) of these linear recurrent regressions for most properties usually exceed 0.999.

Chemical variables that obey Equation 24.5 include normal boiling point, critical parameters, density, refractive index, viscosity, surface tension, dielectric permittivity, solubility, distribution coefficients in heterophase systems of solvents, dissociation constants of organic acids, and many others. Keeping in mind the variety of these properties, most publications [1–9] are devoted to chemical applications of recurrences, but the subject of two of them is the approximation of chromatographic variables [4,7].

24.2 PRINCIPAL MATHEMATICAL PROPERTIES OF RECURRENT RELATIONS

Recurrent equations are a special class of mathematical objects that are used most extensively for discrete functions of integer arguments. It seems important to note that the number of atoms in the molecule is the example of just integer argument; therefore, all properties of organic compounds can be considered as functions of the integer argument. It follows that recurrences can be used to approximate the properties of homologues that differ from each other by homologous differences (most important among them being methylene fragment CH_2). Surprisingly, the feasibility of using such equations for chemical objects remained unknown up to 2005.

Apart from discrete functions of integer arguments, recurrent equations can be used to approximate continuous functions, provided their values correspond to sequences of equidistant argument values:

$$Y(x+\Delta x) = aY(x) + b, \quad \Delta x = \text{const} \tag{24.6}$$

Examples of continuous arguments are such important chemical variables as temperature ($x = T$), pressure ($x = P$), concentration ($x = C$), and so on. The chromatographic retention parameters of the analytes to be separated can be approximated with Equation 24.6 as well.

In mathematics, a recurrence (synonyms: recurrent or recursive relations) is known as a difference equation that defines a numerical sequence recursively. It means that each term of this sequence is defined as a function of the preceding term [10–12]:

$$x_i = f(x_{i-j}), \quad 1 \le j \le i \tag{24.7}$$

A recurrent function can be considered as a partial case of autoregression [13]:

$$x_i = \sum a_j x_{i-j} + \varepsilon_i \tag{24.8}$$

where a_j and ε_j are numerical coefficients.

A linear recurrent equation (24.5 and 24.6) like any recurrence has an algebraic (nonrecurrent) solution that has a function $Z(x)$. This solution can be easily found, for instance, with the use of standard computer Maple software or, if necessary, confirmed by mathematical induction:

$$Z(x) = ka^x + b(a^x - 1)/(a-1) \tag{24.9}$$

This solution is a series, because $(a^x - 1)/(a - 1) = a^{x-1} + a^{x-2} + \cdots + a + 1$, hence $Z(x) = ka^x + ba^{x-1} + ba^{x-2} + \cdots + ba + b$. Thus, the recurrent approximation of any numerical sequence can be regarded as an approximation by polynomials of not a constant but a variable power ($x = n_C \ne \text{const}$), that is, the highest degree of such polynomials is equal to the serial number of a variable in the row. It is the most important feature in the approximation of physicochemical properties of organic compounds: for every object the degree of approximating function depends on the position of the compound within a homologous series, namely, it is equal to the number of carbon atoms in the molecule, $x = n_C$.

A few noticeable consequences on the application of the approach under consideration can be derived from this solution. At first, if $a \equiv 1$ and $b \ne 0$, Equation 24.9 transforms into a relation of simple arithmetic progression, $Z(x) = k + bx$. At the same time, if $0 < a \ne 1$ and $b \equiv 0$, this equation converts into an expression of geometric progression, $Z(x) = ka^x$. Hence, in the general case (at arbitrary values of coefficients a and b), recurrent equation (24.9) combines the mathematical properties of both kinds of progressions and can be used for linear approximation not only of them both, but of any of their combinations in variable proportions. It demonstrates the high approximating "power" of recurrences for the approximation of various physicochemical properties of homologues.

Recurrent relations allow revealing the existence of limiting values of both discrete (A) and continuous $Y(x)$ properties. If the coefficients a of recurrences (24.5) or (24.6) obey the condition $a < 0$, then the values A or $Y(x)$ tend to non-infinite limits at the hypothetical tending of arguments to infinity ($n \rightarrow \infty$ or $x \rightarrow \infty$):

$$\lim \left[A \text{ or } Y(x) \right]\Big|_{n \text{ or } x \rightarrow \infty} = b/(1-a) \quad (\text{only at } a < 1) \tag{24.10}$$

By definition, if any set of numerical values belongs to a geometric progression, their logarithms form an arithmetic progression. If recurrences are suitable for approximation of them both, it means that these relationships can be selected for approximation not only of any set of numerical values $\{A_i\}$, but of their logarithms $\{\log A_i\}$ as well. This consequence seems to be highly important in chemistry, because it permits using the same equations for the approximation of physicochemical variables presented in different scales, namely, pressure (P) and its logarithms (log P), partition coefficients (K_p), values of log K_p, and so on. In accordance with that, chromatographic retention times can be approximated using recurrences both in their native (t_R) and logarithmic (log t_R) forms equivalently. Generalizing this rule, we can conclude that if recurrent relations are applicable for approximation of any variables A or $Y(x)$, the same recurrent relations appear to be applicable in the approximation of any monotonous functions of these variables.

The last, but not the least, remark: every point on the plots of the nonrecurrent dependencies $A(n)$ or $Y(x)$ corresponds to a single A or Y value, while every point on the plots of recurrent dependencies (24.5) or (24.6) corresponds simultaneously to two "contiguous" values A and Y. It means this is a special requirement to the accuracy of data for mathematical processing, because every erroneous or unreliable value distorts not one, but two points of the recurrent dependence.

In addition, recurrences possess some interesting "chemical" features [1–9] that are not directly connected with their mathematical properties. These features are more important for the chemistry of homologues and, hence, should not be considered here.

24.3   CHROMATOGRAPHIC APPLICATIONS OF RECURRENCES

24.3.1   RECURRENT APPROXIMATION OF THE DEPENDENCE OF CHROMATOGRAPHIC RETENTION TIMES ON THE NUMBER OF CARBON ATOMS IN HOMOLOGUE MOLECULES

The approximation of the dependence of retention parameters of homologues on the number of carbon atoms in their molecules, $t_R(n_C)$, is sometimes treated in practice as the problem of estimating sorbate retention parameters from the data on the few preceding homologues. Before discussing the applicability of recurrences for these purposes, note that the equation determining the form of the $t_R(n_C)$ dependence (Equation 24.1) is well known and it is one of the basic relations in the theory of chromatography [14–16].

The logarithms of adjusted retention parameters, log ($t_R - t_0$), are proportional to the number of carbon atoms in the molecule because of the additivity of the Gibbs energy of sorbate/sorbent or sorbate/stationary phase interactions [15–17]. The well-known system of such chromatographic invariants as Kovat's retention indices (RIs) for isothermal conditions of chromatographic separation [18] is based on this principle and the corresponding Equation 24.1.

The algorithm of standard nonrecurrent calculations using Equation 24.1 implies first the conversion of net retention times, t_R, into adjusted values, $t'_R = (t_R - t_0)$, followed by calculation of their logarithms, log($t_R - t_0$), and, finally, approximation of the obtained set of transformed data by LSM. The holdup time of chromatographic system, t_0, can be determined experimentally or estimated from retention times of three consecutive homologues of any series (e.g., by Peterson and Hirsch method [19]). Equation 24.1 is so commonly accepted in contemporary chromatography that the search for some alternative method(s) of solving the problem under consideration often seems nearly paradoxical. However, note that this theoretically correct equation is comparatively rarely used for obtaining real estimations of retention times for certain reasons. To explain these reasons, calculation by Equation 24.1 should be illustrated for at least one example.

Example 24.1

The retention times of n-alkanes C_5–C_{10} were determined on a capillary wall-coated open tubular (WCOT) column with nonpolar polydimethyl siloxane OV-101 under isothermal conditions (110°C, author's experimental data).

n in C_nH_{2n+2}	t_R, min	n in C_nH_{2n+2}	t_R, min
5	6.33	8	7.37
6	6.50	9	8.41
7	6.80	10	10.31

Let us estimate the retention times of n-alkanes C_{11}–C_{13} under the same conditions by extrapolation using these data.

Before calculating with Equation 24.1, we must evaluate t_0; the algorithm of Peterson and Hirsch [19] gives the value of 6.11 min. The parameters of linear regression (24.1) and the estimated retention times of n-alkanes C_{11}–C_{13} are listed in Tables 24.1 and 24.2. For the first target analyte C_{11}, the accuracy of the extrapolation with the "classical" Equation 24.1 can be considered appropriate [the error, $\Delta t_R = t_R(\text{calc.}) - t_R(\text{exp.})$ equals −0.22 min], but for more "remote" compounds C_{12} and C_{13}, the errors reach −0.59 and −1.88 min, correspondingly, which is unacceptable. These errors are observed in spite of the high value ($r = 0.9999$) of the correlation coefficient of regression (24.1). Thus, we can conclude, that this "classical" equation well *interpolates* experimental data but cannot be regarded as satisfactory for *extrapolation* of retention parameters.

In other words, there is mathematical explanation of poor applicability of Equation 24.1 for $t_R(n_C)$ data processing: the optimal approximation of $\log(t_R - t_0) = f(n_C)$ does not mean the optimal approximation of initial non-transformed variables, $t_R = f(n_C)$.

The simplest linear recurrent relation that can be used for approximation of net retention times instead of Equation 24.1 is as follows:

$$t_R(n+1) = at_R(n) + b \tag{24.11}$$

TABLE 24.1

Parameters of "Classical" Equation (24.1) and Alternative Recurrent Relation (24.11) for Estimation of Gas Chromatographic Retention Times of n-Alkanes C_{11}–C_{13} from the Data on n-Alkanes C_5–C_{10}

Parameter	Nonrecurrent Equation (24.1)	Recurrent Equation (24.11)
t_0, min	6.11 (estimated)	No estimation required
a or a'	0.257 ± 0.002	1.848 ± 0.005
b or b'	-1.950 ± 0.013	-5.20 ± 0.03
r	0.9999	0.9999
S_0	0.008	0.016

TABLE 24.2

Retention Times of *n*-Alkanes C_{11}–C_{13} Calculated with Nonrecurrent Equation (24.1) and Recurrent Relation (24.11) in Comparison with Experimental Values

n in C_nH_{2n+2}	t_R(exp.), min	t_R (Equation 24.1)	Difference, Δt_R	t_R (Equation 24.11)	Difference, Δt_R
11	13.86	13.64	**−0.22**	13.85	**−0.01**
12	20.31	19.72	**−0.59**	20.40	**+0.09**
13	32.59	30.71	**−1.88**	32.50	**−0.09**

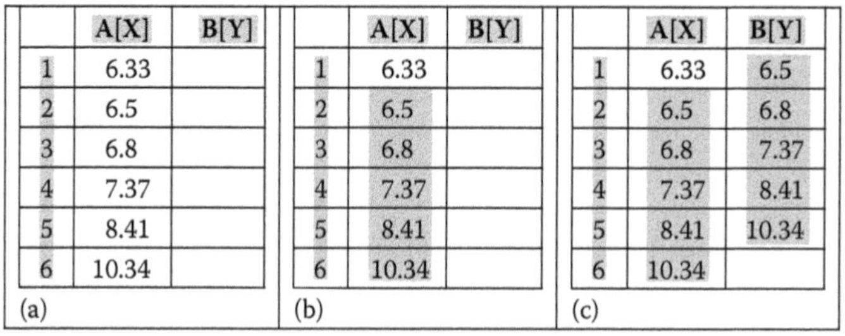

	A[X]	B[Y]			A[X]	B[Y]			A[X]	B[Y]
1	6.33			1	6.33			1	6.33	6.5
2	6.5			2	6.5			2	6.5	6.8
3	6.8			3	6.8			3	6.8	7.37
4	7.37			4	7.37			4	7.37	8.41
5	8.41			5	8.41			5	8.41	10.34
6	10.34			6	10.34			6	10.34	
(a)				(b)				(c)		

FIGURE 24.1 Schematic illustrating the recurrent calculations using standard software Origin (see comments to Example 24.1 in the text). The part of the column **A[X]** marked in gray corresponds to its copy in the column **B[Y]** by shifting one line up.

In accordance with general mathematical properties, the recurrences (see previous) are applicable for approximation of not only the sets of any numerical values $\{A_i\}$, but the sets of their logarithms $\{\log A_i\}$ as well. This property permits us to exclude preliminary conversion of net retention parameters t_R into adjusted (corrected) parameters ($t_R - t_0$) followed by calculation of their logarithms $\log (t_R - t_0)$, as it is necessary when using a "classical" approach based on Equation 24.1 without any loss in accuracy of approximation.

The features of calculations using Equation 24.11 seem to be reasonable to suggest the application of the Origin software. The initial set of one-dimensional data, namely t_R values, should be entered into the column **A[X]** (Figure 24.1a). After that, these data from all lines excluding the first one should be copied (Figure 24.1b) into the column **B[Y]** by shifting one line up (Figure 24.1c). This copying creates the two-dimensional data file corresponding to the first-order recurrences (24.5) or (24.6). The following steps of data processing imply the standard options of the Origin software, namely, plotting the function **B(A)** and calculating the parameters of linear regression $t_R(n + 1) = at_R(n) + b$: $a = 1.847 \pm 0.005$, $b = -5.20 \pm 0.003$, $r = 0.99999$, $S_0 = 0.008$, where r is the correlation coefficient and S_0 is the general dispersion (synonymous: the sum of residuals).

The parameters of linear recurrent regression (24.11) and the retention times of *n*-alkanes C_{11}–C_{13} estimated from the same initial data set without any additional data processing (recalculation into corrected retention times by subtracting holdup time t_0 followed by conversion into logarithms) are given for comparison in Tables 24.1 and 24.2. Graphic illustration of the linearity of the recurrent dependence for retention times of *n*-alkanes C_5–C_{10} is given in Figure 24.2. So long as the correlation coefficient is close to 1, we cannot distinguish the plot of this dependence from a straight line. The accuracy of extrapolated retention times in this case is incomparably higher than that for the preceding version of calculations and ranges from −0.09 to +0.09 min (which is comparable with the widths of chromatographic peaks on capillary columns).

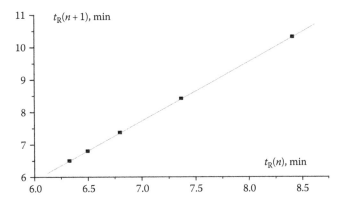

FIGURE 24.2   The linearity of the recurrent dependence of retention times of *n*-alkanes C_5–C_{10} in isothermal conditions of GC separation (Example 24.1).

The extreme simplicity of evaluations using recurrent relations should be especially underlined. The evaluation of retention time of the first target analyte (*n*-undecane) using the t_R value for the previous homologue (*n*-decane) in accordance with Equation 24.11 requires the simplest arithmetic calculations:

$$10.31 \times 1.848 - 5.20 = 13.85 \quad (\text{experimental value is } 13.86 \, \text{min [see Table 24.2]})$$

To evaluate the t_R value for *n*-dodecane, the same arithmetic calculations with the same numerical coefficients should be repeated twice:

$$10.31 \times (10.31 \times 1.848 - 5.20) - 5.20 = 20.40 \quad (\text{experimental value } 20.31 \, \text{min})$$

Similar triplicate sequence of calculations gives the t_R value for the last homologue under consideration (*n*-tridecane):

$$10.31 \times (10.31 \times [10.31 \times 1.848 - 5.20] - 5.20) - 5.20 = 32.50 \quad (32.59 \, \text{min})$$

It follows that substantial simplification of calculations is combined with a substantial increase in the accuracy of the results when recurrent equations are used for interpolation of chromatographic data. Such precision of evaluations in combination with their simplicity seems to be beyond the reach of the other calculation algorithms.

In the following examples, the scheme of recurrent calculation remains the same, which permits us to make shorter the comments required in every case.

A similar recurrent dependence holds for the retention parameters of homologues in RP HPLC in isocratic regimes, because they are also determined by the additivity of the Gibbs energies of sorbate/sorbent interactions. However, this area of application is probably less important, because, in practice, in the presence of several homologues in a sample, gradient rather than isocratic elution conditions are preferably used. Nevertheless, if necessary, the same recurrent approximation can be used in RP HPLC without any restrictions. The applicability of recurrences in RP HPLC for the solution of similar problems can be illustrated shortly by an example taken from a monograph [14].

Example 24.2

The retention times of n-alkyl esters C_3–C_6 of 2,4-dinitrobenzoic acid (homologous series test mixture) were determined on an RP HPLC column under isocratic conditions (for detailed conditions, see [14]):

n in C_nH_{2n+1}	t_R, s	n in C_nH_{2n+1}	t_R, s
3	131.2	5	196.2
4	158.4	6	250.5

Evaluate the retention time of the next homologue, 2,4-dinitrobenzoic acid n-heptyl ester, under the same conditions.

The application of Equation 24.11 to these data gives the following parameters of the linear recurrent regression: $a = 1.418 \pm 0.013$, $b = -27.9 \pm 2.1$, $r = 0.9996$, $S_0 = 0.6$. It is not necessary to illustrate the dependence with $r = 0.9996$ using a graphical plot because it looks like a straight line. After that, it is very easy to evaluate the retention time of the target analyte using the t_R value for the previous homologue (C_6 ester) with the simplest arithmetical calculations similar to those mentioned in Example 24.1:

$$250.5 \times 1.418 - 27.9 = 327.3 \quad \text{(experimental value is } 328.4\,\text{s [14]})$$

Thus, the error of recurrent evaluation is only -1.1 s, or 0.3%.

Of course, the number of examples illustrating the advantages (simplicity and accuracy) of recurrent approximation of chromatographic retention times can be essentially increased, but it cannot change the general character of resulting conclusions. No exceptions in the correctness of approach under consideration have been found up to present among different examples available.

The more complex problem, which requires special consideration, is the estimation of GC retention times of homologues under linear temperature programming (TP, as well as retention times in RP HPLC with gradient elution), because in this case the dependencies $t_R(n_C)$ are not so simple as in Equation 24.1 [20,21].

24.3.2 RECURRENT APPROXIMATION OF THE TEMPERATURE DEPENDENCE OF ANALYTES' RETENTION TIMES IN GAS CHROMATOGRAPHY

The temperature dependence of the retention times of analytes in GC, $t_R(T)$, is a problem often encountered in GC separation. As in the previous problem considered, the equation determining the form of the $t_R(T)$ dependence is well known [15,17,22]; most frequently, a two-parameter relation of the Antoine-type equation (24.2) is used.

As with Equation 24.1, theoretically correct equation (24.2) is comparatively seldom used in real analytical practice. This must be so because of some important reasons. Before discussing the possibility of the application of recurrent relationships for approximation of $t_R(T)$ dependencies, an important mathematical feature not mentioned earlier should be communicated. Hyperbolic dependencies of the type $y = a'/x + b'$ theoretically cannot be approximated by arithmetic or geometric progressions. It means that simple recurrent relationships theoretically cannot be used for approximation of the values of hyperbolical functions over the whole range of argument variations. However, if the set of data is considered over a local range of argument values fairly remote from the boundary values, this dependence can be approximated by a first-order recurrent equation with fairly high accuracy [7].

In solving the problem of estimating the GC retention parameters of a sorbate from the data obtained at other temperatures, several values in limited temperature intervals are considered as a rule. It follows that the use of recurrent equations is then quite justified.

Example 24.3

The retention times of 1-methylnaphthalene were determined on a WCOT column with OV-101 under isothermal conditions at 100°C, 110°C, and 120°C (author's experimental data):

T, °C	t_R, min
100	28.35
110	21.06
120	15.27

The retention time of this compound at 130°C should be estimated.

Calculations with Equation 24.2 (similar to those with Equation 24.1) require estimating the t_0 values, which are not necessarily equal to each other at different temperatures. The t_0 value at 130°C should be determined experimentally. The parameters of linear regression (24.2) are listed in Table 24.3 and the results of t_R evaluation at 130°C in Table 24.4. Note that calculation of linear regression parameters according to only three pairs of (x_i, y_i) data and subsequent use of the equation obtained for experimental purposes is a method that cannot be considered optimum, and the accuracy can be fairly low, as is in the case under consideration. Moreover, the graphic interpretation of Equation 24.2 indicates its clearly observed nonlinear character for the chosen compound (1-methylnaphthalene), as shown in Figure 24.3. Hence, the two-parameter Antoine-like equation (24.2) is not correct for approximation of $t_R(T)$ dependences for all possible analytes. It follows that this version of calculations in chromatography should be considered inapplicable to the precise extrapolation of retention data.

An alternative to Equation 24.2 recurrent relation is as follows:

$$t_R\left(T + \Delta T\right) = a t_R\left(T\right) + b, \quad \Delta T = \text{const} \tag{24.12}$$

TABLE 24.3
Parameters of Antoine-Like Equation (24.2) and Alternative Recurrent Relation (24.12) for Estimation of Gas Chromatographic Retention Time of 1-Methylnaphthalene at 130°C from the Data at Temperatures 100°C, 110°C, and 120°C

Parameter	Equation 24.2	Equation 24.12[a]
t_0, min	6.15 (100°C), 6.15 (110°C), 5.99 (120°C)	No estimation required[a]
a or a'	2.67 ± 0.18	0.809
b or b'	−5.6 ± 0.5	−3.081
r	0.997	1.0[b]
S_0	0.02	0[b]

[a] The t_0 value at 130°C was conventionally set equal to the t_0 at 120°C; generally, this parameter should be independently determined experimentally.
[b] Because only two points were used in calculations.

TABLE 24.4

Retention Time of 1-Methylnaphthalene at 130°C Calculated with Equation 24.2 and Recurrent Relation (24.12) in Comparison with Experimental Value

T, °C	t_R (exp), min	t_R (Equation 24.2)	Difference, Δt_R	t_R (Equation 24.12)	Difference, Δt_R
130	13.65	16.50	+2.85	13.74	+0.09

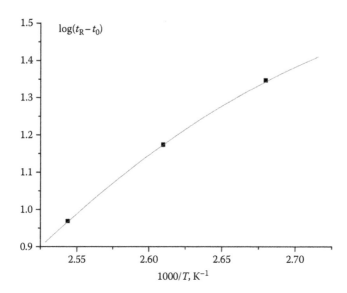

FIGURE 24.3 Nonlinear character of the Antoine-like dependence $\log (t_R - t_0) = a'/T + b'$ for 1-methylnaphthalene (Example 24.3).

Three t_R values at 100°C, 110°C, and 120°C produce only two pairs of their values. For linear regressions determined by only two points, $r \equiv 1$ and $S_0 \equiv 0$ by definition. Theoretically, the impossibility of obtaining accurate statistical estimates of the results prohibits the use of such a method for calculations but may be used for the unique possibilities of recurrences. The parameters of Equation 24.12 and the estimated retention time of 1-methylnaphthalene at 130°C are presented in Tables 24.3 and 24.4, correspondingly. The accuracy of recurrent estimate is +0.09 min, which is much better than that with standard $t_R(T)$ approximation.

The conclusion that can be drawn from this example is the same as that from the previous example. Substantial simplification of calculations is combined with a significant increase in the accuracy of results when recurrent equations are used.

24.3.3 RECURRENT APPROXIMATION OF THE DEPENDENCIES OF SORBATE RETENTION TIMES ON THE CONCENTRATION OF AN ORGANIC SOLVENT IN AN ELUENT IN RP-HPLC

The selection of the optimal regression equation for the dependence $t_R(C)$ of sorbate retention times on the concentration of organic solvents in mobile phases (eluents) in RP HPLC is more complex than that of the $t_R(n_C)$ and $t_R(T)$ dependencies in GC. Moreover, in contrast to the mentioned tasks, there is no unique equation for describing this version of chromatographic separation, which is a consequence of more complex mechanisms of sorbate/sorbent interactions in RP HPLC. Within comparatively narrow ranges of variations in the concentrations of organic mobile phase modifiers (according to the existing estimations, these ranges should be no more than 30%), the logarithms of retention factors $k = (t_R - t_0)/t_0$ are described by Equation 24.3, $\log k' = \log k'_1 - SC$ [23,24], where $\log k'_1$ is the conventional value of the logarithm of the retention factor for eluent containing no organic modifier, C is the volume fraction of organic modifier in an eluent, and S is the coefficient interpreted as a characteristic of the elution ability of the organic solvent.

Since $k' = (t_R - t_0)/t_0$, this equation can be rewritten in a form similar to that of Equation 24.1, namely, $\log (t_R - t_0) = a'C + b'$. If nonlinear $t_R(C)$ dependencies should be approximated over wider ranges of the concentrations of organic solvents in eluents, preference should be given to the use of second-degree polynomials [24]:

$$\log (t_R - t_0) = a'C^2 + b'C + c' \tag{24.13}$$

However, it should be taken into account that polynomials provide appropriate interpolation of data, while extrapolation can be accompanied by unpredictable errors in the results.

Because of limitations of two-parameter equations (24.2) and (24.3), the $t_R(C)$ dependencies for various compounds can be different even at equal eluent compositions. This feature can be exemplified by two compounds of phenol origin: phenol (**I**) and (2R,3R)-3',4',5,7-tetrahydroxyflavanonol (**II**, trivial names dihydroquercetin or taxifolin) with the following retention parameters:

$C(CH_3CN)$, % v/v	$t_R(I)$, min	$t_R(II)$, min
10	23.8	68.6
15	16.3	25.3
20	11.5	11.5
25	8.6	6.7
30	6.7	4.5

The comparison of two sets of $t_R(C)$ data shows that the retention times of taxifolin (**II**) at $C < 20\%$ exceed retention times of phenol (**I**), while at $C > 20\%$, it is *vice versa*. Such changes (inversion of elution order) in the sequences of elution of various compounds (usually of different chemical origin) are well-known effects in RP HPLC. Most interesting is that the mathematical form of optimum equations (the criterion of optimum is the maximal value of correlation coefficient, r) for approximating $t_R(C)$ is different for various sorbates. For phenol (**I**), Equation 24.14 is applicable, whereas for taxifolin (**II**), this dependence is obviously nonlinear, and the best results are obtained using an alternative equation of linear regression:

$$\log (t_R - t_0) = a' \log C + b' \tag{24.14}$$

It follows that the $t_R(C)$ dependencies for the two mentioned compounds are different under equal conditions of chromatographic separation. Additional examples reveal sorbates for which other equations are optimal. For instance, the set of retention parameters for phenol can be approximated by another "nonstandard" hyperbolic equation $t_R = a'/C + b'$ with the parameters $a' = 0.39 \pm 0.01$, $b' = 0.61 \pm 0.02$, $r = 0.998$, and $S_0 = 0.2$, which gives accuracy only insignificantly worse than that obtained using Equation 24.3. Of course, the sets of retention parameters for both sorbates correspond well to quadratic dependence (24.13).

By analogy with previously considered problems of approximation, the $t_R(C)$ dependence can be described by a recurrent equation:

$$t_R\left(C+\Delta C\right)=a t_R\left(C\right)+b, \quad \Delta C = \text{const} \tag{24.15}$$

Hence, irrespective of the chemical origin of analytes, recurrent relationship (24.14) can be used to approximate the concentration dependencies of the retention times of sorbates in RP HPLC with correlation coefficients no less than 0.999. For a more convincing comparison, it may be noted that the correlation coefficient for linear two-parameter recurrent dependence (24.15) for phenol ($r = 0.99997$) has the same order of magnitude as the correlation coefficient of three-parameter quadratic dependence (24.13) ($r = 0.99992$). The parameters of all regression equations are presented in Table 24.5.

The possibilities of calculating t_R values corresponding to arbitrary, but not necessarily equidistant ($\Delta C = 5\%$), concentrations of organic solvent in an eluent are based on the algebraic solution (24.9) of the first-order recurrent equation (24.6). It seems reasonable to observe them, for example, on one of the compounds considered (phenol). The coefficients a and b necessary to solve Equation 24.9 obviously correspond to the coefficients of recurrent equation (24.15), and the auxiliary parameter $k = [C_1 - b]/a$ (for phenol $k = 35.8$ at $C_1 = 10\%$). The x value substituted into (24.9) for any concentration C is

$$x = x_1 + \left(C - C_1\right)/\Delta C \tag{24.16}$$

As a result, for instance, for phenol retention times at acetonitrile concentrations in an eluent from 5% to 15% we obtain the set of values with steps of 1% presented in Table 24.6. Three of them (at $C = 5\%$, 10%, and 15%) can be compared with the experimental data, which confirms their complete coincidence within the number of significant digits specified.

If necessary, the same calculation procedure can be used in the evaluation of $t_R(T)$ values for arbitrary (not equidistant) values of temperature in GC.

TABLE 24.5
Parameters of "Classical" Equation (24.12), "Nonstandard" Equation (24.14), Polynomial of Second Degree (24.13), and Alternative Recurrent Relation (24.15) for Estimation of Retention Times of Phenol (I) and Taxifolin (II) in RP-HPLC Depending on the Concentration of Organic Solvent (CH_3CN) in Eluent

	Equation 24.12 or 24.14		Polynomial (24.13)		Recurrence (24.15)	
Parameter	(I) (Equation 24.12)	(II) (Equation 24.14)	(I)	(II)	(I)	(II)
a or a'	-0.032 ± 0.001	-2.84 ± 0.05	$(3.0 \pm 0.2) \times 10^{-4}$	$(15.2 \pm 0.2) \times 10^{-4}$	0.629 ± 0.005	0.331 ± 0.009
b or b'	1.65 ± 0.02	4.68 ± 0.06	$(-4.38 \pm 0.09) \times 10^{-2}$	-0.128 ± 0.009	1.31 ± 0.09	2.7 ± 0.3
c	—	—	1.756 ± 0.008	2.94 ± 0.08	—	—
r	-0.998	0.9996	0.99997	0.9993	0.99992	0.9991
S_0	0.02	0.02	0.002	0.02	0.06	0.02

TABLE 24.6
Retention Times of Phenol (I) at Various Acetonitrile Concentrations in the Eluent ($5\% \leq C \leq 15\%$) Estimated Using the Algebraic Solution (24.9) to Recurrent Equation (24.15)

C, % v/v	t_R (Equation 24.9)	t_R(exp)
5	35.8	36.2
6	32.9	
7	30.3	
8	27.9	
9	25.8	
10	23.8	23.8
11	22.0	
12	20.4	
13	18.9	
14	17.5	
15	16.3	16.3

24.3.4 RECURRENT APPROXIMATION OF HOLDUP TIME

24.3.4.1 Description of the Problem

The retention time of a component not retained by stationary phase or sorbent of the chromatographic column (t_0, "dead time," holdup time, void time, etc.) is considered as one of the principal parameters characterizing chromatographic separation processes [21,25,26]. It can be determined in the following way [21]:

$$t_0 = \int \left[F(L) \right]^{-1} dL \qquad (24.17)$$

where $F(L)$ is the linear flow of mobile phase at the column section dL, the integrating ranges are from 0 up to L, where L is the length of the column.

The knowledge of t_0 is necessary for excluding the influence of parameters of chromatographic systems on the retention characteristics of analytes. One of the forms of its presentation is retention coefficients (capacity factors, $k' = (t_R - t_0)/t_0$) widely used in HPLC [27,28].

In GC, the conversion of experimentally measured retention times, t_R, into adjusted values, $t_R' = t_R - t_0$, is used preferably. One of the reasons for this is the wide presentation of GC data in the form of RIs. At the isothermal condition of GC separation, the logarithms of adjusted retention parameters of homologues ($t_{R,n}'$) of any series are proportional to the number of carbon atoms in their molecules (n_C):

$$\log t_{R,n}' = a' n_C + b' \qquad (24.18)$$

So far, in most known RI systems, the standard values $RI_n = 100 n_C$ are postulated, and Equation 24.18 can be transformed into the following mathematically equivalent form:

$$RI_n = a' \log \left(t_{R,n} - t_0 \right) + b' \qquad (24.19)$$

Hence, the accurate calculation of RIs implies experimental determination or theoretical pre-calculation of t_0 values.

Chemometrics in Chromatography

Experimentally, the t_0 values are usually chosen to be equal to the retention times of methane, argon, or air injected into the GC column. Fairly often, they can be equaled to retention times of low boiling organic solvents [29,30]. In HPLC, t_0 values are most often accepted to be equal to retention parameters of inorganic salts [31]. In all cases, these experimental operations are highly simple. This fact should be taken into account as the main criterion of the usefulness of any algorithms of t_0 precalculation: they can be accepted only at the maximal simplicity of the calculations. Hence, it is not surprising that historically the first approach proposed as early as 1959 (Peterson and Hirsch method) [19] (some its modifications [32,33] are known) has acquired the most popularity. The evaluation of t_0 at isothermal conditions of GC analysis requires the values of retention times of three consecutive homologues:

$$\text{if } \log\left(t'_{R,2}\right) - \log\left(t'_{R,1}\right) = \log\left(t'_{R,3}\right) - \log\left(t'_{R,2}\right),$$

$$\text{then } t_0 = \frac{\left(t_{R,1}t_{R,3} - t_{R,2}^2\right)}{\left(t_{R,1} + t_{R,3} - 2t_{R,2}\right)} \tag{24.20}$$

Most often, n-alkanes are chosen as the easily available set of compounds for t_0 evaluation, but homologues of other classes can be used as well, such as methyl alkanoates [34].

The large variety of alternative methods is based on the use of relation (24.20). Its linearization in different ways permits both the evaluation of t_0 values and the correction of RIs with this parameter [33–50]. The interpretation of the physicochemical sense of the parameter t_0 in accordance with these approaches is the retention of the hypothetical homologue with zero number of carbon atoms in the molecule. The advantages and restrictions of various methods are summarized in several reviews [51–56].

Overviewing the problem of t_0 evaluation, two important, nearly paradoxical moments should be mentioned. First, most of the mentioned algorithms include rather complex calculations and, sometimes, require application of nonstandard software. Of course, it is not the principal moment, but such exhaustive complexity *de facto* restricts the use of the corresponding algorithms in analytical practice. Just due to this reason, the simplest method of Peterson and Hirsch [19] retains its importance up to present day. Second, the majority of the methods proposed to solve this problem are applicable only at the isothermal conditions of GC separation as well as Equation 24.20 itself. At the same time, most of the separation regimes used in contemporary analytical practice imply TP. The evaluations of parameter t_0 in TP conditions are considered only in a few publications [57–59].

However, the whole procedure of calculations appears to be unduly complex. To simplify the solution of this problem, the simplest empirical approximate relation was proposed [60]. It is based on the difference of retention times of only two successive homologues:

$$t_0 \approx 2t_{min} - t_{min+1} \tag{24.21}$$

where
  t_{min} is the retention time of the homologue with the minimal number of carbon atoms in the molecule (n), which is registered on the chromatogram as an individual peak
  t_{min+1} is the retention time of the next homologue with ($n + 1$) number of carbon atoms in the molecule

Relation (24.21) provides more or less appropriate t_0 evaluation only for selected TP regimes, if values t_{min} and t_{min+1} both correspond to the range $k \ll 1$. However, as it was demonstrated in the same paper [60], the influence of t_0 on RI values at $k > 1$ becomes so small that it can be neglected.

The importance of correct t_0 evaluations in TP conditions for fast and superfast chromatography [61] was demonstrated by Blumberg [21].

Hence, despite many publications considering the evaluation of parameter t_0, two main problems still exist: (1) to simplify the existing algorithms of calculations, and (2) to expand the methods of t_0 precalculation on regimes of linear TP.

24.3.4.2 Application of Recurrences in Evaluation of Holdup Time at Isothermal Conditions of GC Analysis

Table 24.7 includes the retention times of n-alkanes C_5–C_8 at three isothermal regimes. Experimental values of holdup time in the same regimes were accepted to be equal to the retention times of methane (the last column of this table). The use of "classical" equation (24.20) of Peterson and Hirsch's method for t_0 evaluation naturally leads to precise results: the differences between precalculated and experimental t_0 values do not exceed standard deviations of the last of them, that is, $\pm s(t_0)$.

If we operate with retention times of N consecutive homologues for calculation of parameters of recurrent linear regression, we obtain $(N - 1)$ points. Hence, to provide the statistical evaluation of these parameters (correlation coefficients, r, and sum of residuals, S_0), we need to select t_R values for at least four members of a series. If we use recurrent relations in the form (24.11), that is, in the ascending order of retention times, $t_R(n + 1) > t_R(n)$, the values of coefficients a appear to obey the inequality $a > 1$. In accordance with principal properties of recurrences (see earlier), it means the absence of the limiting values of the variable considered, that is, $t_R \to \infty$, which is in complete accordance with the physicochemical meaning of retention times of homologues in isothermal conditions of GC analysis. Nevertheless, to evaluate t_0 using this kind of recurrence, we can calculate retention times of preceding homologues; hence, we should select retention times in the descending order $[t_R(n - 1), t_R(n - 2), ..., t_R(1), t_0]$ using the relation reciprocating to (24.11), that is, $t_R(n) = [t_R(n + 1) - b]/a$. The calculations continue step by step until we obtain the final t_0 value. This algorithm corresponds to the interpretation of t_0 values as retention times of hypothetical homologues with zero number of carbon atoms in the molecule [46–48]. Thus, continuing with Example 24.1 (see earlier), starting from the t_R value for n-pentane (6.33 min), we obtain the following sequence of t_R evaluations for previous homologues:

$$t_R(4) = (6.33 + 5.20)/1.847 \approx 6.24 \text{ min}$$

$$t_R(3) = (6.24 + 5.20)/1.847 \approx 6.19 \text{ min}$$

$$t_R(2) = (6.19 + 5.20)/1.847 \approx 6.16 \text{ min}$$

$$t_R(1) = (6.16 + 5.20)/1.847 \approx 6.15 \text{ min}$$

$$t_R(0) = t_0 = (6.15 + 5.20)/1.847 \approx 6.14 \text{ min}$$

It should be noted that not only the last evaluation corresponds to the experimental t_0 value (6.14 ± 0.03 min, see Table 24.1) within the range of its standard deviation $\pm s(t_0)$, but the two previous values as well. It is not surprising because these are the retention times of low-retained methane $[t_R(1)]$ and ethane $[t_R(2)]$.

To simplify and rationalize the calculations with the use of recurrences, it seems reasonable to extrapolate retention times of homologues not in the ascending but in the descending order. Hence, Equation 24.11 should be transformed into another form:

$$t_R(n-1) = at_R(n) + b \tag{24.22}$$

TABLE 24.7

Comparison of the Precision of Different Evaluations of Holdup Time at Isothermal Conditions

	Evaluation of Holdup Time, min					
Temperature	Retention Times of n-Alkanes C_5–C_8, $t_R(5) \div t_R(8)$, min[a]	Equation 24.20	Equation 24.21	Consecutive Recurrent Evaluations for Retention Times of n-Alkanes C_4–C_0: $t_R(4)$, $t_R(3)$, $t_R(2)$, $t_R(1)$, $t_R(0) = t_0$[b]	$t_0 = \lim[b/(1-a)]$, $t_R(5) \div t_R(8)$	Experimental Values $t_0 \pm s(t_0)$, min[a]
$T_0 = 100°C$ (I)	6.30; 6.51; 6.91; 7.98	6.07	6.09	6.19(4), 6.13(3), **6.10**(2), **6.09**(1), **6.08**(0)	**6.07** (1.0000)[c]	6.08 ± 0.02
$T_0 = 110°C$ (II)	6.33; 6.50; 6.80; 7.37	6.11	6.16	6.24(4), 6.18(3), **6.16**(2), **6.14**(1), **6.13**(0)	**6.13** (0.9998)	6.14 ± 0.03
$T_0 = 130°C$ (III)	6.56; 6.66; 6.85; 7.19	6.45	6.46	6.51(4), **6.48**(3), **6.47**(2), **6.46**(1), **6.46**(0)	**6.45** (0.9999)	6.46 ± 0.02

Note: Evaluated t_0 values corresponding to the experimental data within the ranges $\pm s(t_0)$ are printed in bold.

[a] Averaging of three experimental values.

[b] Coefficients of Equation 24.22 are (I) $a = 0.521 \pm 0.001$, $b = 2.908 \pm 0.010$; (II) $a = 0.538 \pm 0.010$, $b = 2.83 \pm 0.07$; (III) $a = 0.549 \pm 0.008$, $b = 2.91 \pm 0.06$.

[c] Values of correlation coefficients (r) are indicated within parentheses here and hereinafter.

The values of coefficients $a < 1$ and b of this equation for different isothermal regimes I–III are presented in the footnotes to Table 24.7, while the corresponding correlation coefficients $r > 0.999$ are indicated in the column for t_0 evaluations. In the application of Equation 24.22, the possibility to evaluate t_R values for previous homologues $t_R(n-1)$, $t_R(n-2)$, …, $t_R(1)$, t_0 consecutively remains, and this is provided in Table 24.7. However, due to the condition $a < 1$, the calculation of the limiting values $t_0 = \lim t_R(n)|_{n \to 0}$ becomes possible without any intermediate stages:

$$t_0 = \lim t_R(n)\big|_{n \to 0} = b/(1-a), \quad (a < 1) \tag{24.23}$$

Differentiating this relation gives the equation for evaluation of standard deviation of calculated t_0 values, namely:

$$s(t_0) \approx t_0\left[\left(s_a/a\right)^2 + \left(s_b/b\right)^2\right]^{1/2} \tag{24.24}$$

As a result of recurrent approximation of retention times of few consecutive homologues and calculating the limits (24.23), we obtain t_0 values that show a good agreement with both experimental data and t_0 evaluations using a previously known relation (24.20). Thus, the verification of the new algorithm for the evaluation of holdup time confirms its high precision, but the simplicity of calculations exceeds that of previously known approaches. It unites the methods based on both the linearization of the dependence (24.1) and the algorithm proposed by Peterson and Hirsch [19] that can be confirmed by following observations.

If we consider the simplest recurrent dependence (24.22) for three t_R values only, we can write

$$t_2 = at_3 + b$$

$$t_1 = at_2 + b$$

TABLE 24.8

Reproducibility of Holdup Time Evaluations at Isothermal Conditions (Experimental t_0 Value Is 6.14 ± 0.03 min) Using Different Sets of n-Alkanes

Set of Data for n-Alkanes, $t_R(n_C)$, min	Parameters of Equation 24.23				$t_0 = \lim[b/(1-a)]$ (Equation 24.25)
	a	b	r	S_0	
7.37(8), 6.80(7), 6.50(6), 6.33(5)	0.538 ± 0.010	2.83 ± 0.07	0.9998	0.006	**6.13**[a]
8.41(9), 7.37(8), 6.80(7), 6.50(6)	0.541 ± 0.006	2.82 ± 0.04	0.99995	0.006	**6.14**
10.34(10), 8.41(9), 7.37(8), 6.80(7)	0.542 ± 0.002	2.81 ± 0.02	0.99999	0.005	**6.14**
13.86(11), 10.34(10), 8.41(9), 7.37(8)	0.545 ± 0.002	2.78 ± 0.03	0.99999	0.009	**6.11**
20.40(12), 13.86(11), 10.34(10), 8.41(9)	0.541 ± 0.003	2.82 ± 0.04	0.99999	0.018	**6.14**
31.92(13), 20.40(12), 13.86(11), 10.34(10)	0.558 ± 0.008	2.55 ± 0.18	0.99991	0.099	5.77 (unacceptable)
20.40(12), 10.34(10), 7.37(8), 6.50(6)	0.295 ± 0.000	4.325 ± 0.006	1.00000	0.004	**6.13**
20.40(12), 10.34(10), 7.37(8)	0.295	4.317	—	—	**6.12**
20.40(12), 8.41(9). 6.50(6)	0.159	5.160	—	—	**6.14**

[a] Evaluated t_0 values corresponding to the experimental data within the ranges $\pm s(t_0)$ are printed in bold.

This system should be solved relative to the coefficients a and b:

$$a = (t_2 - t_1) / (t_3 - t_2)$$

$$b = (t_1 t_3 - t_2^2) / (t_3 - t_2)$$

The substitution of these coefficients in Equation 24.23 gives the following:

$$t_0 = \lim t_R(n)\big|_{n \to 0} = b / (1 - a) = (t_1 t_3 - t_2^2) / (t_1 + t_3 - 2t_2) \tag{24.25}$$

Hence, the formula f or evaluating the limiting value t_0 using coefficients of recurrent relations appears to be exactly equivalent to the equation proposed by Peterson and Hirsch (24.20) for calculating t_0 by using retention times of three consecutive homologues.

To prove the validity of the recurrence approach, t_0 evaluations obtained with different sets on n-alkanes should be compared. Table 24.8 gives the results at the isothermal regime II using t_R data for n-alkanes C_5–C_8, C_6–C_9, C_7–C_{10}, C_8–C_{11}, and C_9–C_{12}. All t_0 values are well consistent with the experimental t_0 value 6.14 ± 0.03 min. Only the attempt to use the most "remote" set of data as that for C_{10}–C_{13} n-alkanes gives inappropriate t_0 evaluation. Additionally, it should be noted that fairly precise t_0 evaluations are provided using retention times for sets of nonconsecutive n-alkanes, for example, C_6, C_8, C_{10}, C_{12}; C_8, C_{10}, C_{12}; and even C_6, C_9, C_{12}, and so on. It confirms the simplicity and usefulness of recurrent t_0 evaluations in chromatographic practice.

For additional illustration of recurrent approximation of retention times of homologues in evaluation of holdup time at the isothermal conditions of GC separation, let us consider the example taken from the literature.

Example 24.4

In the monograph of R.E. Kaiser and A.J. Rackstraw ([14], pp. 59, 67), the evaluation of parameter t_0 using retention times of n-alkanes C_6–C_{10} is considered: 1880 s (C_{10}), 1090 s (C_9), 676 (C_8), 460 (C_7), and 346 s (C_6). As a result of **five steps** of iterated data processing, the final value $t_0 = 221.7$ s was calculated.

The same problem can be solved in a much simpler way using recurrences. Using the standard Origin software (even its obsolete versions) requires entering data as illustrated in Figure 24.1, followed by calculating the parameters of linear regression (24.22):

$$a = 0.5237 \pm 0.0004, \quad b = 105.4 \pm 0.5, \quad r = 1.0000, \quad S_0 = 0.5$$

These values of coefficients permit us to evaluate the limiting t_0 value $105.4/(1 - 0.5237) \approx 221.3$ s (Equation 24.23) with standard deviation ± 1.1 s, which is equivalent to the value reported in the monograph [14].

The number of examples based on published data can be increased significantly, but in all cases it confirms the correctness of the recurrent algorithm discussed. Besides that, recurrences can be used in the evaluation of holdup times at TP regimes, but this problem seems to be more complex; hence, it is considered in a special publication [4].

24.4 CONCLUSIONS AND PROSPECTS

To summarize, the dependencies of chromatographic retention parameters on the temperature of the column in GC and the concentration of the organic eluent component in RP HPLC can be written by an identical first-order recurrent equation:

$$t_R(Z + \Delta Z) = at_R(Z) + b, \quad \Delta Z = \text{const}$$

where $Z = T$ or C.

In addition, this recurrent equation can be solved to obtain intermediate values of continuous arguments, that is, for $\Delta Z \neq$ const.

Both in GC (isothermal elution) and in RP HPLC (isocratic elution), the retention times of homologues of any series, differing in the number of carbon atoms in the molecule (n_C), are in accordance with a similar recurrent relationship:

$$t_R(n_C + 1) = at_R(n_C) + b$$

Such general conclusions concerning chromatographic retention patterns are in complete accordance with similar conclusions on the application of recurrent approximation to different physicochemical properties of organic compounds. All of them are based on the unique mathematical properties of recurrences, previously established without the attention of chemists.

The application of recurrent relations made it possible for us to propose a new simple algorithm for evaluating the holdup time. Moreover, the unique possibilities of such a chromatographic quantification method as those of double internal standards are explained just by unique properties of recurrent relations. The selection of the previous and the next homologues of the target analyte as two standards permits us to compensate any losses of these compounds at any stages of sample preparation for analysis [62].

REFERENCES

1. Zenkevich, I.G. 2016. *J. Chemometr.* 30:217–225.
2. Zenkevich, I.G. 2014. *J. Chemometr.* 28:311–318.
3. Zenkevich, I.G. 2012. *J. Chemometr.* 26:108–116.
4. Zenkevich, I.G. 2012. *Chromatographia.* 75:767–777.
5. Zenkevich, I.G. 2010. *J. Chemometr.* 24:158–167.
6. Zenkevich, I.G. 2009. *J. Math. Chem.* 46:913–933.

7. Zenkevich, I.G. 2009. *J. Chemometr.* 23:179–187.
8. Zenkevich, I.G. 2007. *Russ. J. Struct. Chem.* 48:1006–1014.
9. Zenkevich, I.G. 2006. *Russ. J. Org. Chem.* 42:1–11.
10. Cormen , T.H., Leiserson, C.E., Rivest, R.L., and Stein, C. 1990. In *Introduction to Algorithms*, 2nd Edn. London: MIT Press and McGraw-Hill. Chapter 4, Recurrences.
11. Romanko, V.K. 2006. *Difference Equations* (in Russian). Moscow, Russia: Binom.
12. Oleinik, V.L. 2001. *Soros Educ. J.* (in Russian) 7:114–120.
13. Bourke, P. 1998. Autoregression analysis. http://astronomy.swin.edu.au/~pbourke/other/ar/ (accessed May 2016).
14. Kaiser, R.E., and Rackstraw, A.J. 1983. *Computer Chromatography*. Heidelberg, Germany: Alfred Huethig Verlag GmbH.
15. Ettre, L.S., and Hinshaw, J.V. 1993. *Basic Relationships in Gas Chromatography. Reference Book.* Cleveland, OH: Advanstar.
16. J. Cazes, Ed. 2001. *Encyclopedia of Chromatography*. New York: Marcel Dekker, Inc. pp. 835–836.
17. Cazes, J., and Scott, R.P.W. 2002. *Chromatography Theory*. New York: Marcel Dekker, Inc.
18. Kovats, E. 1958. *Helv. Chim. Acta.* 41:1915–1932.
19. Peterson, M.L., and Hirsch, J. 1959. *J. Lipid Res.* 1:132–134.
20. Harris, W.E., and Habgood, H.W. 1966. *Programming Temperature Gas Chromatography*. New York: John Wiley & Sons.
21. Blumberg, L.M. 2010. *Temperature-Programmed Gas Chromatography*. Weinheim, Germany: Wiley-VCH.
22. Grob, R.L., and Barry, E.F. 2004. *Modern Practice of Gas Chromatography*, 4th Edn. New York: John Wiley & Sons.
23. Shoenmakers, P.J. 1989. *Optimization of Selectivity in Chromatography* (in Russian transl.). Moscow, Russia: Mir.
24. Meyer, V. 1999. *Practical High Performance Liquid Chromatography*, 3rd Edn. New York: John Wiley & Sons.
25. Scott, R.P.W. 2010. Dead point: Volume or time. In *Encyclopedia of Chromatography*, 3rd Edn., Vol. 1, J. Cazes, Ed. Boca Raton, FL: Taylor & Francis. pp. 557–558.
26. Ettre, L.S. 1980. *Chromatographia.* 13:73–84.
27. Jinno, K. 2010. Void volume in LC. In *Encyclopedia of Chromatography*, 3rd Edn., Vol. 3, J. Cazes, Ed. Boca Raton, FL: Taylor & Francis. pp. 2430–2431.
28. Alhedai, A., Martire, D.E., and Scott, R.P.W. 1989. *Analyst.* 114:869–875.
29. Parcher, J.F., and Jonhson, D.H. 1980. *J. Chromatogr. Sci.* 18:267–272.
30. Vezzani, S., Castello, G., and Pierani, D. 1998. *J. Chromatogr. A.* 811:85–96.
31. Oumada, F.Z., Roses, M., and Bosch, E. 2000. *Talanta.* 53:667–677.
32. Harvey, J., and Gold, U.S. 1962. *Anal. Chem.* 34:174–175.
33. Riedmann, M. 1974. *Chromatographia.* 7:59–62.
34. Lomsugarit, S., Jeyashoke, N., and Krisnangkura, K. 2001. *J. Chromatogr. A.* 926:337–340.
35. Guardino, X., Albaiges, J., Firpo, G., Rodriquez-Vinals, R., and Gassiot, M. 1976. *J. Chromatogr.* 118:13–22.
36. Grobler, A., and Balizs, G. 1974. *J. Chromatogr. Sci.* 12:57–58.
37. Toth, A., and Zala, E. 1984. *J. Chromatogr.* 284:53–62.
38. Toth, A., Zala, E. 1984. *J. Chromatogr.* 298:381–387.
39. Smith, R.J., Haken, J.K., and Wainwright, M.S. 1985. *J. Chromatogr.* 331:389–395.
40. Ballschmiter, K., Heeg, F.J., Meu, H.J., and Zinburg, R. 1985. *Fresenius J. Anal. Chem.* 321:426–435.
41. Touabet, A., Maeck, M., Badjah Hadi Ahmed, A.Y., and Meklati, B.Y. 1986. *Chromatographia.* 22:245–248.
42. Touabet, A., Badjah Hadi Ahmed, A.Y., Maeck, M., and Meklati, B.Y. 1986. *J. High Resol. Chromatogr. Chromatogr. Commun.* 8:456–460.
43. Wronski, B., Szczepaniak, L.M., and Witkiewicz, Z. 1986. *J. Chromatogr.* 364:53–61.
44. Vigdergauz, M.S., and Petrova, E.I. 1987. *Russ. J. Anal. Chem.* 42:1476–1481.
45. Maeck, M., Touabet, A., Badjah Hadi Ahmed, A.Y., and Meklati, B.Y. 1989. *Chromatographia.* 29:205–208.
46. Quintanilla-Lopez, J.E., Lebron-Aguilar, R., and Garcia-Dominguez, J.A. 1997. *J. Chromatogr. A.* 767:127–136.
47. Lebron-Aguilar, R., Quintanilla-Lopez, J.E., and Garcia-Dominguez, J.A. 1997. *J. Chromatogr. A.* 760:219–227.

48. Garcia-Dominguez, J.A., Quintanilla-Lopez, J.E., and Lebron-Aguilar, R. 1998. *J. Chromatogr. A.* 803:197–202.

49. Aryusuk, K., and Krisnangkura, K. 2003. *J. Sep. Sci.* 26:1688–1692.

50. Pous-Torres, S., Torres-Lapasity, J.R., and Garsia-Alvarez-Cogue, M.C. 2009. *J. Liq. Chromatogr. Relat. Technol.* 32:1065–1083.

51. Wainwright, M.S., and Haken, J.K. 1980. *J. Chromatogr.* 184:1–20.

52. Smith, R.J., Haken, J.K., and Wainwright, M.S. 1985. *J. Chromatogr.* 334:95–127.

53. Smith, R.J., Haken, J.K., Wainwright, M.S., and Madden, B.G. 1985. *J. Chromatogr.* 328:11–34.

54. Kaiser, R.E., and Bertsch, W. 1998. *J. High Resol. Chromatogr. Chromatogr. Commun.* 1:115–120.

55. Dominguez, J.A.G., Diez-Masa, J.C., and Davankov, V.A. 2001. *Pure Appl. Chem.* 73:969–992.

56. Wu, N.S., Wu, G.S., and Wu, M.Y. 2006. *J. Chromatogr. Sci.* 44:244–246.

57. Curvers, J., Rijks, J., Cramers, C., Knauss, K., and Larson, P. 1985. *J. High Resol. Chromatogr. Chromatogr. Commun.* 8:611–617.

58. Golovnya, R.V., and Svetlova, N.Y. 1988. *Russ. J. Anal. Chem.* 43:859–865.

59. Golovnya, R.V., and Svetlova, N.Y. 1988. *Chromatographia.* 25:493–496.

60. Zenkevich, I.G., and Szczepaniak, L.M. 1992. *Russ. J. Anal. Chem.* 47:507–513.

61. Klemp, M.A., and Sacks, R.D. 1991. *J. Chromatogr. Sci.* 29:507–510.

62. Zenkevich I.G., and Makarov E.D. 2007. *J. Chromatogr. A.* 1150:117–123.

25 Chemometrics and Image Processing in Thin-Layer Chromatography

Bahram Hemmateenejad, Elaheh Talebanpour Bayat, Elmira Rafatmah, Zahra Shojaeifard, Nabiollah Mobaraki, and Saeed Yousefinejad

CONTENTS

25.1 SHORT INTRODUCTION TO THIN-LAYER CHROMATOGRAPHY

Chromatography, according to the definition by the International Union of Pure and Applied Chemistry (IUPAC), is a physical separation method in which the components to be separated are distributed between two stationary and mobile phases. Chromatographic techniques can be classified depending on the basis of stationary bed shape or mobile phase. Column chromatography and planar chromatography (thin-layer and paper chromatography) can be noted as the two most well-known bed shapes in chromatographic methods. On the other hand, three major branches of chromatography from the view of mobile phases are (1) liquid chromatography, (2) super critical fluid chromatography, and (3) gas chromatography (GC).

In 1938, Izmailov and Shraiber [1] introduced a novel variant of the chromatographic technique, which was called spot chromatography at that time and is known as thin-layer chromatography (TLC) today. TLC, as a subcategory of planar chromatography, is a type of chromatography that should be carried out in a layer of adsorbent spread on different kinds of supports such as paper or glass plate.

The principle of operation in TLC is alternative switching of a molecule from a sorbed to an unsorbed state. The distribution tendency of species toward the sorbent or mobile phase leads to the separation process. In other words, the more the tendencies toward the sorbent, the lesser are the distances on the plate and vice versa. Factors such as types of the stationary phase (mean particle size, morphology of the stationary particle, and size distribution), activation of the stationary phase (expelling the physically adsorbed water from the surface of the silica gel), and choice of solvent system (adequate purity, ability to dissolve the mixture of sample, low viscosity, a vapor pressure that is neither very low nor very high) can effect thin-layer chromatograms.

TLC can be considered as a simple, cheap, and fast qualitative and quantitative technique. Due to these advantages, TLC has been applied in purity testing of pharmaceuticals and drugs [2–4], and in the identification and determination of active substances in fields such as clinical chemistry [5,6], forensic chemistry [7,8], cosmetology [9,10], and food analysis [11,12]. However, it suffers from some drawbacks such as zone broadening and variable mobile phase velocity. To overcome these limitations, special developments have been suggested. Among these, the technique with forced flow known as overpressured layer chromatography [13], high-pressure planar liquid chromatography [14,15], and rotation planar chromatography [16,17] can be highlighted.

Practically, the small predefined volumes of samples are applied on the TLC plate, either as spots or as narrow bands, in precise positions. After evaporating the sample solvent, the mobile phase is developed in one of these three modes: ascending [17–19], descending [20,21], or horizontal [22–24]. Visualization of the results and their representation in terms of retardation factor (R_f) and resolution (RS) are the final steps. The retardation factor is obtained by dividing the distance between the substance zone and the starting line by the distance between the solvent front and the starting line, and resolution is obtained by dividing the distance between the two spot-center zones by calculating the sum of the radius of the two separated zones.

According to the subject of this chapter, it should be emphasized that the visualization of TLC has been developed over time. The human eye was the first detector of visualization of TLC results. The evolution of detectors has divided them into two categories: (1) general reagent detectors and (2) specific ones. In the general reagent kind, all the components need to be visible to make their detection possible. Therefore, these reagents are capable of fulfilling this visibility. Iodine reagent [25–27] (iodine bound physically to the components turning them into brown spots) and sulfuric acid spray [28] (oxidizing the components and leaving black spots) are examples of such reagents. In specific reagent detectors, just the relevant target component will be visualized such as acid base indicators for carboxylic acids [29], ninhydrin for amino acids [28], and aniline for reducing sugars [30,31].

Alongside these reagents, fluorescence detection and scanning densitometry could also help in spot detection [32–34]. Solutes with intrinsic fluorescence can be detected on TLC plates under UV lamps. On the other hand, if components do not have fluorescence, we can use TLC plates with fluorescent background so that these components may quench the background fluorescence. In scanning densitometry, another common technique for this purpose, comparison between the reflectance light or generated fluorescence of spots and other parts of the TLC plate is performed [35].

Recent improvements in detection methods for TLC are focused on image analysis [36–39]. The required equipment is composed of imaging detectors to acquire images of the TLC plates, light source, and appropriate optics such as filters, which are all embedded in a box, and a computer test software for data acquisition. Since the cost of this analyzing system is much less than TLC slit-scanning densitometry, several research groups have conducted quantitative image analysis of TLC plates [37,40,41]. This technique also facilitates data acquisition for various applications and allows utilization of modern chemometric methods for data analysis.

In the following sections of this chapter, we will introduce the concept of image analysis and its instruments (scanner and charge-coupled device [CCD] camera) and also the role of chemometric methods in this ongoing field.

25.2 IMAGE ANALYSIS

25.2.1 COLOR SPACE

A three-dimensional (3D) model is usually utilized to create a specified color defined as "color space." A color space explains an environment to compute, arrange, represent, or compare colors. The components of the color space environment represent intensity values and thus can be utilized as quantitative tools. It is noteworthy that the difference between color spaces and their definition resulted from the choice of the axes which will be explained in this section briefly. Red, green, blue (RGB) and cyan, yellow, magenta, black (CMYK) color spaces are two more common systems; the former is currently utilized extensively in image analysis. The RGB space uses the additive mixing color because it is based on lights and the CMYK space uses the subtractive mixing color because it is based on pigments.

The RGB space is the simplest color space. Colors in the RGB space are expressed with a mixture of red, green, and blue primary colors. It is based on a Cartesian coordinate system and each of the three coordinates stands for one of the three noted colors. For example, the pure red color is defined by R = 1, G = 0, and B = 0, and similarly for the two other colors. It should also be noted that the black color will be created when all the coordinates are equal to zero. The RGB color space could be illustrated with three red, green, and blue circles named as primary colors.

Each of the secondary colors (cyan, magenta, and yellow) could be created by adding the two appropriate primary colors (Figure 25.1). In this space, the other colors could be created by a combination of primary colors. Adobe RGB and sRGB are the most conventionally used color spaces. Adobe RGB is larger than sRGB.

25.2.2 VIDEO DENSITOMETRY

Densitometry can be considered a technique to detect TLC plate tracks and create analog curves for qualitative and quantitative analysis. Densitometry can be categorized into two general subsets: (1) scanning (opto-mechanical) and (2) video [42]. These techniques are applicable for in situ visualization, documentation, and evaluation of tracks on the developed TLC plate or electrophoresis sheet. Scanning densitometry is the stable and relatively mature technique that can record in situ absorption and fluorescence spectra from the entire range of UV-visible wavelength of 190–800 nm [43]. A densitometer can be equipped with a special detector such as radioisotope detection [44] or interfaced with modern spectroscopic techniques to obtain more selective and sensitive online identification or quantification of separated compounds on TLC plates. Spectroscopic methods including mass [45,46], infrared [47], Raman [48], and nuclear magnetic resonance [49] are some examples of such advanced spectroscopic techniques.

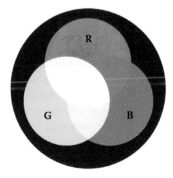

FIGURE 25.1 Representation of RGB color system and defining black and other combinational colors in this system.

The inherent part of scanning densitometry is its capability for sequential scanning through the TLC plate by the predetermined mono- or poly-chromic light. Therefore, the obtained spectral selectivity and accuracy can be utilized to achieve valuable information such as peak purity of chromatographic zone, peak, and compound identification with a desirable figure of merit (sensitivity, selectivity, accuracy, precision, detection limit, and dynamic response range), and resolution [50]. On the other hand, the technique has some disadvantages including being high-cost and time-consuming. In other words, it is not possible to equip all laboratories with these devices because of their high purchase price [51].

Time spent for data acquisition depends on various parameters such as size of TLC plate, desired spatial resolution, wavelength range, type of analyzed sample, and so on. In the case of two-dimensional chromatography, time consumption is one of its greatest limitations for application in mechanical scanning. To overcome these principal limitations, video densitometry was introduced for electrical documentation and evaluation of TLC plates [50].

Video densitometry is a rapidly evolving technique that is based on electronic point-scanning of the total desired area at the same time. Generally, the designed accessory is composed of a suitable external light source if required, a monochromator or filter, and an imaging processor with acceptable resolution for storage of an original (e.g., TLC plate) as a digital image. Most of these sets are housed in a dark package to prevent interference from ambient light. Commonly used imaging processors comprise of a scanner, a CCD camera, and a cell phone. These devices are connected to the computer for adjusting capture setting (e.g., resolution, cropping, exposure time, and enlargement) and color correction (e.g., contrast, brightness, sharpness, saturation, and color balance) with appropriate software to achieve a suitable appearance of the image [52]. A video densitometer cannot be used as a "black box technique" that transfers results into data files [53]. So it needs convenient software for data extraction of a visually captured image. This is the point at which chemometrics finds its entry.

Chemometric methods can be utilized for selecting optimum capturing condition or color correction parameters in the next step [54]. Their versatility, nondestructiveness, and low-cost means for fast data acquisition with a simple instrumental design may be noted as the main attractive parameters for their widespread application in TLC evaluation. Extensive publications in the literature have documented TLC plates with flatbed scanner and CCD camera, so the main focus of the following sections is on these image processors.

25.2.2.1 Flatbed Scanner

Flatbed (desktop) scanner is an opto-electromechanical device that can deliver digital images from both reflective and transparent materials. The principal scanning operation is based on the discrete movement of a carriage at a fixed speed through an object that is placed facedown on the platen transparent glass. The carriage contains a light source, an image sensor, and optic element if required. The light source directs a beam onto the object. For example, white light or infrared radiation [55] can be utilized for special applications such as automatic corrections in transmittance mode. The received light (reflected or transmitted) from the object passes through a set of optic elements (mirror and lens). During this process, the mirror and lens project the collimated light (specula) onto a linear image sensor (LIS), which converts light to analog voltage. At each step of traversing along the scanning direction, a snapshot line of information that is required for image creation is saved in a buffer unit until the scan of the desired object is completed. An analog to digital converter (A/D) is utilized for conversion of analog voltages to digital values. The raw digital data are processed in a standalone driver or by professional image-editing software such as TWAIN interface as a famous and standard software [56,57]. Three commonly used scanners are contact image sensor (CIS), CCD, and LED inDirect exposure (LiDE) scanner. However, all types of scanners use CCD technology in LIS, which is simpler, has lower price, and is more stable in comparison with other technologies of sensing. A typical element layout of a flatbed CCD scanner is shown in Figure 25.2. Performance specification of the scanner and ultimately the captured image quality are evaluated by several factors, which are briefly discussed later [57,58].

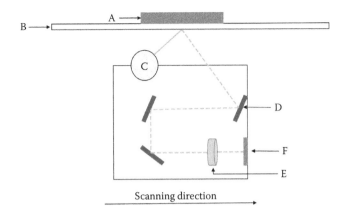

FIGURE 25.2 Element layout of flatbed CCD scanner. The original (A) face down on the platen glass (B) is scanned by the sequential discrete movement of the carriage. The carriage is composed of a light source (C), mirrors (D), a cylindrical lens (E), and CCD linear image sensor (F).

25.2.2.1.1 Resolution

Manufacturers quote resolution of scanners in two terms: optical and interpolated resolution. LIS characteristic (e.g., number of used pixels per inch) and number of steps the carriage moves per inch are the main factors in optical resolution. Optical resolution is limited by LIS specification. The true capability of a scanner for rendering detailed information of the original is evaluated via LIS specification and should be considered in scientific application. By interpolation, a scanner driver provides an option for increasing the number of pixels in an image to greater than optical resolution without adding any further details about the original image (interpolated resolution).

25.2.2.1.2 Optical Density Range (Dynamic Range)

Optical density (D) is a common logarithmic scale of opacity (O). Opacity is referred as the ability of an original to absorb the illuminated light. It is given in Equation 25.1 for the opaque original and in Equation 25.2 for the transparent original:

$$O = \frac{\text{Reflected light intensity}}{\text{Incident light intensity}} \tag{25.1}$$

$$O = \frac{\text{Incident light intensity}}{\text{Transmitted light intensity}} \tag{25.2}$$

Manufacturers address optical density range (ODR) in terms of dynamic range. It shows the ability of the scanner to perceive the subtle discrimination of tones at the end of both shadowed and highlighted areas. It is worth noting that the ODR of the captured image not only depends on the ODR of the scanner but is also affected by the ODR of the original.

25.2.2.1.3 Color Depth

Color depth is referred as the number of used bits for conversion of analog data to color information for a single pixel and for each channel of triple RGB channels by A/D converter. Color depth can be considered as the indirect indicator of the dynamic range and the direct indicator of the fineness of the output tone scale. In addition to the scanner, image-editing software and scanner driver must support the desired color depth of the delivered image to prevent loss of information.

25.2.2.1.4 *Noise Level*

Noise comprises the most considerable proportion of visual errors that take place during the scanning process. Condition of lightening, image sensor, image enlargement, and so on can lead to introduction of noise into data file information. It limits the available dynamic range of the scanner. On the other hand, the impact of noise can be reduced by averaging multiple scans and using specially designed software.

A powerful scanner driver or professional software package via TWAIN interface can provide several options that permit control over the scanning and display parameters. These enable the operator to manipulate the image, such as providing special features for automatic correction, color management, and noise reduction. These additional options reduce scanning error and develop image quality. Tagged Image File Format (TIFF) and Joint Photographic Experts Group (JPEG) file formats are the common forms that are usually used for image storage. TIFF is a better option for quantitative analysis because of its flexibility with higher color bit (8 or 16 bits per channel) without losing information while saving images [59].

Open-source licenses or inexpensive software (e.g., ImageJ, JustTLC, Biostep, and Sorbfil TLC Videodensitomer) [59] and home-written programs [60] have been used to extract chromatographic data form TLC plates. Triple RGB channels or averaging as gray scale has been utilized to describe image information.

Aside from peripheral office scanners, commercial flatbed scanners are available for scientific applications, for example, ViewPix (Biostep, Jahnsdouf, Germany) and ChromImage (AR2i, Clamart, France) scanners. The outstanding features of these scanners include advanced optic system and software, and their applicability in the UV region for TLC analysis.

Despite significant color consistency, dynamic range, color fidelity, and resolution of scanners [56] as a result of their uniform illumination, common commercial peripheral flatbed scanners suffer from some intrinsic limitations such as wavelength region (only visible), ability for only real-time or online analysis, and fluorescence detection. In this regard, CCD technology was introduced to use as an image processor in the field of TLC analysis.

25.2.2.2 CCD Camera

The progress of technology and particularly technical improvement in photography hardware and software has promoted the usage of CCD cameras in various fields of analysis. Since 1970, improvements in the digital world and accessibility of new digital instruments for most people have led to finding ways to import digital devices into science.

CCDs are one of the instruments that have been commercialized and are available from the 1970s [61,62]. Because of the progress in the development of CCDs, they have become one of the most significant technological elements in the digital world, especially in the role of detectors in digital cameras. A CCD plays the same role as photographic film in conventional devices (chemistry photography) with the replacement of the sensitized film with a CCD photon detector [63].

In fact, a CCD camera is the most common way of capturing images in the form of electrical signals. They are widely used in various applications such as television camera, astronomy (spy satellite), and analytical spectroscopy. With rapid progress in the digital world and the spread of its applications in homes, digital cameras will soon be replaced by camera phones in smartphones [64].

It is worth noting that CCDs are known as highly sensitive photon detectors. A CCD consists of many small light-sensitive parts that are known as pixels, which can be used to get an image of the desired scene. Photons of light that come into the pixels are allowed to be converted into electrons. The electrons are collected and read by an imposed electric field and camera electronics, respectively. Obviously, the number of trapped electrons will be directly related to the intensity of light at each pixel [65]. The capturing of photons and their conversion into electrons occur through the creation of electron–hole pairs in metal oxide semiconductors. The CCD itself is primarily made of silicon with different doped elements to make it p-doped silicon oxides.

By applying a positive charge on the electrode, all of the negatively charged electrons close to the electrode are attracted to its surface. On the other hand, extant positively charged holes are repulsed from the positive electrode. Consequently, a "potential well" of negative charge that is produced by incoming photons is stored on the electrode and more light produces more electrons. The number of electrons that are attracted to the potential well increases until the potential well is full near the electrode (potential well capacity). A CCD consists of many potential wells (pixels) that are arranged in rows and columns. The numbers of pixels in the rows and columns determine the size of a CCD (e.g., 1024 pixel high × 1024 pixel wide per 1 in.). In addition, the width or patch of each pixel is defined as CCD resolution.

The charges of each pixel can be collected by transferring their contents to their neighbors and the charge of the last capacitor is converted into a voltage by a charge amplifier. In the final step, a CCD chip is embedded into the CCD camera as an analogue device that amplifies small voltages, decreases noises, and converts the pixel values into a digital format and finally sends the values of each pixel to a personal computer (PC) as the device's output for image displacement [65,66].

Scientific grades of CCDs are dependent on their applicability features such as quantum efficiency (QE), wavelength range, dynamic range, linearity, noise, and power [67]. Because of the importance of these features, each one is introduced here briefly:

Quantum efficiency: The percentage of detectable electrons created from photons is known as the QE and has a significant effect on CCD quality.

Wavelength range: CCDs can support a wide range of wavelengths from 400 nm (blue) to about 1050 nm (infrared). The most sensitive recording of images can be done by a CCD in 700 nm. However, by using a process known as back-thinning, the wavelength range of a CCD can be extended into shorter wavelengths such as the extreme ultraviolet and x-ray [63,64].

Dynamic range and resolution: By directing light onto a CCD, photons are converted into electrons until the saturation of the potential well occurs. The brightness and dimness (dynamic range) of an image depends on the maximum and minimum number of electrons that can be supplied within the potential well. Furthermore, the brightness and faintness of an image can be a measure of CCD resolution.

Linearity: The transformation of incident photons into final digitized electronic outputs should be linear. It is noteworthy that possessing an appropriate linearity is so essential for quantitative image analysis using CCD cameras.

Noises: In order to gain the true quality of an instrument, the signal-to-noise ratio should be determined. The magnitude of the signal can be determined through QE measurement. To keep noise level as low as possible, various kinds of noises in CCD should be considered. Readout noises and dark noise can be noted as two main types of CCD noises.

Dark noise comes from the creation of electron–hole pairs by external sources such as high temperature. Consequently, the full capacity of each pixel is reached in a few seconds and this leads to saturation of the CCD. The dark noises can be significantly reduced by cooling the device.

Readout noises, the ultimate noise limit of cameras, arise from the conversion of electrons trapped in each pixel into a voltage by the CCD output node. The size of the output node can influence the magnitude of the readout noise. It is remarkable how the combination of two noise levels and QE parameters can define the sensitivity of a CCD camera [68].

Separation of chemical species in a quantitative view through digital images is an attractive aspect in analyses. Previously reported methods such as scanners suffer some intrinsic limitation such as limited wavelength region (only visible), slow operation, and thereby their inability for real-time or online analysis. Whereas the ability to obtain digital image of chromatograms through a faster and easier method for recording and preserving the results of a TLC analysis has made the entrance of CCD cameras quick and widespread [63,69].

Consequently, extraction and analysis of the information presented in an image can offer helpful output details in an easier way. The combination of image analysis with processing methods like chemometric techniques can provide good opportunity for handling analytical and biological data.

25.3 APPLICATION OF CHEMOMETRICS IN IMAGE ANALYSIS

According to the nature and huge amount of data in image processing, application of multivariate chemometric methods seems necessary for resolving hidden relationships in data and/or extraction of more useful information. It is well known that chemometrics plays a significant role in obtaining robust results from a large number of data, finding indirect function relationships in the data, and making multidimensional structures [70–74]. Various chemometric methods have been developed for different purposes, but based on the reported applications the utilized approaches in TLC image analysis can be divided into two categories: "factor analysis" for calibration and regression goals and "pattern recognition" methods for data exploratory and classification aims. A brief description about these categories and more applicable methods are represented in Figure 25.3.

25.3.1 FACTOR ANALYSIS

As noted previously, factor analysis in TLC image processing has been applied for two main purposes. One is data reduction and omission of collinearity in large amounts of gathered imaging data and the other is for quantitative analysis and determination of concentration or function of analyzed component(s). In the first approach, the analysis is commonly conducted on the input variables (XX) without using target information (y, i.e., concentration or activity), which is known as "factor analysis without a priori." In the second approach, which is named "factor analysis with a priori," both X and y are required to build a calibration model for prediction and quantification in TLC image analysis [75].

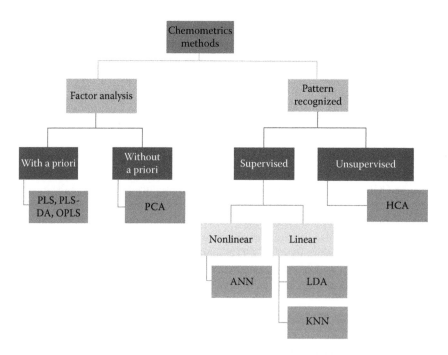

FIGURE 25.3 Classification of chemometric methods utilized in image analysis.

25.3.1.1 Factor Analysis without A Priori

Principal component analysis (PCA)* is the most common technique for multivariate data analysis. In 1901, Pearson introduced the base of the PCA theorem, similar to the previous principal axis theorem in mechanics and mathematics [76], which was a model developed by Hotelling in the 1930s for experimental science [77].

PCA is a statistical procedure that can convert a series of correlated variables into linearly uncorrelated values by orthogonal transformation. Each uncorrelated value is called a principal component [78]. The utility of PCA has been proven in various scientific fields, such as physics, mathematics, and chemistry [79–81]. Different goals can be achieved by the PCA method such as simplifying data by reducing the large original data size, classification, variable selection, outlier detection, prediction, and modeling [82–85].

PCA consists of mathematical decomposition and transformation of original data matrix (X) that contains n samples and m variables that are located in rows and columns, respectively [86], as shown schematically in Figure 25.4.

The outcome of decomposition of matrix (X) is two T (n × k) and P (k × m) matrixes in which k is the significant component causing variation in the X matrix. T is known as "score" matrix and the P matrix is called "loading." The unexplained variances that are not covered by PCA during decomposition, for example, measurement errors, are collected in the error matrix (E). The orthogonality of score and loading vectors, and the normality of the loading vector are two distinct features of eigenvectors.

The product of ath column of T (t_a) and a^{th} row of P (p_a) is named the eigenvector of ath PC. In this regard, the summation of eigenvectors can be regarded as a model of original data (X):

$$X = c_1 p_1^T + c_2 p_2^T + \cdots + c_k p_k^T + E \tag{25.3}$$

The value of each PC that is named an eigenvalue is commonly used as a marker for determining the number of significant components (rank) in a series of samples. Eigenvalues are arranged in a way that the first PC has the largest possible variance. A simple definition of eigenvalue is the sum of squares of the scores, as shown in the following equation:

$$\lambda_a = \sum_{i=1}^{I} t_{ia}^2 \tag{25.4}$$

where λ_a is the ath eigenvalue.

Successive eigenvalues correspond to smaller contents that can be used to evaluate the number of PCs. This number of PCs is the best way to model a data matrix [78,87]. Various algorithms have been suggested for PCA, such as nonlinear iterative partial least squares (NIPALS) and singular value decomposition (SVD) [88,89]. The SVD algorithm is extensively accepted as the most precise

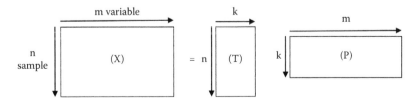

FIGURE 25.4 Schematic of decomposition of a data matrix into score and loading.

* See also Chapter 8.

and stable technique for PCA or a data set [87]. The Matrix Laboratory (MATLAB®) software has a code for SVD (known as svd) that gives three explained SVD outputs (U, S, V) in which singular values and eigenvectors are sorted from the largest to the smallest.

In image analysis, PCA can play diverse roles. At first and as a common approach, it is utilized to compress image information by converting it into a reduced number of uncorrelated PCs while relevant information is retained. In the next step, PCA can be used as an input for exploration of data analysis, image segmentation, modeling, classification, and so on [90]. Because of the known advantages of PCA, this method is applicable in computer visions for feature extraction and also as an auxiliary tool for feature selection in order to identify the most relevant variables in the original data matrix to produce image information [90].

Komsta et al. [91] applied the PCA technique for clustering three different carex species based on two-dimensional TLC digital photos. Since the differences between images can be related to the spots' intensities, differences in backgrounds and spots' location can influence the PCA results. In this regard and to reduce these drawbacks, some preprocessing filters have been utilized before cluster analysis [91]. On the other hand, because TLC images recorded by a CCD camera need denoising and baseline correction, Komsta et al. used a median filter and a rollerball algorithm, and also nonlinear warping with spline function for these goals. Applying these preprocessing techniques and using first principal components were significantly effective to obtain an obvious clustering in the analyzed samples by image analysis.

In 2015, Hawryl et al. separated and classified 11 essential oils obtained from various mentha specimens. They applied two-dimensional TLC for separation and a scanner to record their samples. Furthermore, a chemometric method such as PCA has been used for classification and identification of unknown samples [92].

25.3.1.2 Factor Analysis with A Priori

Partial least squares (PLS)* is the most well-known regression technique in multivariate data analysis that has been developed by Herman Wold [93]. This technique reduces the variables to a smaller set of uncorrelated components that are weighted upon their correlation with response [94,95]. Least squares regression is performed on obtained variables and finally PLS results in a linear model that fits all variables.

There are several algorithms that have been used in PLS regression (PLSR): Kernel algorithm [96], NIPALS algorithm [97], SIMPLS algorithm [95], and HÖskuldsson algorithm [98] can be noted as some of the well-known algorithms for PLS.

PLS is also used in discriminant analysis or classification. In such cases, it is called partial least squares discriminant analysis (PLS-DA) [99]. PLS-DA uses the PLSR algorithm to find latent variables (linear combinations of the original variables) with a maximum covariance with the Y variables [100]. In other words, if we have the independent X variables, we can derive the discriminant plane by PLS. Then projecting new (unknown) observations onto the obtained model can lead to predicting the class of these observations.

Another form of PLS is orthogonal partial least square (OPLS). Here, the systematic variation in **X** that is orthogonal to **Y** or, in other words, variations that are not correlated to **Y** are removed [101]. So, the PLS model can be easily explained and we are able to analyze the uncorrelated variations separately. A small modification in the NIPALS algorithm is enough for obtaining the OPLS algorithm. With OPLS, the detection limit of outliers can be improved and the source of disturbing variation can be identified.

It should be noted that PLS has been used efficiently in combination with TLC and image analysis, and some bold examples will be noted later. Because of the potential of the well-known PLS method, it can be a good and useful tool in TLC image analysis for simultaneous determination of analytes with overlapped signals or determination in the presence of unknown interferences.

* See also Chapter 14.

In 2010, Hemmateenejad et al. [102] offered the solution for peak overlapping by multivariate image analysis, which uses PLS as the calibration method. Images of TLC plates were captured by a digital camera fixed on top of the cabinet, beside two white fluorescent lamps as shown in Figure 25.5. Images were related o the measurement of methyl yellow (MY), cresol red (CR), and bromocresol green (BG) and also for the determination of nifedipine and its photo degradation product. The saved images in JPEG format were imported into MATLAB environment, and the obtained results in each image contains three color intensity matrices of red (R), green (G), and blue (B). As represented in Figure 25.6, the image data were converted to 3D (color value vs. length and widths of TLC sheet) and 2D (color value vs. length of TLC sheet) chromatograms using the written subroutine. As can be seen in Figure 25.6, the significant overlapping can cause difficulty in the determination of analytes in a solution containing their mixture. This problem was solved by smoothing and feeding data to multivariate calibration methods (PLS model).

TLC image analysis was also used to monitor the quantitative progress of organic reactions such as reduction of benzaldehyde derivatives and alkaline hydrolysis of phenyl benzoate [103]. Small aliquots (µl) of starting material of these reactions as well as successive aliquots of the reaction mixture were spotted on TLC plates. Hexane–ethyl acetate (3:1 and 15:1 for reduction reactions and alkaline hydrolysis, respectively) was utilized as the mobile phase to elute the spots [103]. After TLC development, the images of the TLC plates were recorded by the home-made image-recording cabinet (schematically shown in Figure 25.5) to convert the images into 2D and 3D TLC

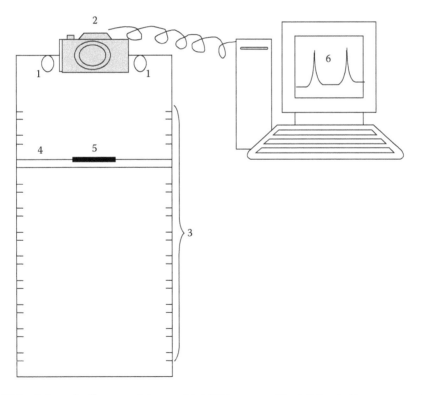

FIGURE 25.5 Schematic diagram of the provided MIA system. The left-hand side represents the cross-sectional area of the imaging cabinet. 1, White radiation fluorescent lamps; 2, digital camera; 3, groves for tuning the distance of TLC sheets from camera; 4, holder for TLC plates; 5, TLC plates; 6, multivariate analysis of the images recorded by a camera. (From Hemmateenejad, B., Mobaraki, N., Shakerizadeh-Shirazi, F., and Miri R., Multivariate image analysis-thin layer chromatography (MIA-TLC) for simultaneous determination of co-eluting components, *Analyst*, 135(7), 1747–1758. Copyright 2010. Reproduced by permission of the Royal Society of Chemistry.)

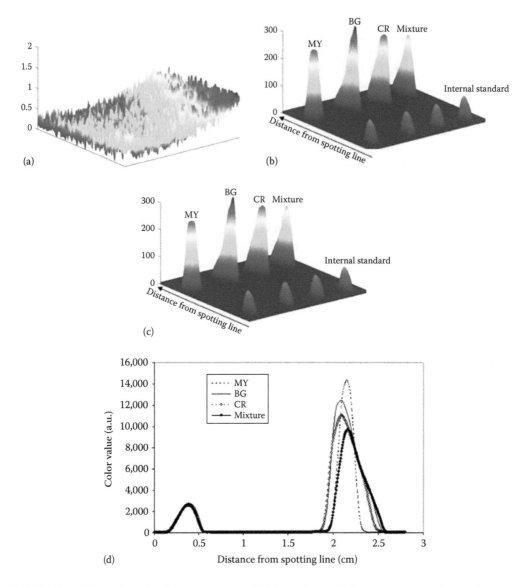

FIGURE 25.6 Three-dimensional chromatograms of (a) the background; (b) pure analytes and a typical mixture of indicators before background subtraction; and (c) after background subtraction. (d) Two-dimensional chromatograms of the indicators and their mixture. The levels of analytes in the individual spots are 300.0 ng and in the mixture, 100.0 ng. (From Hemmateenejad, B., Mobaraki, N., Shakerizadeh-Shirazi, F., and Miri R., Multivariate image analysis-thin layer chromatography (MIA-TLC) for simultaneous determination of co-eluting components, *Analyst*, 135(7), 1747–1758. Copyright 2010. Reproduced by permission of the Royal Society of Chemistry.)

chromatograms. These obtained chromatograms were analyzed as a function of reaction time to obtain conversion percentage of both model reactions (equal to 99.9%) and to estimate the rate constants of these reactions. Results showed that TLC in combination with image analysis can be suggested as a quantitative reaction-monitoring tool and a simple alternative to GC for organic laboratories [103].

This procedure had been also used for simultaneous determination of two isomeric amino acids leucine and isoleucine in 2014 by Hemmateenejad's research group [104]. Linear calibration models were constructed for these amino acids and detection limits were equal to 0.60 and

0.48 µg for leucine and isoleucine, respectively, after optimization of the composition of mobile phase. Because of incomplete separation of the spots of the analytes (transformed into image chromatograms), the chromatograms were analyzed by PLS. This led to simultaneous determination of leucine and isoleucine in their binary mixtures with root mean square error of prediction of 0.008 and 0.009 µg, respectively. The prediction ability of the multivariate model was led to good recoveries for determination of these biological isomers (in the range of 95%–105%) in diluted serum plasma [104].

Trifković et al. [105] presented a method for classification of propolis (a resinous material collected by honey bees) with high-performance thin-layer chromatography (HPTLC) and image analysis. Image capturing was performed at 366 nm with Camag video documentation system, which provided uniform lighting of surfaces, fast scanning, and high optical resolution. HPTLC chromatograms were divided by orange and blue types upon the orange and blue bands that appeared on HPTLC plates. Orange bands represent phenolics, such as pinocembrin, galangin, caffeic acid phenethylester, and chrysin, while blue patterns are caused by p-coumaric acid and ellagic acid. The captured images were processed with *IMAGE J* software and the exported data were used for chemometrics data handling. PCA had been done at the beginning to reduce data dimension, identification of variables, and determination of outliers. Hierarchical cluster analysis (HCA) was performed on groups of similar objects with consideration of all data variability. Propolis types were modeled by using PLS-DA and a mathematical model was obtained for further classification of unknown samples.

In 2015, multicomponent image analysis of Pereskia bleo leaf compounds had been conducted by TLC to predict their antioxidant activity [106]. A digital camera was used to photograph the TLC plates, and these images were preprocessed by Image J software. TLC images were converted into a wavelet signal and then 500 variables were chosen in a way that 70% of total variation in wavelet signals could be described by them. These 500 selected variables were the $\mathbf{X}$ (input) variables and antioxidant activities of samples were the $\mathbf{y}$ (output or response) variables for the OPLS model. The obtained OPLS model from TLC images can be used for fast bioactivity measurement of complex mixtures of Pereskia bleo leaf extract.

25.3.2 PATTERN RECOGNITION METHODS

As shown in Figure 25.3, the pattern recognition methods can be divided based on the nature of the relation between the input variable (X) and the target vector (y) into linear and nonlinear methods, which the former limits to respect only a linear mathematical relationship but the latter has no such limitation [107,108]. On the other hand, pattern recognition methods can also be categorized into supervised and non-supervised groups according to whether they use any information about the classes in data or not. In other words, non-supervised methods operate only on $\mathbf{X}$ data but supervised methods need the class marker information ($\mathbf{y}$) to analyze $\mathbf{X}$. In the following subsection, a brief discussion about the common pattern recognition algorithms in TLC image analysis will be presented with some applications.

25.3.2.1 Linear Supervised Pattern Recognition Methods

Discriminant analysis (DA)* deals with transforming input variables into a definite number of latent variables with maximum class separation. The most common and also the oldest version of DA method is known as linear discriminant analysis (LDA), which was developed by Ronald Fisher [109], and can be categorized as a supervised pattern recognition method. LDA assumes that all of the classes have the same dispersion in the observations, which means the variance and covariance matrices of each category are equal to the others.

---

* See also Chapter 15.

In LDA, determination of a line (onto which the points will be projected) is done in a direction causing maximum separation among classes. The methodology is calculating within-class and between-class scatter matrix, obtaining the ratio of between-class scatter to within-class scatter, and finally maximizing this ratio [110]. Suppose we have n classes that each have a mean vector μ_i and r number of samples; then within-class scatter (S_w) and between-class scatter (S_b) can be computed using the following equations [110]:

$$S_w = \sum_{i=1}^{n}\sum_{j=1}^{r_i}\left(x_j - \mu_i\right)\left(x_j - \mu_i\right)^T \tag{25.5}$$

$$S_b = \sum_{i=1}^{n}\left(\mu_i - \mu\right)\left(\mu_i - \mu\right)^T \tag{25.6}$$

where clearly μ is the mean of observations:

$$\mu = \frac{1}{n}\sum_{i=1}^{n}\mu_i \tag{25.7}$$

LDA is mostly applied in classification problems and also for the reduction of dimension. As an example, effective TLC image analysis has been developed for monitoring and controlling the quality of dietary supplements with the aim of LDA [111]. In this research, dietary supplements from sea buckthorn, bilberry, and cranberry were characterized by visualization of the amounts of polyphenols and antioxidants, and the compounds were observed by UV-visible detection on TLC plates. The procedure was composed of (1) injection of sample spots on the TLC plates, (2) ascending development in a chamber, and (3) capturing their images by a scanner or by photography. The obtained images were transformed into a digitized chromatogram by TLC Analyzer software [111], and the chromatograms were handled by chemometric techniques. First of all, PCA was used for reduction of the original dataset dimension and then LDA was performed to classify unknown samples according to their complex chromatograms.

The k-nearest neighbor (k-NN) method is one of the simplest pattern recognition techniques which was introduced in the beginning of the 1970s as a nonparametric method. The k-NN technique can be efficient for both classification and regression purposes. In both cases, the inputs are based on the voting scheme. Hence, k-nearest neighbors (k is odd number) are selected to determine their distance from an unknown sample. Accordingly, the majority vote of its k-nearest neighbors is assigned as the input.

It is worth noting that the output, which is requested from k-NN, depends on the types of applications. In classification, the output should be able to classify the unknown sample as a class membership since the majority of samples in a class determines the allocation of the unknown sample. While in k-NN regression, the output is the average of the values of its k-nearest neighbors of the unknown sample.

The usefulness and simplicity of the k-NN technique has made it one of the most common methods of classification. The algorithm of classification starts by assigning known objects to their specified classes. Figure 25.7 shows that objects are classified in three classes as training sets based on their similar features.

In order to assign unknown objects (or test set) to the most similar object, the following steps have to be carried out.

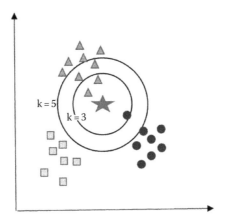

FIGURE 25.7   Principal diagram of the k-NN classification algorithm with two k values equal to 3 and 5.

1. Distance of the unknown object from all members of the training set should be calculated. Distances can be measured by various functions between two samples k and l, such as the Euclidean distance:

$$ED = \sqrt{\sum_{j=1}^{J} \left(x_{kj} - x_{lj} \right)^2}$$
(25.8)

or Manhattan distance:

$$MD = \sum_{j=1}^{J} | x_{kj} - x_{lj} |$$
(25.9)

2. Samples should be sorted in ascending according to their distances. In other words, in this ranking, the smallest distance is the first in order.
3. k training samples nearest to the unknown object are chosen to estimate the nearest class. Usually, k is a small odd number (especially in binary classification to avoid ties). In Figure 25.7, two values of k (k = 3 and k = 5) are selected.
4. A majority of voting is used for classification.
5. Assessment of classification ability is done by evaluating different values of k, for example, 1, 3, 5, and 7. It should be noted that small values of k can lead to larger variances which may result in large bias in the model. Thus, selecting an optimal value for k is essential to avoid bias and obtain good variance in the model. Some common statistical methods such as the well-known cross validation can be used to estimate an appropriate value of k.

The simple algorithm of k-NN introduced it as one of the primary classification techniques and a benchmark in image analysis. By classifying training images according to their feature vectors, unknown image features can be assigned to one of the classes based on their k closest distance. As represented in Figure 25.7, an unknown object is assigned to the triangle group based on both k = 3 and k = 5. In spite of the noted advantages, k-NN suffers from increasing classification error rate when distribution of training samples is not even or the sample size of each class is very different [112].

Sousa et al. used the k-NN technique for classifying oligosaccharides of urine samples in two pathological and normal cases with the aid of TLC images digitized through light reflection using a scanner. They analyzed the oligosaccharide bands based on the lane mean intensity profile in order to avoid large curvature observed in the bands in the 1D Gaussian deconvolution method. Their results showed the effects of band correction at initial band position on improvement of classifier performance [113,114].

In 2009, authenticated Chaihu samples versus many commercial samples were evaluated by two high-performance liquid chromatography evaporative light-scattering detectors and HPTLC analyses of their principal bioactive components (saikosaponins). Pretreatment and pattern recognition of both methods were completed by two chemometric techniques: artificial neural networks (ANNs) and k-NN.

In general, the k-NN classifier was employed as an appropriate and flexible algorithm for processing HPTLC fingerprint in the case of changing R_f values and varying color or saturation of the appeared bands between different TLC sheets [115].

25.3.2.2 Nonlinear Supervised Pattern Recognition Methods

ANN is one of the famous methods in chemometrics [116] with some reported applications and more potential for utilization in image analysis and TLC image analysis [117]. The human's brain has many interesting characteristics that cause the emergence of conscious and sophisticated behaviors including perceptual, control, and cognitive functionality. These characteristics comprise learning ability, adaptivity, ability of generalization, and capability of information processing [118]. A neuron is considered as a biological processor that can process the incoming information (input) from the neighboring neurons, and then propagate the processed information (output) to another neuron through a complex biochemical process. Each neuron is approximately linked to 10^3–10^4 neurons, which makes a highly interconnected network [118].

For a typical neuron, the multiple input signals are gathered from the m neighboring ANNs of the desired ANN. The cumulated input signals are combined based on the predefined rule before introducing them into the ANN processor unit. For this purpose, the weighted sum (net) of m input signals is computed according to the following equations:

$$net = \sum_{j=1}^{m} w_j x_j + \theta = AC + \theta = XW^T \tag{25.10}$$

$$net = \sum_{j=1}^{m} w_j x_j + \theta = AC + \theta = XW^T \tag{25.11}$$

where

$$X = x_0, x_1, x_2, \ldots, x_m \tag{25.12}$$

$$W^T = \theta, w_1, w_2, \ldots, w_m \tag{25.13}$$

in which AC shows the current activation state of the neuron and W^T denotes the transposed weight vector of the neuron. It corresponds to the connection weight from m neighboring ANNs of the desired ANN, which is augmented with θ as the variable bias (θ is a threshold value that is appended to another weight for notation simplicity). X is referred to the input signals from m neighboring ANs, which are augmented with a constant input value x_0. In most cases, x_0 is equal to unity.

To investigate the effects of impinging input signals on the current state of activation node, a transfer (activation) function f is introduced that takes net signals and generates a new value that can determine a new state of AN activation.

$$y_m = f(net) \qquad (25.14)$$

where
    y_m is output value (target value) of AN with respect to the net value that is obtained according to Equation 25.10
    f is a type of transfer function

It means that no output value is generated by the transfer function f unless AC is greater than a certain threshold value (θ). Different types of transfer functions have been used such as sgn (e.g., binary step function), piecewise linear, or smoothly limiting threshold (e.g., sigmoid) [119–122].

Multilayer perceptron (MLP) is the most popular topology in the subclasses of feed-forward network architectures. Among the mentioned architectures, MLP is explained in more detail because this architecture is applied in the analysis of TLC chromatograms.

As can be seen in Figure 25.8, MLP is composed of unidirectional nodes that are arranged in the three layers involved: an input, one or more hidden (processing), and an output. Each of the observations in the training set is received by the nodes of the input layer and is merely transferred to the hidden layer without doing any computational processing, such as the computational processing involved in the other nodes of ANN. Therefore, this layer is not frequently counted as an operator layer in the MLP architecture. The number of nodes in this anterior layer is equal to the number of observations in the training set.

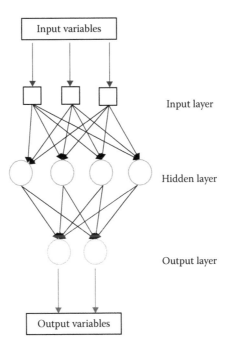

FIGURE 25.8 A typical network of multilayer perceptron.

The most applied learning algorithm in the MLP architecture to train the network is the generalized delta rule or back propagation (BP). It has been employed in various applications such as pattern recognition, control, forecasting, and function approximation [123]. In simple terms, the BP algorithm provides a procedure to approximate the weights of the feed-forward network based on a generalized delta rule.

The potential applicability of ANN in the analysis of the obtained images from TLC can be found in the literature in three main categories including calibration, classification, and optimization. At first, two reported applications in the field of calibration category are described, and then two examples of classification category are provided. Finally, a hybrid of ANN with a genetic algorithm is illustrated for optimization purpose.

Agatonovic-Kustrin et al. investigated the potential applicability of the ANN method to analyze obtained chromatograms from the HPTLC as fingerprint patterns. This method is utilized to quantify the amount of biologically active components, including caffeic acid, chlorogenic acid, and rutin as chemical markers in *Calendula officinalis* extract, which is traditionally applied as a herbal medicinal plant for its anti-inflammatory effect. To obtain an analytical signal for quantitative estimation of chemical markers in extract of *C. officinalis*, different extract samples containing various amounts of chemical markers were applied to the activated NC Silica HPTLC plate. After gradient development of HPTLC plate, the documentation of chromatograms was done using a TLC visualizer camera as the detector. In the next step, the images were captured under UV illumination (336, 254 nm) and white lamps on the plate of HPTLC separately. Unfortunately, only two active components (caffeic acid, chlorogenic acid) were visualized on the HPTLC plate under the described conditions. Two derivative reagents including sulfuric acid and 2-aminoethyl diphenylborinate were utilized to visualize all the marker compounds. In this regard, to make the images of the components of interest detectable, the derivative reagent was sprayed over the HPLTLC plate. Then this process was allowed to be completed in an oven over 30 min. In the next step, the obtained images were imported to software to recognize and create the band chromatograms. The signal intensities of the created band chromatograms were introduced to the ANN. In data modeling, ANN was utilized to correlate the obtained signal intensities of chromatogram bands and the amount of interested chemical markers. In the next step, the developed model was accurately utilized to predict the amount of chemical markers in the *C. officinalis* extract.

The effects of different conditions (factors) on chemical patterns and four chemical markers were investigated. Growth location, conditions of growth, plant age, various parts of utilized plant, and extraction methods can be noted as some of the investigated conditions. The results based on ANN approved that all factors had impressed the chemical profiling but the type of extraction method was the most detrimental factor. Detection limits of 57, 183, and 282 ng were reported for caffeic acid, chlorogenic acid, and rutin, respectively [124].

Another application of ANN for calibration purpose was also reported by Agatonovic-Kustrin et al. [124]. The goal of utilizing ANN was quantification of some chemical phenylpropanoid markers that may be responsible for special properties of *Echinacea*. The studied chemical markers included chlorogenic acid, echinacoside, and chicoric acid. *Echinacea* has widespread applications, which arises from its immunity enhancement property, and antifungal, antiviral, antibacterial, and antioxidative properties.

After development of the TLC plate, without performing any derivatization step, the images of TLC plates were captured under a UV lamp (366 nm). A similar procedure, as mentioned in the previous example, was used to transform the captured images from the produced chemical fingerprints on the TLC plate onto the chromatograms. The quantification of chlorogenic acid, echinacoside, and chicoric acid was performed by ANN. Different kinds of topologies were examined. Finally, the best results were obtained by the topology of MLP with two hidden layers. High correlation coefficients in the prediction of different chemical markers in the test set indicated good ability of ANN to couple with TLC image analysis. The theoretical detection limits (DLs) of 46, 19, and 29 ng were calculated for chlorogenic acid, echinacoside, and chicoric acid , respectively [125].

In most cases, discrimination of natural products from their counterfeits needs sophisticated and expensive instruments such as HPLC due to the high similarity between their chemical profiles. If chemometrics is combined with low-cost methods such as TLC image analysis, a powerful, budget-friendly, and fascinating analytical tool will be generated to identify natural products. In this regard, Yang et al. utilized such a combination to discriminate poplar tree gum and Chinese propolis (bee glue) [126]. They developed a new visualization reagent (vapors of ammonia and hydrogen peroxide) to provide simple detection of polyphenol species as major bioactive components in natural products (under natural light without using any background correction). The applied strategy of the proposed method to identify the samples by chemometric fingerprinting of TLC densitograms is depicted in Figure 25.9.

In the first step, spots of the known samples were visualized on the TLC plate based on the new proposed method and were documented in the JPEG format. In the next step, the region from the initially injected sample to the solvent front was cropped to remove redundant area that was empty of any sample spots. Subsequently, the cropped image was virtually segmented into different lanes in the migration direction. Afterwards, one-dimensional array of each lane was calculated by summing over the gray scale intensity of each row of the lane in the same migration direction. A densitogram was created by plotting the dimensional array data as a function of pixel numbers in the migration direction. This procedure was also utilized to obtain a densitogram of unknown samples. For classification purpose, the obtained densitograms were used as the input of various chemometric tools including similarity analysis,

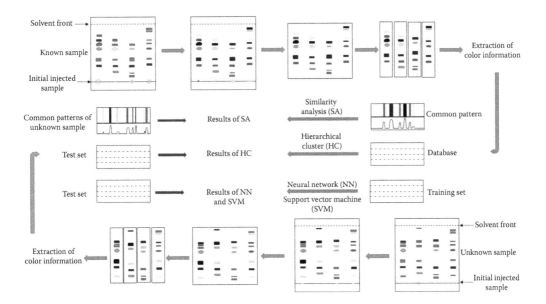

FIGURE 25.9 Schematic diagram analysis of the obtained images from thin-layer chromatography to identify unknown samples by chemometric fingerprinting. Identification of unknown samples is possible by the discrimination models which are constructed by chemometric methods including similarity analysis (SA), hierarchical clustering (HC), and neural network (NN)/support vector machine (SVM). The required inputs for these chemometric methods include common pattern, database of samples, and training set that are produced by the densitograms of known samples. Densitograms are created by the extraction of color information on gray scale. (Based on Tang, T.X. et al., *Phytochem. Anal.*, 25, 266, 2017.)

hierarchical clustering (HC) analysis, and neural network/support vector machine (NN/SVM) analysis, respectively. It should be noted that a common pattern was generated by averaging all the calculated densitograms of the known samples. In the last step, chemometric methods were used for recognition of the unknown samples. The results showed that HC analysis, ANN, and SVM could provide accurate classification to discriminate Chinese propolis from its counterfeit (poplar tree gum).

Rezić et al. [127] proposed that the genetic algorithm and ANN can be successfully combined to optimize the analytical separation conditions in TLC and to predict resolution (R_S). Their proposed method was examined for a mixture containing seven amino acids including alanine, serine, cysteine, leucine, asparagine, threonine, and phenylalanine. Resolution (R_S) was used as an optimization criterion to locate an optimum mobile phase composition.

25.3.2.3 Hierarchical Cluster Analysis as a Linear Unsupervised Pattern Recognition Method

Clustering methods are broadly divided into two distinct categories: (1) the hierarchical clustering method and (2) the partitioning method. Hierarchical clustering is done when the observations are distinct and we are interested in the individuals themselves. However, partitioning is done when the observations by themselves are not distinct and we are not interested in the individuals but rather the groups that they form.

Now again, hierarchical clustering can be divided into two types (Figure 25.10). It can either be formed in an agglomerative manner (bottom-up approach) or a divisive method (top-down approach). The agglomerative manner starts with n clusters and finishes with a single cluster on the top. On the other hand, the divisive method is initiated with a single large cluster, which keeps dividing till n different clusters are obtained [128]. Depending on these two factors, the closest clusters are merged, one at a time, until all clusters merge into a single cluster of n members at the top of the tree [129–131].

Agglomerative hierarchical clustering starts with n distinct clusters at the bottom of the tree of the dendrogram and has to predecide on which distance measure to use. The distance measure obviously is the distance between the members of the observations. It also needs to decide on which linkage for measurement of the cluster distances.

In 2011, Zarzycki et al. utilized an eco-friendly micro-TLC technique for the classification of herbs and spirulina samples. Hierarchical clustering methods were used to classify these samples according to the chromatographic band intensities which were derived from the sample plates [132].

In 2015, Taylor et al. studied the influence of retardation factor variation and nonuniform image background for the similarity measure of TLC with image analysis. They used Citrus herbal medicine as a sample in their experiments. Hierarchical clustering methods were applied to compare the influence of these two factors [133].

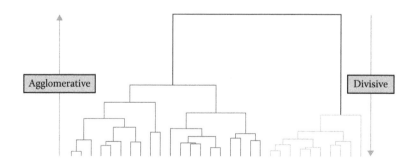

FIGURE 25.10 Hierarchical clustering for n individuals.

25.4 CONCLUSION

TLC plates, owing to their characteristics such as fast data acquisition with a simple instrumental design, versatility, and budget-friendliness, are a good target for multivariate methods to enhance their potential. Three common types of video densitometer including flatbed scanner, CCD camera, and cell phone are utilized to document information of TLC plates for further analysis. In this chapter, the basic principal operations and the parameters that are important in the performance of these devices were briefly discussed. Indeed, controlling these parameters can practically help researchers to capture images with acceptable quality for quantitative analysis, especially chemometrics data treatment.

Because of the limitations such as high amount of backgrounds in the obtained TLC chromatograms, chemometric tools have critical importance in the processing and extraction of information. In case of image analysis, the chemometric exploration of chromatographic data is also a well-established topic. The chemometric techniques applied for TLC can be divided into two groups of factor analysis and pattern recognition. Factor analysis techniques such as PCA, PLSR, PLS-DA, and OPLS have been utilized in TLC measurements. Some methods such as ANN, LDA, k-NN, and HCA have been coupled with TLC image analysis for pattern recognition applications.

REFERENCES

1. Izmailov NA and Schraiber MS. 1938. A new method of separation for the organic compounds. *Farmatsiya* 3: 1–7.
2. Senf HJ. 1969. *Thin Layer Chromatography in Drug Analysis, Microchimica Acta* 57: 522–526.
3. Ferenczi-Fodor K, Végh Z, and Renger B. 2006. Thin-layer chromatography in testing the purity of pharmaceuticals. *Trends in Analytical Chemistry* 25: 778–789.
4. Yousefinejad S, Honarasa F, and Saeed N. 2015. Quantitative structure-retardation factor relationship of protein amino acids in different solvent mixtures for normal-phase thin-layer chromatography. *Journal of Separation Science* 38: 1771–1776.
5. Sharma L, Desai A, and Sharma A. 2006. A thin layer chromatography laboratory experiment of medical importance. *Biochemistry and Molecular Biology Education* 34: 44–48.
6. Bladek J and Neffe S. 2003. Application of thin-layer chromatography in clinical chemistry. *Separation and Purification Reviews* 32: 63–122.
7. Lim AF, Marimuthu Y, Chang KH, Mohamad SNF, Muslim MNZ et al. 2011. Forensic discrimination of lipsticks by thin layer chromatography and gas chromatography-mass spectrometry. *Malaysian Journal of Forensic Sciences* 2: 22–28.
8. Kaur A, Sahota SS, and Garg RK. 2015. Separation of different components of hair cosmetic (hairsprays) using tlc and hptlc. *Anil Aggrawal's Internet Journal of Forensic Medicine and Toxicology* 16: 1–1.
9. Vetrova OY, Kokorina KA, Bazhaeva ZB, Mel'nik YV, and Petrova YY. 2015. Sorption-catalytic determination of rutin, lysine, and collagen in pharmaceuticals and cosmetics. *Pharmaceutical Chemistry Journal* 48: 753–758.
10. Skorek M, Kozik V, Kowalska T, and Sajewicz M. 2015. Thin-layer chromatographic quantification of trans -resveratrol in cosmetic raw materials of botanic origin. *Journal of Planar Chromatography— Modern TLC* 28: 167–172.
11. Starek M, Guja A, Dąbrowska M, and Krzek J. 2015. Assay of β-carotene in dietary supplements and fruit juices by tlc-densitometry. *Food Analytical Methods* 8: 1347–1355.
12. Sherma J. 2000. Thin-layer chromatography in food and agricultural analysis. *Journal of Chromatography A* 880: 129–147.
13. David AZ, Mincsovics E, Papai K, Ludanyi K, Antal I, and Klebovich I. 2008. Oplc comparison of methods for aqueous extraction of sennae folium and tiliae flos plant samples. *Section Title: Pharmaceuticals* 21: 119–123.
14. Loescher CM, Morton DW, Razic S, and Agatonovic-Kustrin S. 2014. High performance thin layer chromatography (HPTLC) and high performance liquid chromatography (HPLC) for the qualitative and quantitative analysis of calendula officinalis-advantages and limitations. *Journal of Pharmaceutical and Biomedical Analysis* 98: 52–59.
15. El-Yazbi AF and Youssef RM. 2015. An eco-friendly hptlc method for assay of eszopiclone in pharmaceutical preparation: Investigation of its water-induced degradation kinetics. *Analytical Methods* 7: 7590–7595.

16. Vovk I, Simonovska B, Andrenšek S, Vuorela H, and Vuorela P. 2003. Rotation planar extraction and rotation planar chromatography of oak (quercus robur l.) bark. *Journal of Chromatography A* 991: 267–274.

17. Van Berkel GJ, Llave JJ, De Apadoca MF, and Ford MJ. 2004. Rotation planar chromatography coupled on-line with atmospheric pressure chemical ionization mass spectrometry. *Analytical Chemistry* 76: 479–482.

18. Chavan M and Vaidya A. 2012. Development of high-performance thin-layer chromatography (HPTLC) technique for evaluation of sunscreen photostability. *Journal of Planar Chromatography—Modern TLC* 25: 122–126.

19. Bhushan R and Tanwar S. 2010. Different approaches of impregnation for resolution of enantiomers of atenolol, propranolol and salbutamol using cu(ii)-l-amino acid complexes for ligand exchange on commercial thin layer chromatographic plates. *Journal of Chromatography A* 1217: 1395–1398.

20. Berezkin VG, Khrebtova SS, Redina EA, and Egorova E V. 2010. A combined version of planar chromatography. *Journal of Analytical Chemistry* 65: 492–497.

21. Quesenberry RQ, Donaldson EM, and Ungar F. 1965. Descending and ascending chromatography of steroids using thin-layer chromatography sheets. *Steroids* 6: 167–175.

22. Vlajković J, Andrić F, Ristivojević P, Radoičić A, Tešić Ž et al. 2013. Development and validation of a tlc method for the analysis of synthetic food-stuff dyes. *Journal of Liquid Chromatography & Related Technologies* 36: 2476–2488.

23. Mandal K, Kaur R, and Singh B. 2014. Development of thin layer chromatographic technique for qualitative and quantitative analysis of fipronil in different formulations. *Journal of Liquid Chromatography & Related Technologies* 37: 2746–2755.

24. Hawryl MA and Waksmundzka-Hajnos M. 2013. Micro 2d-tlc of selected plant extracts in screening of their composition and antioxidative properties. *Chromatographia* 76: 1347–1352.

25. Ramić A, Medić-Šarić M, Turina S, and Jasprica I. 2006. Tlc detection of chemical interactions of vitamins a and d with drugs. *Journal of Planar Chromatography—Modern TLC* 19: 27–31.

26. Bhushan R, Martens J, Agarwal C, and Dixit S. 2012. Enantioresolution of some β-blockers and a β2-agonist using ligand exchange TLC. *Journal of Planar Chromatography—Modern TLC* 25: 463–467.

27. Bhushan R and Agarwal C. 2008. Direct tlc resolution of (±)-ketamine and (±)-lisinopril by use of (+)-tartaric acid or (−)-mandelic acid as impregnating reagents or mobile phase additives. isolation of the enantiomers. *Chromatographia* 68: 1045–1051.

28. Eloff J, Ntloedibe D, and Van Brummelen R. 2011. A simplified but effective method for the quality control of medicinal plants by planar chromatography. *African Journal of Traditional, Complementary, and Alternative Medicines* 8: 1–12.

29. Massa DR, Chejlava MJ, Fried B, and Sherma J. 2007. Thin layer and high performance column liquid chromatographic analysis of selected carboxylic acids in standards and from helisoma trivolvis (colorado strain) snails. *Journal of Liquid Chromatography & Related Technologies* 30: 2221–2229.

30. Zhou B, Chang J, Wang P, Li J, Cheng D, and Zheng P-W. 2014. Qualitative and quantitative analysis of seven oligosaccharides in morinda officinalis using double-development hptlc and scanning densitometry. *Bio-medical Materials and Engineering* 24: 953–960.

31. Reiffová K and Nemcová R. 2006. Thin-layer chromatography analysis of fructooligosaccharides in biological samples. *Journal of Chromatography A* 1110: 214–221.

32. Feng Y-L. 2001. Determination of fleroxacin and sparfloxacin simultaneously by tlc-fluorescence scanning densitometry. *Analytical Letters* 34: 2693–2700.

33. Stroka J and Anklam E. 2000. Development of a simplified densitometer for the determination of aflatoxins by thin-layer chromatography. *Journal of Chromatography A* 904: 263–268.

34. Xie H, Dong C, Fen Y, and Liu C. 1997. Determination of doxycycline, tetracycline and oxytetracycline simultaneously by tlc-fluorescence scanning densitometry. *Analytical Letters* 30: 79–90.

35. Hahn-Deinstrop E and Leach RG. 2007. *Applied Thin-Layer Chromatography*, 2nd edn. Weinheim, Germany: Wiley-VCH.

36. Phattanawasin P, Sotanaphun U, and Sriphong L. 2011. A comparison of image analysis software for quantitative tlc of ceftriaxone sodium. *Silpakorn University Science and Technology Journal* 5: 7–13.

37. Tie-xin T and Hong W. 2008. An image analysis system for thin-layer chromatography quantification and its validation. *Journal of Chromatographic Science* 46: 560–564.

38. Sakunpak A, Suksaeree J, Monton C, and Pathompak P. 2014. Development and quantitative determination of barakol in senna siamea leaf extract by tlc-image analysis method. *International Journal of Pharmacy and Pharmaceutical Sciences* 6: 3–6.

39. Tozar T, Stoicu A, Radu E, Pascu ML, and Physics R. 2015. Evaluation of thin layer chromatography image analysis method for irradiated chlorpromazine quantification. *Romanian Reports in Physics* 67: 1608–1615.

40. Phattanawasin P, Sotanaphun U, and Sriphong L. 2008. Validated tlc-image analysis method for simultaneous quantification of curcuminoids in curcuma longa. *Chromatographia* 69: 397–400.

41. Bansal K, McCrady J, Hansen A, and Bhalerao K. 2008. Thin layer chromatography and image analysis to detect glycerol in biodiesel. *Fuel* 87: 3369–3372.

42. Wagner H, Bauer R, Melchart D, Xiao P, and Staudinger A. 2011. *Chromatographic Fingerprint Analysis of Herbal Medicines*, 2nd edn. Berlin, Germany: Springer.

43. Bernard-Savary P and Poole CF. 2015. Instrument platforms for thin-layer chromatography. *Journal of Chromatography A* 1421: 184–202.

44. Sherma J and DeGrandchamp D. 2015. Review of advances in planar radiochromatography. *Journal of Liquid Chromatography & Related Technologies* 38: 381–389.

45. Cheng S, Huang MZ, and Shiea J. 2011. Thin layer chromatography/mass spectrometry. *Journal of Chromatography A* 1218: 2700–2711.

46. Poole CF. 2015. *Instrumental Thin-Layer Chromatography*. Amsterdam, the Netherlands: Elsevier.

47. Cimpoiu C. 2005. Qualitative and quantitative analysis by hyphenated (hp)tlc-ftir technique. *Journal of Liquid Chromatography & Related Technologies* 28: 1203–1213.

48. Csehati T and Forgacs E. 1998. Hyphenated techniques in thin layer chromatography. *Journal of the Association of Official Analytical Chemists* 81: 329–332.

49. Gössi A, Scherer U, and Schlotterbeck G. 2012. Thin-layer chromatography–nuclear magnetic resonance spectroscopy—A versatile tool for pharmaceutical and natural products analysisa. *Chimia International Journal for Chemistry* 66: 347–349.

50. Poole CF. 2003. Thin-layer chromatography: Challenges and opportunities. *Journal of Chromatography A* 1000: 963–984.

51. Spangenberg B, Poole CF, and Weins C. 2011. *Quantitative Thin-Layer Chromatography: A Practical Survey*. Berlin, Germany: Springer.

52. Poole CF. 2003. *The Essence of Chromatography*. Amsterdam, the Netherlands: Elsevier.

53. Reich E and Schibli A. 2007. *High-Performance Thin-Layer Chromatography for the Analysis of Medicinal Plants*. New York: Thieme Medical Publishers.

54. Kompany-Zareh M and Mirzaei S. 2004. Genetic algorithm-based method for selecting conditions in multivariate determination of povidone-iodine using hand scanner. *Analytica Chimica Acta* 521: 231–236.

55. Andrews P. 2009. *Advanced Photoshop Elements 7 for Digital Photographers*. Oxford, U.K.: Focal Press.

56. Blitzer H, Stein-Ferguson K, and Huang J. 2008. *Understanding Forensic Digital Imaging*. Burlington, MA Academic Press.

57. Dempster J. 2001. *The Laboratory Computer: A Practical Guide for Physiologists and Neuroscientists* London, U.K.: Academic Press.

58. Bilissi M. 2011. *Langford's Advanced Photography: The Guide for Aspiring Photographers*, 8th edn. Oxford, U.K.: Focal Press.

59. Milojković-Opsenica D, Ristivojević P, Andrić F, and Trifković J. 2013. Planar chromatographic systems in pattern recognition and fingerprint analysis. *Chromatographia* 76: 1239–1247.

60. Djozan D, Baheri T, Karimian G, and Shahidi M. 2008. Forensic discrimination of blue ballpoint pen inks based on thin layer chromatography and image analysis. *Forensic Science International* 179: 199–205.

61. Tompsett MF. 1970. Charge coupled 8-bit shift register. *Applied Physics Letters* 17: 111–115.

62. Tompsett MF, Amelio GF, Bertram WJ, Buckley RR, McNamara WJ et al. 1971. Charge-coupled imaging devices: Experimental results. *IEEE Transactions on Electron Devices* 18: 992–996.

63. Lahuerta Zamora L and Pérez-Gracia MT. 2012. Using digital photography to implement the mcfarland method. *Journal of the Royal Society Interface* 9:1892–1897.

64. Baker IM, Beynon JDE, and Copeland MA. 1973. Charge-coupled devices with submicron electrode separations. *Electronics Letters* 9: 48–49.

65. Janesick JR. 2001. *Scientific Charge-coupled Devices*. Bellingham, WA: SPIE Press.

66. Holst GC and Lomheim TS. 2011. *CMOS/CCD Sensors and Camera Systems*. London, U.K.: SPIE Press.

67. Giles MJ, Ridder TD, Williams RM, Jones AD, and Denton MB. 1998. Selecting a CCD camera. *Analytical Chemistry* 70: 663A–668A.

68. Aikens RS, Agard DA, and Sedat JW. 1989. Solid-state imagers for microscopy. *Methods in Cell Biology* 29: 291–313.

69. Biesemeier A, Schraermeyer U, and Eibl O. 2011. Quantitative chemical analysis of ocular melanosomes in stained and non-stained tissues. *Micron* 42: 461–470.

70. Brereton RG. 2003. *Chemometrics: Data Analysis for the Laboratory and Chemical Plant*. Chichester, U.K.: John Wiley & Sons, Ltd.

71. Lavine BK and Workman J. 2013. Chemometrics. *Analytical Chemistry* 85: 705–714.

72. Hemmateenejad B, Karimi S, and Mobaraki N. 2013. Clustering of variables in regression analysis: A comparative study between different algorithms. *Journal of Chemometrics* 27: 306–317.

73. Hemmateenejad B and Yousefinejad S. 2009. Multivariate standard addition method solved by net analyte signal calculation and rank annihilation factor analysis. *Analytical and Bioanalytical Chemistry* 394: 1965–1975.

74. Mahboubifar M, Yousefinejad S, Alizadeh M, and Hemmateenejad B. 2016. Prediction of the acid value, peroxide value and the percentage of some fatty acids in edible oils during long heating time by chemometrics analysis of ftir-atr spectra. *Journal of the Iranian Chemical Society* 13: 2291–2299.

75. Gendrin C, Roggo Y, and Collet C. 2008. Pharmaceutical applications of vibrational chemical imaging and chemometrics. *Journal of Pharmaceutical and Biomedical Analysis* 48: 533–553.

76. Pearson K. 1901. On lines and planes of closest fit to systems of points in space. *The London, Edinburgh, and Dublin Philosophical Magazine and Journal of Science* 2: 559–572.

77. Hotelling H. 1933. Analysis of a complex of statistical variables into principal components. *Journal of Educational Psychology* 24: 498–520.

78. Brereton RG. 2009. *Chemometrics for Pattern Recognition*, Chapter 3. Chichester, U.K.: John Wiley & Sons, Ltd.

79. Bro R and Smilde AK. 2014. Principal component analysis. *Analytical Methods* 6: 2812–2831.

80. Hemmateenejad B and Elyasi M. 2009. A segmented principal component analysis-regression approach to quantitative structure-activity relationship modeling. *Analytica Chimica Acta* 646: 30–38.

81. Yousefinejad S and Hemmateenejad MAR 2012. New autocorrelation qtms-based descriptors for use in qsam of peptides. *Journal of the Iranian Chemical Society* 9: 569–577.

82. Wold S, Esbensen K, and Geladi P. 1987. Principal component analysis. *Chemometrics and Intelligent Laboratory Systems* 2: 37–52.

83. Hemmateenejad B. 2004. Optimal qsar analysis of the carcinogenic activity of drugs by correlation ranking and genetic algorithm-based PCR. *Journal of Chemometrics* 18: 475–485.

84. Hemmateenejad B. 2005. Correlation ranking procedure for factor selection in pc-ann modeling and application to admetox evaluation. *Chemometrics and Intelligent Laboratory Systems* 75: 231–245.

85. Shojaeifard Z, Hemmateenejad B, and Shamsipur M. 2016. Efficient on–off ratiometric fluorescence probe for cyanide ion based on perturbation of the interaction between gold nanoclusters and a copper(ii)-phthalocyanine complex. *ACS Applied Materials & Interfaces* 8: 15177–15186.

86. Hemmateenejad B, Yousefinejad S, and Mehdipour AR. 2011. Novel amino acids indices based on quantum topological molecular similarity and their application to qsar study of peptides. *Amino Acids* 40: 1169–1183.

87. Gemperline P. 2006. *Practical Guide to Chemometrics*, 2nd edn. Boca Raton, FL: CRC Press.

88. Shlens J. 2003. A tutorial on principal component analysis: Derivation, discussion and singular value decomposition. Online Note, 2: 1–16. http://www.cs.princeton.edu/picasso/mats/PCATutorial-Intuition Jp.pdf. Accessed on March 2, 2017.

89. Halko N, Martinsson P-G, Shkolnisky Y, and Tygert M. 2011. An algorithm for the principal component analysis. *SIAM Journal on Scientific Computing* 33: 2580–2594.

90. Prats-Montalbán JM, de Juan A, and Ferrer A. 2011. Multivariate image analysis: A review with applications. *Chemometrics and Intelligent Laboratory Systems* 107: 1–23.

91. Komsta Ł, Cieslla L, Bogucka-Kocka A, Jozefczyk A, Kryszeri J, and Waksmundzka-Hajnos M. 2011. The start-to-end chemometric image processing of 2d thin-layer videoscans. *Journal of Chromatography A* 1218: 2820–2825.

92. Hawryl M, Świeboda R, Hawryl A, Niemiec M, Stępak K et al. 2015. Micro two-dimensional thin-layer chromatography and chemometric analysis of essential oils from selected mentha species and its application in herbal fingerprinting. *Journal of Liquid Chromatography & Related Technologies* 38: 1794–1801.

93. Wold H. 1966. *Estimation of Principal Components and Related Models by Iterative Least Squares*. New York: Academic Press.

94. Geladi P and Kowalski BR. 1986. Partial least-squares regression: A tutorial. *Analytica Chimica Acta* 185: 1–17.

95. De Jong S. 1993. Simpls: An alternative approach to partial least squares regression. *Chemometrics and Intelligent Laboratory Systems* 18: 251–263.

96. Lindgren F, Geladi P, and Wold S. 1993. The kernel algorithm for pls. *Journal of Chemometrics* 7: 45–59.

97. Wold SSM and Eriksson L. 2001. Pls-regression: A basic tool of chemometrics. *Chemometrics and Intelligent Laboratory Systems* 58: 109–130.

98. Höskuldsson A. 1995. A combined theory for pca and pls. *Journal of Chemometrics* 9: 91–123.

99. Barker M and Rayens W. 2003. Partial least squares for discrimination. *Journal of Chemometrics* 17: 166–173.

100. Mobaraki N and Hemmateenejad B. 2011. Structural characterization of carbonyl compounds by ir spectroscopy and chemometrics data analysis. *Chemometrics and Intelligent Laboratory Systems* 109: 171–177.

101. Trygg J and Wold S. 2002. Orthogonal projections to latent structures (o-pls). *Journal of Chemometrics* 16: 119–128.

102. Hemmateenejad B, Mobaraki N, Shakerizadeh-Shirazi F, and Miri R. 2010. Multivariate image analysis-thin layer chromatography (mia-tlc) for simultaneous determination of co-eluting components. *Analyst* 135: 1747–1758.

103. Hemmateenejad B, Akhond M, Mohammadpour Z, and Mobaraki N. 2012. Quantitative monitoring of the progress of organic reactions using multivariate image analysis-thin layer chromatography (mia-tlc) method. *Analytical Methods* 4: 933–939.

104. Hemmateenejad B, Farzam SF, and Mobaraki N. 2014. Simultaneous measurement of leucine and isoleucine by multivariate image analysis-thin layer chromatography (mia-tlc). *Journal of the Iranian Chemical Society* 11: 1609–1617.

105. Ristivojević P, Andrić FL, Trifković JĐ, Vovk I, Stanisavljević LŽ et al. 2014. Pattern recognition methods and multivariate image analysis in hptlc fingerprinting of propolis extracts. *Journal of Chemometrics* 28: 301–310.

106. Sharif KM, Rahman MM, Azmir J, Khatib A, Sabina E et al. 2015. Multivariate analysis of prisma optimized tlc image for predicting antioxidant activity and identification of contributing compounds from pereskia bleo. *Biomedical Chromatography* 29: 1826–1833.

107. Javidnia K, Parish M, Karimi S, and Hemmateenejad B. 2013. Discrimination of edible oils and fats by combination of multivariate pattern recognition and ft-ir spectroscopy: A comparative study between different modeling methods. *Spectrochimica Acta Part A: Molecular and Biomolecular Spectroscopy* 104: 175–181.

108. Yousefinejad S, Aalizadeh L, and Honarasa F. 2016. Application of atr-ftir spectroscopy and chemometrics for the discrimination of furnace oil, gas oil and mazut oil. *Analytical Methods* 8: 4640–4647.

109. Fisher RA. 1936. The use of multiple measurements in taxonomic problems. *Annals of Eugenics* 7: 179–188.

110. Kim TK and Kittler J. 2005. Locally linear discriminant analysis for multimodally distributed classes for face recognition with a single model image. *IEEE Transactions on Pattern Analysis and Machine Intelligence* 27: 318–327.

111. Sima IA, Sârbu C, and Naşcu-Briciu RD. 2015. Use of TLC and UV–visible spectrometry for fingerprinting of dietary supplements. *Chromatographia* 78: 929–935.

112. Haralick RM. 1979. Statistical and structural approaches to texture. *Proceedings of the IEEE* 67: 786–804.

113. Kamel M and Campilho A. 2005. *Image Analysis and Recognition*. New York: Springer-Verlag.

114. Sousa A and Aguiar R. 2004. Automatic lane and band detection in images of thin layer chromatography. *Image Analysis and Recognition*. 3212: 158–165.

115. Tian RT, Xie PS, and Liu HP. 2009. Evaluation of traditional Chinese herbal medicine: Chaihu (bupleuri radix) by both high-performance liquid chromatographic and high-performance thin-layer chromatographic fingerprint and chemometric analysis. *Journal of Chromatography A* 1216: 2150–2155.

116. Yousefinejad S and Hemmateenejad B. 2015. Chemometrics tools in qsar/qspr studies: A historical perspective. *Chemometrics and Intelligent Laboratory Systems* 149: 177–204.

117. Shakerizadeh-Shirazi F, Hemmateenejad B, and Mehranpour AM. 2013. Determination of the empirical solvent polarity parameter e t (30) by multivariate image analysis. *Analytical Methods* 5: 891–896.

118. Jain AK, Mao J, and Mohiuddin KM. 1996. Artificial neural networks: A tutorial. *Computer* 29: 31–44.

119. Patnaik PR. 1999. Applications of neural networks to recovery of biological products. *Biotechnology Advances* 17: 477–488.

120. Zupan J and Gasteiger J. 1991. Neural networks: A new method for solving chemical problems or just a passing phase? *Analytica Chimica Acta* 248: 1–30.

121. Kröse B and Smagt V. 1996. *Introduction To Neural Networks*, 8th edn. Amsterdam, Netherlands: University of Amsterdam.

122. Krenker A, Kos A, and Bešter J. 2011. Introduction to the artificial neural networks, in *Artificial Neural Networks: Methodological Advances and Biomedical Applications*, ed. K. Suzuki, pp. 1–18. Rijeka, Croatia: INTECH Open Access Publisher.

123. Mehrotra K, Mohan C, and Ranka S. 1997. *Elements of Artificial Neural Networks*. Cambridge, MA: MIT Press.
124. Agatonovic-Kustrin S and Loescher CM. 2013. Qualitative and quantitative high performance thin layer chromatography analysis of calendula officinalis using high resolution plate imaging and artificial neural network data modelling. *Analytica Chimica Acta* 798: 103–108.
125. Agatonovic-Kustrin SL School CM, and Singh R. 2013. Quantification of phenylpropanoids in commercial Echinacea products using TLC with video densitometry as detection technique and ANN for data modelling. *Phytochemical Analysis* 24: 303–308.
126. Tang TX, Guo WY, Xu Y, Zhang SM, Xu XJ et al. 2014. Thin-layer chromatographic identification of chinese propolis using chemometric fingerprinting. *Phytochemical Analysis* 25: 266–272.
127. Rolich T and Rezi I. 2011. Use of genetic algorithms and artificial neural networks to predict the resolution of amino acids in thin-layer chromatography. *Journal of Planar Chromatography* 24: 16–22.
128. Rokach L and Maimon O. 2005. *Data Mining and Knowledge Discovery Handbook*, 2nd edn. New York: Springer-Verlag.
129. Theodoridis S and Koutroumbas K. 2009. *Pattern Recognition*, 4th edn. London, U.K.: Elsevier.
130. Rui X and Wunsch DC. 2009. *Clustering*. Hoboken, NJ: John Wiley & Sons, Inc.
131. Gan G, Ma C, and Wu J. 2007. *Data Clustering: Theory, Algorithms, and Applications*. Philadelphia, PA: Society for Industrial and Applied Mathematics.
132. Zarzycki PK, Zarzycka MB, Clifton VL, Adamski J, and Głód BK. 2011. Low-parachor solvents extraction and thermostated micro-thin-layer chromatography separation for fast screening and classification of spirulina from pharmaceutical formulations and food samples. *Journal of Chromatography A* 1218: 5694–5704.
133. Taylor P, Tang T, Guo Y, Li Q, Xu X et al. 2015. The influence of image correction and chromatogram alignment on similarity measure of one-dimensional tlc by means of image analysis. *Journal of Liquid Chromatography & Related Technologies* 38: 1279–1285.

Index